Preface

The purpose of this tutorial collection is to introduce readers to digital private branch exchanges (PBXs).

The organization of this tutorial consists of six principal chapters, a general introduction, concluding remarks, a glossary, and an annotated bibliography. Each chapter has an introduction that provides an overview of the principal issues and a brief summary of each selection and why it is included. The selections in each chapter are ordered in difficulty from low to high.

The emphasis of Chapter 1, ''What PBXs Do, and How They Do It,'' is on the general design and workings of PBXs, the various roles PBXs are capable of playing, and an introduction to the PBX market and the major players.

Because PBXs most often support office environments, Chapter 2, ''PBXs as Office Service Centers,'' addresses the possible roles a PBX may play in optimizing white collar productivity, managing information, and integrating office systems from multiple vendor sources.

Planning for and procuring PBXs are complex processes involving different groups with different skills. Chapter 3, ''A Modern PBX in Detail: NEC's NEAX2400 IMS as Example,'' presents a complete system description of a modern PBX with the intent of emphasizing the numerous technology decisions and level of specification required in PBX procurement.

The advent of desktop computers and on-line information services has greatly changed the office environment from even a decade ago. Interconnecting personal computers (PCs) and/or connecting PCs or terminals to information services has spawned local area networks (LANs). In Chapter 4, ''PBXs as LANs,'' PBXs are presented as an appropriate technology to deliver LAN services.

Although conventional and largely centralized architectures dominate the present PBX market, distributed and highly modular architectures combining powerful and low-cost microprocessors and high speed extended buses—often employing fiber optics—appear to be the direction of the future. Chapter 5, ''Distributed Architectures,'' presents a number of these distributed designs.

The last chapter, Chapter 6, is entitled ''Positioning for ISDN.'' Integrated services digital networks (ISDN), if they become economical and ubiquitous in the next decade, will result in a quantum leap in the availability of information services. Many business users will access these services through equipment attached to a PBX, and present PBX vendors are scurrying in several directions at once: to deliver the services in an unintegrated or semi-integrated manner immediately, and to be ready to provide them in an integrated fashion as soon as final standards, central office availability, and common carrier network changes converge.

The intended audience for this tutorial are graduate engineers and others with experience in computer or telecommunications systems that require an introduction to digital PBXs and an understanding of the directions in which PBXs are evolving.

The reader's expectations should include enough introductory information to understand what PBXs are, what they do, how they do it, and their principal components; information on what functions PBXs can perform in a modern office; an insight into the principal areas, the relative complexity, and the level of detail to be encountered when faced with PBX technical specifications, such as in a procurement situation; what the use of PBXs of the future may be like; and, last, the role that PBXs will play in future ISDNs.

The reader's level of achievement or education assumed is probably best loosely drawn: either graduate engineers or those with equivalent technical background or individuals whose work experience has imparted a familiarity with computer or telecommunications systems.

Ed Coover
Chevy Chase, Maryland, USA

Acknowledgments

Tutorials are intrinsically collaborative and this one is no exception. But where the reprint contributors are clearly identified, the behind-the-scenes contributors are not. What follows is a partial, and probably insufficient, response to this deficiency.

Mike Kane, Bill Chandler, Pat Carney, Paul Tsuchiya and David Lee, colleagues at the MITRE Corporation, influenced my thinking on PBXs. Beth Roth of MITRE was extremely helpful with the bibliographic aspects. Dan Curry of MITRE provided careful and witty criticisms of several drafts. Jon Butler of the IEEE Computer Society provided sage advice and helped preserve the tutorial's conception. Finally, Denise Felix and Margaret Brown of the IEEE Computer Society were instrumental in fielding the conception.

E.R.C.

Table of Contents

Preface iii

Acknowledgments v

Introduction 1

Chapter 1: What PBXs Do and How They Do It 5

The ABCs of the PBX 7
    L.F. Goeller Jr. and J.A. Goldstone (Business Communications
    Review Manual of PBXs, 1983)
The Evolution of the SL-1 PBX 17
    D. Bir, J. Eng, and R. Hoo (Telesis, 1984 one, pages 20-27)
An Inside Look at the New Rolm CBX II Architecture 26
    H.W. Johnson (Business Communications Review,
    January-February 1984, pages 9-19)
Private Branch Exchanges: The Best Time to Shop May Be Right Now 37
    D. Levin (Data Communications, August 1987,
    pages 100-122)
An Overview of Recent Developments in the Designs and Applications

of Customer Premises Switches 49
    N. Janakiraman (IEEE Communications Magazine,
    October 1985, pages 32-45)
System 75: Communications and Control Architecture 63
    L.A. Baxter, P.R. Berkowitz, C.A. Buzzard, J.J. Horenkamp,
    and F.E. Wyatt (AT&T Technical Journal, January 1985,
    pages 153-173)
System 75: Switch Services Software 79
    W. Densmore, R.J. Jakubek, M.J. Miracle, and J.H. Sun
Just When Your Thought It Made Sense to Get Rid of Centrex... 91
    C.A. Klinck (Data Communications, March 1986,
    pages 118-129)

Chapter 2: PBXs as Office Service Centers 103

Surveying the PBX Path 105
    E. Horwitt (Business Computer Systems, May 1985,
    pages 56-65)
Voice-Data PABX Makes Data Analysis Easy for E-Tech 110
    R.D. Creadon (Telephony, March 2, 1987, pages 28-32)
Voice-Data Integration in the Office: A PBX Approach 115
    E.R. Coover (IEEE Communications Magazine,
    July 1986, pages 24-29)
Some Architectural Considerations for Local Area Networks 121
    R.L. Sharma (Proceedings: 8th Conference on Local
    Computer Networks, October 1983, pages 6-13)
Helping Data Managers Find Their Voice 129
    D. Thomson (Data Communications, May 1985,
    pages 111-123)
Analog vs. Digital 136
    E.R. Coover and M.J. Kane (Telephony, August 10, 1987,
    pages 76-81)
The Role of Information Management System in Business Communication 139
    O. Enomoto and S. Kadota (IEEE Communications Magazine,
    July 1986, pges 37-43)
Use Integrated PBXs and X.25 in Today's Networks; Don't Wait for ISDN 147
    B.C. Sagaser (Data Communications, May 1986, pages 253-261)

Chapter 3: A Modern PBX in Detail: NEC's NEAX2400 IMS as Example 153

NEAX2400 Information Management System (IMS) General Description 155
   (NEC Corporation, November 1986)

Chapter 4: PBXs as LANs 213

New Niches for Switches 216
   F.H. Harris, F.L. Sweeney Jr., and R.H. Vonderohe
   (Datamation, March 1983, pages 109-220)
An Introduction to Integrated Voice/Data PBX Systems 221
   K.J. Thurber and H.A. Freeman (Office Automation Systems,
   1986, pages 209-215)
The Heat Is on for Phone Switches That Do a Lot of Fast Shuffling 228
   M.W. Patrick (Data Communications, March 1985,
   pages 227-236)
PBX Trends and Technology Update: Following the Leaders 234
   E.E. Mier (Data Communications, September 1985, pages 82-96)
Meridian DV-1: A Fast, Functional and Flexible Data Voice System 243
   C. Murray and W. Carr (IEEE Communications Magazine,
   December 1986, pages 36-42
Who Needs a Lan? 250
   S. Mehta (LAN Magazine, June 1987, pages 24-28)
Enhanced Data Switches May Outshine LAN, PBX Alternatives 254
   N.J. Muller (Data Communications, May 1987, pages 185-193)
The Evolution of Data Switching for PBX's 259
   B. Bhushan and H. Opderbeck (IEEE Journal on Selected Areas
   in Communications, July 1985, pages 569-573)

Chapter 5: Distributed Architectures 265

Notes from Mid-Revolution: Searching for the Perfect PBX 267
   E.R. Coover and M.J. Kane (Data Communications, August 1985, pages 141-150)
New Wave Coming in Data-Voice Switching 272
   F.G. McKay (Telephone Engineer & Management, October 15, 1984, pages 71-74)
The Next Generation in Business Communication 277
   W.P. Karavatos (Telecommunications, August 1983, pages 28-35)
Ztel, Inc. and CXC Corp. Announce Voice/Data Switches 281
   Ronald A. Frank (Business Communications Review, May-June 1983, pages 41-44)
Ericsson MD110 Intelligent Network General Description 297
   (Ericsson Corp., November 1987)

Chapter 6: Positioning for ISDN 301

ISDN—New Vistas in Information Processing 303
   W.V. Tang (Proceedings of the International Conference, ISDN, Volume 1)
Transition to the AT&T-IS Integrated Private Network Architecture 309
   G.M. Anderson (IEEE Journal on Selected Areas in Communications,
   July 1985, pages 600-605)
Network of the Future 315
   T.E. Browne (Proceedings of the IEEE, September 1986, pages 1222-1230)
Customer Installations for the ISDN 324
   G. Robin (IEEE Communications Magazine, April 1984, pages 18-23)
CCITT Recommendations on the ISDN: A Review 330
   M. Decina (IEEE Journal on Selected Areas in Communications,
   May 1986, pages 320-325)
The Evolution of Telecommunications Technology 336
   R. Vickers and T. Vilmansen (Proceedings of the IEEE, September 1986,
   pages 1231-1245)

Design of an Integrated Services Packet Network 351
    J.S. Turner (IEEE Journal on Selected Areas in Communications,
    November 1986, pages 1373-1379)

Concluding Remarks 359

Glossary 361

Annotated Bibliography 377

Author Biography 383

Introduction

Current Appeal

The last 15 years have seen an expanding interest in private branch exchanges (PBXs). Several reasons for this interest are most conspicuous.

Deregulation has played probably the premier role. The controversy accompanying communications deregulation served to spotlight available features and what they cost. Under plain old telephone service (POTS), each telephone line was associated with a physical pair of copper wires terminating in the central office. Even Centrex, the telephone companies' most advanced business-oriented service, was cumbersome in use and required telephone companies intervention to activate and change features and service assignments. Under deregulation, Centrex service rapidly escalated in cost. Within many organizations, few people were even aware of communications costs, much less the ability to manipulate them. With the advent of computerized PBXs, a software-controlled switch was interposed between the user's telephone set and the central office and a mind-numbing number of options, including data transmission, was offered to the end user and the company.

Technology has also played an important role. Production economies of scale and cheap digital components have greatly dropped the cost of PBXs. Purchased systems with few features now cost around $600 to $700 per user. Software control and the nearly universal transition to digital switching technologies in PBXs have opened the door to an arrray of features ranging from switching (and often converting the form of) computer data, storing voice messages, optimizing long-distance calling by a complex set of variables, implementing elaborate chargeback routines, and even controlling aspects of the building environment. From the technologist's perspective, the modern digital PBX offers a software-defined environment, a flexible and usually fault-tolerant hardware interface, and ubiquitous connectivity to the user.

Voice-data integration has also played a role. All modern PBXs support data switching, usually in the form of using all or part of a 64 Kbps voice channel. Typical levels of support are 9.6 or 19.2 Kbps for asynchronous traffic, and 56 or 64 Kbps for synchronous traffic. For a typical office with terminals and/or personal computers (PCs), a modern PBX provides a means to connect to electronic mail, shared word processing, or information services with relatively low costs and complexity.

Predictably, bureaucracies have been less flexible in taking advantage of the integrated capabilities. Companies still exist that parrot "if it's voice it's AT&T, if it's data it's IBM, and if it's word processing it's Wang," but such intellectual economy is becoming rarer. Also rarer are the separate voice and data bureaucracies that acted as if the other did not exist up to the point when overhead trays and underground conduits could not accept one more wire, regardless of type.

Increasingly, pushed by new efforts to contain costs and drawn by the publicity and promise attending integrated services digital networks (ISDNs), voice and data bureacracies are being placed under a common management structure, often with "information" and "management" in its title. The perception of the information resource manager, or IRM, as the position is often called, is very different from the voice and data managers. Set on reducing or controlling costs and, where possible, instituting new or improved services, this new emphasis is especially critical of previously contrived bureaucratic territorialisms. In-plant wiring is seen as a corporate strategic resource and its dual use for voice and data, if technically possible, a predisposition.

Within this new organizational context, and faced with demands to economically network a large number of PCs, and bring long-distance expenses under tight control, the modern PBX emerges as a leading technological contender.

Major Problems

Nonetheless, even when the bureaucracies are responsive and malleable, PBX procurements, at least in part because PBXs can subsume functions provided by several existing systems, often become stressful events. First, fear of change is high, and the prospect of a company's telephone system's not working, or working very badly, is chilling. Second, the functional flexibility of modern digital PBXs normally exceeds the knowledge and experiential bases of any single individual. Whether expertise is assembled in-house or via consultants, there are nontrivial problems in evaluating PBXs functionally (will they do a particular job better than alternatives?) and procuring them (which system is most cost-efficient for a particular environment?). Third, successful systems can change their environments in ways that are not always perceptible in advance. As a result, "success" can come back to haunt the implementator in the form of rapid growth and altered work styles that are unsustainable within the system responsible for the "success."

Solutions

Because the chief problem with PBXs is managerial—how does one deal with the complex decisions involved in an item absolutely central to most companies' business—the solutions are largely managerial as well. When faced with strategic decisions regarding PBXs, some companies place their trust in a particular vendor or follow the recommendation of consultants. For many, however, a PBX evaluation committee is often the answer. These committee members need experience in the following areas:

- Information management, where basic decisions are made regarding the utility of PBX-capable services such as voice or text mail, building monitoring, switched data support, network bridges and/or protocol conversion facilities, modem pools, and the like;

- Voice operations, especially regarding policy toward feature sets to be assigned to various staff levels, call-forwarding sequences, and managing telephone moves and changes;

- Data operations, particularly regarding connectivity and performance requirements;

- Maintenance and capacity/configuration planning;

- Cost management, with emphasis on optimizing long-distance resources, finding a middle ground on call restriction, and implementing the chargeback scheme;

- Physical plant, where experience is needed in the details of extant conduits, wiring, space, air conditioning, and emergency power and where changes need be made;

- System engineering, wherein numbers and types of trunk lines are engineered to meet preset grades of service or special needs of the organization; and

- System acquisition, where the various options regarding lease and purchase can be evaluated in accordance with institutional guidelines.

Future Trends

A lasting effect of deregulation was to initiate a struggle for turf between two of the United States' largest corporations, AT&T in communications and IBM in computer systems. Given the technological convergence of computers and communications over the last half century, it was probably inevitable that these two industry giants would collide, but deregulation ensured it. AT&T has used their strength in networks and long distance, as well as corporate alliances with Olivetti, EDS, and Sun Microsystems, to push their PCs and 3B minicomputers and encourage multi-vendor shops and communications-intensive distributed processing. IBM, for its part, has enhanced its data strengths in mainframes and PCs to move into voice switching by acquiring Rolm, with

T1 network switching by acquiring Network Equipment Technologies, and has aggressively pitched its capabilities to perform voice/data network management with its NetView product.

Although the large scale of the IBM-AT&T conflict tends to be distracting, many of the important technological changes in PBXs have come from other sources. Before acquisition by IBM, Rolm pioneered with distributed systems linked by a network operating system. Mitel and Ericsson and others who initially concentrated on the low end of the market have led the way in compact and energy-efficient packaging. InteCom has carved out a sizeable niche in the upper end of the market with its advanced data handling capabilities. NEC, the Japanese manufacturer, has led the market in low-cost production. And Northern Telecom has been a leader in responding to the challenge to support PCs and to functionally combine the telephone and PC.

Most analysts who have looked to the future of the PBX market have concluded that there are too many players, too many companies in the marketplace. Given the conventional definition of the market (larger businesses and institutions) and PBX economics ($500–$1100 a line), that conclusion is probably true. But given the current cost tilt against Centrex, the increasingly complex long-distance environment, the low cost of digital electronics, and the increase in services such as electronic mail and facsimile, it is not wild-eyed to contemplate low-end, PC-like PBX devices and their proliferation through public schools, homes, jails, libraries, motels, churches, and small businesses of every kind. Not uncoincidentally, many of the future PBX trends alluded to (compact size, low cost, air-cooled, an array of smart, software-driven services, easy to use administrative features, easy PC connection) also point in that direction.

Aspects to Be Covered

In the chapters that follow, the readings assembled address many of these concerns. Chapter 1, "What PBXs Do, and How They Do It," is intended to introduce the technology and technical issues to the unfamiliar. Chapter 2, "PBXs as Office Service Centers," examines the multi-function role of the PBX in an office environment and reexamines the revitalized Centrex offering. Chapter 3, "A Modern PBX in Detail: NEC's NEAX2400 IMS as Example," is not intended to plug NEC's PBX but rather demonstrate, in a reasonably compact form, the various aspects of PBX technology and the sheer range of issues on which decisions must be made. Chapter 4, "PBXs as LANs," focuses on the various data options offered by modern digital PBXs and casts a sidelong look at data (only) PBXs. Chapter 5, "Distributed Architectures," examines a number of radical designs whose influence is being felt both in the repackaging and repositioning of older designs. Chapter 6, "Positioning for ISDN," looks at what elements in the present PBX environment are likely to change with ISDN availability. A final section, "Concluding Remarks," examines a number of design directions likely to be seen in future PBX products.

Finally, a word about the un-inevitability of technological change and progress. Among engineers there runs a not-so-silent credo of technical positivism, a feeling that the world would both make sense and work well were it not for low-tech managers, blundering politicians, parsimonious bean counters, lying salespeople, and the like. The truth is that things do not often turn out as technologically prophesied. With this in mind, there have been inserted, at various spots in this tutorial, papers that are intentionally nonconforming. Thus, the reader will find articles that cite the advantages of Centrex over PBXs, direct the manager to extend ISDN-type services now rather than wait for ISDN standards, suggest that a data-only PBX may be the best LAN for office environments, opinion that the best PBX design is a hybrid switch combining circuit and packet switching, describe the radically distributed PBX architecture of a company that has since seen Chapter 11, and explain to the reader why the current scheme for ISDN is wrong-headed, the point being not only that these nonconformists are an important part of the digital PBX story but that they may point the way in the future.

Chapter 1: What PBXs Do and How They Do It

What PBXs Do...

Private branch exchanges (PBXs) are switches that connect circuits. Connecting circuits in conjunction with particular hardware and software provides different services. For instance, voice services include not only the obvious intra- or inter-site, local, and long-distance switching connections but a large number of supplementary functions. Some of these functions such as call forwarding, paging, or conferencing are visible to the user. Others are aspects of overall system administration, such as establishing the "profile" of user privileges, implementing moves and changes, or capturing station message detail records (SMDR) in order to analyze usage and chargeback costs.

Although voice services typically provide the bulk of PBX usage, generally 80 percent or more, other services are increasingly being added to the PBX repertoire. Connecting computer terminals to different hosts is common, and the PBX often provides the additional functions such as port selector, protocol conversion, in bound and out bound modem pooler, and/or network bridge. In these roles the PBX functions as a local area network (LAN).

Another service offered frequently with PBXs is electronic mail. This is available in two predominant forms. Text mail systems include those that access a mail processor via a terminal connected with the PBX, a message center where forwarded calls are answered by operators who enter the messages in a computer, or a combination of the two. Oftentimes the text service is expanded to include access to the institution's telephone directory, announcements of general interest, and staff schedules. Voice mail systems, where spoken messages are stored in a compressed digital format, are more similar in operation to the home answering machines and may complement text-oriented systems.

Other services, such as facsimile transmission, piggyback on the voice services, using the switch's least-cost routing facilities to reduce the cost of the service. Building monitoring systems, both for the environment (heat/air conditioning) and for the security (fire, personnel access) are supported by today's PBXs and are likely to increase in popularity with integrated services digital networks (ISDN).

...and How They Do It

The modern digital PBX consists of several major elements working together: Various kinds of line cards connect both user telephone and central office trunks to the PBX's time-division multiplexed (TDM) bus; the TDM bus, a large band-width, shared transmission medium, provides time "slots" to the communicating devices; and one or more larger processors run the switch's control system, higher level call processing routines, and applications software that act on the call data.

In digital switches the signals are manipulated as digital pulses. Where analog devices are interconnected with the PBX, the analog signals, such as voice or modulated data signals, must be converted to digital signals. This A-to-D conversion, as it is called, is performed by coder-decoders, or codecs, which may be part of a line card or integrated into individual telephone instruments. Line cards usually contain one or more microprocessors and may serve multiple telephone instruments or data devices.

They play a distributed processing role vis-a-vis the central processor(s) and provide services such as conversion, noting an off-hook status, and collecting call digits. Common types of line cards include those for analog or digital central office trunks, T1 trunks (a group of 24 digitial trunks), analog telephones, digital telephones, and interfaces for asynchronous or synchronous data devices.

Within this complex system of components, performance limitations are imposed both by switch design and by economic decisions relative to the operating environment. One is said to be "blocked" when one cannot complete a call, but the reasons for the blockage can be numerous. At the most basic, there may be only one phone in a particular area and it may be in use. Also, a busy number is a form of blockage. During an emergency or under extremely heavy use, the PBX may not have the capacity in its multiplexed bus to complete all the calls placed, or the sheer number of attempts may cause it to slow down. Beyond the common busy signal, the most likely blockage to be visible to users is contention for a limited number of long distance trunk lines. Depending on the institution's policies and its PBX implementation, the inability to access a wide area telephone service (WATS) or foreign exchange (FX) circuit may result in a busy signal, queuing, an explanatory message advising to try later, or an automatic overflow to the long distance supplier's toll network.

Modern PBX systems are designed to be modular and provide, in numerous configurations, services to a wide variety of constituencies. PBXs range in size from just above key systems (around 12 lines) up to the size of a medium central office (around 20,000 lines). Several have bus capacities that allow them to be internally nonblocking above 18,000 lines.

Many offer special software packages for hotels or for automatic call distribution (much used by airlines and mail order businesses) where calls are queued until an operator is free. Still others have software suites that allow them to perform other functions altogether, such as tandem switching (switching high capacity trunks in an electronic tandem network or ETN). The point is that modern PBXs can be configured to deliver a wide variety of services economically; the typical problem, alluded to before, is assembling the staff expertise necessary to describe the requirements to the PBX vendor accurately and to evaluate the inherent tradeoffs in the many alternatives likely to be proposed.

Note should also be taken of PBX packaging, because it is an important aspect that is slowly changing. Most current PBX's employ one or more standard 19-inch racks, locate the power supplies on the bottom, use redundant sets of fans to dissipate heat, and provide maintenance access to the switch via large doors. The larger and/or high-cost systems often resemble mainframe computer centers with their raised floors and large amounts—and often duplexed—forced air cooling. In the raised floor installations, all wiring, including telephone, power, and inter-cabinet connectors, run under the floors; in the installations where the switch sits on the floor, usually in the basement, the telephone connections are overhead in cable racks and terminate at the cable ducts or main distribution frame.

There are considerable reliablility and environmental costs associated with the inefficiencies inherent in such traditional packaging. Most switch failures are due to power outages, and although batteries and motor generators can bridge the typical power outage of less than a half hour, an extended outage can create a situation where the lack of power to run the air conditioners forces the PBX to be shut down to avoid overheating. As a consequence, in large or otherwise critical environments, substantial add-on costs will be incurred for raised floors, air-conditioning, batteries, motor generator, and power control and protection equipment.

In the area of packaging, smaller systems and the new, distributed systems (Chapter 5) are leading the way. Oriented toward installation in offices, they are incorporating the same downsizing seen in computer systems such as the NCR Tower, DEC MicroVAX, and IBM PS/2 Model 80. They combine universal slots (any card can be fit into any slot), less space, less power consumption, built-in UPS, and a wide temperature operating range. To a large extent they embody cost consciousness toward office floor space, recognizing that it is an expensive resource. The newer, downsized systems allow the manager to "shrink" the plant, or nonrevenue generating portion, of the office space.

The papers included in this chapter have been ordered by difficulty, with the easier ones first. The Goeller and Gold-stone selection, "The ABCs of the PBX," reprinted in *Datamation* but taken from their book, *Business Communications Review Manual of PBXs,* introduces the reader to basic PBX functions and terminology. Their emphasis is on the "life cycle" of a call. The Bir, Eng, and Hoo paper, "The Evolution of the SL—1 PBX," recounts the progressive software and hardware evolution of Northern Telecom's SL-1, introduced in 1975. Particular note should be taken of the development process (see box) describing the process whereby new features are added. In a similar vein, Johnson, in "An Inside Look at the New Rolm CBX II Architecture," details how Rolm has evolved a 16-bit single-node PBX with 75 Mbps bus introduced in 1975 into a 32-bit multinode PBX with a 295 Mbps bus while maintaining upward compatibility. Special note should be taken of the central role played by the high speed bus in today's modular architectures.

The next two selections are overviews. Levin's recent paper, "Private Branch Exchanges: The Best Time to Shop May Be Right Now," provides a detailed look at the products in the very crowded low end of the PBX market. Here, the emphasis is on the various configuration options (See Figure 2 in the paper) available with PBXs and how the numerous suppliers and the very competitive market "drives" the products. Janakiraman, in "An Overview of Recent Developments in the Designs and Applications of Customer Premises Switches," reviews the entire field of customer premise equipment and how the forces of "office automation," technological advance, ISDN, and deregulation have influenced both the design and marketing of the newer PBX systems. Note in particular the gross cost per line tendencies presented in Figure 1 in that paper. These cost trends have been the major force in greatly expanding the PBX market.

The two most difficult papers in this chapter recount the design of AT&T's System 75. Together they are meant to provide insight, at a fairly low level of detail, on not only how a modern PBX works but what are the large number of choices open to the designers.

Baxter et al., in "Communications and Control Architecture," focus on concepts, terms, and control architecture of the System 75. Densmore et. al., in "Switch Services Software," describe the software environment and present, in considerable detail, the anatomy of a call and the large number of cooperating processes. Note how much more complex the call process appears here than in the early Goeller and Goldstone paper.

The final selection is unabashedly nonconformist. Kinck, in "Just When You Thought It Made Sense to Get Rid of Centrex…" describes the numerous techniques by which the Bell Operating Companies (BOCs) are adding PBX-like features to central office-based Centrex services to stay competitive. At least by implication, the Kinck paper questions the "historical inevitability" of the customer premise-based solution advanced by Janakiraman.

THE ABCS OF
THE PBX

Leo F. Goeller Jr. and J.A. Goldstone

Private branch exchanges aren't just for communications types anymore. Here's part two of our four-part series on these important devices.

THE ABCS OF THE PBX

by Leo F. Goeller Jr. and Jerry A. Goldstone

A PBX or private branch exchange is a switchboard for business telephones. It differs from the phone company's central office (CO) switch in two very important ways. First, it must in general have someone to say, "XYZ Company, good morning" and then complete the call. Then, if the call reaches the wrong party, some means must be provided to transfer that call to another extension.

At the smaller end of the spectrum, key telephone systems compete with PBXs. A key system may have 50 or more telephones associated with it and many advanced features. There is a philosophical difference, however, between a PBX and a key system: a PBX has a relatively high proportion of calling among telephones it serves, while a key system has most of its traffic with the outside world and is used very little for internal calling. In much larger sizes, Centrex competes, usually by using specially modified central office equipment.

A PBX system (Fig. 1) is composed of four basic parts: switching matrix, control, user terminals, and trunks. The PBX proper includes the matrix and control, along with interface units, usually called line and trunk circuits, that terminate the transmission facilities that extend to user telephones or other switches. In addition, one usually finds "service circuits" in a PBX to apply ringing and call progress tones (busy, reorder, etc.), and to assimilate caller signaling information from telephone dials and tone pads, converting this information to something the control can use.

No matter how complex a modern PBX may become internally, it still works

PHOTOGRAPH BY STEVE COOPER

Excerpted with permission from the *Business Communications Review Manual of PBXs*, published by Business Communications Review, Hinsdale, Ill.

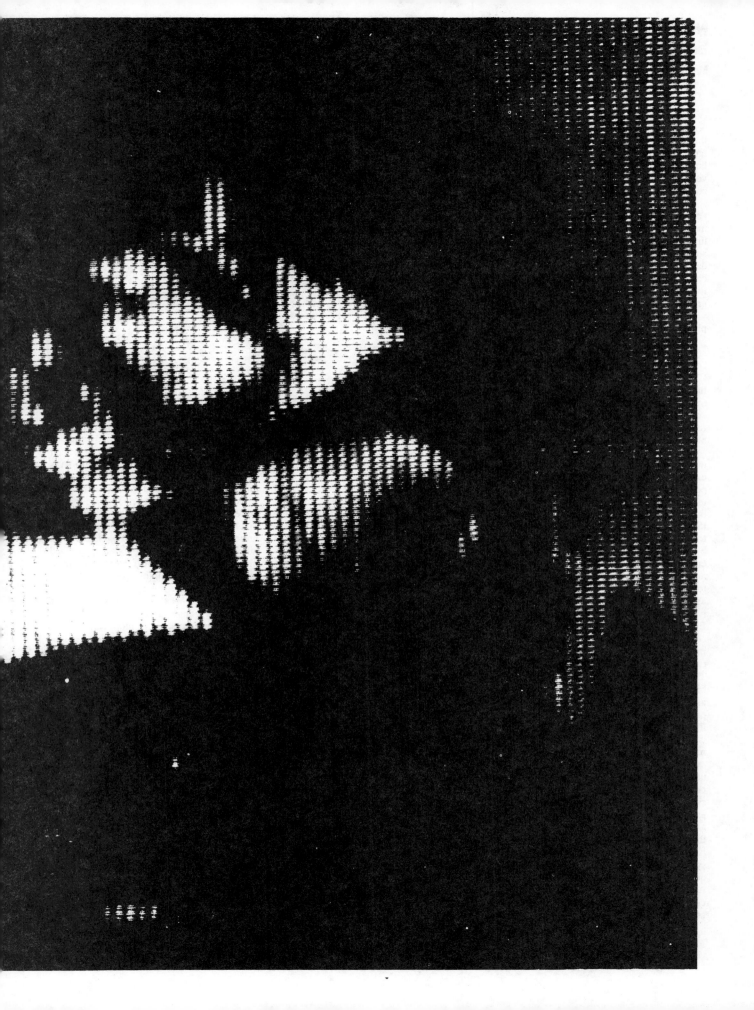

No matter how complex a modern PBX may become internally, it still works very much the way the old manual switchboards did.

very much the way the old manual switchboards did. In the so-called cord board, each line and trunk terminates in a suitable interface circuit that controls one or more signal lamps to alert the attendant and passes the voice path on to a jack that permits easy interconnection.

When a user picks up his phone, the PBX supplies power to the instrument and, by monitoring this flow of power (usually with a line relay), knows to light a lamp to tell the attendant that assistance is needed. The attendant has in front of him a number of pairs of cords that can be used to make connections to lines and/or trunks. These cords perform the precise functions that a modern PBX performs with its switching matrix. The attendant plugs one cord of a particular pair into the jack associated with the calling line and operates a switch that connects his headset to the cord circuit. He then says something like "Number, please" so that the caller knows the attendant is ready to respond to his communication needs. A modern system, of course, uses a dial tone to perform the same function, even though many such systems have replaced rotary dials with tone pads.

The caller then gives the attendant the number he wishes to reach. If it is another extension on the PBX, the attendant takes the other cord of the pair and makes a "busy test" on the called line to make sure it is not busy on another call. If the called line is idle, the attendant then completes the connection to the appropriate jack.

Either the attendant or automatic circuitry detects an answer, terminates ringing, and leaves the calling and called-station users to converse. If either of them needs further assistance, he flashes his switch-hook. That is, he depresses and releases his switch-hook. Each of the two cords in the pair making the connection has a lamp associated with it. Momentary depression of the switch-hook causes the lamp to come on momentarily. This flash signals the attendant to reconnect his headset to the cord circuit to see what assistance is required. When the call is completed and the parties hang up, the cord lamps light, letting the attendant know that the call is over and the cords can be pulled down and made available for another call.

There are still a few manual PBXs in use. They serve their customers well and, because they consist primarily of a very small switching matrix (the cords) and simple line and trunk circuits, they occupy little space and are relatively inexpensive. But perhaps their main advantage, in addition to low cost, is their control. They have the most advanced system control available today, one that is quite flexible and very smart: a human being. A human attendant at a manual switchboard can provide almost all the modern features associated with the new, computer-con-

trolled PBXs. Attendants, available in common to all lines and trunks, meet all the requirements of what is called "common control" in modern switching systems. (Common control can be contrasted with "distributed control" in step-by-step electromechanical systems where each switch has its own control equipment and control is distributed over the entire switching matrix. Common control equipment, shared by all parts of the system, eliminates a great deal of duplication and, when done properly, reduces costs and improves operations.)

However, people are getting more and more expensive: they like to go to lunch in the middle of the day, they want to go home at night, and they require training. Humans being human, the trend today is to effect as many of the functions of the attendant as possible with automatic equipment, often in the form of a computer acting as a common control.

MODERN PBX FEATURES

A modern PBX still detects the flow of power when the telephone user picks up his instrument. It then makes a connection automatically through the switching matrix to a digit receiver that returns dial tone. The caller dials or keys the called number into the digit receiver (sometimes called a register, a decoder, or something similar). The system responds by making sure the call is permitted and then rings the called line. Upon detection of answer, the calling and called parties are interconnected via the switching matrix for conversation; the system monitors for switch-hook flashes that indicate the need for some additional service, or hang up to free the portions of the switching matrix used on this call for future calls.

A major point of complexity, far

more important in PBXs than in most central office switches, lies in terminating the call to some line other than the one requested. Hunting is usually available and widely used in PBXs so that if the boss's line is busy, the call will be routed to his secretary. Modern PBXs have added a variety of call-forwarding features that can be invoked and canceled by the station user. Sometimes privacy features require the return of a special tone to indicate that the called party does not wish to be disturbed. Pickup is a relatively new feature that allows any station in a previously defined pickup group to come off hook and snatch away a call that is ringing unanswered at another station in the group. Camp-on and call waiting may have calls stacked up waiting for an existing conversation to end; automatic callback may have one or more callers waiting for an existing call to be completed so that the system can call them back and then complete to the called party. Obviously, a call encountering camp-on, call waiting, or callback at the terminating line will be dealing with a situation that has several levels of "busy" to contend with. A simple test to see if the line is busy is no longer sufficient. From all this, it is easy to see that the terminating half of the call setup procedure can be quite complex.

Trunk calls are a little more intricate. An outgoing call starts off exactly the same way as an extension-to-extension call, but now the control can complete its part of the job by connecting to any one of several trunks to the local central office. Further, it can, if it wants to, do this long before the caller has finished dialing. Traditionally, callers on a PBX dial 9, get dial tone from the CO, and then dial the telephone number of the called party. This approach was very important in the early automatic PBXs that were constructed from

DATAMATION

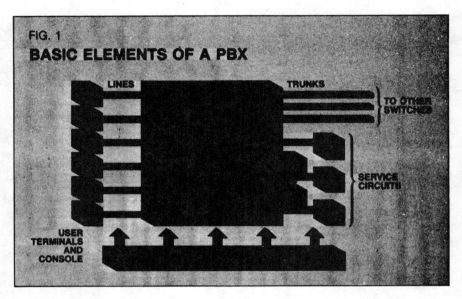

FIG. 1
BASIC ELEMENTS OF A PBX

LINES

TRUNKS

TO OTHER SWITCHES

SERVICE CIRCUITS

USER TERMINALS AND CONSOLE

CHART BY CYNTHIA STODDARD

With the present PBX and central office designs, it is imperative that the CO use ground start circuits.

relays and other electromechanical components. It let the PBX complete its part of the job quickly and easily and turn all the hard work over to the CO. The CO would then store the called number, set up the connection, handle the billing, etc.

The problem with this was the propensity of PBX station users to make personal calls that kept showing up in the bill from the phone company as message units, short-haul tolls, or even long-haul tolls. This led to a central office feature called toll diversion, which automatically terminated all toll calls or diverted them to the switchboard attendant. But because many COs did not have this feature, the restrictor was developed. A restrictor was a box full of switching components that attached to each trunk of the CO and could be programmed with a list of approved calling regions. As time went on, however, it became evident that the restrictor function should be incorporated in the PBX design.

A PBX that incorporates restriction knows which extension is originating the call. Thus the PBX can automatically perform restricting and routing functions based on the "class of service" assigned to the specific extension without the caller having to take any action. External restrictors must treat all calls alike since they have no way of identifying the caller, and external call routers, quite popular during the last 10 years, require the user to identify himself by dialing in an additional group of digits.

Two ways of handling outward calls became possible: cut through and register sender. With memory and control capability being relatively inexpensive in modern PBXs, the register-sender approach has much to recommend it. Many manufacturers, however, still cling to the cut-through approach. Note carefully the difference: with cut through, the system connects as soon as possible to the CO and, in certain circumstances, monitors digits as they go past, while with register-sender operation, the system takes in the entire number and performs restriction, routing, and whatever else may be appropriate and then continues with setting up the call.

In setting up an outgoing call, the PBX sometimes doesn't know when the CO is ready to receive digits, and it almost never knows when the called party answers. Further, it may not know when the call is ended by the called party hanging up. In cut-through operation, the user hears CO dial tone and, ultimately, the called-party answer (or busy tone or whatever); this relieves the PBX of the need to do anything. But the PBX must monitor the internal extension very carefully since this, in many instances, is the only source of information about adding features or ending the call. If the internal extension has been put on hold, the outside party can sometimes hang up without being detected.

When call detail recording equipment is added to or built into the PBX, the inability of the system to determine when a call is answered is a severe problem in making accurate toll bills. On the other hand, machine detection of hang-up is fairly simple if "ground start" trunks are obtained from the CO; ground start trunks also work well to provide an automatic start dialing signal in register-sender operation, although, for a variety of reasons, detection of dial tone directly has much to recommend it.

INCOMING CO CALLS DIFFER Traditional incoming calls from the CO are quite different from the intra-PBX and outgoing calls we have examined so far. It must be recalled that a PBX trunk is, as far as the CO is concerned, a station line just like the one that goes to a telephone. Thus when the CO wants to complete a call to somebody served by a PBX, it will send ringing down the line (viewed from its end) or trunk (when viewed from the PBX end). Ringing (which is a large, high-power signal at 20Hz and 86 volts) operates a ring-up relay or similar device in the PBX's trunk circuit to make some kind of indication to the system that a CO call is coming in. This indication is forwarded to the console position where the attendant can see and respond to it.

When the attendant signals the system control that the call is to be accepted, ringing is tripped and the call is connected to the console. Tripping ringing causes the CO to put its equipment in the talking state and start charging. Thus, the calling party pays for the call if it reaches the PBX attendant.

The attendant obtains the called extension number or, more often, the name of the called party, and instructs the control to manipulate the switching matrix to complete the call. The control applies ringing toward the called extension, generally returns audible ringing (ringback) to the calling party, and watches for answer. It is highly desirable that a PBX monitor this ringing situation and return the call to the console after a timed interval if no answer is obtained. It is also desirable that the PBX be able to monitor the trunk circuit for abandon in case the calling party hangs up. Ground start trunks are required for this to be possible. As on any call, the system must monitor the internal party for switch-hook flashes and hang-up.

In many systems, the same CO trunks used for incoming calls are used for outgoing dial 9 calls. This poses a problem that must be understood if serious trouble is to be avoided. The trouble concerns seizure of the same trunk simultaneously by both the CO and the PBX. This situation puts the PBX caller, who has just dialed 9, in direct contact with an incoming call that is almost certainly intended for someone else.

Simultaneous seizure can be minimized at the PBX if the trunk circuit involved is made busy to dial 9 calls as soon as the CO has seized the trunk from the far end. This cannot, in general, be done if all the CO does is apply ringing. Ringing, in most central offices, is on for two seconds and off for four. Thus, if the CO seized the trunk at the start of the silent interval in the ringing cycle, up to four seconds could elapse before the PBX trunk circuit would know. In a busy hour, four seconds is eternity and many dial 9 calls would have a crack at the "idle" trunk. Thus, once again ground start trunks are necessary. They tell the PBX immediately when the trunk is seized from the CO and thus allow the PBX to direct outgoing calls to other circuits. Trunks used in the outgoing direction for dial 9 and incoming for completion to the attendant are usually called combination trunks. With the present PBX and CO designs, it is imperative that the CO use ground start circuits.

Direct inward dialing (DID) is rapidly making Centrex unnecessary, and most modern PBXs offer it. As mentioned above, DID is not particularly difficult at the PBX. Any PBX that can handle dial repeating tie trunks is ready to go immediately. All that is needed is for a CO to be able to send dial pulses; this is harder to come by. In some metropolitan regions, New York City in particular, tandem offices in the public network bypass the local central offices and connect directly to DID PBXs. This lets each PBX be treated as a small central office, and everything works fine.

There are, of course, some practical details. All the old-fashioned step-by-step (SXS) PBXs were quite fast in that they could accept dial pulsing on a tie trunk as soon as the trunk was seized at the distant end. Unfortunately, most modern COs and PBXs are much slower. After detecting seizure, they must find a digit detector to attach to the trunk, or arrange their internal operation to examine the trunk circuit fast and often enough to catch all the dial pulses as they come in. This may take a while. Thus, the distant end must be held off until the PBX is ready. The technique used is called "wink start." The CO (or tandem office) is psyched up to watch for a momentary off-hook signal to be returned. At the end of this one fifth of a second interval of off-hook, the telco end knows it can start sending dial pulses.

Some modern PBXs are always watching their DID trunks (and tie trunks) for incoming dial pulses, and thus are as fast as SXS systems. However, they sometimes have to return a wink start signal anyway to make the CO happy. Thus, we have two problems here: we must know if a PBX is able to send wink start to satisfy the CO, and if it must send wink start to fend off the CO until the PBX is ready to receive digits.

The current trend in PBX design is away from the relatively inexpensive 500/2500-type sets.

When digits are sent to a PBX in a DID situation, the telco sends them at the slowest possible rate: 10 pulses per second, with something more than half a second between trains of pulses that constitute a digit. Thus, to send a four-digit extension number to a PBX, the CO will require at least four seconds. If Touch-Tone had been used, about a half a second would have been needed. There is even a standardized form of dial pulsing, long used between common control switches in the public network and between PBX attendants and common control COs, in which the time could be cut to two seconds. But, for reasons that are unknown, the slowest form

of dialing is used. This is particularly amusing in that the Bell System is installing, as rapidly as possible, a new digital method of signaling called CCIS (Common Channel Interoffice Signaling) that may, among other things, cut down signaling time in the public network. The time saved will, of course, be balanced by the slow pulsing into DID PBXs.

A DID call seizes the trunk, gets wink start (if required), and sends the PBX the extension-identifying digits it needs. The PBX then completes the call. That is, the PBX rings the called extension, sends audible ringing (or, if appropriate, busy tone) to the calling party, and monitors for answer. When the

call is answered, the answer signal is returned to the CO and the talking connection is established. Note that charging does not start on a DID call until the called party (or somebody else) answers. The telephone company, to be sure it gets paid for the call, will not make the trunk work in both directions until the answer signal is received. The trunk must work in one direction so that the caller can hear audible ringing or busy tone—but it won't work in both directions until after answer has taken place and charging begins.

THE USER INTERFACE

Telephone sets are the interface between the telephone system and the users. Consoles, in modern equipment seldom more than glorified telephone sets, fall into the same category. Additional interfaces, provided for maintenance and information exchange, are often standard teletypewriters or data terminals although sets or consoles with special displays are sometimes provided. Needless to say, the instrument and the switching system must be able to work with one another if the system is to support communications.

Telephone sets must, in general, convert acoustical energy to electrical energy and vice versa to permit voice communications. For nonvoice services, analogous requirements exist. In addition, the station user must be able to signal toward the system to "place his order," and the system must be able to signal toward the caller to let him know what the system is doing to carry out his instructions. The system must also signal toward the called party to encourage him to answer his phone and, in some instances, to tell him which line he is supposed to answer.

The basic mechanism of the 500-type telephone set, the one in present use throughout the country, was perfected just about the time transistors were announced. Thus, station apparatus has not, as yet, taken any particular advantage of developments in solid-state physics. The 2500-type telephone set, identical to the 500 except for the use of a DTMF (the generic name for AT&T's Touch-Tone trademark) pad for signaling rather than a rotary dial, does, it is true, use transistors for generating the tones. But that is sort of an add-on application.

The 500/2500-type telephone does use some rather early solid-state components to compensate for the distance between the central office and the telephone. The curious result is that the closer to the switch a telephone is, the more it upsets transmission when connecting to trunks. Or, to put it another way, the majority of PBX telephone sets are located where they will harm overall network transmission.

It happens that something less than 20% of all telephones are served by Centrex

DEPT. OF FUNNY SMELLS

EMERGENCY

CARTOON BY SIDNEY HARRIS

or PBX systems, while the rest are served directly by central offices on (mostly) a single line basis. It appears, however, that the great majority of toll calls are placed from PBX and Centrex telephones, and these calls are, of necessity, made during the business day when rates are high. Thus, improving transmission at PBX telephones would seem to be one of the easiest ways to improve overall transmission on a nationwide basis. And the easiest way to improve transmission would be to make PBX systems four-wire end-to-end, including the telephone sets.

The current trend in PBX design is away from the relatively inexpensive 500/2500-type sets. The SL-1 PBX from Northern Telecom was the first to go in this direction, and Bell followed quickly with sets for Dimension. Danray also had electronic sets from the beginning. These sets require three or four pairs between the telephone set and the switch, but unfortunately all except Danray use only one pair for voice transmission. One can understand the Dimension using two-wire telephone sets, since Dimension is a two-wire switch. And one can even understand the SL-1 using two-wire telephone sets even though the switch is four-wire internally; after all, Northern Telecom is a telephone-oriented company and tradition is of considerable importance. What is really hard to understand, however, is the new Rolm electronic telephone set. The Rolm switch is four-wire, and the people who designed it were determined to innovate. But their new microprocessor controlled telephone set, the ETS 100, uses two-wire transmission.

Only Tele/Resources and Danray have, to date, seen the logic of four-wire telephone sets. The T/R System 32 is four-wire from station to station, as is the Danray. The only time they are vulnerable to transmission echo is when they connect to two-wire facilities, as when they go to off-premises stations or, more important, to CO trunks.

Conventional 500- or 2500-type telephone sets have a microphone or transmitter for converting acoustical energy to electrical energy to be transmitted via wires. Similarly, they have a receiver for making the opposite conversion. The two are connected via a network, which permits a single pair of wires to be used for transmission in each direction when two separate devices are obviously required at the telephone set to interface a caller's ear and mouth.

The "network" serves an additional purpose. It provides "sidetone." That is, it allows a certain amount of the spoken energy to be fed back to the ear of the speaker (without delay). That is necessary because we are used to hearing ourselves speak (cover both ears and speak aloud to detect the impact of not hearing yourself); if we don't, we change our speaking patterns. As it happens, the

sidetone circuit feeds back slightly less sound than we hear through the air; this tends to make us talk a little louder, and provides better volume for the person on the far end of the connection.

The switch-hook, the next device we encounter, makes a path for current from the PBX or CO switch through the telephone set when the caller picks up the handset. The handset, of course, contains the transmitter and receiver, and, when not in use, sits in a cradle on the telephone base. While placed in the cradle, it operates a switch that turns off power; when lifted, the switch closes and completes the path needed for current to flow.

Note that the PBX or CO must always monitor a line for its on-hook or off-hook status. This is called supervision. The switch must know if the user is originating a call, answering a call to his phone, or terminating a call that is already in progress. In recent years, the switch-hook "flash" has regained the importance it had in manual systems. A flash is a momentary on-hook. The system detects it and knows that it is not a hang-up followed by the origination of a new call.

The purpose of the flash is to tell the system that the user wants to send an additional command, a "feature code." Particularly if the system uses DTMF, but often when dial pulsing is used as well, a digit receiver must be connected to unscramble the user's new command. Use of the switch-hook flash and as many as 20 feature codes is common in modern PBXs.

TELEPHONE SET FEATURES

Special telephone sets designed to work directly with the PBX combine PBX capabilities with user convenience. These sets, available from Northern Telecom, the Bell System, and others, usually have a group of buttons that can be either line pickup keys (as in conventional key telephone sets) or feature keys. What usually happens is depression of a selected key tells the system which key is depressed. The system looks up in a table in memory to find out what is going on, and carries out the appropriate action. If a line select key is pushed, the system connects an incoming call for the line to the tel set via the switching matrix; on an outgoing call, charging and restriction are based on the class marks of the line selected. If a feature key has been pushed (the hold button, for instance, although we now have many other possibilities), the system does whatever is required. If visual cues are required by the user, the PBX control causes an appropriate signal to be returned to light or blink the appropriate lamp, to cause the tone ringer to sound, or to provide some other suitable signal.

These modern electronic sets have a

number of advantages. First, they are much easier for the user when contrasted with flashing the switch-hook, dialing a number of feature codes, and identifying a variety of call progress tones. Second, they usually need only three pairs, and three pairs will work on any kind of phone from a simple single line set to a 30-button call director. Finally, they permit complete program control of changes in lines picked up, features selected, etc. This kind of control can often be handled from remote locations via a dataset in a dial-up connection or by the customer's communication manager. The saving in OCC (other charges and credits) for moves and changes can be considerable in an active location, and the saving in aggravation and time when the communication manager can effect the change directly rather than trying to honcho an order through a reluctant vendor can hardly be appreciated until it is experienced.

One problem must be noted, however. Single line sets usually require different line cards in the PBX than do these modern sets. This shows that more than a program change is required when one goes from a single line to a multiline set. Further, it is evident that when one changes from a five-button set to a 30-button set, the sets themselves will have to be interchanged even though programming can do most of the rest. In any event, the identification required on each button must be made at the set and not with the program. Even so, the modern sets can be cost effective when properly used.

It should be possible to switch data over public and private voice networks like any other signal. Just how readily this can be done was demonstrated by Danray and taken up by Northern Telecom upon its purchase of Danray. One simply puts an RS232C interface on the electronic telephone set (there are, apparently, better or less expensive interfaces, but 232 exists in vast quantities) and plugs in a terminal or a computer. Danray uses the power pair (one of the three pairs required by their electronic set) to carry off 9,600 bps full-duplex data to a separate auxiliary data switch. The voice telephone is used to set up the connection, but once the data path is set up, the voice equipment is free for make or receive voice calls. This kind of simultaneous voice/data operation seems to be highly favored by those who normally use terminals, particularly over some version of alternate voice/data where the voice phone is tied up with the data connection.

Northern Telecom uses the same approach with the SL-1 PBX, but without a separate data matrix. The SL-1 has a digital time-division switching matrix, and can switch data directly by simply omitting the analog to digital conversion in the data line circuit. Each telephone has two appearances on the

No amount of processor duplication will save the system in the event of a power failure.

switching matrix: one for voice and one for data. Again, the voice phone is free to make and receive calls once it has the data connection established. With the long holding times of data calls (normally found with timesharing in industry), some traffic problems may develop here—but the idea seems to be quite good.

Rolm has done something different. A single voice path through the matrix is submultiplexed into as many as 40 data paths (depending on the speed of the switched data). In addition to adding switching capability with negligible degradation of the system's voice handling capability, this approach appears to be quite inexpensive compared with even relatively low-speed modems.

Many people feel that a telephone set

in a business area should have a full alphanumeric keyboard and display and a handset, along with an array of feature and line pickup buttons. This relatively dumb terminal could then interface with the switch, which could include software to permit the set to do whatever is required: be a timesharing terminal, a word processing station, an electronic mail system, or whatever. The cost of modern electronic components would seem to point in this direction.

Consoles are provided with PBXs to permit human assistance when required to complete calls. Early consoles interfaced trunks directly and, working through the trunk appearance on the switching matrix, effected call completion. Today the trend is to use something very much like an electronic telephone set as a console; the switch con-

nects calls needing assistance to the console as to any other telephone, but provides a bit more information by driving suitable displays. Consoles, like electronic telephone sets, can be arranged to have digital signaling paths to the system control to interchange information directly without using the voice path.

SYSTEM CONTROL METHODS

Most modern PBX switches today use a processor of some sort for control. Stored program approaches are now standard, but this hardly tells the whole story. We are confronted with minicomputers and microprocessors, RAM and ROM, backup tapes and disks, and separate standalone mechanisms for call data recording and processing. Like any small computer-controlled industrial system, a PBX is operated by its processor. The processor, backed with program instructions, inspects all lines, trunks, service circuits, and other portions of the system, first for supervisory information (on-hook and off-hook) and then for any additional information that may be needed to make things work (dialed addresses, feature codes, signals from feature/line buttons on electronic telephones, alarm and trouble warning information, etc.). Further, maintenance or other control information may also be coming in.

The control responds to externally generated signals and to internal signals that result from processing information in accordance with data stored in memory. That is, it checks class marks and other instructions and sets up paths through the switching matrix between lines and/or trunks, records information in memory for billing and traffic purposes, and, in general, carries out the functions of an operator at a manual switchboard.

Processors are usually not duplicated for redundancy in small systems; they almost always are in large systems. The breakpoint appears to be at about 400 lines. Duplication for reliability requires additional complexity in that both sets of memories have to be kept up to date, and the computers have to be able to decide which one is in charge. In some systems with duplicated processors, standard business programs (payroll, inventory, etc.) can be run on the standby machine; other systems use the standby machine to handle other functions (such as Rolm's electronic mail) until it is needed. With or without duplication, the system may very well have extra time and memory that could be devoted to store and forward data, etc.

No amount of processor duplication will save the system in the event of power failure. In such an instance, a backup battery is required to keep the system running (or, at least, to save the memory) until the power is restored. If batteries are not provided, some

THE
DYNAMIC BUILDING

BIO-DYNAMIC PLAN	312
COMMUNICATION DYNAMICS	402
CONTROL DYNAMIC DESIGNS	105
CREATIVE DYNAMICS, INC.	107
DYNAMIC ASSOCIATES	210
DYNAMIC MANAGEMENT GROUP	220
DYNAMIC SYSTEMS, LTD.	320
DYNAMIC THERAPY INST.	115
MEDICO-DYNAMICS, INC.	405
MOTIVATIONAL DYNAMICS	120
MULTIMODAL DYNAMICS, LTD.	206
PERSONAL DYNAMICS, INC.	305
REGIONAL DYNAMIC DESIGNS	415
SPECTRUM DYNAMICS	202
VITAL DYNAMICS ASSN.	318

One of the main advantages of computer control of switching systems is the ease with which new features can be installed.

means may be required to reprogram memory when the power comes back up. Systems using a volatile random access memory (RAM) lose their program when power is cut off, so the system must be able to reload from some reasonably permanent memory when the power is restored. Often a magnetic tape or disk contains the backup program. The backup memory does not always keep track of the calls in progress or the station-programmed features such as call forwarding. Thus, reload sometimes does not get back to the starting point.

One of the main advantages of computer control of switching systems (or other industrial equipment) is the ease with which new features can be installed, services can be upgraded, and changes can be made in the course of regular activity. Note, however, that not all stored-program processor controlled systems permit easy changes. Read only memories (ROMs) are not always changeable in the field—sometimes they have to be unplugged and replaced, while in other instances, they must be altered in a special programming device. ROM memory is quite reliable—it can't be changed by unauthorized personnel very easily, and it is non-volatile so that it is ready to go as soon as power comes back after a failure. But in most instances, the greater flexibility of making changes from a maintenance terminal, the console, or a special telephone set is of much greater importance. Usually some sort of compromise is used: program instructions that do not change are stored in ROM, while parameters that define special features and translations are keyed in, stored in memory that is easily changed but wiped out with power failure, and duplicated on the backup tape for reload with power restoral. There are almost as many variations as there are PBX systems on the market, and we haven't seen the end of it yet.

Switching matrices can be cataloged under three general headings: space division, frequency division, and time division. In space division, which includes all the older electromechanical systems and some electronic systems, the control finds a path through one or more sets of switches between the line and trunk, sets up that path, and the path, a physical connection that can be traced from point to point, is used by the caller for the duration of the call. When the call is over, the various switches are released for use by other callers.

Frequency division is seldom used, while time division is taking over the field. Unfortunately, time division can be carried out in many different ways, some analog and some digital. And even the digital techniques permit a wide variety. For example, pulse code modulation (PCM) and delta modulation are incompatible with each other, and no two

forms of a delta modulation system are compatible.

ANALOG VERSUS DIGITAL

There is a great deal of discussion about the merits of digital versus analog switching. Some people feel that a switch controlled by a digital computer is a digital switch, and they point out quite correctly that most of the features of interest to users are a direct result of stored program control. But this is not what is generally meant when one speaks of digital switching. Digital vs. analog in this context refers to the way the switching matrix operates, and has little to do with the control equipment. Further, it is true that many of the features of interest to users today can be carried out quite well with a switch that has a matrix that works on analog principles.

Analog transmission means that the signal transmitted is a direct analog of the actual signal. In telephony, the actual signal is compressions and rarefactions of air—changes of air pressure that are interpreted as speech. These changes of pressure fall on the microphone or transmitter in the telephone handset, and the microphone converts pressure variations to current variations. Note that the current variations are a direct analog of the pressure variations, increasing and decreasing in exactly the same way. Note also that the power in the current variations is appreciably larger than the power in the sound pressure; there is gain or amplification available in the telephone set. This is the only way long distance telephony could have been practical 25 years before the invention of the vacuum tube.

The electric current can be transmitted over wires, and it can operate more modern equipment to make another analog of the original signal which permits transmission via radio beam. In all these instances, a view of the signal on an oscilloscope would look just the same as if the scope could read sound pressure directly.

Amplification is very important in analog transmission. A small incoming signal must control something that makes a large but directly analogous outgoing signal. The signal will be attenuated by a long cable, but can be amplified again to get back to its proper size. Unfortunately, amplification can't tell the difference between the original signal and that signal plus any noise that may have been picked up along the way. Each time amplification is used, more noise is added to the signal. Ultimately, the signal is submerged in noise. Even a very small variation in the original signal will change it, and all such changes will add up.

A digital signal cannot be continuously variable as can a pressure wave or the analogous current variations produced by the

microphone. A digital signal can only take on one of a finite number of values at any instant. It does not have to be amplified, however. The digital signal is measured, its value is determined, and a new signal is made just like the old one. All the noise is stripped off and lost.

This technique is called regeneration and is very old. It was used in telegraph systems before the telephone was invented. Telegraph pulses would be distorted by long wires, and the dots and dashes couldn't be distinguished; then somebody figured that, instead of using one long wire with a key at one end and a sounder and a huge battery at the far end, a number of short wires could be used, end to end, each powered with a smaller battery. The first wire would have a key on one end for sending, and a relay and battery at the far end for receiving. The relay would follow the key properly, and its contacts could act as a key in the next circuit to regenerate a new signal. This regeneration could go from circuit to circuit, outwitting noise and distortion. In a digital telephone system, we convert a voice signal to something that looks like a telegraph signal. Then we can use regeneration rather than amplification and be free of noise and distortion.

The characteristic of digital modulation techniques is that all of the pulses are the same in both amplitude and width. All the system has to do, then, is determine whether a pulse is present or not. A series of standard pulses (or absence of pulses) in successive time intervals can define the amplitude of a voice signal at any instant. This requires more pulses than analog modulation techniques, but simplifies pulse detection and retransmission.

Can a customer tell the difference between an analog and a digital switch? Not on a voice call. Make a connection through relays, reed switches, crossbar, or SXS switches, or with a variety of electronic techniques including the various digital modulation schemes discussed above, and a caller cannot tell which system has been used. On a per-call basis, there is no difference. Why, then, go to all this trouble? Why not stick with SXS, or even stop with 8,000 samples per second where each is of variable height or width?

There are reasons for a PBX to be digital. The public telephone network is about half digital today in an 8-bit PCM format. If these 8-bit words or bytes of PCM are kept intact when they go through switches and trunks, new frontiers open up. It just happens that ASCII uses seven bits, often plus a parity or check bit. Voice channels normally handle 8,000 eight-bit bytes per second; if they can be switched together in built-up connections ultimately extending from one telephone to another, there are possibilities for data transmission that make all present data systems,

The purchase of a digital PBX to obtain all the advantages of the digital future is a bit premature at the present.

including packet networks, seem roughly comparable to the postal service.

THE ULTIMATE PAYOFF

This is the ultimate payoff. With an all-digital network served by digital switches handling digital user stations, voice and nonvoice communications can be mixed as needed on a per-call basis. Note that nonvoice communications are particularly easy to handle: on that day in the future when there is an all-digital network, modems will not be needed. A data signal will just bypass the A/D converters required for the speech part of the telephone and get directly on the bit stream with information precoded to fit. When any dial-up connection can handle 8,000 ASCII characters/sec, new opportunities for business communications not yet even considered will be possible.

Unfortunately, there is a long way to go before such a future will be possible. Most intertoll trunks will remain analog until fiber optics becomes generally available for long-haul PCM circuits. Microwave, satellite, and coaxial cable can handle many more analog circuits than digital at the same cost or bandwidth so they will not vanish easily. Further, protocols required by various data machines to permit them to talk to one another vary from data system to system; indeed, in data communications, one of the principal functions of a node or switch is to convert from one system to another to permit connections between otherwise incompatible equipment. Finally, digital telephone systems need a variety of other signals for synchronization, supervision and control; these tend to make a universal voice/nonvoice system hard to achieve. Thus, the purchase of a digital PBX to obtain all the advantages of the digital future is a bit premature at the present time.

What can a digital switch do for a user today? A few things, some of which may be quite important in special applications. First, distributed switching is now possible with several modern switches. One can take parts of the switch, analogous to line groups in a SXS system, and put them near the telephones. Using T-Carrier techniques, these remote units can connect something like 100 extensions to the main part of the switch with two pairs; this minimizes the three-pair wiring from the remote units to the electronic telephone sets, and the overall economies are considerable in a large building or in a campus-type environment.

The second thing a digital switch can do is switch digital signals internally in digital form, without modems. A digital switch is needed, although Danray, using an analog switch, pioneered the effort as described above. A small interface box is required to take the computer or terminal's digital signal and convert it slightly to conform to the needs of the power pair, which is the channel to the switch, but this little box is relatively inexpensive compared to a high-speed modem. In the newest PBXs, A/D conversion takes place in the telephone set and not at the line card in the switch. Then the data signal can enter the digital world immediately, without the need to utilize power pairs for access.

A modern PBX, from the equipment point of view, is much simpler than the older electromechanical systems. Most of the sophistication is built into the program that instructs the control, and modern digital circuitry has simplified most of what is left. The switching matrix tends to be quite small, and new opportunities are available for convenient design of user terminals. Trunks circuits, going mostly to obsolete (two-wire analog) central offices and to analog long-haul tie trunks, have to meet the outside world on its own terms, but even they tend to take advantage of PBX control and component sophistication.

A modern PBX can do many things the older PBXs could not do, and can open the way to doing things that are done today by completely separate systems. We are right on the edge of a whole range of developments that will change the way business is conducted. But we are not there yet. Thus, one should not rush to buy the current state of the art just because it is more modern. One should understand just what is involved, and make sound decisions on the basis of rational information. And above all, be ready for change. ✻

Leo F. Goeller Jr. is an independent consultant with Communication Resources in Haddonfield, N.J. He is also on the Board of Members of the Business Communication Review.

Jerry A. Goldstone has been the editor and publisher of *Business Communications Review* for the past 12 years. Copies of the BCR manual can be obtained from the company by writing to 950 York Rd., Suite 203, Hinsdale, IL 60521.

CARTOON BY JACK CORBETT

THE EVOLUTION
OF THE SL-1 PBX

Dinker Bir, Jim Eng, and Rod Hoo

The evolution of the SL-1 PBX

Dinker Bir, Jim Eng, Rod Hoo

Late last year Northern Telecom introduced enhanced models of the SL-1 PBX (private branch exchange). These models represent the latest stage in the SL-1's evolution. Over the course of this evolution, the SL-1's hardware has incorporated new technology that has become available. As a result, the SL-1's per-line power consumption has dropped; the floor area, or footprint, that it occupies has been reduced; and manufacturing costs have been reduced, which has enabled the SL-1 to remain competitive in price.

As the hardware has evolved, so has the software. The SL-1 was introduced with about 100 000 lines of software code. Today its software base contains more than a million lines of code, which represent about 500 man-years of development work. Most software development has been for optional features. The SL-1 now has feature packages that support the needs of businesses, North American military forces, hospitals and medical clinics, and other types of organizations. Features in these packages provide convenience calling services, reduce the cost of communication service, make better use of private networks, and meet special user needs. Software development has also created the international

software package, which enables the SL-1 to operate in countries that use tones, ringing, and signaling schemes different from North American ones. The recently introduced international private networking features enable the SL-1 to support private networks in most countries in the world (*Telesis* 1983, no. 2).

As part of the hardware and software evolution, the SL-1 has integrated voice communication with data communication. The integrated voice and data switching (IVDS) group of features includes interface modules to computer terminals, data line cards, and high capacity interfaces to computers. (See page 28.) The terminal interface modules are connected by standard telephone twisted pairs to the SL-1. Twisted pairs, high capacity links, or both are used to connect the SL-1 to one or more computers. The combination of transmitting data over twisted pairs and connecting terminals and computers to each other through the SL-1 brings several important benefits. Terminals can be located wherever there are twisted pairs – in other words, almost anywhere in a typical office building. Terminals can be moved easily from one part of a building to another – relocation does not require that wiring be reinstalled. And terminals

that have had access to only one computer can gain access through the SL-1 to numerous computers.

As the SL-1 evolved, its performance improved. The average per-line power consumption of the new SL-1 models is 20 percent less than that of previous ones. The footprint of the SL-1 has been reduced between 25 and 50 percent, depending on the number of lines supported, by the most recent hardware enhancements. And the mean time between system level failures has increased to 20 years for a duplicated central control system.

The SL-1 was the first fully digital switching system that BNR designed and Northern Telecom manufactured. Close to 8000 SL-1s are now in service in more than 40 countries. Together these systems serve the equivalent of three million voice lines. The four most recently introduced models of the SL-1 are the SL-1S, SL-1MS, SL-1N, and SL-1XN. All models of the SL-1 have the same hardware and software architectures, use the same software, and support the same user features. They differ only in their hardware packaging and in the number of lines that they support. The SL-1S supports 30 to 120 lines; the SL-1MS supports up to 400 lines; the SL-1N supports up to 1500 lines; and the SL-1XN supports up to 5000 lines.

The SL-1N and SL-1XN are the products of a recent development program undertaken by BNR's Mountain View, California laboratory in association with Northern Telecom's Santa Clara, California and Belleville, Ontario manufacturing plants. In this development program the central

Since the SL-1 PBX was introduced, its software base has increased 10 times in size. The most recent models of the SL-1 consume 20 percent less power than the models of the previous generation and they occupy 25 to 50 percent less floor space, depending on the number of lines supported. These improvements are represented by the colored areas in the disk pack, the electrical plug, and the equipment cabinet.

Reprinted from <u>Telesis</u>, 1984 one, pages 20-27. Copyright © 1984 by Bell Northern Research. All rights reserved.

control, switching network, and peripheral equipment (Figure 1) were redesigned. Some of the enhancements in this program have also been applied to the SL-1S and the SL-1MS.

The development program introduced new very large scale integrated (VLSI) components that include 64 kilobit memory chips and the W05, a filter codec chip designed by BNR and Northern Telecom. These components have been instrumental in reducing the SL-1's footprint and power consumption. The development program has also enabled the SL-1 to be engineered for nonblocking service in data switching applications and to meet new electromagnetic and radio frequency interference regulations. (These regulations do not apply to SL-1s already in service.)

Since the SL-1 was introduced, it has incorporated all new enhancements without changes being made to its hardware and software architectures. This is no accident: the SL-1 was originally designed to be able to incorporate such changes. As a result, SL-1 PBXs already in service

are able to add new hardware and software features as these become available. The ability of the SL-1 to add new developments is a demonstration of one of the fundamental concepts of Northern Telecom's OPEN World. (OPEN stands for Open Protocol Enhanced Networks.) That concept, called continuity, is that a communications system should be able to evolve without becoming obsolete. The concept of continuity has been and will continue to be applied to the SL-1 and other Northern Telecom communications systems.

Hardware evolution
The SL-1's hardware enhancement program had three parts: common equipment enhancement, switching network enhancement, and peripheral equipment enhancement.

The common equipment enhancement consisted of redesigning the memory and central processing unit (CPU) circuit packs. The cost of 64 kilobit random access memory (RAM) chips had dropped sufficiently to permit their use on the memory packs. As a result, the capacity

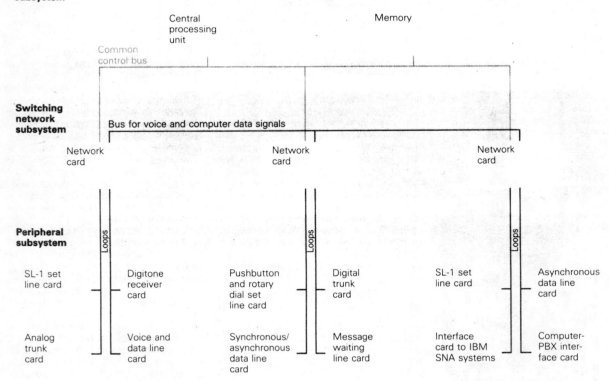

Common control subsystem

Central processing unit

Memory

Common control bus

Switching network subsystem

Bus for voice and computer data signals

Network card

Network card

Network card

Peripheral subsystem

Loops

Loops

Loops

SL-1 set line card

Digitone receiver card

Pushbutton and rotary dial set line card

Digital trunk card

SL-1 set line card

Asynchronous data line card

Analog trunk card

Voice and data line card

Synchronous/ asynchronous data line card

Message waiting line card

Interface card to IBM SNA systems

Computer- PBX inter- face card

Figure 1. The common equipment for the SL-1 PBX is divided into three functional subsystems: the common control, the switching network, and the peripheral modules. The main components in the common control are the central processing unit, which controls the operation of the system, and the memory, which holds the operating programs for the system and user data such as features and class-of-service information. Control signals between the common control components and the network circuit packs travel over the common control bus. The network subsystem is where actual connections are made. This subsystem consists of network cards and a bus that carries system user traffic – voice signals and computer data – between these cards. In the most recent models of the SL-1, each network card has two multiplex loops; each loop serves a number of peripheral modules. One hundred and sixty ports can be supported through peripheral modules and one loop. The modules in the peripheral subsystem support both voice and data communication. Various types of peripheral modules provide service circuits and serve as interfaces to station sets, data communication equipment, and trunk facilities.

of three earlier memory packs – 192 kilowords of memory – is accommodated on a single new memory pack. A new type of microprocessor was introduced to the CPU for the MS and N models. It has enabled the number of circuit packs needed for the CPU to be reduced from three to one. The new microprocessor has also increased the processing speed of the CPU by 60 percent over the equivalent earlier models of the SL-1. The combined reduction in the number of circuit packs for the memory and CPU has enabled these units to be housed on one shelf in the SL-1N rather than on the two shelves used in the equivalent earlier

model. Similarly, the space required for the memory in the XN has been reduced from two shelves to one.

The enhancements to the switching network were made primarily to improve the support of data communication on the SL-1. Both parts of the network enhancement concern the loop, which is the group of paths between a network card and the peripheral modules that it serves. (This loop should not be confused with the subscriber loop.) The SL-1 loop contains 32 channels. Two channels are reserved for control signaling between the CPU and the peripheral modules. The other 30 provide communication paths between the switching network card and the peripheral modules that it serves. (These modules can support up to 160 lines.) In earlier versions of the network card, consecutive channels on the loop were paired. For example, channel 2 carried one direction of a

call, channel 3 the other. The SL-1 has provided a high level of voice service with this arrangement: it can be engineered for a blocking probability of 0.01.

The traffic requirements for data communication differ significantly from those for voice communication because the holding time of the average data call connection is substantially longer than that for a voice connection. Two developments have improved the SL-1's support of data service. First, the number of loops on a network card was increased from one to two. Second, the requirement that consecutive channels in the loop be paired was removed by increasing the size of the memory within the network card. These developments increased the traffic capacity of the loop by 110 percent. As a result the SL-1 can be engineered non-blocking for data service.

The peripheral enhancement program reached five main objectives. One, the footprint of the SL-1 was substantially reduced. Two, the per-line power consumption was reduced. Three, a new transmission standard recommended for digital PBXs by the U.S. Electronic Industries Association (EIA) was met. Four, new U.S. Federal Communications Commission (FCC) standards for radio frequency interference (RFI) and electromagnetic interference (EMI) were met. Five, engineering the power supply for an SL-1 being placed in service has been made easier.

Key to reaching the first three objectives was the W05, a new filter codec chip that was custom-designed by BNR and Northern Telecom for use in the SL-1 and other systems. This chip can perform analog-digital conversion according to both A-law and μ-law conversion rules. Thus the W05 can be used for both North American and international service.

The W05 has replaced thick film hybrid components used on earlier versions of the five most commonly used peripheral cards. The number of ports on the line cards for SL-1 station sets and for rotary dial (500) and pushbutton (2500) station sets was increased from four to eight. This year the number of ports on the message waiting line cards and two types of trunk cards will also be doubled. Also contributing to the increased density of the 500/2500 line card are new smaller relays and transformers.

The W05 filter codec has reduced the SL-1's per-line power consumption. When a line is supporting a call, whether voice or data, the W05 consumes 100 mW of power. When the line is idle, the W05 powers itself down to consume only 15 mW of power.

The W05 meets the transmission standards for digital PBXs that were recently recommended by the EIA. These standards cover return loss, longitudinal balance, gain variation, idle channel noise, and other transmission characteristics.

The FCC's new EMI and RFI requirements were dealt with at both the circuit pack and system levels. At the circuit pack level EMI and RFI were minimized through design practices: CMOS (complementary metal oxide semiconductor) components were used wherever they would not adversely affect the performance of the circuit pack; traces on the pack that carried rapidly changing signals were isolated; and multilayer circuit boards were used for the backplanes for peripheral and network cards.

Figure 2. Recently the U.S. Federal Communications Commission introduced new regulations that govern electromagnetic interference (EMI) and radio frequency interference (RFI) emissions. To meet these regulations, the SL-1 PBX has been provided with a new cabinet. The cables that extend from the cabinet have the potential both to introduce noise and interference into the system and to carry noise and interference out. To prevent such noise and interference transmission, filtered connectors, as shown above, are used on the new cabinet. In addition, the cabinet has been plated with zinc and has gasketing fastened to its doors and side panels; the zinc and gasketing together ground out EMI and RFI emitted by the SL-1 equipment.

Table 1. Basic business features on the SL-1 — a sample

Cost control
Class of service restriction
Restricted access to trunk groups
Code restriction
Automatic route selection
Authorization codes

Time-saving
Speed calling
Autodialing
Voice calling
Attendant recall
On-hook dialing
Handsfree conversation

Call answering
Call waiting
Call forwarding
Call forward, no answer
Call pick-up
Several directory numbers on station sets
Call park
Message center

Busy party access
Ring again
Call override
Paging access
Busy verification

Multilocation
Tie trunks
Tandem trunks
Off-premise extensions
Remote peripheral equipment

System attendant
Alarm lamps
Automatic timed recall
Barge in
Busy verify
Busy field lamp
Night service
Position busy
Incoming indicator

Executive
Call selection
Voice call
Private line
Conference
Call transfer
Dictation trunks
Call screening by secretary

These design practices tackled the EMI and RFI problem at the source – the circuit pack. They gave the designers confidence that work at the system level would guarantee that the SL-1 would meet the new regulations – which it has. Work at the system level aimed at keeping EMI and RFI that is emitted by circuit packs within the cabinets that hold them. A new cabinet was designed for this purpose (Figure 2). It is coated with zinc, which conducts away emitted EMI and

RFI. The edges around the doors are filled with wire mesh, which grounds out EMI and RFI. And filter connectors are used for cables that leave the cabinets.

The final part of the peripheral equipment enhancement program reduced the powering restrictions. New relays for the 500/2500 line cards can use unregulated power at 52 V rather than regulated power at 48 V. Other improvements, such as the W05, have further reduced power consumption. The extra regulated power is used for other purposes, which makes the SL-1 easier to engineer for service.

Software

The SL-1 was originally introduced with software that provided basic telephone service and supported a group of user features. These features, today considered essential for a digital PBX, include call waiting, toll restriction, direct inward dialing, and remote administration and maintenance; the original features for the SL-1 station set include autodialing, call forwarding, speed calling, and ring again.

Most of the development on the SL-1 since its introduction has been to create software features (Box 1). Today more than 300 basic features are available for the SL-1 set, attendant console, 500 and 2500 station sets, and the system as a whole (Table 1). In addition to these general purpose features are features that meet the special needs of businesses, military forces, hotels and motels, and hospitals and medical clinics.

In North America, SL-1 features are available on two generic feature packages.

One of these generic packages provides basic and specialized features for SL-1 PBXs used in hotels and motels. The other North American generic package is a general one that contains both general purpose features and specialized features for the user groups mentioned above. A third generic package contains the software needed for international service as well as user, system, and networking features.

When an organization obtains an SL-1, it chooses the features it wants from the appropriate generic package – that is, it does not have to obtain all the features in the package. Northern Telecom prepares a magnetic tape with the desired features and sends it to the organization's SL-1. The tape is placed in a unit on the SL-1 that loads the features into the common equipment memory. Fea-

Box 1. Software development for new SL-1 features

Since the SL-1 PBX was introduced, its total software base has grown from 100 000 lines of code to more than a million lines. New software is made available in two major and several minor releases each year. The features on each release generally take 6 to 12 months to develop. Development of a new release passes through five main stages. One, the design requirements for the new release are determined. Two, the new features for the release are developed. Three, the new release is tested by the BNR group that designed the new features. Four, the new release is tested by an independent BNR verification group. Five, the new release is tested by Northern Telecom staff and goes through field trial before it is made publicly available. The many reviews and tests within this development process ensure that the high level of quality for SL-1 software is maintained.

Determining new feature requirements

The commercial specification
–describes basic requirements for new features in the new release
–produced by a Northern Telecom marketing group

↓

Systems requirements
–drawn up for the group of new features if it is especially complex
–produced by BNR's networks technology division

↓

The preliminary feature specifications
–produced from the commercial specification and the systems requirements by BNR designers
–describe how each feature is to operate

↓

Review of the preliminary specification
–BNR designers and Northern Telecom marketing staff agree upon final changes to the preliminary specification

Developing the new features

The design specification
–created for each new feature by the designer to whom it has been assigned
–describes the interaction between the new feature and other SL-1 modules
–written in pseudocode

↓

Review of the design specification
–conducted by senior BNR designers
–evaluates the impact of each new feature on the processing speed of the SL-1's CPU, on system memory, and on traffic flow
–ensures that the new feature conforms to design and architectural philosophies for the SL-1

↓

Feature software
–written in the SL-1 programming language by the designer
–is modular in structure; modules are loosely coupled

↓

Review of the code logic
–performed by other members of the new release design team

↓

Compiling of the feature code
–performed by the designer
–feature is now ready for testing

tures are assigned to station sets through a terminal connected to the SL-1's common equipment.

Feature groups

Hotel/motel: This generic package was one of the first groups of specialized features to be developed. The software base on which it was developed was an early version of the general North American generic package. Today the hotel/motel package contains about 20 specialized features for hotels and motels. One of these features prevents hotel guests from making long distance calls from the telephone in their room after they have checked out. Another feature enables the telephone number to match the room number. Guests can dial directly between rooms; they do not need to have the connection made by the hotel's SL-1 operator. Guests can use their

telephone to set the time at which they wish to be woken up. An SL-1 system that serves a large hotel complex can support up to 31 attendant consoles.

International: The international generic package, like the hotel/motel package, was developed in 1978 on a base of the general North American generic package. Added to this base were the features the SL-1 needed for international operation – new tones and ringing cadences, a signaling interface to R2 public networks, periodic pulse metering, and an interface for 2.048 megabits per second (Mb/s) trunks to peripheral modules at remote locations.

Tones and ringing cadences vary between countries. For example, in one country a telephone may ring for one second every four seconds; in another it

may give two short rings every three seconds. The international tones and ringing software was written in a way such that different signaling schemes are easily accommodated.

A second development, equally important in the international marketplace, was the interface for connections to public networks that use the R2 signaling scheme of the International Telephone and Telegraph Consultative Committee (CCITT). By supporting both North American and R2 signaling scheme interfaces, the SL-1 is able to operate in most countries in the world.

Associated with the R2 interface is periodic pulse metering, which is used to assess both local and long distance service charges in many countries. The interface to 2.048 Mb/s digital trunks enables

Testing by the new release design group

Testing of individual features
–operation of each feature on its own is tested by the feature's designer
–then all new features are added to a duplicate of the current SL-1 software release

Regression testing – first series
–ensures that the existing features on the current software release are still functioning correctly

New feature testing
–checks the operation of the new features on the current release base

New release operation
–operation of the new release as a whole is verified
–once verified the new release is passed to an independent BNR software verification group

Testing by the BNR verification group

Regression testing – second series
–checks the operation of the current release software a second time
–performed with the testing and traffic simulation system, which was designed by BNR (*Telesis* 1983, no. 2)

Design verification testing
–ensures that the original design requirements for the new release have been met
–new software release then passes to Northern Telecom

Testing by Northern Telecom

Acceptance testing
–conducted by technology group in Northern Telecom's subscriber switching division

Field trial
–conducted on an in-service SL-1 at an actual customer site

Final approval
–is required from Northern Telecom's corporate quality control group before the new release can become publicly available

New release made available for use on new and in-service SL¹ PBXs

the SL-1 to provide service through remote peripheral modules to telephones and terminals at distant locations.

General North American generic
The main groups of specialized features in the general North American generic package are automatic call distribution, Electronic Switched Networking, Automatic Voice Networking, the integrated message service, and the hospital and medical clinic features.

ACD: Automatic call distribution (ACD) was developed for applications such as airline reservation service in which large numbers of incoming calls must be distributed evenly among service agents. ACD routes each incoming call to the first available service agent. If all service agents are busy, the call can be routed to a recording that asks the caller to wait for service. ACD monitors the level of incoming traffic, which enables the user organization to determine just how many agents are needed to provide proper service.

ESN: The Electronic Switched Network (ESN) enables convenience calling, automatic route selection, cost-control, and maintenance and diagnostic features to be provided over a whole private network rather than only for an individual SL-1. SL-1s and SL-100 PBXs, which can support up to 30 000 lines, serve as the nodes in an ESN network. The PBXs served by a node can be SL-1s, SL-100s, or other types of PBXs. Among ESN's major features are the following:

- uniform network dialing
- enhanced network routing
- on- and off-hook queuing of calls through the network
- 16 class-of-service levels
- compatibility with AT&T's Electronic Tandem Network (ETN)
- centralized administration and maintenance for the network
- network directory service
- network traffic analysis
- processing of call detail records

ESN simplifies the process of calling through a private network by enabling all locations in the network to use the same dial plan. Network users can be assigned to one of 16 service classes, each of which has certain calling privileges and restrictions associated with it. For example, users who belong to a certain class may be prevented by their host SL-1 or SL-100 from making long distance calls through the public network. The service classes enable calling privileges to be as-

signed to those people who really need them. This helps the using organization to control its communication costs.

The operation of the network can be monitored through the customer management center (CMC). When problems arise in an area of the network, the CMC identifies them, which enables corrective action to be taken immediately. The CMC also monitors traffic through the network; it can produce reports of the traffic patterns and detailed call records for all parts of the network.

Autovon: This software package meets the special requirements of the Autovon military network, which serves the U.S., Canada, and parts of Europe and the Far East. The most important requirement takes the form of the priority dialing scheme. Autovon calls are classified into five levels of precedence: routine and four levels of ascending priority. When an Autovon user makes a call, he dials an extra digit that indicates the priority level of the call. If a priority call is blocked when it tries to pass through the network, the SL-1 will examine the priority level of the calls on the busy trunks. If one of these calls is of lower priority, the SL-1 will disconnect it in favor of the higher priority call. The people conducting the call that was disconnected are informed by a prerecorded message that a higher priority call was given the connection. (For more about Autovon, see *Telesis* 1983, no. 2.)

IMS: The integrated messaging system (IMS) is a centralized messaging service that is connected to the SL-1. Incoming calls made to busy lines or that go unanswered are routed automatically by the SL-1 to the IMS attendant. This attendant has a terminal that is connected to the IMS computer. The terminal displays the called number and the name of the person called. It also displays any information previously provided by the called person. For example, the called person may have indicated to the IMS that he is in an all-day meeting. The IMS attendant passes on any such information to the caller and enters any messages from the caller into the computer. If the caller is also served by the SL-1, he can leave simple predetermined messages by dialing numbers on his station set. To let the called person know that the IMS has a message waiting for him, the SL-1 turns on an indicator lamp on his station set. To receive the message, the called person calls the IMS attendant, who can read the message back to him or forward a printed copy of it.

IVMS: The integrated voice messaging system (IVMS) provides a messaging service that does not require a system attendant. An IVMS can support up to 1000 telephone users on an SL-1; the

SL-1 can support several IVMSs. The IVMS enables callers who cannot reach people served by the SL-1 to leave a voice message. The IVMS answers the incoming call automatically, guides the caller through the simple process of leaving a voice message, records and stores the message, and notifies the called SL-1 user that a message is waiting. Like the IMS, the IVMS enables people calling from SL-1 sets or 2500 sets to leave brief predetermined messages such as, please call. The called person uses an SL-1 set or a 2500 set to retrieve and listen to the message. The set used can be served by the SL-1 through which the message was left or it can be served by another SL-1, another type of PBX, or a central office switching system.

Hospital/medical clinic: This package is among the most recently developed feature groups. One of its features, call override, enables a call in progress to be disconnected in favor of an emergency call. An automatic answerback feature enables calls to operating rooms and other sterile areas to be answered automatically by a handsfree module attached to an SL-1 set. The SL-1 PBX can be connected to the hospital's or clinic's paging system. This permits hospital staff to use their telephone to page other staff through the whole hospital or in predetermined parts of it.

The future holds further evolution in store for the SL-1. Upgrading of the hardware will continue as will development of new software features. The major direction of development will be along the line set down by Northern Telecom's OPEN World program. As part of this program BNR is now developing fully digital terminals that will be used in conjunction with the SL-1 and other Northern Telecom communications systems. These terminals will combine voice, text, and graphics communication; they will communicate with their host systems by way of twisted pairs that carry data at rates of several megabits per second. This will enable the terminals to support new office services, some of which will be provided by a server system now being developed for the SL family of business communications systems. The new services will give users of the voice and data terminals transparent access to information stored in computers made by different manufacturers. Much of the development work for these services will be an extension of recent agreements between Northern Telecom and several computer manufacturers.

Dinker Bir holds degrees in mathematics and physics and in electrical engineering from Bombay University in India. In 1964 he moved to England, joining GEC as a hardware designer for a crossbar switching system. While at GEC he obtained an MSc in electrical engineering from the University of London. Dinker moved to the U.S. in 1970. During the next eight years he was a senior software designer at Automatic Electric Laboratories and at ITT; at both companies he was primarily concerned with call processing software for stored-program-controlled switching systems. Dinker joined BNR as a senior designer for the SL-1's international software development group, which he went on to manage. Between 1982 and 1984 he managed enhancement of the SL-1 network, development of the digital trunk interface, and development for the general North American software generic package. Recently Dinker joined Northern Telecom as the director in charge of SL-1 planning for Canadian and international service.

Jim Eng received his bachelor's degree in electrical engineering in 1972 from the University of Windsor (Ontario) and his master's degree in the same discipline in 1973 from the University of Waterloo (Ontario). He then joined BNR in Ottawa as a software designer, working primarily on call processing for the SL-1's attendant console. In 1977 he joined the development team for the DMS-10 central office switching system, managing a group that designed call processing and administrative software. Three years later he

transferred to BNR's Mountain View, California laboratory. There he has managed development of the Electronic Switched Network (ESN) features for the SL-1 and hardware and software enhancements for the SL-1 system in general and the peripheral equipment in particular.

Rod Hoo holds bachelor's and master's degrees in electrical engineering, which he obtained from the University of Hawaii in 1968 and 1969 respectively, and an MBA, which he received from the University of Santa Clara, California in 1978. After obtaining his engineering degrees, he worked at McDonnell Douglas Astronautics and at GTE-Sylvania as a designer and project leader in the fields of digital signal processing and microcomputer system design. In 1976 he joined BNR in Mountain View, California, where he has primarily worked on software development for the SL-1. He first worked on enhancements to the SL-1's CPU; then he led exploratory development for data communication on the system. Since 1978 he has managed development of ESN, IMS, IVMS, North American business features, Autovon, and exploration of voice messaging. He is now concerned with value-added services and new interfaces in the SL-1 to other manufacturers' computers.

Howard W. Johnson, Rolm Corp.

AN INSIDE LOOK AT THE NEW ROLM CBX II ARCHITECTURE

In 1975 ROLM Introduced its digital PBX, the Computerized Branch Exchange (CBX). At that time, customers were seeking replacements for their electromechanical PBXs and their old-fashioned cord boards that required an operator to connect calls by hand.

From its inception, designers recognized that a digital PBX could play a key role in transferring and transmitting data as well as voice communications. With the proliferation of word processors, personal computers, terminals and other digital devices in the early 1980s, digital PBXs began to serve as the voice and data communications hub of the office. To a large extent then, the long awaited "office of the future" is here now. Today, about 10 percent of all office workers have data devices on their desks.

The key requirements for business communications systems in the second half of this decade and on into the 1990s are exactly those two required in the past: the ability to act as the communications controller for the office and to easily incorporate new technical advances. The volume of data communications between devices in the office and between devices in the office and external sources will increase exponentially. PBXs must be equipped to handle not only voice communications at every desk but data communications as well. Unlike telephones, which are used sporadically, terminnals or personal computers are oftentimes used throughout the day. Thus, the PBX must possess sufficient data communications capability to permit all data devices to be simultaneously active. In addition, the transmission speed for such devices will surely exceed the 9,600 bits per second which is, for practical purposes, today's upper limit. Although some suppliers believe that a PBX that can handle 64,000 bps will be adequate for future desktop devices, Rolm believes that to achieve the kind of performance needed in the second half of this decade and in the 1990s, it will be necessary to support data rates of at least hundreds of thousands of bits per second from the PBX to many desktop devices.

Another key trend in the marketplace is the need for the communications controller to deal with a much wider universe of communications needs. Within the office this means that a PBX must be capable of providing communication between all devices in the customer's establishment, regardless of vendor. Thus, a PBX must be able to accommodate standards such as asynchronous, synchronous, X.25, and T1/D3 that are now evolving. The ability to connect to the IBM world, using existing protocols, such as 3270 and SNA, and allowing compatibility with future ones, such as IBM's proposed token ring, will be crucial here.

In addition, an organization with more than one building should still be able to use a PBX as its communications controller. Whether for a university or large corporation, one PBX system, with distributed processing, should be able to control the communications of many sites separated by thousands of feet.

If we are to live in a "global village," the PBX must be able to connect to a global business communications network as well. Coaxial cable, microwave, satellite, and fiber optic links to other systems and private and public data bases and networks will become increasingly important. Brokerage houses, banks, major industrial corporations, and other Fortune 500 companies will want access to these data bases and networks from any site in the world.

A business communications system, then, must follow flexible growth, provide easy-to-use devices, integrate voice, data text and even video applications, and permit access to the outside world through networks.

Rolm Corporation recently announced a new product called CBX II™ to meet these needs. Major engineering advancements on four separate fronts have been incorporated into a new business communications architecture that will provide total switching capability at high bandwidths and line sizes. This architecture is the basis of CBX II, the successor to the Rolm CBX telephone and data switching product line. CBX II is typified not only for the large increases in system capacity that it affords, but for the modular, extensible nature of its architecture, and the important fact that the new capabilities can be retrofitted into any existing Rolm CBX. Customers no longer need to worry that they might "run out" of vital communications systems bandwidth (communication capacity) or that they might not have sufficient growth capacity in their CBX system.

CBX II Physical Structure

The CBX II, which provides both voice and data switching capacity for a variety of applications, accepts data, asynchronous or synchronous, at speeds up to 64,000 bits per second (bps). Voice connections may be either analog or digital. All voice and data connections are brought to the CBX II system over single twisted-pair wiring and terminated on replaceable electronic circuit cards, called line cards. The line cards are arranged in racks, called interface shelves, of up to thirty-two cards each. Six shelves make up a cabinet, and three side-by-side cabinets make up a CBX II switching node. Every CBX II node contains a master processor, which can be either the Rolm 8000 (16-bit) or 9000 (32-bit) processor. Each interface shelf has its own backplane capable of switching 74 Mbps. A special circuit card, or expander, on each shelf passes information from the shelf backplane to a system backplane, called the ROLMbus™, which is used for exchanges of information between shelves. There are two choices of system backplanes: ROLMbus 74 and ROLMbus 295. ROLMbus 295 provides four times the connection capacity of ROLMbus 74, and is an option for high-performance applications.

Large systems are built by joining up to fifteen nodes in a distributed processing network. Between nodes, there are several possibilities for voice/data communication. For small systems the Digital Intertie, which multiplexes several calls onto one digital T1 link is used. For larger systems, CBX II uses a fiber-optic inter-node link (INL) that can carry up to 295 Mbps between nodes. For installations of more than 4,000 lines there is an inter-node network (INN), which can service more than 10,000 users providing non-blocking voice and 64 Kbps data to every desk.

CBX Philosophy

The CBX II project started with six primary objectives, two which were design guidelines, and four which were functional objectives. The design guidelines were to build a modular product, and to maintain full compatibility with previous products. The four functional objectives were to: (1) increase the bandwidth of each node; (2) develop extremely high bandwidth connections between nodes; (3) retain and enhance distributed

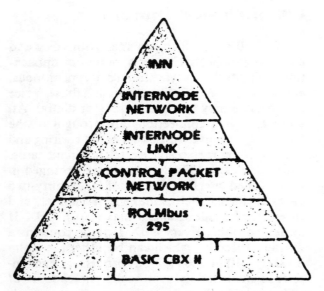

Figure 1. The hierarchical structure of the CBX II architecture.

control of the system by multiple processors communicating over a packet network; and (4) design in new modes of data handling to accommodate future functional enhancements.

Modularity

Large PBX systems that are modular and expandable benefit customers in many ways. From an engineering standpoint, a modular design ensures fault isolation, self-testing and reliability. A more direct benefit is that it is not necessary to configure a system with all the layers of the hierarchy.

If a customer's switching needs can be filled by some subset of the hierarchy, then the upper layers do not need to be installed or can be installed later as the need arises. In some systems, ultimate control over the entire system is vested in the top layer, so that even for modest applications the entire package must be purchased.

The ROLMbus 295, INL and INN modules are examples of hierarchical layers in CBX II that are not necessary for small systems (see Figure 1).

Compatibility

Since it is directly compatible with previous CBX systems, CBX II retains the powerful software feature set and innovative distributed control functionality common to the Rolm CBX family. CBX II also uses telephony and data in-

terface circuit cards common to the entire Rolm CBX family, so no installed system has been obsoleted. Even line cards in the new architecture are the same as those in older systems.

Bandwidth

Bandwidth, the fundamental communication capacity inside any PBX switching node, is the logical means by which information (data, telephone conversations, FAX, video) is moved from place to place. Units for measuring information transfer capability, or bandwidth, in a digital system are bits per second. When fully configured, CBX II has a total capacity exceeding 4.4 billion bits per second. That's enough bandwidth to switch 3,000 high-quality video channels, or 23,000 data channels of 192,000 bps each — more bandwidth than any other system today.

Inter-Node Connections

The switching bandwidth of ROLMbus 295 would be wasted if there were no means of carrying it to another node. The CBX II uses an internode link (INL), which comes in basic modules of 73.728 Mbps (plus parity and framing). Four INL Modules between a pair of nodes extend the full node bandwidth of 295 Mbps to each. The link can use either coaxial cable for short distances or fiber optics for distances up to 20,000 feet. Up to twelve links can be placed on every CBX II node. INL hardware, available in a redundant configuration, can be used for voice, data, or any combination.

For very large systems, the inter-node network (INN) expands the connection possibilities of each node.

Control Flexibility

It is sometimes difficult to balance the two naturally opposing system requirements of centralized control and modularity. While it is advantageous to have one extremely fast, centralized processor to handle the entire system, it is also beneficial to have each section of the system control itself. The CBX II system uses distributed processors connected via a packed switched network to achieve both goals. Each node retains the intelligence it needs for local call processing, while the network of controllers can share common processing tasks and centralize some duties (such as call detail recording). This control flexibility will be the key in providing new and innovative software features and functions. The distributed packet network will allow Rolm CBX II processors to link up with specialized external processors to configure cost-effective integrated communications solutions.

CBX II Architecture

There are ten major parts to the CBX II architecture. They are ROLMlink™, the Interface Shelf, Expanders, ROLMbus, Turnaround, Time Division Multiplexing (TDM) Controller, Control Processing Unit (CPU), Inter-Node Link, Inter-Node Network and the Control Packet Network. These are listed roughly in terms of their status on the hierarchical tree.

ROLMlink

ROLMlink is not specific to any one product, but is a pervasive new way of connecting most types of terminal devices to the CBX. ROLMlink provides a basic communications channel of 256 Kbps bidirectionally from the CBX to the terminal equipment. The digital communications channel back to the switch is carried on ordinary voice-grade twisted pairs of wire at distances up to 3,000 feet. The link protocol is synchronous, with the line interface card in the CBX acting as master for 16 ROLMlink channels. The data transmission is divided into distinct frames, with each frame containing a fixed amount of bandwidth for voice, data, and control. The link adds its own

parity check to each stream independently. The link operates over a single twisted pair in a bidirectional mode. Both sides transmit simultaneously, and the signals are separated by specialized data hybrids at each end of the line. Voice-grade twisted pair was chosen because of its low cost, ease of installation and servicing, and the fact that many buildings are already wired with it. No special shielding is required for a ROLMlink since the link meets all applicable FCC standards regarding emitted radiation. ROLMlink can be bundled into twenty-five and one-hundred pair cables with no loss in performance.

The ROLMphone family of digital desktop devices (ROLMphone© 120, 240 or 400) shown in Figure 2 uses ROLMlink and is compatible with both the previous CBX and the new CBX II architecture.

The ROLMphones use an integrated codec chip to both sample and digitize the analog speech wave forms from the handset. The same chip is used to drive the handset ear piece. By itself, digitized voice may be transmitted long distances without distortion or noise effects. Not only does the ROLMphone communicate voice, but any ROLMphone can be configured to accept and transmit data at speeds up to 64 Kbps *simultaneously* with voice conversations. For data connections, the user's data equipment plugs into a telephone socket which accepts data according to the appropriate physical standards. From there, the data can be switched through the ROLM integrated voice/data network to any

Figure 2. The ROLMphone 120, 400 and 240, from left to right.

properly configured data port, be it another ROLMphone, a desktop data terminal interface (DTI), or a Cypress™ Personal Communication Terminal.

The digitized voice signals and the user's data stream, along with control information vital to the operation of the phone, are all multiplexed together into the common ROLMlink digital data channel. The ROLMphone 400, for instance has a 60-character display and forty soft-function keys, in addition to the dialing pad. Key stroke capture, off-hook sensing, and display handling are performed by an on-board microprocessor, which communicates with its counterpart inside the CBX through the ROLMlink control channel. This pre-processing greatly reduces the real-time operating burden on the CBX II main processor, allowing it to devote more time to high-level routing and management decisions. The processors at both ends of the ROLMlink are continually checking each other to maintain dependable service and participate in fault isolation and determination, to reduce service time. Transient errors are likely to be caught by the parity checking hardware integral to the link, and immediate retransmissions will prevent any service interruption to the user. This feature is built into every ROLMlink and does not require additional user support.

Intershelf/Expander

A main function of the switch is to collect the data from all cabinet shelves and send it back to the appropriate places. The Expander Card in a CBX II (see Figure 3) acts as a bus master for the shelf bus, and is the only device capable of driving the shelf address and enable lines. The Expander exchanges data and control information with other shelves via the ROLMbus.

ROLMbus 74

The bandwidth (74 Mbps) and clock speed (4.608 Mhz) of ROLMbus 74 are identical to an interface shelf. The TDM Controller (Time Division Multiplexing Controller) for ROLMbus 74 contains a complete connection table for all system events and continuously supplies the Expanders with addressing information, which is forwarded by the Expanders to the shelves. This addressing information specifies which devices are to communicate in what time slots. The Expanders are relatively simple circuits that can buffer data

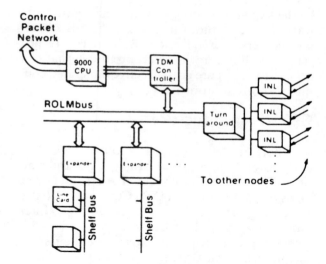

Figure 3. The Expanders act as the interface between the individual shelf busses and the ROLMbus.

bidirectionally from the shelf to the ROLMbus, and drive the shelf address and card select lines based on information from the TDM Controller.

ROLMbus 295

ROLMbus 295 has a bandwidth of 295 Mbps, four times that of the ROLMbus 74, and can be installed as original equipment in new systems, or retrofitted in any previous CBX or CBX II. ROLMbus 295 provides substantially greater service in those applications where it is needed. To install ROLMbus 295 in CBX II, the Expanders and TDM controller are changed, and a new circuit card called the turnaround is plugged in. ROLMbus 295 uses the same flat ribbon cable harness used by ROLMbus 74. ROLMbus 295 is classified as a *undirectional* bus. It is also a *traveling wave* bus. Because the total bus length is longer than the wave length of one clock period, several successive signals can actually be present on the bus at one time. The signal pulses are pipelined through the bus in such a way that they do not interfere with each other.

CPU

The CPU, TDM Controller, ROLMbus, and one Expander for each shelf all go together to form the control subsystem. The ROLMbus can be either a type 74 or 295 and the CPU can be either the 8000 (16-bit) or 9000 (32-bit) processor. This entire control subsystem is completely and independently mirrored by a redundant system.

Each shelf has two Expanders. Only one of the control subsystems is active at a time, the other is a "hot standby," ready to take over at any instant, and constantly running diagnostics and self tests. The active CPU also constantly runs self tests and diagnostics on all the components of the system.

In a CBX II 9000 processor environment with several nodes, each CPU has a packet network interface called the Control Packet Network for communicating with its peer processors in other nodes, and also for talking to specialized applications processors. The Control Packet Network is an ideal place to connect external processing power such as mainframe resources, or remote diagnostic computers for off-site servicing. Once a sound network has been established the growth potential for the system becomes practically unlimited.

The Control Packet Network is carried by the INL between remote nodes, capitalizing on the fiber optic technology and high bandwidth available over those connections. This is an excellent example of synergistic relationship that exists between the Rolm CBX II and the LAN world.

The Control Packet Network is used for node-to-node system traffic, and not for user device communications or voice traffic.

Inter-Node Link (INL)

The INL hardware must interface through a turnaround, and is available only on systems equipped with ROLMbus 295. The INL is designed to permit extremely high bandwidth connections between nodes. It is most often used to perform a data swapping operation, and is, therefore, able to effectively double the ROLMbus bandwidth. Data going out from the ROLMbus can be buffered in the INL subsystem, passed through a time slot interchange, and sent across a 73.5 Mbps serial (coax or fiber optic) link to the destination node. Data coming back from other nodes through the INL may be subjected to another time slot interchange before being placed on the local ROLMbus.

INL links may be installed in a redundant mirrored fashion. There is a turnaround for each half of the mirrored ROLMbus 295, and each turnaround interfaces to both of the redundant INL networks. The INL equipment can be used in a local sharing arrangement. Up to 12 fully redundant INL links can be installed on any CBX II node. A fiber optic INL using all glass 100 uM cable can be run as far as 20,000 feet.

Inter-Node Network (INN)

What if 12 redundant INL links aren't enough? What if the user wants fifteen nodes, all fully inter-connected with 294 Mbps of non-blocking service between each pair? Research indicates that such requirements may exist in the forseeable future. To meet this requirement, Rolm has developed an inter-node network (INN) that can absorb and switch the full 294 Mbps from each of 15 CBX II nodes. The INN is a slave type device, and responds to requests from the community of CBX II nodes which arrive over the Control Packet Network. The INN is connected to each CBX II node with a full complement of four INL links.

CBX II Growth Plan

A user might start with a single cabinet version of a CBX II using the ROLMbus 74 and 8000 processor and expand it to include more shelves and up to three cabinets. As the total traffic or feature load grows, the ROLMbus 295 can be installed, followed by the 9000 processor. The next step is to add the Control Packet Network, a second node, and the INL connecting them together. More INLs and nodes will suffice until the requirement grows beyond a few thousand users. For very large systems, the INN becomes appropriate, and the user can expand to 15 nodes.

CBX II Meets Needs of Today and Tomorrow

The distributed architecture and flexible hierarchical structure give the customer a system tailored exactly to individual needs, not tailored for some hypothetical "average" customer. The CBX II architecture can grow consistently from 16 lines to more than 10,000 users by adding successive layers of the hierarchy — up to 4.4 billion bits per second of non-stop performance, connecting users eight miles apart (or further with T1 links and microwave equipment). Distributed processing and fully redundant switching capability keep problems local, eliminating shutdowns.

More Than You Probably Want to Know About the CBX II

The Rolm CBX II is both a product and an architecture; the latter being the basic capabilities which are implemented by the specific product. The key architectural element is a high-speed, parallel, time-division-multiplexed bus, called the ROLMbus™.

Voice and data devices and resources are connected, typically by twisted-pair telephone wiring, to appropriate interface cards, all connected to the ROLMbus. A central processor runs the software that defines the intelligence of CBX II, and it allocates via commands to the time division controller, the bandwidth of he ROLMbus, providing appropriate connections between the various interfaces.

There are two types of ROLMbus available, ROLMbus 74, which provides 74 Mbps of capacity, is the standard configuration and provides the information capacity and capability required for today's voice and data devices.

Because the ROLMbus 74 is used in a bidirectional manner, it is not possible to increase the system clock speed without distorting the signal pulses. If the signals could be made to travel along the cable in only one direction, they could be pipelined through at a much higher rate without distorting. That is the basic idea for ROLMbus 295. With a bandwidth of 295 Mbps, ROLMbus 295 has the designed-in flexibility to handle demanding future communication requirements.

At first glance into a digital PBX, one will see the interface shelf. The shelf supplies a basic interface to the CBX II switching capabilities. The shelf definition includes card dimensions, connector types, power, grounding, shielding and other mechanical specifications designed to make the interface circuit cards reliable and easy to service.

In the Rolm system, each interface card has two connectors. One is the actual line interface connector and goes off to various types of telephone and data service equipment. The second connector goes to the "shelf bus," which is a motherboard used to bring voice samples, data and control information to and from the card. The total bandwidth potential of this connector is 74 Mbps, although no one card would be expected to make sense of such an enormous flow of information.

Any bus cycle can be used either to transmit voice, data or control information at any time. The "Expander" is hardware that is contained on each shelf, serving as the bus master, which sequences the flow of the bus cycle so that each voice or data connection is sampled at an even rate. The flexibility of using any bus cycle for any type of operation means that CBX II can be adapted to many different uses.

The motherboard format includes a 16-bit bidirectional data bus, a 10-bit address bus, and a special "enable" line to each card. The enable line serves three basic purposes: (1) cards do not have to be configured with a special physical address when they are plugged in, as the enable line performs that function for them; (2) the reliability of the system has been enhanced relative to letting each card do its own address decoding, because the enable line interface is simpler on the card; and (3) because the card select decoding is done on the Expander (the bus master) instead of on each interface card, the enable line is actually replacing five coded card select lines, which saves connector pins on each interface card.

The bus cycle rate is 4.608 Mhz. Addresses (and enables) appear on the bus synchronously with data, but one cycle earlier. By convention, 832 of the possible 1024 address patterns are used in conjunction with the enable lines. This is a subtle, but important point, as it is the key to the new CBX II broadcast and shared-access modes of operation. The remaining 192 patterns are called non-enabled commands and can be used for broadcast and multi-access. The 832 commands are more than enough to program intelligent interface cards, such as the ROLMphone interface group.

For a typical data transfer, the Expander first outputs a card enable and 10-bit address indicating which word it wants. In the next shelf, the enabled card places its data on the shelf bus. From there the Expander transfers the data out to the ROLMbus. At the destination shelf, the corresponding Expander captures the data word off of the ROLMbus and places it on the destination shelf from which it may be read by the intended interface card.

The Expander is the first point on the hierarchical ladder at which redundant (i.e., duplicated) hardware can be supplied. The entire common control structure of a node can be supplied with mirrored redundancy including the CPU, ROLMbus and Expanders. Two Expanders sit on each shelf, one for each side of the common control system. Only one common control side is active at a time; the other side spends its day checking for errors.

The Expanders form what is called a fully connected redundant interface between the non-

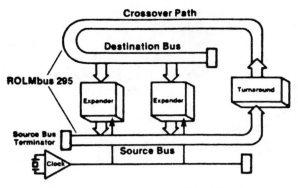

Figure A. The two halves of the ROLMbus.

redundant shelf bus and the redundant ROLMbus system. The two Expanders are completely independent and may be replaced without a service interruption.

Figure A illustrates the connection of Expanders onto the bus, which is split into two halves: a source half and a destination half. The turnaround card is used in conjunction with the Inter-Node Link (INL) equipment, but from the perspective of the ROLMbus 295, it is just a repeater. Clock signals are generated by the turnaround and propagate from left to right along the bus. Each expander uses the ROLMbus 295 clock for all timing, so that while the Expanders are all "out of phase" with each other, they maintain the correct phase relationship with the data, which travels along the ROLMbus 295 in the same direction as the clock.

Up to 17 devices can be connected directly to the ROLMbus 295. A data word traveling on the ROLMbus 295 would be split out by the source shelf Expander onto the source half of the ROLMbus 295 and progress to the right until it is absorbed by the turnaround. The turnaround then repeats the word back onto the destination bus from which it may be captured by the destination shelf Expander.

Because ROLMbus 295 goes four times faster than the interface shelves, it has time to service four shelves during each shelf cycle. This means that four shelf transfers can take place simultaneously, multiplying the CBX II internal bandwidth by a factor of four.

Inside the ROLMbus 295 Expanders, a complete connection table for all regular voice and data connections affecting the shelf is stored (see Figure B). The Expander uses the information in its connection table to generate addresses for the shelf. This means that address information is not required to flow on a regular basis between the controlling CPU and the Expander. That saves a great deal of bandwidth on the ROLMbus because the addressing information for a one word data transfer is much longer than the data itself.

The ROLMbus taps are not unidirectional, so some waves actually do propagate to the left along the source bus. These reverse waves are harmlessly absorbed at the source bus terminator at the left end of the ROLMbus. The Expander transmitters are linear current sources so that they can transmit correctly even when other reverse waves are passing by. There can be as many as three sets of forward and reverse waves traveling on the bus at any one time. The turnaround repeats all the signals only to the destination bus and they propagate on the crossover link back to the left end of the ROLMbus and then back from left to right. This is necessary to preserve the correct phase relationship between clock and data at every Expander. Only the turnaround is allowed to transmit on the destination bus, and since the turnaround is located at the end of that bus, waves can propagate only in one direction on the destination bus.

All interfaces to the ROLMbus have been designed to sustain local power outages, failed components and other local disasters without affecting the operation of other devices on the ROLMbus. The ROLMbus 295 always uses parity.

The turnaround is needed only on systems equipped with the high performance ROLMbus 295. The turnaround has the duty of absorbing information from the source bus and putting it back onto the destination bus. But nothing in the architecture requires the turnaround to put back the same information it pulled off. The capacity of the switch effectively can be doubled by doing just that!

Suppose a call is in progress between a local node with phone "X" and a remote node with phone "Y." When "X" transmits a voice sample it flows out to the local source bus and down to the turnaround. If the turnaround could capture that sample and send it directly to the remote node, then

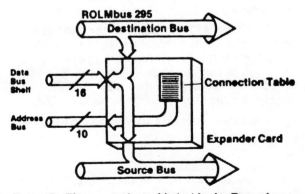

Figure B. The connection table inside the Expander.

33

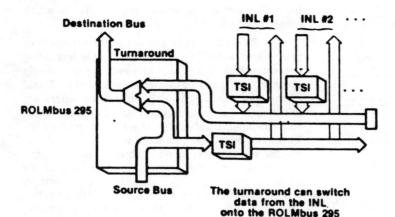

Figure C. Signal flow paths inside the turnaround.

the local destination bus could lie idle for the remainder of the cycle (because the sample certainly isn't needed anywhere else in the local node).

The empty destinations bus cycle could be filled in with a sample from "Y," if the turnaround had such a sample. This is a way to do a full conversation using only a single timeslot, but it only works if the turnaround knows exactly what to do.

Figure C shows the signal flow paths inside the turnaround that allow it to communicate with the INL Hardware. The INL exchanges information with other nodes and is capable of supplying data for the turnaround to perform this efficient swapping operation. The net result is that a two-way conversation takes up only one timeslot if the call is being placed to another node, instead of two timeslots like an ordinary intra-node call. In multi-node systems, many of the calls are between nodes, and the efficiency of the turnaround/INL combination is almost twice as high as tie lines or digital interties.

The ROLMbus 74 TDM Controller (TC) contains the full system connection table and acts as bus master for the entire ROLMbus 74, generating and driving all the address and select lines systemwide.

ROLMbus 295 vests more of the cycle by cycle control with the individual Expanders, freeing the main CPU and TC to perform more polling, testing and system configuration duties. The ROLMbus 295 TC can handle more than three million bits per second of control information in packets of 64 bits.

The TDM Controller attaches to the ROLMbus

295 exactly like an Expander. Control commands from the TC propagate through the ROLMbus 295 system just like regular data. The TC is responsible for three activities: loading the connection tables on every Expander, configuring the turnaround and INL hardware, and communicating with all the various line card groups. The TC does not contain its own connection table, and does not participate in the normal connection oriented data flow.

The TC uses the ROLMbus 295 special control field line (CFL) to signal its activities. When the TC has a message to send it raises the CFL line and broadcasts a packet of 64 bits of control information. The CFL is wired as a seventeenth data bit that only the TC is allowed to use. The packet contains addressing control and data information to do operations like loading the Expander connection table, or reading the status of a line card. The TC is careful to never set up regular connections in the timeslots that it uses for control packages. In most respects the TC behaves very much like an Expander.

There are two important points about the control packet strategy: First, any number of timeslots can be used for control. The control bandwidth could be expanded to a full 295 Mbps if necessary. The current TC version is configured in hardware to use a maximum bandwidth of slightly more than three bits per second, about one percent of the total system capacity. Second, there can be multiple TCs inside one node (as long as they use non-overlapping timeslots). This provides yet another important direction of expansion possibilities for the CBX II architecture.

Further software developments eventually will open up even more configuration possibilities. For instance, the INL could support a chain of CBX IIs with INL connections between adjacent pairs

stretching out for tens of miles. Combining two INN subsystems might pave the way for configuring systems of twenty or thirty nodes, even though such huge installations are difficult to imagine today.

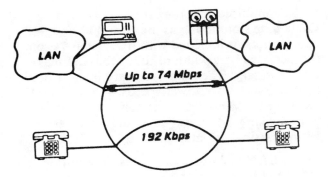

Figure 4. Supermultiplexing capability.

External computing facilities could be interfaced with the Control Packet Network; such arrangements are already under study. That could provide for more comprehensive on-site billing and recordkeeping. Automatic call distribution systems could be integrated with the terminal handling equipment for airline reservations and other applications. Text processing applications may be integrated with voice systems, enhancing the value of both. A telephone could be used to access computer files (using voice responses from the computer and the DTMF pad on the telephone). A computer terminal could be used to screen calls, or to help file and retrieve voice recordings.

New Data Communications Capabilities

Many people believe that a PBX can only be used to set up low-speed point-to-point data connections. While that may be true of some systems, the CBX II architecture will allow it to expand and grow as office automation matures.

Super-multiplexing is one of three new switching modes included in the CBX II ROLMbus 295 for future expansion. Blocks of bandwidth can be allocated to provide high-speed switching services. Bidirectional high speed data streams of up to 37 Mbps can be configured (see Figure 4). Super-multiplexing works well for providing bridge and gateway functions between local area networks (LAN), such as Ethernet or the IBM token ring network. For multiple LANs spread out over a large geographical area, the CBX II and its INL hardware can act as a high-speed backbone network for the LANs. Suppose a site had several LANs located three miles apart. The CBX II, with INL links, could be used to provide a basic high-speed data connection from

place-to-place with each LAN having a bridge interface to the CBX.

Video transmission is possible with super-multiplexing. Out of 295 Mbps per node, a mere 1.5 Mbps for high quality video could hardly be missed. In addition, enormous amounts of burst traffic between computers, like magnetic tape transfers, could be easily handled with super-multiplexing.

The broadcast mode allows one device to transmit to a number of listeners simultaneously as shown in Figure 5. This mode uses non-enabled card commands to instruct a number of cards to read data from a single source simultaneously. If an Expander has a non-enabled command programmed into its connection table, all the cards on that shelf sensitive to the command will respond. The broadcast mode may prove to be extremely effective for solving automatic program load problems. Many intelligent workstations could use the broadcast feature for obtaining system update information. Another use for this mode is to synchronize events in different parts of the system by sending out global timing information. Mail distribution and document circulation are other potential applications.

Broadcast groups can be configured at any speed up to the full interface shelf bandwidth of 74 Mbps.

Shared-Access mode also uses the non-enabled commands. This mode provides common broadcast bandwidth to a number of cards and allows the cards to arbitrate among themselves for access. The only difference between shared-access and plain broadcast is that in broadcast an enable command is typically sent to the source device,

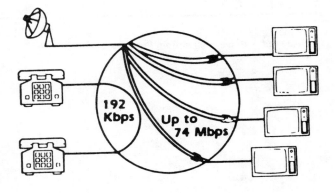

Figure 5. Shared access capability.

while in the shared-access mode it is not. In shared-access mode, the non-enable command is broadcast to the whole group. From the CBX II perspective, any device in the group can transmit when it wants. By combining super-multiplexing with shared-access, a number of one megabit shared communication channels can be configured. CBX II can partition the ROLMbus 295 bandwidth into many non-interacting segments of super-multiplexed shared-access groups — broadcast groups, bidirectional super-multiplexed channels and ordinary data or voice connections. This capability is called "dynamically allocatable bandwidth."

The shared-access mode can be used to implement a packet switching service directly embedded in CBX II. As long as the devices doing the transmitting have an agreement not to transmit simultaneously, every station can hear every other station. The access control could be via token passing between the sharing devices or access control could be statistical, as in Carrier Sense Multiple Access methods. Or a new TC type card could be plugged in to the ROLMbus 295 and programmed to poll devices and grant transmit requests, providing centralized arbitration.

Sub-multiplexing, the capability to split up a single time slot among as many as 40 low-speed data devices, is included with ROLMbus 74, but the increased bandwidth of ROLMbus 295 makes it unnecessary. The trend is toward higher-speed connections — not lower speeds — and CBX II and the ROLMbus 295 can supply every desk with non-blocking 64 Kbps data and voice traffic without using sub-multiplexing.

As LAN technology matures, in addition to providing backbone and gateway functions, the CBX can act as a center for administration and maintenance of the various LAN networks that will coexist with the CBX at the customer site. It can function to configure, allocate, and monitor vital LAN resources such as satellite links, printers, diagnostic equipment, and other network servers. The need for these services is virtually independent of the type of LAN, and is fundamentally the same for CSMA broadcast bus networks, token passing ring networks, and even cellular mobile radio installations. The CBX will continue to be a point of departure for external communications, such as X.25 networks, just as it is today.

DR. HOWARD W. JOHNSON is a senior member of the Rolm Corporation Technology and Advanced Development Group. He earned his BSEE, MEE and Ph.D degrees from Rice University. He joined Rolm in 1979 and completed his graduate studies under a Rolm fellowship. Dr. Johnson is a member of Tau Beta Pi, Sigma Xi, IEEE ASSP Society and a winner of the ACM computer art contest.

Reprinted with permission from *Business Communications Review* 950 York Road, Hinsdale, IL 60521. Tel: (312) 986-1432

David Levin, Netcomm Inc., New York, N. Y.

Private branch exchanges: The best time to shop may be right now

Growing competition in the maturing PBX arena is a boon, especially for users with 16-to-800-line configurations.

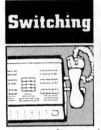

Switching

Vendors of private branch exchanges face a classic challenge: Supply exceeds demand. Their dilemma began in the guise of good times during the Bell divestiture years of 1984 and 1985, when demand for PBX and key telephone products artificially swelled. But by mid-1985, inflated inventories resulted in PBX price cuts of 20 to 30 percent. Stiff competition still persists, and knowledgeable users can turn this situation to their advantage.

Telecommunications switching vendors have adopted two new product-marketing strategies:

■ *Differentiate the core product line from that of the competition.*

■ *Focus on new product opportunities in integrated voice and data PBXs.*

Stressing the product line's unique qualities is essential to the PBX vendor's survival, and niche boundaries are taking shape. Low-end telephone equipment (fewer than 16 extensions with fewer than six trunk lines) is rapidly becoming a retail item. AT&T sells its product through Sears Roebuck & Co., which is competing with discount electronics stores peddling such Japanese brands as Panasonic.

The cost per line of these low-end products will soon fall to less than $300, making it difficult for traditional interconnect suppliers to compete in this category.

At the other end of the PBX product spectrum (more than 800 stations), IBM has quietly sacrificed marketshare held by its Rolm subsidiary in favor of aligning the subsidiary's activities more closely with those of its parent. Rolm's sales and marketing staff are being integrated into IBM's staff, and hopes are that IBM will soon emerge with significant enhancements to the Rolm PBX products. Some observers even say that the Rolm logo will disappear in the near future.

Many suppliers of high-end PBXs are retrenching

through new distributor agreements. AT&T, for example, has agreed to have Sonecor (Southern New England Telephone) resell its System 75 and System 85 products, but not its smaller System 25. The company has also spun off Intelliserve, a shared-tenant resale operation, into an independent organization.

Northern Telecom recently sold its western distribution operations to PacTel and named Centel Business Systems its midwestern distributor. Centel provides the majority of distribution for Northern Telecom in the southern states. L. M. Ericsson has completely withdrawn from distribution in the United States; Nippon Electric Company is selling its products through RCA; and Siemens purchased Tel Plus, a large nationwide interconnect company. What were two promising start-ups, Ztel and CXC, have all but disappeared. GTE recently merged with Fujitsu and now markets Fujitsu's 10,000-line switch as the Omni SIV. GTE plans to begin manufacture of the product in Anaheim, Calif., by year's end. These developments echo the consolidation taking place in the PBX arena as companies draw lines around their targeted markets.

Integration

The second new marketing tactic—to focus on new product opportunities in integrated voice and data PBXs— is the logical extension of a maturing PBX marketplace. Low-end telephone equipment has become a do-it-yourself retail distribution product. New high-end PBX installations are few and far between. The result is the emergence of a new mid-range PBX category.

Figure 1 illustrates the forces redefining the mid-range PBX. The large PBX suppliers, including ATTIS, Rolm, and Northern Telecom, have introduced smaller versions of their existing switch products. The small PBX and key telephone equipment suppliers including AT&T, TIE/communications, and Mitel have developed larger versions of their products.

1. Targeting the middle. *Mid-range private branch exchanges have emerged as low-end gear becomes a do-it-yourself retail product and high-end installations dwindle.*

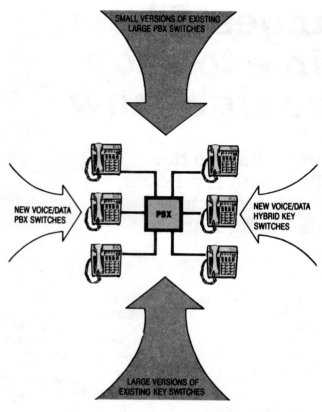

SMALL VERSIONS OF EXISTING LARGE PBX SWITCHES

NEW VOICE/DATA PBX SWITCHES

PBX

NEW VOICE/DATA HYBRID KEY SWITCHES

LARGE VERSIONS OF EXISTING KEY SWITCHES

PBX = PRIVATE BRANCH EXCHANGE

Additionally, these suppliers have introduced completely new voice and data PBX and key switches, and several vendors, including Northern Telecom and Wang, have introduced new software that substantially improves their products.

Mid-range PBXs

Size of the switch hardware and functional operation are the two parameters defining the mid-range PBX category. In terms of size, the mid-range PBX begins with as few as 16 stations and extends to approximately 800 stations. One could argue that a 16-station PBX is considerably different than an 800-station PBX. While this is true, the buyer of the 20- or 30-station PBX knows that it must accommodate considerable growth; otherwise, the user will never realize the potential life of the product. The mid-range PBX may be linked to a network of PBXs and thus require the sophistication of a larger switch.

Figure 2 illustrates the four ways in which a mid-range PBX can function.

■ 1. In its most common form, a PBX operates as a standalone private branch exchange (Fig. 2a). It will, typically, support four types of trunk lines: a combination of incoming and outgoing trunks, direct inward dial trunks, Wide Area Telephone Service (WATS) trunks (and equiva-

lents from common carriers other than AT&T Communications), and tie-line trunks (usually two- and four-wire leased lines supporting E and M supervisory signaling). A fifth type of trunk interface for digital T1 (1.544-Mbit/s) channels is supported by some, but not all, of the mid-range PBX switches.

When functioning as a standalone PBX, the majority of incoming calls are handled through an attendant console. The attendant operator greets the calling party and connects the call to the desired extension. Multiple attendant consoles are usually supported in the mid-range PBX. The attendant console has the highest privilege level of any station attached to the PBX, including station and trunk status displays. The attendant console operator can override a do-not-disturb station as well as interrupting a connection already in progress. The attendant console is usually connected to the PBX using a 25-pair twisted-pair cable.

Analog telephone instruments, including 2500-type equipment, are usually supported by installing an analog station card within the PBX. Data adapters are available for analog telephone instruments that permit attachment of data devices, including personal computers. Most mid-range PBX switches support asynchronous and synchronous data devices at speeds up to and including 19.2 kbit/s. Some mid-range PBX devices support asynchronous communications at the additional speed of 56 kbit/s, and some support Netbios microcomputer communications at more than 1 Mbit/s.

All manufacturers of mid-range PBX switches offer a proprietary line of digital telephone instruments with integrated data adapters. Data devices, including the asynchronous ASCII terminal shown in Figure 2a, are connected to the data port on the telephone instrument. In-house data processing equipment may be connected to the PBX to enable local as well as remote users to connect to these facilities. IBM mainframes connect to the PBX by attaching data modules to the front-end processor ports. Data modules are typically connected to ports on digital station PBX cards, and minicomputer terminal communications ports are connected in a similar fashion.

■ 2. A mid-range private branch exchange can function as a standalone key telephone network, in contrast to a standalone PBX (Fig. 2b). When functioning as a key telephone network, each individual telephone instrument must be capable of accessing each and every trunk line used by the customer. In this scenario, analog 2500-type telephone instruments are not supported because they do not have the capability of accessing multiple lines. Moreover, they cannot understand the proprietary control signaling used by the digital station cards within the central PBX switch.

When configured as a key telephone network, the mid-range PBX does not have a console, since by definition every telephone instrument has the same ability to access every trunk. The PBX may be programmed so that only one station rings audibly while the remaining stations ring visually. The audible/visual ringing instrument can then function like a console. The designated attendant operator would answer incoming calls as they rang, put the calling party on hold, "intercom" the party being called, and then wait until someone picked up the held trunk—as indicated

2. The four-way mid-range PBX. *The private branch exchange can operate as a standalone device with most calls handled via an attendant console (A), as a standalone key* *telephone network (B), and as a voice and data switching device operating either behind the Centrex service (C) or behind another large PBX (D).*

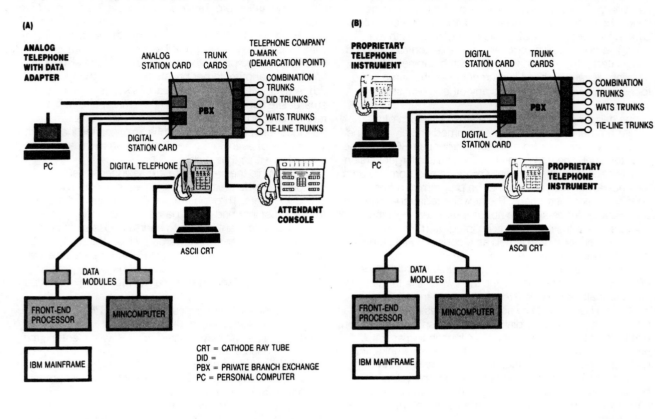

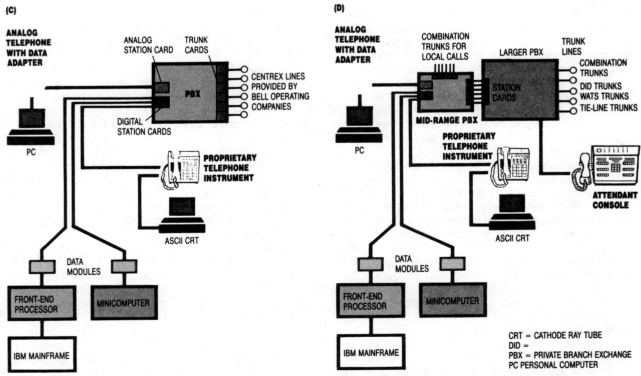

CRT = CATHODE RAY TUBE
DID =
PBX = PRIVATE BRANCH EXCHANGE
PC = PERSONAL COMPUTER

CRT = CATHODE RAY TUBE
DID =
PBX = PRIVATE BRANCH EXCHANGE
PC PERSONAL COMPUTER

by the trunk line visually changing from flashing to being steadily lit.

While the multiline digital telephone instrument may function in the same way as a console for most routine tasks, the console is considerably different. The console typically requires a 25-pair cable connecting it with the PBX. The console also requires a console control card and associated port(s) within the PBX switch cabinet. The multiline digital telephone instrument uses the same two-, four-, or six-wire connection as any other digital station instrument.

The mid-range PBX, which is a smaller version of a bigger brother PBX, may be programmed entirely from the console or, optionally, from an asynchronous ASCII/teletype data terminal. The mid-range PBX, which is a larger version of a key service unit, is typically programmed from the data terminal device only and cannot be programmed from a specific station instrument. The newer integrated voice and data PBX switches also are not programmed from the console, since their console is an integrated terminal/telephone designated as a console within the control program software.

When functioning as a key service unit, the mid-range PBX supports fewer trunk types than when it is functioning as a standalone device. Direct inward dial trunks are not supported because a key service unit does not have the requirement to translate dialed digits into specific extension numbers. While the mid-range PBX functioning as a key telephone network supports WATS and tie-line trunks, the user must select these facilities manually rather than through an automatic route-selection capability. Many PBX products support tie lines through an off-premises station capability rather than through a true tie-line interface with E and M supervisory signaling.

■ 3. The mid-range PBX can function as a voice and data switching device behind the Centrex service (Fig. 2c). Since each extension line in Centrex service is an individual leased-line connection back to the Bell operating company's central office switch, single-line analog telephone instruments are supported. When using a mid-range PBX behind Centrex service, the user must ensure that the supervisory signaling of the PBX does not conflict with the supervisory signaling of the Centrex service. Typically, the signaling is disabled in software either in the central office or in the PBX.

When functioning behind Centrex service, the mid-range PBX supports its proprietary digital and analog telephone instruments. Data devices may be connected to a digital telephone instrument with integrated data adapter or to an analog telephone instrument with external data adapter. Front-end processor ports and minicomputer ports may be connected to the mid-range PBX when functioning behind Centrex. Only Centrex lines provided by the local BOC are used in this environment.

Centrex software typically provides automatic route selection, and long-distance trunk services terminate at the BOC's point of presence. Some products, most notably the Rolm VSCBX, have considerable difficulty when trying to compete cost-effectively in a behind-Centrex situation. For example, a 50-station voice with data user behind-Centrex configuration is difficult to accomplish using the Rolm VSCBX. This configuration requires 50 trunk inter-faces, 50 voice station interfaces, 50 data station interfaces, and 50 data-host port interfaces.

The Rolm salesman running this configuration through his autoquotation program quickly discovers an upgrade to a CBX-II. The VSCBX can easily accommodate between 50 and 75 voice and data users in a non-Centrex environment with considerably fewer trunk lines to worry about.

■ 4. A mid-range private branch exchange can also function behind another larger private branch exchange (Fig. 2d). A user with a large headquarters site alongside a smaller site might elect to equip the smaller site with a mid-range PBX and connect it to the main PBX in the headquarters site. This approach minimizes the load on the larger PBX of users in the smaller site communicating with resources within their building.

Users within the smaller site are assigned extension numbers outside the range of the user extension numbers in headquarters. Through the use of direct inward dial trunks, most inbound traffic passes through the large PBX without an attendant having to handle it. Users calling the main listed number are answered by the attendant, who transfers the call to the mid-range PBX at the satellite site. Optionally, the mid-range PBX uses consoles to provide message-center functions as well as facilitating moves and changes.

When functioning behind another larger PBX, the mid-range PBX interfaces its trunk interface cards with station interface cards within the larger PBX. As with Centrex, the user must ensure that the supervisory signaling of the PBX does not conflict with the supervisory signaling of the larger PBX. Analog and proprietary digital telephone instruments are both supported in this environment.

The mid-range PBX typically uses the long-distance facilities of the larger PBX. Local combination trunks may also be shared, although some local trunks should be connected to the mid-range PBX. Local calls need not unnecessarily use resources of the larger PBX. The automatic route selection within the mid-range PBX can easily divert local calls to trunks attached to the originating PBX. Without trunks of its own, the mid-range PBX cannot support outside dialing in the event the larger PBX fails.

Size spectrum of the mid-range PBX

Table 1 shows the number of stations in the mid-range PBX spectrum. The definition of a mid-range PBX is shown as a configuration from as few as 16 to as many as 800 stations. Most mid-range PBX switches focus on a specific segment of the mid-range while many cover the other parts of the spectrum through available upgrades.

The AT&T System 75 is a good example of a mid-range PBX. Prior to the introduction of the System 25, many users purchased System 75 PBX switches for as few as 20 to 25 stations. System 25 was introduced in June of 1986 and quickly became a leading product for AT&T. While the AT&T System 25 covers a range up to 200 stations, many users with less than 50 stations still opt for the more sophisticated and more costly System 75.

AT&T now offers several product lines that overlap considerably in the available configurations. The traditional small Merlin key service units support four trunks and 10 stations. The larger Merlin key service units support a range of configurations, from those consisting of 10 trunks and

Table 1: The mid-range PBX product size spectrum

NUMBER OF STATIONS

Scale: 0 10 20 30 40 50 60 70 80 90 100 150 200 300 400 500 600 700 800 900 1000 1500 2000 2500

DEFINITION OF MID-RANGE PBX	16 — 800
ATTIS DIGITAL MERLIN	8 — 72
ATTIS SYSTEM 25	24 — 200
ATTIS SYSTEM 75	40 — 600 (ATTIS SYSTEM 75XE)
NORTHERN TELECOM	24 — 600 (NORTHERN TELECOM SL-1ST)
NORTHERN TELECOM	16 — 99 (NORTHERN TELECOM DV-1)
ROLM REDWOOD	16 — 144
ROLM VSCBX	24 — 300
ROLM CBX-II (SINGLE NODE)	100 — 800
MITEL SX-50	16 — 100
MITEL SX-200 DIGITAL	40 — 336
MITEL SX-2000	200 — 700 (SX-2000S); 700 — 2500 (SX-2000SG)
TIE DATASTAR	16 — 192
TIE MORGAN	16 — 144
TIE ULTRACOM	16 — 160 (TIE ULTRACOM DCX)
TIE M-1000	40 — 600
TIE MERCURY	60 — 256 (TX-250); 256 — 432 (TX-450); 432 — 1920 (TX2000)
GTE-FUJITSU OMNI SI	60 — 256
GTE-FUJITSU OMNI SIII	200 — 2048
WANG WBX	16 — 200
ITT 3100	40 — 300 (ITT 3100LX); 300 — 500 (3100XL)

30 stations to those having 30 trunks and 70 stations. The recently introduced digital Merlin equipment, called the Merlin II, is available in configurations of between eight and 72 stations. Three basic station instruments are offered, including five-, 10-, and 34-button models. All stations use the same station card as well as the AT&T universal four-pair wiring regardless of added modem and speakerphone modules. Each Merlin telephone instrument includes a speaker but no microphone. The speaker may be used for voice announce functions and on-hook dialing. A true speakerphone requires a separate and costly unit that connects to the 10- or 34-button instruments.

The Merlin offers a high quality of voice, especially in multiparty conference calls. Every party is clearly audible, even in six-way conference calls. Feature module 5 offers automatic route selection as well as call detail recording output. The digital Merlin II PBX offers most of the data functions available on the AT&T System 25 including asynchronous data switching at speeds up to 19.2 kbit/s. AT&T recently boosted the Merlin product line with a new 24-trunk, 48-station Spirit 2448 key service unit. Additional Spirit models are expected in a three-trunk, eight-station model, and in a six-trunk, 16-station model.

The AT&T System 25 covers a station range from 24 stations to 200 stations with a maximum number of 240 universal ports. The System 25 utilizes a universal bus, universal card slots, and a universal addressing scheme. The three control cards include central processing unit, memory, and service, which may be placed in any three slots. The remaining slots are available for trunk and station cards.

AT&T uses erasable programmable read only memory (EPROM) chips on the control cards to retain programming instructions. When the PBX is initially powered, it self-configures to a default configuration. Program upgrades are accomplished by unplugging old memory packs and replacing them with new memory packs.

The System 25 uses many of the same stations as the Merlin key equipment. A user can employ older Horizon 10-button MET instruments as well as analog 2500-type single-line instruments. There are two types of station features: fixed and assignable. Fixed-station features include transfer, conferencing, and hold. These features are available on all multiline instruments. Assignable station features are programmed by the administrator and include speed dialing and call coverage. Five station instruments are offered. The 34-button instrument includes 32 programmable keys as well as a standard message-waiting indicator. The 10-button instrument offers eight programmable keys, while the 5-button set has three programmable keys. The 34-button version is also available with a built-in speaker.

Last on the list is an instrument called the Hands-Free Answer Set that has a 10-button set with speakerphone. The System 25 uses a 34-button instrument as the console with the addition of a DSS/BLF (Display Station Status/Busy Lamp Field). This station can then handle from 20 to 200 stations with additionally supported software features including return busy calls, return no-answer calls, message-waiting indication, and call release.

The AT&T System 25 supports only asynchronous data communications at speeds up to 19.2 kbit/s. The PBX switch consists of a single card slot chassis occupying

fewer than six cubic feet. While AT&T claims the installed per-station cost of a System 25 is from $550 to $750, at least one user has found the cost to be in the $750-to-$1,000 range.

Leading mid-range PBX

The leading AT&T mid-range PBX is the System 75. The product's available configurations span from as few as 30 or 40 stations to a present maximum of 800 stations. AT&T has been very successful in selling this PBX to large organizations with multiple branch sites as well as medium-sized businesses. As the flagship product from AT&T, it is defining the mid-range PBX category as the product to target against. While the rumor continues to circulate that a 1,600-station System 75 is already in beta test, AT&T denies it—presumably fearing that customers will be reluctant to buy the more expensive System 85 switch in the 800-to-1,600 station size.

AT&T recently introduced a repackaged System 75 XE, an expandable edition of its mid-range PBX product line intended for small- and intermediate-sized businesses. The System 75 XE is available in configurations from 40 to 600 stations with Version 2 software station features. It may be configured for automatic call distribution tasks as well as with optional software for hotels. The PBX may be interfaced with AT&T's AUDIX voice mail equipment, and it supports direct inward dial trunks using special trunk interface cards.

The System 75 XE supports asynchronous data at speeds up to 19.2 kbit/s as well as synchronous data at speeds up to 56 kbit/s. Data, control information, and digitized voice are combined in the digital telephone instrument. This traffic is communicated using AT&T's proprietary protocol, the DCP (Distributed Communications Protocol), between the instrument and the System 75 XE PBX.

A System 75 XE can function as a node on an Information Systems Network (ISN) and as an end-point node on an Electronic Tandem Network (ETN). The System 75 XE supports a DS-1-framed T1 interface and the Digital Multiplexed Interface (DMI) for direct attachment to a minicomputer. The advantages of the System 75 XE compared with its System 75 predecessor models include lower start-up costs, reduced pricing, and improved modularity of the cabinet.

Software for the System 75 has evolved since the switch's initial release in April of 1984. The System 75 XE is available with either Version 2 software or Version 3 software and provides the same features and functionality offered by its bigger brother. AT&T has reduced switch prices as well as instrument prices making the System 75 an attractive PBX solution. AT&T is willing to negotiate generous credits toward ATTCOMM long-distance charges as a way of discounting PBX equipment, providing an additional economic incentive.

Northern Telecom entries

The all-time largest selling non-Bell PBX is the Northern Telecom SL-1. Northern still markets this switch in configurations of between 32 and 5,000 stations. While the switch architecture has remained basically the same except for the recent Meridian data bus enhancement, Northern continues to reduce the electrical consumption and to improve the packaging.

At the beginning of this year, Northern Telecom announced the Meridian SL-1S, serving from 30 to 160 stations. This PBX supports the usual Northern Telecom Meridian data services, including support for asynchronous data up to 19.2 kbit/s and for synchronous data up to 56 kbit/s.

The SL-1S comes in a three-shelf floor-standing chassis with the main wiring distribution frame physically inside the PBX, prewired to the interface cards. This packaging of prewired distribution blocks within the PBX cabinetry saves a significant amount of installation time. When the PBX arrives, the installer simply cross-connects the station cables and the trunk interfaces.

In contrast to the universal signal bus used in the AT&T System 25, the Northern Telecom SL-1S uses a segmented signal bus. Each segment of the multiplexer signal bus consists of 32 time slices. Two are used for control signaling while the rest are used for voice traffic. The master network controller allocates time slices to a specific carrier shelf. Time slices to a specific carrier shelf are only a portion of all available time slices. If a carrier shelf fails, only part of the PBX network fails. This is not true of universal bus structures.

The SL-1S was barely six-months old when Northern announced the SL-1ST, a modular package which consists of four tiers of card cages. The basic single-tier SL-1ST supports up to 80 stations with 12 trunks and one attendant console. The second tier permits the PBX to grow to 240 stations, while the third tier increases the size to 400 stations. A fourth tier, which will be available by summer's end, will permit expansion of low trunk intensive applications to 600 stations. Northern Telecom has introduced the concept of putting the tone receiver electronics onto a daughter board, instead of a dedicated printed circuit board. Also introduced was a new 16-port station card for 2500-type analog single-line telephone instruments. The SL-1ST is shipped with a preprogrammed software database to facilitate installation.

The Northern Telecom SL-1ST supports digital trunk interfaces including T1 and DMI digital transmission channels. The PBX may function as an end-point in an electronic tandem network. Software is available for using the PBX as an automatic call distributor. Standard software provides call detail recording output and basic automatic route selection. Comprehensive automatic route selection is optional.

The SL-1ST supports a wide variety of data communications interfaces and services. Protocol converter gateways are available for IBM 3270 and System 36/38 terminals and printers. Gateways for X.25 networks are supported along with modem pooling.

A coax elimination and switching capability is available that allows the connection of 3270-type terminals to a coax interface module (CIM). The CIM is connected to a port of a data station card within the SL-1ST PBX using one twisted-pair cable. The IBM 3274 cluster controller ports are connected to a multichannel coax system (MCCS), which in turn connects each port to a data station card using one pair per controller port. This facility allows a user with a 3278 terminal to contend for different cluster controllers,

and their associated hosts, using the SL-1ST as a port-contention switch.

Whereas the majority of PBX vendors offer mid-range PBXs that switch asynchronous data at 19.2 kbit/s and synchronous data at 56 kbit/s, Northern Telecom is unique in its offering of a 2.5-Mbit/s microcomputer desktop interface using twisted-pair cable. Its LANstar interface allows connection of a microcomputer that is a Netbios-compatible local area network interface. Novell's Netware has run successfully on a Meridian SL-1ST PBX.

Northern Telecom enhancements

Serving up to 100 telephones and 48 data devices, Northern Telecom's DV-1 combines circuit switching for voice applications and packet switching for data applications using two 20-Mbit/s signal buses. A variety of modules are available for applications processing and voice and data integration applications. The company has disclosed sales of only 1,000 DV-1 stations since the product was introduced in 1985. The typical DV-1 workstation is a model M4020 station that is an asynchronous terminal with telephone set physically integrated into the workstation. All of the trunk interfaces, station interfaces, power modules, network services modules, and applications processors are packaged to allow self-maintenance by the user.

Northern recently enhanced the audio/graphics conference-calling capabilities to allow up to 24 DV-1 users to be linked using voice and X.25-based data transmission channels. Users can share any specific user's terminal screen in addition to participating in a properly amplified audio teleconference. Northern is also beefing up its software offerings with telemarketing and message center applications. The message center can activate the message-waiting light on AT&T System 75 and System 85 products in addition to Northern PBXs.

Northern Telecom's difficulty in selling its DV-1 product is primarily due to the company's mistakenly marketing it as a minicomputer replacement with voice and data telecommunications capabilities. The DV-1 has limited processing power, which has led, unjustly, to the product's bad reputation. Northern recently dropped a majority of its value-added resellers, who have failed to sell the product because many customers were using it as a replacement for an IBM System/36. Northern is refocusing its distribution efforts toward regional Bell holding companies and major telephone companies that sell the SL-1 and SL-100 products.

IBM/Rolm PBX offerings

Rolm recently merged its marketing and sales support operations into those of parent IBM. Since then, the only major introduction from Rolm has been its Redwood product. The digital Redwood Model 1 functions only as a key service unit using one card cage cabinet to support up to 16 trunks and 48 stations. The Redwood Model 2 uses up to three cabinets with a capacity of 47 trunks and 144 stations. It may be configured as either a PBX or a key service unit, allowing the PBX to function in any of the four ways previously described.

The Redwood uses a distributed architecture to reduce the risk of failure and to facilitate simple repairs. The PBX uses several eight-port digital switching subsystems that communicate using a nonblocking, time division multiplexed signal bus. Each station and trunk in the PBX is assigned a specific time slot in the multiplexer bus so that the switch is always nonblocking. The Redwood uses standard pulse code modulation (PCM) techniques as well as sampling rates of 8,000 samples per second. This is different from the 12-bit linear coding technique and 12,000 per second sampling rate used in Rolm's VSCBX, CBX II, and VSCBX products. Rolm has preserved its 256-kbit/s one-pair Rolmlink interface supporting their Rolmphone station instruments.

The Redwood was announced as a voice switch with no data communications capabilities. Its digital architecture will easily accommodate data and, contrary to company denials, data support is expected in the near future. The overlap between the newer Redwood and the older VSCBX indicates that the latter may be phased out soon. Were it not for the 50 percent premium cost of a typical 80-user CBX-II configuration compared with an equivalent VSCBX configuration, the VSCBX probably would have been withdrawn earlier.

Release 2 of Redwood software has enhanced networking and management capabilities. Networking among multiple Redwood switches has been improved through support of direct inward system access (DISA). A basic electronic tandem network may be created using Redwood PBXs, tie-line trunks, DISA trunks, and combination in/out trunks. Central management of multiple remote Redwoods is possible using a personal computer. And the personal computer may also be used for PBX administrative functions, including station moves, changes, additions, and even diagnostics for remote switches. At the beginning of the year, Rolm announced the CBX II 9000AE—an upgrade to Rolm's largest switch with Release 9004.2 software. The new product includes a faster processor with more memory. The CBX II 9000AE also uses the Rolmbus 295, a 295-Mbit/s multiplexer bus architecture, as opposed to the more commonly found 74-Mbit/s signal bus. Installed CBX II switches may be field-upgraded to the newer product. Rolm's Cypress and Juniper voice and data workstations are now supported, functioning as automatic call distribution (ACD) "agent" workstations and supervisory workstations.

Rolm is marketing a Rolmphone digital telephone to be used with personal computers. The Rolmphone 244PC does not use an interface card within the microcomputer. Any personal computer serial port supporting asynchronous communications may be connected to the Rolmphone 244PC. The telephone instrument requires a Rolmphone station card in the PBX. This product supports asynchronous and synchronous communications at speeds up to 19.2 kbit/s. The Rolmphone 244PC interface was designed to support IBM's recently announced asynchronous communications device interface (ACDI). This interface allows customers to design and write their own applications using screen-based dialing and coordinated telephony/data processing for customer service departments.

Mitel's SX-200/SX-100 family of PBX products was one of the most popular PBX product lines during the early 1980s with a present installed base of 40,000 switches. The analog SX-200 serves up to 200 stations including 2500-type sets as well as Mitel's Superset line of proprietary instruments. Several recent market research surveys indicate that Mitel more than doubled its marketshare in the 100-to-400-station PBX market during 1985.

Mitel goes digital

The 14-line Superset 4 is widely acknowledged as a leader in terms of price, performance, and user interface. The Superset 4 instrument has a liquid crystal display that labels six function keys. The displayed choices change as the user changes station functions. During late 1986 Mitel announced a completely digital 336-port SX-200D as well as a second cabinet upgrade kit enlarging existing switches to 480 ports. The Mitel SX-200D uses the digital network interface circuit, a proprietary transmission protocol, to provide simultaneous voice and data communications using a single pair of twisted-pair cable and a single station port. This is a significant advantage when compared to most other switches, which require two ports for voice and data support, and it enables the Mitel products to handle more traffic than other switches with a similar port capacity.

Mitel now markets Superset 4DN and Superset 3DN—digital models with additional data communications capabilities. Synchronous and asynchronous data transmissions up to 19.2 kbit/s are presently supported. Useful telephony features may be used for data communications calls, including verified account code access, internal queuing, and speed dialing. The associated modem-line feature allows alternate voice and data communications using a single dial-up connection and the swap key on a Superset instrument.

Mitel's second digital PBX offspring is the SX-50, a 160-port PBX that includes such features as automatic route selection, modem-based data communications capabilities, messaging features, cost control software, and networking capabilities. The SX-50 uses a proprietary digital crosspoint chip to provide a nonblocking switching matrix. Each chip has eight input and eight output links, and each link supports 32 channels of time division multiplexed or pulse code modulated voice or data. There are 65,536 crosspoints on a single silicon chip allowing smaller PBX packaging. A custom filter codec chip also helps to make the cabinet very compact.

The SX-50 has the capacity for up to 10 trunk and station interface cards. Trunk cards are available for all Mitel SX products with four or eight ports per card. Station cards are available with eight or 16 ports per card. The networking capabilities of the SX-50 include support for tie-line trunks with E and M signaling, external call forwarding, direct inward system access (DISA), and satellite-PBX operation.

Mitel offers an interesting messaging facility between Superset 4 instruments connected to the SX family of PBX switches. Preprogrammed call-back and networkwide administrative messages are available through the Superset instrument as well as the capacity for user-programmed messages. A user programs the telephone to an option called "in a meeting," and the message may be read at the attendant console or at any Superset station calling the person who is in a meeting. Call-back messages can easily include the caller's name, extension, and time of call.

Mitel offers a new liquid crystal display-based attendant console for its SX line of PBX products. The first display line on the console provides traditional status and call information, including source and destination data. The second line displays English-language prompts for five function keys controlling call-answering. The console includes 21 fixed-function keys, of which seven are programmable.

The repackaged Mitel SX-2000 PBX

Mitel recently announced a repackaging of its SX-2000SG Integrated Communications System (ICS). The SX-2000S is a 768-port version of the company's existing 5376-port switch. Mitel has installed more than 500 of the SX-2000SG switches since introducing them in 1984. The PBX architecture consists of centralized processors to provide overall control with distributed microprocessors to handle switching, messaging, and supervisory functions. A dedicated packet switch is used for interprocessor communications. A high-capacity voice-and-data switching matrix allocates a 2-Mbit/s link to each card in the PBX. Mitel believes that the real-time parallel processing and software specialization among many processors provide better horsepower. The Mitel architecture separates message and switching control information, which is consistent with Integrated Services Digital Network (ISDN).

While the SX-2000S retains all of the functions of the SX-2000SG, the new product provides improved packaging, a lower cost per line through denser interface cards, and other features. The SX-2000SG requires 17 cards for the main controller function, whereas the SX-2000S requires just one card. CMOS semiconductors reduce not only the switch's power requirements but also the size of the power converters.

The single cabinet configuration that supports 193 ports may be expanded to a two-cabinet, four-shelf configuration that supports 758 ports. Mitel recommends the SX-2000S for configurations of 200 to 500 stations with 20 percent trunking. The SX-2000S supports both digital central office trunk interfaces and digital private line interfaces, including T1 channels on a single card. This company is one of the few PBX manufacturers offering a digital private networking package. It is called Mitel Superswitch Digital Network (MSDN).

The Superset 7 is a screen-based terminal with an integrated telephone instrument used as the attendant console and administrative workstation for the SX-2000S PBX. The Superset 7 supports screen windows for call-handling information, a digital keypad, a QWERTY keyboard, and function keys that are labeled on-screen by the PBX application program. Features such as dialing from a directory screen and message center capabilities are provided.

TIE/communications' confusing array

Most users are confused when they face the abundance of products from TIE/communications. Adding to the muddle, TIE sells similar products with different packaging through regional Bell operating companies, supply houses, and interconnect dealers. The company briefly surpassed AT&T as the leader in key service unit sales, but it overestimated the ability of the regional Bell operating companies to sell

large volumes. The sluggish market growth in the key telephone market along with severe product price erosion have continued to plague TIE.

After focusing on PBX products in the under 200-station category that have data communications capabilities, TIE has brought inventories and costs under control. It cut staff by 50 percent to 1,500 people with the associated consolidation of facilities. While the cost structure has improved, it is still difficult to differentiate TIE's product from that of the other 65 some-odd vendors.

TIE's first venture into the new voice and data arena was the Data Star, a digital nonblocking PBX using mu-law 255 pulse code modulation techniques in a time/space division switching matrix network. The architecture consists of distributed microprocessors for control along with redundancy for most of the critical electronic circuitry. Nonvolatile bubble memory is used to store the PBX's programming. The Data Star becomes more economical than the competition as the percentage of data users increases beyond 25 percent. The product is sold by RBOCs and selected dealers. A single-cabinet Data Star configuration has a capacity of 32 trunks and 96 voice-with-data stations. An expansion cabinet allows the PBX to grow to 64 trunks and 192 stations. Time slots within the switch may carry either voice or data. The Data Star supports asynchronous data communications up to 19.2 kbit/s with support for synchronous communications up to 56 kbit/s promised in the near future. Internal modem pooling cards are available, which are Bell 212A-compatible with the Hayes Smartmodem dialing emulation.

Data Star supports a variety of telephone instruments, most of which are proprietary. Analog 2500-type single-line instruments are supported only through special off-premises extension station cards. The executive display set includes 15 programmable feature keys, 14 function keys, and a one-line visual alphanumeric display. The display shows the call status, calling party's name, call duration, called number, time, and date. The executive display set has a built-in four-function calculator as well as a two-way speakerphone.

The remaining instrument choices of the Data Star station include:
- Multibutton key set available with or without data support.
- Single-line featurephone.
- Electronic single-line set.

The multibutton set offers the same features as the executive display set without the visual display. The single-line featurephone includes 13 feature function keys and is recommended for most locations. The electronic single-line set has four function keys in addition to the 12-button keypad. TIE offers Voice Star, a voice store-and-forward facility, as an enhancement to the Data Star PBX product. Packaged separately, but physically integrated, Voice Star supports up to 1,000 users, 32 messages per station, 16 ports, and 33 hours of storage. Voice Star can provide automated attendant services, synthesized voice announcements, and the equivalent of direct inward dial capabilities without the cost of DID trunks. Voice mailbox features allow Voice Star to be used as a messaging facility similar to Rolm's Phonemail product.

The largest TIE PBX products include the Mercury PBX product line and the M-1000. Mercury consists of three packages: the TX-250 with 288 ports, the TX-450 with 432 ports, and the TX-2000 with 1920 aggregate ports. This product represents the efforts of the company's Canadian operations in enhancing technology acquired from Plessey. Up to 10 Mercury PBXs may be connected to support up to 10,000 lines. Mercury competes against the largest switches from AT&T, Rolm, and Northern Telecom. TIE's limited experience in this marketplace is reflected in the low sales volume of Mercury PBXs.

The new Meritor M1000 represents a repackaging of the Mercury technology in a smaller configuration that is sold through TIE's Mercury dealers. Data Star modules support 16 trunks, whereas M1000 trunk modules support 32 trunks. Each M1000 cabinet contains 160 ports supporting a total of 128 stations and 32 trunks. A four-cabinet M1000 supports 640 ports. This product is a direct competitor of the AT&T System 75XE. A networking software package is available for supporting 6,400 ports spread over 10 locations.

The M1000 uses data terminal line cards, terminal interface modules (TIMs), and TIE's Datalink software to support data communications at transmission speeds up to 19.2 kbit/s. Each M1000 cabinet supports nonblocked data connectivity for up to 64 data ports, which yields a total of 256 data ports.

Three types of TIE M1000 station instruments are available. The 28-button Digiphone includes a liquid crystal display, 28 programmable function keys, and a 16-button keypad. An optional 64-button direct station selector may be attached, making the set function as though it were a console. The nine-button Digiphone uses a 12-digit keypad. The least expensive M-set has no function keys.

TIE's Morgan PBX alternative

A digital voice and data PBX, TIE's Morgan is sold through supply houses in order to avoid conflict with the Data Star, which is distributed through the RBOCs and interconnect companies. The Morgan is sold preprogrammed. A personal computer is used to make programming changes by entering data into spreadsheets, which are edited to check for validity. According to Ray Dellecker, a TIE senior manager in system management, "All instruments are controlled by a single card type. If the phone types actually installed differ from the program, they automatically reconfigure and go into service."

The Morgan PBX product is technologically more advanced than TIE's Data Star PBX. The Morgan provides three 64-kbit/s pulse code modulation transmission channels for each station using a single twisted-pair cable. On the trunk side of the PBX, double bandwidth is available for potential upgrade to ISDN. The Morgan uses a 16-bit processor with a microfloppy diskette for memory backup. TIE's Morgan is available in two configurations. The smaller cabinet supports up to 72 stations and 24 trunks while the larger cabinet supports the full 144 station capacity with 48 trunks.

Only proprietary digital instruments are used with the Morgan. Every instrument includes support for data communications using an RS-232-C port on the back of the telephone instrument. The improved PBX packaging, single-card design for all stations, preprogrammed software configuration, and improved manufacturing costs allow TIE

to claim a per-voice-and-data-station cost of less than $500, uninstalled and including instruments.

TIE's Ultracom DCX rounds out TIE's digital PBX product line. The Ultracom supports a maximum of 160 stations and 48 trunks in a nonblocked configuration with a total of 320 available time slots. Station instruments are similar in design to those used with the Morgan and Data Star, but they differ in the way the instruments are packaged.

TIE's Ultracom DCX is sold by interconnect dealers and promoted primarily as a hybrid key service unit supporting integrated voice and data. The Ultracom PBX overlaps with the Morgan and Data Star in terms of its configuration size. While the three products are sold through different distribution channels, users perceive only that all three products compete. No wonder they cannot easily compare the strengths and weaknesses of TIE's PBX products with other vendors' products.

GTE joins forces with Fujitsu

In April, Fujitsu GTE Business Systems was born—but it was 80 percent Fujitsu and only 20 percent GTE. The new venture sells the Omni SI in the 60- to 256-station size, the Omni SIII in the 200- to 2,000-station size, and the Omni SIV serving up to 10,000 stations. The large switch was developed by Fujitsu to compete against a Northern Telecom SL-100 and will soon be manufactured in California. The two smaller PBX switches are GTE products. Fujitsu GTE sells the Omni S-series against a wholly owned Fujitsu subsidiary called Fujitsu Business Communications with its Focus PBX product. The Omni series uses conventional PCM techniques at the station and trunk interfaces. Station cards support eight stations each, while trunk cards support four trunks per card. The dual bus, packetized data architecture ensures that voice and data communications cannot contend for the same resources. Eight ports are supported per data station interface card. Recently, a 16-port station card was introduced.

The Omni series of PBXs uses 2500-type analog instruments as well as proprietary digital FeatureComm V/VI instruments. FeatureComm instruments have visual displays, eight or 16 function keys, and support for simultaneous voice and data communications using a single pair of cables. GTE uses a proprietary technique of packetizing data internally within the switch rather than using an applications processor. Each individual station instrument has an integrated packet assembler/disassembler for data, a codec to digitize voice, and circuitry to interleave voice with data packets.

The station instrument generates 11-byte packets, one byte of which contains addressing information. All data communications speed and protocol conversion is accomplished in the station instrument. Error checking begins in the station with a cyclic redundancy check (CRC) technique. CRC checking occurs up to five times within the switch to allow the vendor to claim a bit error rate of one error in 1 billion bits.

The Omni SI supports a single T1 interface while the SIII version supports up to 16 T1 transmission channels. (Plans include increasing the capacity of the Omni SIII to 2,500 stations.) According to Tony Brown, national account manager at Fujitsu-GTE in New York, the data communications features that are new to the Omni SIV provide "the

VENDOR	PRODUCT	MAXIMUM STATIONS	MAXIMUM TRUNKS	ASYNC DATA UP TO 19.6K	SYNC DATA UP TO 9.6K	SYNC DATA UP TO 56K	NETBIOS DATA AT 1 MBIT/S	T1 INTERFACE
ATTIS	DIGITAL MERLIN	72	30	A	N/A	N/A	N/A	N/A
	SYSTEM 25	200	104	A	N/A	N/A	N/A	N/A
	SYSTEM 75XE	600	150	A	A	A	N/A	A
NORTHERN TELECOM	SL-1ST	600	48	A	A	A	A	A
	DV-1	99	99	A	A	A	A	N/A
ROLM	REDWOOD	144	47	N/A	N/A	N/A	N/A	N/A
	VSCBX	300	50	A	A	A	N/A	A
	CBXII	800	150	A	A	A	N/A	A
MITEL	SX-50	100	48	B	B	B	N/A	N/A
	SX-200D	336	100	B	B	B	N/A	N/A
	SX-2000S	700	150	A	A	A	N/A	A
TIE	DATASTAR	192	64	A	B	B	N/A	N/A
	MORGAN	144	48	B	B	B	N/A	N/A
	ULTRACOM	160	48	A	B	B	N/A	N/A
	M-1000	600	128	A	B	B	N/A	N/A
	MERCURY	1,920	250	A	B	B	N/A	A
GTE-FUJITSU	OMNI SI	256	64	A	A	B	N/A	A
	OMNI SIII	2,048	800	A	A	A	N/A	A
WANG	WBX	200	200	A	N/A	N/A	N/A	B
ITT	3100XL	576	150	A	B	B	N/A	N/A

A — AVAILABLE   B — ANNOUNCED BUT NOT AVAILABLE   N/A — NOT ANNOUNCED   *AS OF JULY 1, 1987

processing power of a central office switch with the features of an end-user private branch exchange." As for cabinets, the Omni SI comes with a 4-foot enclosure whereas that of the SIII is 6 feet.

Telenova/Wang's PBXs

Wang recently acquired 40 percent of Telenova, and together they market the Telenova 1/Wang Business Exchange (WBX). Telenova/Wang PBX Release 2, announced in late 1986, increased the station capacity to 200, doubled the bandwidth to 512 time slots, and added an 80286 processor to the existing 8086 for more processing power. A second cabinet houses the additional interface cards required for the second hundred users. Says Suzanne Matick, Telenova's marketing communications manager: "You will see an increase in port size [beyond 200], but we'd prefer not to say when."

The Wang WBX architecture uses intelligent station devices that multiplex digitized voice and data signals over two pairs of twisted-pair cable. The switch's control processor group consists of a module processor, up to 16 Mbytes of random access memory, and a 10-Mbyte Winchester disk storage unit. Each of the switch's communications group includes a dedicated time division multiplexer bus supporting up to 120 full-duplex, nonblocking logical ports. The time division multiplexer bus is divided into 2,048 dynamically allocatable 9-kbit/s time slots. Groups of time slots are normally dynamically allocated to 256 logical time

slots of 72 kbit/s each. One time division multiplexer bus is dedicated to carrying voice and data signals. A second TDM bus transmits control messages between the module processor and the interface units. Interface units with logically defined ports can plug into any card slot.

Each communications group bus supports 18 interface card slots. A single cabinet configuration includes a single communications group. The upgraded configuration includes two communications groups and a group network interface. The group network interface combines the functions of the control processor group with a time slot interchanger. The upgraded configuration includes twice the number of card slots and twice the number of time slots.

Interface cards pass data between the control processor group and the station devices. Interface card port densities may change as silicon technology and packaging improve in this type of PBX implementation. Current interfaces include a four-port trunk interface and an eight-port voice and data station card with 16 logical port connections. The WBX uses a service unit processor with 50 logical ports to perform basic telephony functions. The Telenova/Wang PBX product supports only asynchronous data communications through the data adapter device (DAD) at speeds up to 19.2 kbit/s. Support for T1 has been announced as a future option, not currently available. Plans for supporting synchronous transmission have not been disclosed.

Telenova seems to have adopted the Mitel Superset 4

3. Looking ahead. With a 30 percent share of the market, AT&T's System 75 and System 25 would lead in 1990 sales of mid-range PBXs.

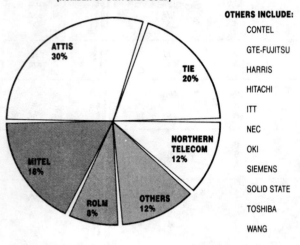

**1990 PERCENTAGE MARKET SHARE
(NUMBER OF SWITCHES SOLD)**

ATTIS 30%

TIE 20%

NORTHERN TELECOM 12%

OTHERS 12%

ROLM 8%

MITEL 18%

OTHERS INCLUDE:

CONTEL

GTE-FUJITSU

HARRIS

HITACHI

ITT

NEC

OKI

SIEMENS

SOLID STATE

TOSHIBA

WANG

approach to its station instruments. Only one station instrument type is offered, and it includes a two-line liquid crystal display, five software-defined function keys, and eight additional function keys. The function keys' definitions depend upon the functions available at any time.

A recent enhancement provides secretarial coverage for more than nine stations using an asynchronous cathode ray tube (CRT) terminal. The terminal acts as a busy lamp field and station status monitor indicating whether a station is busy, active, or inactive. The terminal lists all extensions being covered, the person's name, and whether the station is forwarded. Single-button access is available to any covered extension.

ITT's enhanced System 3100
By releasing new eight-port station cards, ITT recently expanded its 384-port System 3100LX to a 576-port System 3100XL. The ITT System 3100XL consists of up to eight stackable cabinet modules. A new software release supports four-digit dialing between System 3100 PBXs.

New 11-button and 19-button feature instruments with displays were also released. Support for asynchronous data communications at speeds up to 19.2 kbit/s is now supported using three pairs of cable. One cable pair carries voice, one carries data, and the third carries control signals. Support for synchronous data communications at speeds up to 64 kbit/s has been announced but not yet delivered.

ITT sold the last of its direct sales operations to distributors during late 1986. At the same time, the company substantially raised its repair prices and lengthened its repair turnaround time. It has approximately 150 interconnect distributors and no longer sells directly to end users.

Reviewing data requirements
Table 2 summarizes the PBX switches reviewed in this article. For each product, the switch's maximum station capacity is indicated along with the maximum trunk capac-

ity. Support for various data communications services has been divided into three states: available, announced but not yet available, and not announced. Included in the five measures of data communications shown are support for: asynchronous data transmission up to 19.2 kbit/s; synchronous data transmission up to 9.6 kbit/s; synchronous data communications up to 56 kbit/s; personal computer Netbios interfaces at no less than 1 Mbit/s; and T1 interfaces.

Only the Northern Telecom Meridian PBX products provide Netbios microcomputer interface support. Only Rolm's Redwood fails to support asynchronous data transmission at speeds up to 19.2 kbit/s. Many PBX vendors announce data communications capabilities that are not delivered or are delivered considerably later than promised. A user is urged to carefully check customer references when verifying vendors' claims concerning various data communications capabilities.

Coming attractions
Figure 3 shows the projected marketshare for mid-range PBX switches in 1990. With a 30 percent share of the market, AT&T is expected to emerge the leader primarily through sales of its System 75 and, to a lesser degree, System 25 PBX products. TIE is expected to attain the second leading position with 20 percent. Its strength should be in the low end of the mid-range PBX market, with configurations of fewer than 200 stations.

Mitel should emerge as the third leading mid-range PBX supplier. Currently the company has the largest installed base in the under-100-station market. Mitel's new SX-50 and digital SX-200 are sound PBX offerings in the under-300-station market segment. Moreover, the new SX-2000SG provides all of the features that are presently offered in large PBX switches—and the Superset station interface is very user friendly.

Northern Telecom will continue exerting pressure by marketing to existing customers that are adding sites or upgrading their PBXs. With a projected marketshare of 12 percent, Northern is expected to surpass Rolm in sales of mid-range PBX switches, while Rolm is expected to garner 8 percent of this market. Rolm's Redwood is at a serious competitive disadvantage without data communications support. The Rolm VSCBX is aging and needs updating or replacement.

The remaining vendors are expected to make up approximately 12 percent of the mid-range PBX marketplace in 1990. These include: Contel, GTE-Fujitsu, Harris, Hitachi, ITT, NEC, Oki, Siemens, Solid State, Toshiba, and Wang/Telenova. The product life cycle for mid-range PBX products is shrinking. Vendors are continually improving the capabilities and features of their products in response to the increasingly competitive marketplace. As would be expected in such circumstances, the user benefits by a wide spectrum of PBX product alternatives, better price/performance, in addition to a rich array of data communications capabilities. ■

David P. Levin is president of Netcomm, a data communications engineering and management consulting firm. He has worked for the Chase Bank, Bankers Trust Company, and Donovan Data Systems. He earned his B. S. and M. B. A. from Rensselaer Polytechnic Institute.

An Overview of Recent Developments In the Designs and Applications of Customer Premises Switches

Natesa Janakiraman

Some of the recent developments in the customer premises switch area are traced and the forces at play—technology, office automation, ISDN developments, and deregulation—are highlighted

It has been widely recognized that the Customer Premises Switch—that is, the Private Automatic Branch Exchange (PABX), as well as the Local Area Network (LAN) type architecture—is not only the center piece in the "office of the future" but also one that plays an important role in the evolving public and private integrated voice/data networks. In this context, the influencing factors in the evolution of Customer Premises Switches are many. Among these, technology, office automation, Integrated Services Digital Network (ISDN) developments, and deregulation of the communications industry are important.

Recent trends in technology have made distributed processing very cost effective. Ever increasing chip densities and processing power have enabled autonomous distributed switch modules with declining cost per line. Further, the advent of fiber optic technology has made unprecedented data rates possible in the customer premises, in addition to providing an alternative to the Telephone Company (TelCo), in terms of a bypass medium between the customer premises and the terrestrial/satellite gateway for private and public networks.

The emergence of ISDN and office automation has given a new thrust to Customer Premises Switch development. With personal computers expected to attain ubiquity very soon and the need for integrated voice/data services with total non-blocking characteristics, great changes in the architecture of new switches and significantly improved features and performance are anticipated. In fact, we may say that ISDN has already made its presence in the customer premises in some form, and office automation is well on its way to maturity.

The third factor, which is perhaps of far-reaching significance, is the deregulation-divestiture environment. This has created a Centrex-PABX conflict with switch vendors—and, ironically, the unregulated subsidiaries of the divested Regional Bell Operating Companies (RBOC's)—trying to promote Customer Premises Switches and accelerate the replacement of Centrexes. Increases in access charges are driving the Inter Exchange Carriers to bypass TelCo facilities and interconnect Customer Premises Switches to their gateways using alternative media, wherever possible.

The above mentioned environments have created a vigorously competitive industry presenting the users with a myriad of products and services with varying degrees of price, features, and performance. This article reviews some of the recent developments and trends influenced by these environments in the area of Customer Premises Switches. Table I, provides a comparison of a number of relatively recent customer premises switches and sets the stage for the discussion in the remainder of this paper.

EHO282-4/89/0000/0049$01.00 © 1985 IEEE

TABLE I
A COMPARISON OF SOME CUSTOMER PREMISES SWITCHES

Product	Capacity for Fully Equipped System: # ports, aggregate bandwidth	Call handling	Topology Switching Scheme	Subscriber Loop for Integrated Voice/Data Transmission: No. of Pairs	Bit Rate (kbps)	Protocol	Data Rates Handled: ASync (kbps)	Sync (kbps)	Interfaces: T1	X.25	Comments
AT&T IS SYSTEM 85 II	7000 ports	Both blocking (1 in 500 and 1 in a million) and non-blocking configurations available.	Star. Time-Space-Time	2	160	Proprietary Digital Communication Protocol (DCP) using one signalling channel and two information channels.	Up to 19.2	Up to 19.2	Yes	Yes, via Data Communication Interface Unit (DCIU)	Uses 501cc bit slice processor. Several Systems 85 II's can be configured in a Distributed Communication System (DCS) supporting up to 25,000 stations in 12 geographically dispersed locations anywhere in the country.
CXC ROSE	12,288 ports. 33 Mbps for circuit switching. 10 Mbps for Ethernet. 16 Mbps optional for token ring (See Comments)	Non-blocking for intra-nodal calls. Some blocking internodal.	Ring. Time Division switching CSMA/CD and token passing	2	192	Proprietary scheme using two 64 kbps information channels and one 32 kbps signalling channel.	Up to 19.2	Up to 64 kbps for voice and data. 128 kbps for data alone.	Yes	Yes	Handles voice, data and compressed-video in 8 kbps increments. 192 ports node. 61 nodes maximum.
ERRICCSON MD-110	20,000 ports	Non-blocking for intra-nodal calls. Inter-nodal blocking can be eliminated by increasing the number of T1 links between nodes.	Star. Time-Space-Time	1	256	Proprietary scheme using time compression multiplexing method.	Up to 19.2 for standalone terminals. Up to 9.6 kpbs for integrated voice-data.	Up to 56 kbps standalone. 9.6 kbps for integrated voice/data.	Yes	Yes, by means of external hardware.	MD110's architecture consists of autonomous Line Interface Modules (LIM's) interconnected by a Group Switch (GS). LIM's and GS are connected by PCM links.
INTECOM IBX S/80	8,192 ports. Can have up to 512 Mbps in a Local Area Network Configuration.	Totally non-blocking.	Star. Time-Space-Time	2	144	Proprietary scheme.	Up to 19.2.	Up to 57.6.	Yes	Yes	Intercom Switch has extensive data handling capabilities in addition to regular PABX voice features. Can be configured as a Local Area Network (LANMARK). Interface to other LAN's. Speed protocol conversion are also provided as an integral part of the switch.
NEC NEAX-2400	23,186 ports.	Non-blocking for up to 736 ports. Some blocking (with 32 ccs/port) for bigger systems.	Star. Time-Space-Time	2	256	Proprietary scheme. Two 64 kbps for voice and data, 32 kbps for control. Remaining 96 kbps reserved for future use.	Up to 19.2	Up to 56.	Yes	Yes	NEAX-2400 is expandable from 184 ports. Main processor operates in a Load Shared redundancy mode. VLSI intensive, uses bubble memory.
NTI SL100	30,000 ports.	Totally non-blocking.	Star. Time-Space-Time	1	160	Proprietary "Data-Path" protocol using Time Compression Multiplexing.	Up to 19.2	Up to 64.	Yes	Yes. Provided by means of external hardware.	SL-100 is based on NTI's DMS-100 Class 5 family. Has extensive voice data features. Has Remote Switch Unit capability. CCITT #7 Signalling System planned.
ROLM CBX-II	11,520 ports. Individual data circuits can be set up at up to 37 Mbps. (See Comments)	Totally nonblocking.	Star. Time-Space-Time	1	256	Proprietary Time Compression Multiplexing.	Up to 19.2	Up to 56 kbps for integrated voice/data. (See Comments for high-speed data)	Yes	Yes	Using "Dynamically Allocatable Bandwidth Concept" devices requiring faster communication channels can be assigned more than one 192 kbps channel. It is possible to have data connections up to 37 Mbps.
ZTEL PNX	30,000 ports. 10 Mbps rings for Time Division Multiplexed and token passing traffic.	Non-blocking intra-nodally. Internodal blocking can be eliminated using multiple rings.	Time Division Multiplexed ring and IBM token passing ring.	2	192	Proprietary Ztel channel has one voice channel, one data channel and two control channels.	Up to 19.2	Up to 56	Yes	Yes	CXC-ROSE and Ztel PNX are similar in many respects. Each node (Ring Processor) handles up to 2,000 lines. Each circuit board has a maintenance processor. Application software can be distributed among nodes. Load shared redundancy employed.

Technology Related Developments

Distributed versus Central Processing

One of the important characteristics of new Customer Premises Switches is extensively distributed switching, processing and related functions. This is enabling truly modular systems with:—minimal startup costs,—incremental costs growing linearly with the amount of additional capacity (number of lines added), and—non replacement of existing hardware/software.

What can be termed the "acid-test" for distributed processing architecture is the cost per line behavior. In a truly distributed switch structure, the cost per line is nearly a constant, while varying degrees of distribution of intelligence require higher initial investment for the necessary centralized hardware/software units. Figures 1(a) and (b) illustrate this point. The switches compared in the figures were of about 10,000 lines of maximum capacity PABXs with varying degrees of distributed processing. It should be emphasized further that the type of technology used significantly affects the power, floor space and air-conditioning requirements. These variables, often ignored by designers as well as users, can increase recurring costs by an order of magnitude—See Figures 1(c) and (d). In fact, it was calculated that for a 3000 lines capacity switch, the cost of power alone can vary from $3,000 to $30,000 per year depending on the switch design. With regard to air

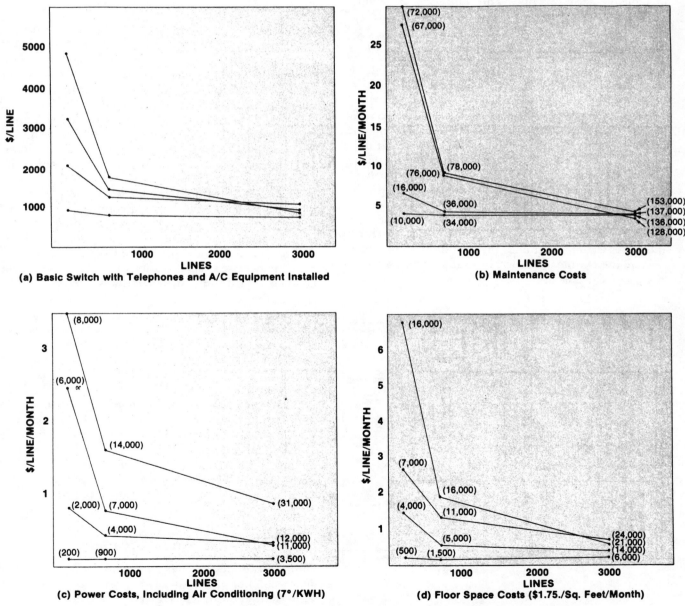

Fig. 1. *Cost per line behavior of PABX's. Bracketed numbers represent cost per year per switch.*

conditioning, it is a design objective not to require special air conditioning. In fact, natural convection cooling is used in one of the new switches [1].

The cost, power, and space behavior as influenced by technology is hypothetically depicted in Fig. 2. Ideally, these variables have the behavior of a shifted hyperbola. The curves shown depict switches designed for capacities ranging from N_1 to N_2 lines. Relative cost effectiveness of a switch for a given capacity is determined by the closeness of the cost curve to the asymptote, that is, the horizontal line denoting incremental cost per line I. It may be noted that the closer a point on a curve is to the horizontal asymptote, the smaller is the start-up cost per line, S/N. From observations of switch evolution, it is obvious that technology has been diminishing the start up cost as well as the incremental cost per line. This implies increasingly VLSI intensive products.

geographical distribution

Fully distributed architectures do not only allow functional distribution in switching modules, but also lend themselves to physical or geographical distribution of these modules. The advantage of geographical distribution is that there is no need for a big common switching room, which may cost a few thousand dollars a year (see Fig. 1d) in leasing costs. It should be emphasized that there will still be some requirement for a common central space to house application processors. One disadvantage, however, is that maintenance and troubleshooting become inconvenient with maintenance software in one location and switching modules scattered around the building. This may require the maintenance person to move from place to place to fix problems.

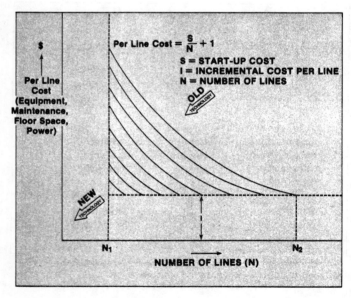

Fig. 2. *Technology influence on distributed processing*

From a design point of view, geographical distribution implies small physical dimensions—not larger than an office desk—so that these modules can be situated neatly, for example, on every floor of an office building. The modules must also be capable of being easily moved. This also calls for aesthetics in design so that the cabinets have a furniture-like pleasing appearance in an office environment. Such geographically distributed nodes must have smaller capacity (200–300 lines) to avoid significant backup power, large space, and special AC requirements. They should preferably be designed for natural convection cooling; alternatively forced air cooling with fans should meet office noise requirements.

Redundancy, Reliability, and Diagnostics

Another area that modern technology has influenced is redundancy, reliability, and diagnostics. Load shared redundancy, that is, in the event of a component failure other components share the system load and enable operation in a somewhat degraded fashion, is becoming a strong design objective. In addition, the tools for detection and isolation of problem areas are becoming increasingly sophisticated. Ztel's PNX switch, for example, incorporates dedicated maintenance processors in each circuit board to monitor, report, and remedy failures. These processors can divert traffic away from faulty subsystems once they detect the source of malfunction. In addition, for sophisticated troubleshooting, remote diagnostic centers are established by many vendors, from which fault isolation and detection can be performed. It is not uncommon to see that a manufacturer's specification for the mean time between catastrophic switch failures exceed the lifespan of the switch itself, usually 7–10 years.

Ring structures are particularly vulnerable to catastrophic failures. Thus, special reliability related design considerations, such as loop redundancy and automatic loop switching, are essential. The reliability of ring structures has been the subject of serious study. Contributions to the IEEE on the token ring LAN Standard and work undertaken in IBM Zurich Labs and the Nippon Telephone and Telegraph Company (NTT) have demonstrated how such rings can be made reliable. Although this issue can be satisfactorily addressed theoretically, the actual reliability related performance can only be learned after these systems are installed and operational over a period of time.

Technology and Product Planning

Significant design changes due to the availability of new technology have already happened to some extent, and will continue. However, sweeping design changes—granting that they are technologically and economically feasible—can only happen slowly and on a small scale. For manufacturers who have a large base of older generation switches, the business strategy warrants gradual upgrade of an older switch into a new and advanced switch. Design considerations under those circumstances will be influenced by the need to provide compatibility with existing hardware and software in the customer base. Also, manufacturers of large public network switches in order to efficiently use their development resources may opt for a PABX design borrowed largely from the CO (central office) switch design. Thus, protection of installed base and optimal use of internal corporate resources play a significant role in product planning and in deciding the architecture, feature, and price considerations. In any event, we may conclude that the impact of technology is more immediate in the customer premises than in central offices where billions of dollars already invested present considerable inertia to react to rapid changes in technology.

Office Automation—ISDN Related Developments
PABX—LAN Dichotomy in the Office of the Future

The evolution of PABX's has been occurring over the past five decades or so. This evolutionary history has often been divided into different "generations," each marking a phase in the switch design. Table II gives a description of these generations. In addition to the three generations described in Table II, a new "fourth-generation" has been added recently. This fourth-generation is distinguished by a ring structure interconnecting autonomous nodes. The topological difference between these two generations is depicted in Fig. 3.

It has been claimed that the fourth-generation ring structured switches are the only fully distributed architectures with a Local Area Network (LAN) as an integral part of it, and truly integrated voice (circuit-

TABLE II
CHARACTERISTICS OF PBX GENERATIONS*

| Call Control | FIRST | SECOND | | | THIRD |
	Early	Middle	Late	
In	Manual	Manual	Manual	DID	DID
Out	Manual	Dial	Dial	Dial	Dial
Intercom	Manual	Dial	Dial	Dial	Dial
Switching	Mechanical	Electro Mechanical	Electro Mechanical	Electronic	Electronic
	(Plug/cord connection or toggle switches)	(Step by step)	(Step by step, crossbar)		
	Space	Space	Space	Time	Time
	Blocked	Blocked	Blocked	Blocked	Non-Blocked
Topology					
Centralized	x	x	x	x	
Distributed					x
Control					
Centralized	x	x	x	x	
Distributed					x
Comments	Entirely manual operation	Operator involvement in supervision and control of exception activities (1)	Push button consoles. Automated control of exception activities except incoming calls.	Stored program control	Stored program control. Inherently integrated voice/data

(1) Control of incoming calls, call transfer, alternate routing, providing outside lines to restricted stations etc.
*Adapted with permission from "Intra-Site Communications Networks", Final Report, Perspective Telecommunications Group, Paramus, NJ 07652, March 1982.

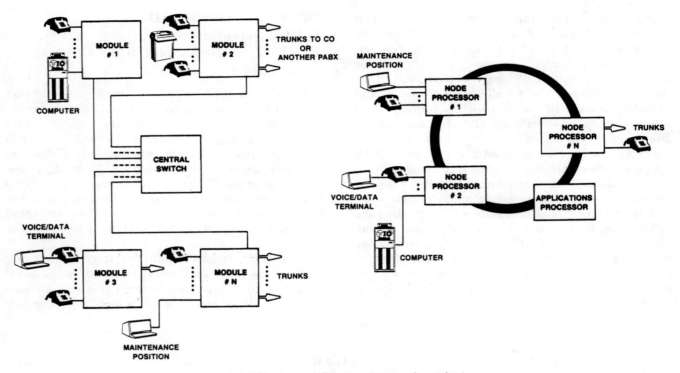

Fig. 3. *The star and the ring structured topologies*

switched) and data (circuit and/or packet-switched) transmission. However, the "generation gap" between the third and fourth generations is not as great as it may first appear, since some of the very well designed star-structured PABX's (see Fig. 1a) have essentially flat cost per line meeting the criterion for distributed processing, and satisfactorily address the issues involved in interfacing with LAN's and packet switched networks. These switches provide up to 56 kbs circuit switched data capability with non-blocking switching networks. This may be compared with a 10 Mbs ring in which certain degrees of internodal blocking for 64 Kbs voice is to be expected.

The discussion on third and fourth generations is really a debate on whether PABX or LAN will survive in the future. In this context, it appears that the coexistence of LAN's and the PABX's is inevitable at least through the 1980s. The reasons are: 1) voice dominates customer premises traffic (90 to 95 percent by various estimates) and the LAN's do not satisfactorily support the voice bandwidth requirements for a large speaker population, 2) the present day PABX's do not meet high bandwidth data transfer applications requirements, and 3) the desire for total integration, that is, voice in LAN or data in PABX, in the office environment is still not widespread in the user set due to a large number of older generation switches in place. While the PABX-LAN dichotomy is debated, each side claiming superiority over the other, designers of LAN's and PABX's are slowly removing the constraints in their products. For example, (see Table I for a compar-

ison of the capacities of different switches) Intecom has 512 Mbs of data rate capability integrated in their PABX referred to as LANMARK and, as referred to before, ROLM's CBX II with the optional "ROLM bus 295" can support 768 full duplex voice/data communication channels with each channel accommodating up to 192 kbs [2, 3]. The LAN type ROSE architecture [4], on the other hand, has up to 33 Mbs for circuit switched voice increasing to some extent its voice handling capability. It appears that in the long run, LAN's with multiple fiber optic rings for voice and data will be a formidable competitor in the customer premises.

Digital Connectivity to the Station Instrument in the ISDN Environment

The concept of the Integrated Services Digital Network (ISDN) is to harmonize the existence of a myriad of digital networks and services by providing common interfaces and end-to-end digital connectivity. The ISDN is conceived to be of three parts: the user network (which is customer premises equipment related), the local network providing access to the end-office consisting of the TelCo loop-plant and the ISDN end office, and the transit network consisting of a variety of circuit and packet switched networks [5]. While a great deal of work remains to be done by way of standardization, there has already been a significant amount of effort in defining and recommending the user network

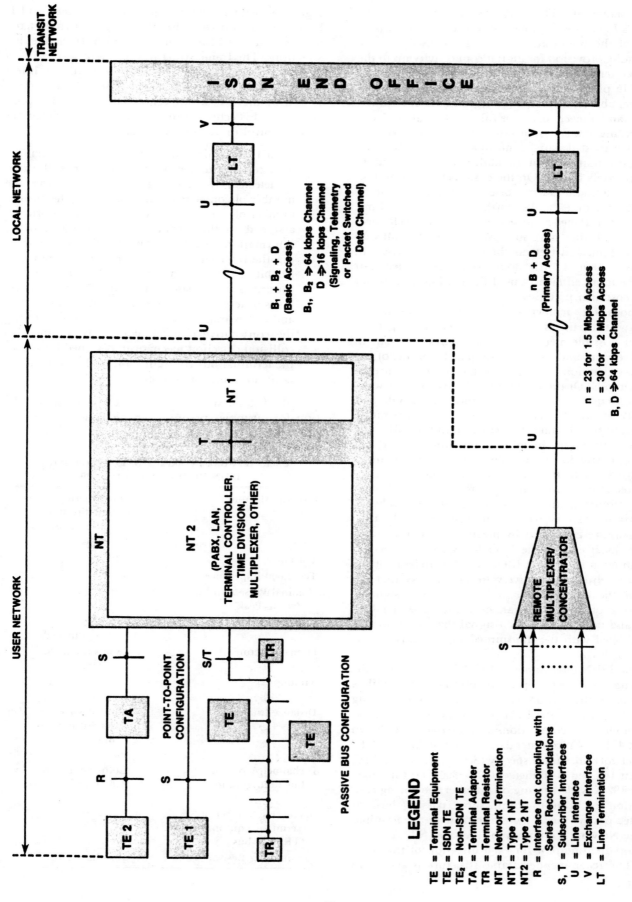

Fig. 4. *User and local network interfaces in ISDN*

LEGEND

TE = Terminal Equipment
TE₁ = ISDN TE
TE₂ = Non-ISDN TE
TA = Terminal Adapter
TR = Terminal Resistor
NT = Network Termination
NT1 = Type 1 NT
NT2 = Type 2 NT
R = Interface not compling with I Series Recommendations
S, T = Subscriber Interfaces
U = Line Interface
V = Exchange Interface
LT = Line Termination

55

related standards. These will soon appear in the CCITT's I Series Recommendations [6].

One of the important aspects of the user network issues is the provision for digital connectivity up to the end user terminal with ability to use twisted pairs already in place. In the use of twisted pairs, one must assure reliable transmission of digital data at required bit rates and a mechanism for full duplex capability.

Regarding transmission rates, it is well known that the 26 Ga twisted pair in multipair TelCo cables has significant transmission capability —1.5 Mbs up to approximately a mile. On the customer premises, taking into account such noise sources as elevators, dimmers, machines, and so forth, it is required to limit the bit rate to some extent. Some vendors have claimed, however, that they have successfully tested 1.5 Mbs up to a few thousand feet on the customer premises. In any case, there is little doubt that the twisted pair could be more efficiently used than it has been so far on the customer premises.

Methods for achieving full duplex transmission using twisted pairs is a subject in itself. Briefly, there are four possible methods: use of 4 wires, one pair in each direction, and use of single pair by means of frequency separation, time separation, or echo cancellation methods. Of these methods, the use of four wires and the frequency separation method in a single pair are rather straightforward. In the time separation method, bursts of data are interleaved in time allowing transmission in one direction at one time interval and in the opposite direction at the consecutive time interval. This alternate transmission of data bursts in opposite directions is popularly referred to as the "ping-pong" protocol. In the echo cancellation scheme, transmit and receive sections at each end of the two-wire loop are connected by means of a hybrid transformer. Each end of the loop has an echo canceller with an adaptive digital filter. The coefficients of the filter are adjusted adaptively in such a way that the sum of the filter output and the received signal—consisting of far-end signal, near and far-end reflections, and noise—produces a signal that is completely uncorrelated with the transmitted signal, thus removing echo.

In the ISDN environment, CCITT recommends the four-wire subscriber loop in the user network, while a two-wire configuration is assumed for the existing TelCo loop in the local network. Different configurations of terminal connections and interfaces are shown in Fig. 4. ISDN recommendations on the "S" and "T" user network interfaces specify (See Tables III and IV) between 150 and 1000 meters (500–3300 feet) for a data rate of 192 Kbs, depending on whether the wiring configuration is point-to-point, short passive bus, or extended passive bus. The wiring between the terminal and the Network Termination Interface (NT), consists of two pairs, one pair for each direction of transmission. The recommended frame structure is shown in

Fig. 5, the details of which can be found in the CCITT Recommendation I.430. Briefly, there are two B channels (B_1 and B_2) 64 Kbs each and a 16 Kbs D channel for signaling. The frame structures in the forward (TE to NT) and the reverse (NT to TE) directions are different. The D bits received from the terminal are retransmitted by the NT on the echo channel called D-echo channel. This echo channel is used for the D channel access control in the following manner:

An active terminal monitors the D-echo channel for a certain number of consecutive ones. If a zero is detected, the terminal restarts counting the number of consecutive one-bits. When the current number of ones is equal to or exceeds the priority assigned to the terminal, it starts transmitting information on the D channel and starts comparing the received D-echo channel. If the transmitted bit is the same as the echo bit, the terminal continues its transmission. If the D-echo channel bit is different, then the terminal detects collision and ceases transmission. It returns to monitoring the consecutive number of ones as before. This collision detection mechanism designed for multiple terminal bus configuration is also suitable for point-to-point connections.

In the U.S., despite the 4 wire CCITT recommendation for customer premises wiring between terminal

TABLE III
USER NETWORK INTERFACE DATA ACCORDING TO ISDN RECOMMENDATION I-430

Physical Medium	:	One twisted pair for each direction of transmission
Bit Rate	:	192 Kbps
Tolerance	:	±100 ppm
Test Load Impedance	:	50 ohms
Nominal Pulse Amplitude (Zero to Peak)	:	750 mV
Line Code	:	Pseudo ternary with 100% duty cycle
Frame Structure	:	48 bits frame (See Fig. 5)
Timing	:	Derived from network clock
Distance Limitations	:	Short Passive Bus 150m
	:	Extended Passive Bus 500m
	:	Point-to-Point 1000m
D channel Access in Bus Configuration	:	Collision detection by monitoring echo channel.
Maximum Number of Terminal Equipment (TE) in a Bus	:	8
Terminating Resistor	:	100 ohms ±5%

TABLE IV
ISDN USER/NETWORK INTERFACE FUNCTIONS

Network Termination 1 (NT1)

Functions broadly equivalent to Layer 1 (physical) of the Open System Interconnection (OSI) model. They include:

- Line Transmission and Termination;
- Layer 1 line maintenance functions and performance monitoring;
- Timing;
- Power transfer;
- Layer 1 multiplexing;
- Interface termination including multidrop termination employing layer 1 contention resolution.

Network Termination 2 (NT2)

Functions broadly equivalent to Layer 1 and higher layers of the OSI model. These functions include:

- Layers 2 and 3 protocol handling;
- Layers 2 and 3 multiplexing;
- Switching;
- Concentration;
- Maintenance functions; and
- Interface termination and other layer 1 functions.

Terminal Equipment (TE)

Functions include Layer 1 and higher layers of OSI model. They are:

- Protocol handling;
- Maintenance functions;
- Interface functions;
- Connection functions to other equipments.

Terminal Equipment Type 1 (TE1)

Includes TE functions with an interface that complies with the ISDN User/Network Interface Recommendations.

Terminal Equipment Type 2 (TE2)

Includes functions belonging to TE but with an interface that complies with Non-ISDN Interface Recommendations (e.g., the X Series Recommendations) or interfaces not included in CCITT Recommendations.

Terminal Adapter (TA)

TA allows a TE2 terminal to be served by an ISDN User/Network Interface. TA functions broadly belong to layer 1 and higher layers of the OSI model.

equipment and the PABX, a number of PABX manufacturers have preferred a single pair for voice and data using the "ping-pong" approach with different bit rates and frame structures. These designs have bit rates up to 256 kbs over a twisted pair (for voice as well as data transmission) between the digital telephones or terminals and the PABX ports. At the present time, most data speed requirements at the terminal end do not require more than 56 kbs. This, in addition to 64 kbs of voice and (for example) 8 kbs for signaling, constitutes the 128 kbs data rate in each direction. Since subscriber loops on customer premises are in most cases within a few thousand feet, such a data rate has been found to operate satisfactorily over the twisted pair. Simultaneous voice and data transmission is achieved by methods which are proprietary in nature, and generally use the "ping-pong" approach using "bursts" or "packets" sent and received during consecutive time intervals. Figure 6 is representative of two of the voice/data frame structures made known by the manufacturers; the ROSE switch [4] has 192 kbs data rate in either direction using two pairs.

Between the PABX and the computer there are two interfaces receiving increasing industry acceptance in the U.S. The first one, called Computer-to-PBX Interface (CPI) developed jointly by Digital Equipment Corporation and Northern Telecom [7] uses the standard DS1 rate transmission facility for 24 data channels of 56 Kbs each. The signaling is by means of the "A" and "B" bits. The second alternative has been proposed by AT&T and is called Digital Multiplexed Interface (DMI) [8]. DMI uses the same T1 transmission facility but provides 23, 64 Kbs clear channels for data and one 64 Kbs channel for signaling which complies with ISDN recommended format for primary access. It is expected that DMI will be upgraded to full ISDN compliance soon. Both of these schemes have a large number of licensees, and chip sets are planned or beginning to be available for these interfaces.

Summarizing the user network interface developments, it is seen that the U.S. is not necessarily in step with CCITT recommendations. However, these interfaces are "Pre-ISDN" interim versions, and very soon it is expected that ISDN recommendations—as opposed to the multiplicity of proprietary protocols—will be complied with by the U.S. manufacturers.

In the local network—that is, between the customer premises equipment and the end office, it is interesting to note that developments parallel to those noted above are taking place. While there are no CCITT recommendations regarding what is known as the "U" interface, this is a subject which will be studied during the 1985–1988 CCITT study period. However, in Europe many ISDN field trials have demonstrated ISDN end office connectivity using both "ping-pong" [9] and echo cancellation methods [10] in the 2-wire TelCo loops. In the U.S., as an early step in the direction of ISDN, 56 kbs switched data over subscriber loops using the existing twisted pair for alternate voice and data, called Circuit Switched Digital Capability (CSDC) [11, 12], has been proposed by AT&T. Telephone Operating Companies on their part, have also been interested in offering circuit switched data. For example, Illinois Bell and Pacific Bell, have already announced circuit switched data services along the lines of CSDC. As a parallel development, a sister service to CSDC, called Local Area Data Transport (LADT) for dial-up voice and data transmission up to 4800 bps, is already operational in Southeastern Florida [13].

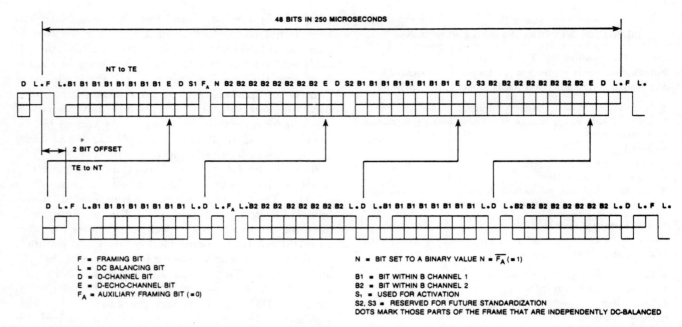

F = FRAMING BIT
L = DC BALANCING BIT
D = D-CHANNEL BIT
E = D-ECHO-CHANNEL BIT
F_A = AUXILIARY FRAMING BIT (=0)

N = BIT SET TO A BINARY VALUE N = $\overline{F_A}$ (=1)

B1 = BIT WITHIN B CHANNEL 1
B2 = BIT WITHIN B CHANNEL 2
S_1 = USED FOR ACTIVATION
S2, S3 = RESERVED FOR FUTURE STANDARDIZATION
DOTS MARK THOSE PARTS OF THE FRAME THAT ARE INDEPENDENTLY DC-BALANCED

Fig. 5. Frame structure at reference points S and T

The factors in the single versus multipair issue on the customer premises, namely the wiring costs as well as the proprietary nature of the protocols involved, are of relevance both from the user's as well as the manufacturer's point of view. It is true that the use of single pair instead of multiple pair reduces the cost of wiring telephones and data terminals by up to approximately $100 per station. This is assuming an average cable run of 200 feet per station costing approximately $.50 per foot. However, in modern buildings, multiple pairs are already wired from the individual stations to the switch room or at least up to the wire closet on each floor. In the case of older buildings, it is often very expensive to trace the existing wires during the installation of a new switch. They are usually removed or disconnected completely and new cables with multiple pairs are installed. Thus, wiring costs do not appear to be a criterion, at least for now, for the use of single pair. Even if it is an issue, and we assume that single pairs save money, this has to be weighed against the increase in the cost of digital interfaces to handle simultaneous voice and data. Manufacturers, however, have a strong incentive to develop single pair voice/data capability for their switches for an entirely different reason. It has been estimated that for every dollar earned from the switching equipment there are four dollars to be earned from the telephones, workstations, hardware/software interfaces for plain data as well as several value added services. Thus, a proprietary protocol in the subscriber loop for voice/data literally locks up a customer for proprietary telephone sets and data interfaces.

In closing, we might say that the customer premises switches are well poised toward an era of Integrated Services Digital Networks. This is clear from the design of station instruments, simultaneous voice and data transmission over subscriber loops, T1 interfaces, and the provision of voice and text mail, packet switching, word processing and Local Area Networking capabilities. Moreover, some of the manufacturers are planning or providing common channel signaling (CCITT #7) capability in their new switches in anticipation of its widespread use.

Deregulation–Divestiture Related Developments

Centrex versus PABX in the Post Divestiture Era

An important side effect of the Bell System divestiture is the uncertainty of the role of Centrex services. The reaction to this side effect by the Regional Bell Operating Companies (RBOCs), switch vendors, and customers has lately become the subject matter for an intense debate.

Some RBOC's (for example, U.S. West, Ameritech, and Pacific Tel) are determined to keep Centrex as their "flagship" vehicle. The main reason attributed to such a decision is that an RBOC would otherwise be left with significant idle CO capacity, and the lost revenue can only be compensated by charging residential telephone customers more; a measure very hard to get approved by the utility commissions.

To make Centrex services a success, these companies have petitioned for rate reduction for Centrex lines—for it has often been said that post divestiture access charges will prove to be a death-knell for Centrex—and

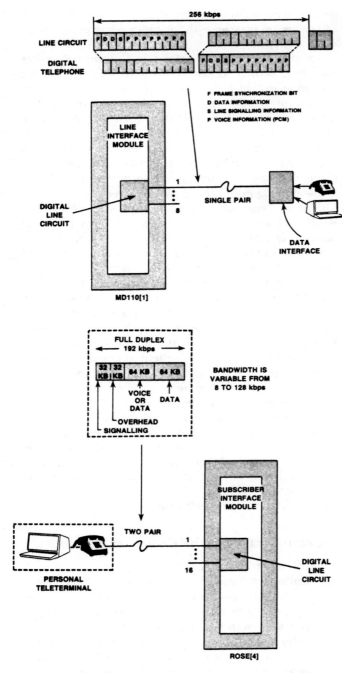

Fig. 6. *Two implementations of frame structures for voice and data transmission on subscriber loops*

are planning to provide enhanced services such as Station Message Detail Recording (SMDR), data handling capability up to 56 kbs, modem pooling, customer moves and changes, least cost routing, and so on. To provide office of the future capabilities adequately, a remote switch option as a part of the Centrex offering at the customer premises for LAN type applications, value added services such as voice/text mail, is also on the drawing board.

The RBOC's with unregulated subsidiaries (for example, Pacific Tel)—a benefit of deregulation—are promoting Centrex Services, as well as planning to sell, install, maintain, and operate customer premises switches. These companies have realized that they need to carefully target disjoint market segments to avoid the parent company and its unregulated subsidiary working at cross purposes. There are other operating companies that would not want to emphasize Centrex services, but would like to be dominant customer premises switch vendors. For example, SONECOR, a subsidiary of Southern New England Telephone (SNET), is aggressively positioning itself to be a top provider of customer premises switches and to be a single point of contact for customers for installation, maintenance, network integration, and customer training.

The switch vendors are excited about the Centrex replacement market. The replacement has already started and is likely to accelerate. It has been estimated that in the 1984–86 time frame, approximately four million Centrex lines may be replaced [14]. Contractual agreements between vendors, RBOC's, and independents have already run into hundreds of millions of dollars for the sale of customer premises products. Larger vendors (such as, AT&T, NTI) who sell both customer premises and CO switches have, in addition, agreed to provide Centrex-based products to interested operating companies.

The customer, on his/her part, understands the Centrex-PBX issue as not only a price performance issue, but also a corporate economy one. For some customers, the carefree 24-hour professional service provided at the CO offers a great feeling of security; while for others, there is no alternative—they cannot afford to have their own customer premises switches. The remaining customers, in the words of Eugene Lotochinski [15], should on the one hand look at whether the TelCo will provide required features in a cost-effective manner, and on the other hand they should evaluate lease versus buy options for the customer premises switches and determine if they really want to run a "telephone company" on their premises. The result is that some customers would prefer Centrex services based on required features and economics while others would decide to have an on premise switch owned or leased.

In conclusion, it appears that a small percentage of present day Centrex systems will remain for a long time. However, it is doubtful that Centrex, in spite of modernization, can effectively compete with the customer premises switches in the office of the future market.

TelCo-Bypass Capability and Networking Requirements

Deregulation of the communication industry, and technology, have provided opportunities to use communication media other than the lines and trunks provided by the local telephone companies. This, usually referred to as TelCo bypass, is becoming attractive depending on the traffic (number of circuits) and the bypass distance. Among alternative voice/data communication media available for bypass, point-to-point/point-to-multipoint microwave, fiber, and the little utilized potential of the TV cables in place and important.

In high population density metropolitan areas, where a company is spatially distributed among different buildings, the interconnection of different customer premises switches should take advantage of the several communication media available (microwave, TV cables, and so forth). The solution is site dependent and often requires dissimilar interfaces, including physical connections, digital frame formats, timing, and signaling schemes to the by-pass medium. It has been found * that an economical solution to this multimedia situation is to design a low-cost generic, "intelligent" node that can be placed between incompatible interfaces to convert the incoming information (voice, data, video) into a suitable format compatible with the outgoing medium. In addition, the node can monitor the performance of the media providing real time diagnostics, network usage accounting, and management reports. Further, it can dynamically allocate bandwidth to reflect specific user requirements, provide optimum routing capability, the ability to switch to different routes in the situations of malfunction, provide traffic statistics, and so on.

Like the customers who are interested in economics, the Other Common Carriers (OCC's) are also interested in bypassing the TelCo to interconnect the customer premises equipment to their tandem switch or the satellite gateways in an economical way. This awareness has been intensified further due to the proposed increase in TelCo access charges and private line rates after the Bell System breakup. The OCC's have special interest in reaching out to the customer premises, where their revenue is generated. In fact, it is possible that the satellite gateway is integrated, to some extent, with the customer premises switches in private networks. This will be accelerated by declining earthstation costs and increased sophistication in switch design. Realizing the OCC market potential, some PABX manufacturers are already offering tandem switching features, the TSX M/80 of Intecom, Inc., for example, in their product line, so as to enable the concentration of traffic from several geographically dispersed locations. To serve isolated locations, some of the distributed processor switches offer Remote Switching Unit (RSU) options. In such situations, a

design with fully distributed autonomous modules offers advantages over the centralized structure with respect to cost.

Another phenomenon related to Customer Premises Switches, which is only of recent origin, is the "teaming-up" of real estate companies and communication companies. In this arrangement, one of the partners is typically a real estate developer with properties across the continent while the other is a communications carrier, communications product vendor, or a subsidiary of a telephone operating company. Such partnership agreements have often exceeded $100 million—notable examples being the agreements between SBS Real Estate Communications Corporation. (SBS RealCom) and Ameritech Communications Inc., and between Olympia and York, the nation's largest office developer and United Telecommunications, the second largest independent telephone company.

This trend is creating opportunities to provide "total" communication systems for high-rise buildings and office parks with several corporate tenants. The Customer Premises Switch in such an environment should be able to function as a totally separated partition for each tenant for internal communications, while the trunk facilities are shared by all the tenants. In addition, a full range of enhanced services such as video teleconferencing, electronic mail, high-speed facsimile, energy management, and security systems can be provided. Tenants will be able to achieve economies of scale in equipment, transmission, administration, and maintenance. While individually, a tenant may find owning a facility such as teleconferencing very expensive, an entire office park will be able to share such a service very cost effectively.

Human Factors in the Customer Premises

Until recently, human factor engineering in communications products has generally been of very low priority for the product planners. To cite a simple example, the designer's lack of interest in human factors may be seen in the single piece telephones with the manually activated switch that people often forget to deactivate to on-hook status, resulting in loss of important incoming calls. It is often mentioned that most of the users, except the trained secretaries, were using only one or two of the features available in the "Feature-phones" and this too was rather reluctantly and often unsuccessfully (for example, call transfer) done. The reason is that a casual user of telephone features does not want to remember the push-button sequences and often shows lack of interest in referring to a booklet or feels frustrated and intimidated. At a recent conference, a large number of participants—intimately associated with the communication industry—when unexpectedly asked by a speaker to draw the layout of a DTMF keyboard, were unable to place the numerals and alphabets in the right place. In this con-

---

*This product, designed and built by Cohesive Network Corp., a Campbell, California firm is currently undergoing field trials.

text, one may note Klapman's [16] observation, "Executives will not use a system that complicates their lives, no matter how technologically advanced it may be." This is, in fact, true for any human being coming into contact with a machine.

The situation is changing, however. There is an increasing amount of discussion about people not using the features provided in the station instruments and the reasons for it. The requirement that the learning time for features and functions must be almost zero is becoming the designer's objective. The basic beliefs behind the revolutionary LISA and its sister product MacIntosh Personal Computers (PC's), namely that a learning time requirement of a few hours rather than a few days, extensive graphics with single button "mouse" for pointing, ease of switching from one program to another on the same screen, are indicative of the importance placed on man-machine interfaces. The emphasis on "teaching people about machines" is changing to "teaching machines about people."

Some examples of the human factor awareness in the customer premises are evident in the design of station instruments, electronic mail, PC's/Workstations, namely, re-labelable feature buttons, pressure sensitive switches in single piece telephones, extensive graphics with split screen capabilities, menu driven programs, and so on. It must be said that there is still a long way to go in human factors design. As R. W. Lucky [17] notes, most of us are still amateurs in this business. Subjective tests of prototypes are still not widespread in this industry. In this context, an area where significant user acceptance experiments are required is in voice mail. To be sure, voice mail has a personal touch as contrasted with the text mail, and most people seem to prefer it. However, its effectiveness for "long" messages in terms of accuracy, ease of reference, and recollection is not clear. The ease of invoking the voice mail, storage and retrieval, and attended cost effectiveness, combined with user preferences, seem to require further examination.

Conclusion

Customer premises products are experiencing exploding growth rates driven by market conditions, technological developments and, in general, the deregulation of the communications industry. Industry analysts predict that in the next few years, at least a quarter of the customer premises switch manufacturers will go out of business, while another quarter will be taken over by the remaining dominant manufacturers. To know the fast-paced developments in this industry, the alignments and realignments among operating companies, interconnect companies, Inter Exchange Carriers, computer and switch manufacturers, one only needs to glance at some of the trade magazines. The events are happening so fast and are so volatile that this author's knowledge on the subject has already become obsolete. This transient environment is likely to persist for at least another 2–3 years before the dominant players are identified.

Acknowledgment

The author is indebted to Chet M. Day of Bell Communications Research, Inc., Per O. Dahlman of Satellite Business Systems, John F. Malone of the Eastern Management Group, and George M. Pfister of the Perspective Telecommunications Group whose views have significantly influenced the contents of this paper. In addition, the author is grateful to several anonymous reviewers for their valuable criticism of this paper.

References

[1] Ericsson Information Switching Systems: MD110 General Description, Issue 1, February 1983.

[2] "ROLM introduces the CBX-II system with large voice/data capabilities," *Communication News*, p. 261, Dec. 1983.

[3] R. Raffensperger, "The broadband switch: a powerful bid for control of the automated office," *Telephony*, pp. 32–41, March 1984.

[4] "CXC announces its voice/data PBX-ROSE," *Nielson Dataquest Research Newsletter*, May 27, 1983.

[5] D. J. Kostas, "ISDN overview," *Telephone Engineer and Management*, Dec. 1, 1984.

[6] Maurizio Decina, "CCITT recommendations on the ISDN: a review," *Telephony*, Dec. 3, 1984.

[7] P. C. Janca, "The computer-to-PBX interface," *Business Communications Review*, pp. 24–29, March–April 1984.

[8] A. R. Severson, "AT&T's proposed PBX-to-computer interface standard," *Data Communications*, pp. 157–162, April 1984.

[9] S. Giorcelli, et al., "Experiment of ISDN facilities in a Proteo UT10/3 local exchange," XI International Switching Symposium, Florence (Italy), Session 21B, paper 2, pp. 1–7, May 1984.

[10] ISDN Field Trial in Italy, Ericsson Review, No. ISDN, 1984

[11] "Inside AT&T's new 56-Kbit/s switched digital service," *Data Communications*, pp. 53–58, Nov. 1983.

[12] AT&T Switched Network Compatibility and Performance Specifications for 2-wire Connection to the Digital Public Switched Network, AT&T Communications, Publication 61320, April 1984.

[13] R. J. Grellner, "Developing a gateway to the information age," *Telephony*, pp. 66–72, Dec. 1983.

[14] The Eastern Management Group, "PBX markets/competitors: a ten year strategic analysis," Nov. 1982.

[15] E. B. Lotochinski, "A balanced view," speech delivered at CRISIS '84: PBX versus Centrex, a conference presented by The Eastern Management Group, Nov. 9–10, 1983.

[16] R. N. Klapman, "Enhanced communications in an executive office," *IEEE Journal on Selected Areas in Communications—special section on experimental teleterminals,* vol. SAC-1, no. 2, pp. 327–332, Feb. 1983.

[17] R. W. Lucky, Guest Editorial on Experimental Telecommunications Services and Terminals, loc. cit., p. 309.

Natesa Janakiraman received the B.E. degree from the University of Madras, Madras, India, in 1970, the M.E. degree (with distinction) from the Indian Institute of Science, Bangalore, India, in 1972, and the Ph.D. degree from Carleton University, Ottawa, Ont., Canada, in 1980.

From 1972 to 1976, he was a Scientific Officer at the Defence Electronics Research Laboratories, Hyderabad, India. From 1976 to 1980, he was a Canadian Commonwealth scholar. From December 1980 to July 1983, he was with GTE Business Communication Systems, McLean, VA. He is now with Satellite Business Systems, McLean, VA. His experience includes digital telephone systems as well as circuit and packet switched data communication systems. Dr. Janakiraman represents SBS in the T1X1.1 subcommittee on Common Channel Signaling and is actively involved in the ISDN developments. ∎

AT&T Technical Journal
Vol. 64, No. 1, January 1985
Printed in U.S.A.

System 75:

Communications and Control Architecture

By L. A. BAXTER,* P. R. BERKOWITZ,* C. A. BUZZARD,[†]
J. J. HORENKAMP,* and F. E. WYATT*

(Manuscript received July 11, 1984)

The System 75 office communication system uses a unique communications
and control architecture that provides great flexibility and a minimum of
overhead for small configurations while growing smoothly to larger line sizes.
A distributed communication network provides 64 kb/s connectivity for both
voice and data. It consists of a pair of time division multiplexed buses and
intelligent port circuits. Flexible conferencing and gain adjustment are supported as an integral part of the network. The control complex supports an
operating-system-based software structure.

I. INTRODUCTION

The System 75 office communications system hardware architecture
consists of a control complex and a communications network, which
are connected by a pair of Time Division Multiplexed (TDM)[‡] buses,
as shown in Fig. 1. Part of the bandwidth of the TDM buses is used
as a control channel between the control complex and the intelligent
port circuits in the communications network.

One of the main architectural features of System 75 is the distributed
switching network. This was chosen to allow as much complexity as
possible to be transferred to the port boards, thereby reducing the
amount of common circuitry required and minimizing the getting-
started cost (often referred to as the intercept cost). At the same time,
aggressive use of VLSI devices in the port circuits allows the per-port
cost (slope) to remain low. This combination of low intercept cost and
moderate slope allows the System 75 network to remain cost-effective
over a wide range of line sizes.

Another key architectural feature is the universal slot concept. All
circuit pack slots in System 75 port carriers are identical (with the
exception of a unique address for each slot). Every port slot has the
same interfaces to the TDM buses, I/O access to outside devices, and

---

* AT&T Information Systems Laboratories, an entity of AT&T Information Systems,
Inc. †AT&T Information Systems Laboratories; present affiliation, Bell Communica-
tions Research, Inc.
‡Acronyms and abbreviations used in the text are defined at the back of the *Journal*.

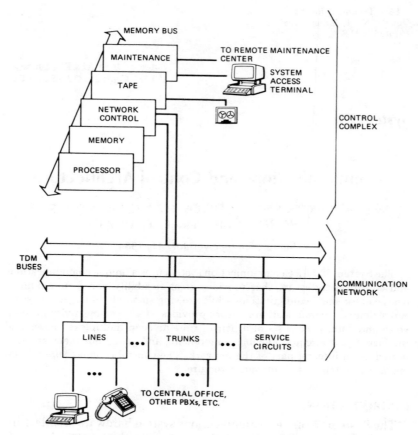

Fig. 1—System 75 communication and control architecture.

power supplies. This freedom to plug any port circuit into any slot allows customers to flexibly configure their system, so that it is optimized to their particular needs.

This highly distributed and modular architecture has allowed System 75 to meet the following design objectives:

1. Provide a digital network which efficiently supports both voice and data communication.

2. Serve customers up to four hundred lines with a single-cabinet hardware configuration.

3. Provide support for an operating system-based software structure.

4. Maintain high-reliability operation with complete self-diagnosis and alarming.

5. Utilize a flexible architecture so that future needs may be easily accommodated, and the system may be gracefully upgraded as technology advances.

6. Provide the above functions at a competitive price.

The following sections describe the architecture in more detail.

II. COMMUNICATION NETWORK

2.1 TDM buses

System 75 has two parallel TDM buses, each of which is 8 bits wide and runs at 2.048 MHz. Functionally, the dual-bus structure is equivalent to a single 512-time-slot bus. Separating the bandwidth into two physically distinct buses has two advantages. First, it lowers the speed of each bus, which eases the timing requirements on VLSI interface devices. Second, the redundancy provided by two buses improves system reliability. If one bus fails, the architecture permits continued operation at reduced capacity on the other bus.

The buses are implemented as printed paths on the backplane. The geometry of these printed paths has been carefully designed to maintain the proper characteristic impedance. Several carriers may be daisy-chained together within a cabinet. Each bus path has a resistive termination at each end.

A novel current-source bus transceiver was designed for this application. Up to 100 port boards may be plugged into the bus in a simple party-line fashion. These transceivers have been specifically designed with low signal levels and controlled rise times to minimize radiated noise. They are designed to allow nondisruptive insertion and removal of boards with the system power on and to isolate port boards from the bus during failure conditions.

Voice signals on the buses are encoded in μ-255 PCM (Pulse Code Modulation) format[1] for domestic systems, while A-law PCM could be provided for international applications. Data signals utilize the Digital Communications Protocol (DCP).[2] Several time slots are reserved for tone distribution and for the control channel.

2.2 Control channel

The first five time slots on each bus are reserved for a control channel between the control complex and the port circuits. In essence, the control channel is the backbone for a network of microprocessors. It provides a communication path between the control complex and the microprocessor on each port circuit pack (commonly referred to as the angel). On each port board a custom VLSI device known as the SAKI (Sanity and Control Interface) provides address recognition, buffering, and synchronization between the angel and the five control time slots. The control channel is active on only one bus at a time. It can be moved to the other bus in the event of a bus failure.

The control channel operates strictly in a polled mode, with the network control (often referred to as the archangel) as master, and the angels as slaves. The first time slot of each frame (TS0) carries the control address, while the following four time slots (TS1 to TS4) contain control data. The archangel grants bus usage to a particular angel or group of angels by transmitting a specific address in TS0. The direction of transmission during the control data time slots (TS1 to TS4) is dependent on the message type.

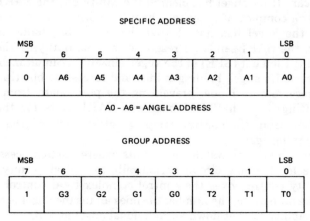

Fig. 2—Addressing modes of TDM bus control channel.

65

Each port slot in System 75 contains seven address pins that are hard-wired to define a unique address. During initialization, the angel reads in this address and writes it to the address-detection portion of the SAKI. This seven-bit address fixes a limit of 127 angels (plus one null address) in the archangel's address space. This restriction is well above the physical limitations on the number of port circuit packs in a single cabinet.

There are two modes in which the archangel can address an angel, as shown in Fig. 2. To differentiate between these two modes, the SAKI inspects the Most Significant Bit (MSB) of the control address which appears in TS0. If the MSB = 0, then the archangel is addressing the specific angel whose address is given in the remaining seven bits. If the MSB = 1, then the archangel is addressing the group of eight angels whose address matches bits 3 through 6 of TS0. In this case, the three Least Significant Bits (LSB) indicate the type of scan or command.

2.2.1 Group addressing

The group address mode is used in two ways: to collect status information (called short-scanning) and to send certain commands to a group of eight angels simultaneously (group commands). In a short scan, each angel in the addressed group responds with a single bit of information in TS2. The bit-position assignments on the TDM bus are determined by a binary decoding of the lower three bits of the angel address. (In other words, the angel whose lower three address bits are 000 responds on TDM bus bit 0, etc.) Using short scans, the archangel can gather status information from a full complement of angels an order of magnitude faster than if each angel were polled directly. Thus, short scanning reduces the latency period before a stimulus is reported to the control complex.

The archangel obtains two types of status information via short scans. Activity scans determine which angels have messages waiting for uplink transmission. Those angels will be individually polled to collect the messages. Sanity scans collect state-of-health information from the angels.

The sanity control circuitry in the SAKI gives the control complex the ability to identify and isolate insane angels quickly. When the archangel sends a sanity scan to a group of eight angels, each SAKI in the group checks the sanity of its angel by verifying that its angel has updated a special sanity bit latch in the SAKI since the previous sanity scan. If its angel has cleared the sanity bit, the SAKI notifies the control complex of its angel's health by driving its bit low during TS2. If the angel has not cleared the sanity bit, indicating angel insanity, the SAKI sends no response in TS2, notifying the Switch Processing Element (SPE) of the angel's insanity. Simultaneously, the SAKI resets its angel, holding it idle, and forces the bus transceivers into a receive-only mode, preventing the port board from errantly transmitting onto the TDM bus. The SAKI waits for the restart instruction from the control complex before allowing the angel to begin running again.

Unlike traditional watchdog circuits where each processor must update a local timing circuit periodically to indicate sanity, System 75's sanity control gives the control complex total control over the sanity scanning rate, the number of times an insane angel is restarted, and the ability to shut down an individual angel at will.

Each SAKI also protects the control channel by monitoring transmission onto the TDM bus during the control time slots. When it detects transmission during a control time slot by anyone other than itself, it disables its angel and Network Processing Elements (NPEs) and waits for the control complex to send the restart command.

2.2.2 Specific addressing

When the control complex wants to send a message to a specific angel, or retrieve a message from an angel that gave a positive response to an activity scan, the archangel must use the individual addressing mode. In this mode, a message may span a number of frames.

Messages sent across the control channel use a well-defined format known as the Control Channel Message Set (CCMS). The CCMS provides a combination of stimulus and functional messages that are common across all types of ports. Downlink (network control to port circuit) messages allow the control complex to control the ringer and LEDs on stations; seize, release, and outpulse on trunks; set up and tear down network connections; execute various maintenance tests, etc. Uplink (port circuit to network control) messages allow the port to report state changes, such as switchhook and button pushes on stations and seizure of incoming trunks. Control channel messages are protected by a checksum, and are retransmitted in the event of an error.

2.3 Network processing element

2.3.1 Switching functions

In conventional digital switches, each port is permanently assigned a time slot on which to talk and another on which to listen. A centralized mechanism called a Time Slot Interchanger (TSI) is used to enable, reorder, and transfer time slots from talking to listening ports. Conventional TSIs require additional centralized equipment to perform gain adjustment or form conferences, features that require arithmetic processing of voice samples. However, intelligent TSIs have been designed to perform these operations.[3]

Whether intelligent or not, a centralized TSI that is sized to accommodate the full capacity of the communications system represents a cost burden on small-size customers who pay for more capacity than they need. The alternative of providing a family of TSIs optimized for several sizes entails extra development effort and complicates growth in the field.

The communications network in System 75 solves these problems. Each port board carries with it a modular piece of the network in the form of a VLSI chip, the NPE.[4] Each NPE serves four ports and is resident on each of the port boards. Two are used on each of the eight-port boards with the exception of the digital line, which uses four NPEs to switch the two information channels of each of its eight ports. The distributed network architecture and absence of a centralized TSI allows customers to buy just the right amount of network for their needs and permits smooth growth as needs expand, while the VLSI technology provides a low per-port cost even though each port has its own dedicated TSI and voice processing logic.

The NPE provides the functions of time-slot assignment for listening and talking, gain adjustment, and eight-party conferencing.[5] (System 75 actually features a six-party conference limit with the remaining two conferencing slots reserved for tones.) The NPE contains over 18,000 transistors, with half of them making up a novel memory network for control and processing functions. As an indication of its complexity, a Transistor-Transistor Logic (TTL) breadboard of this device required six 8- by 13-inch circuit boards.

2.3.2 NPE operation

Figure 3 is a diagram of the operation of one of the NPE's four channels. A network of memory arrays, the associative conference buffer,[6] is used both as a control store, written and read by the angel, and as a buffer for PCM samples from the time division bus. Memory locations are loaded by the angel with time-slot numbers for specifying a talking slot and up to seven listening time slots. Companion memory locations are loaded with a gain or loss value to be applied to samples from listening time slots received from the bus. A talk enable/disable bit can also be stored for the talking time slot. The locations holding the time slot number also act as a content addressable memory by comparing their content against a time-slot counter and individually controlling a sample transfer on the specified time slot. The sample transfers consist of placing a talking sample from a station onto the bus and storing up to eight listening samples from the bus in a sample buffer memory array. The sample buffer holds the samples until they are accessed with their respective gain values to form a conference sum.

An active port is usually allowed to talk on one time slot and listen to from one to seven others. An idle port uses no time slots. When multiple listen time slots are selected, their samples are converted to linear PCM, multiplied by the stored gain values, summed together in an accumulator, and then restored to μ-law for delivery to the station. Of course, data samples are passed through the NPE without any of the conference processing which would corrupt the data.

A simple two-party connection occupies two time slots: each port talks on one and listens to the other. An N-party conference uses N time slots. Tones may be broadcast on a single time slot and received by an unlimited number of ports. In System 75 the seven single-frequency components of the Dual-Tone Multifrequency (DTMF) signaling tones are broadcast continuously. Each port requiring access to a DTMF tone for dialing out forms a brief conference between the two appropriate single frequencies without interfering with other ports similarly doing so.

2.3.3 Conferencing algorithms

The forming of a gain-adjusted conference sum can be thought of as a sequence of arithmetic operations. (In the following discussion, conversions between linear and μ-law PCM are neglected, since they are required in all cases.) Each recipient of the conference sum must hear the composite of the other conferees' samples minus his or her own. The sample received by the kth member of an N-party conference is:

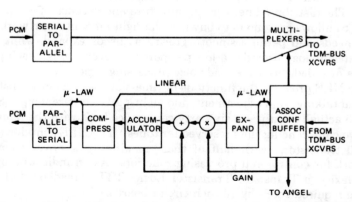

Fig. 3—Architecture of one of the network processing element's four channels.

$$R_k = \sum_{i=1}^{N} g_{ik}T_i - g_{kk}T_k,$$

where

    R_k = Receive sample for port k
    T_k = Transmit sample for port k
    g_{ik} = Gain coefficient from port i to port k
    T_i = Transmit sample for port i
    g_{kk} = Gain coefficient from port k to itself.

There are conceptually two algorithms for generating a conference sum. One method is to hold the g_{ik} constant for all k. Then:

$$R_k = \sum_{i=1}^{N} g_i T_i - g_k T_k,$$

where g_i = Transmit gain for ith port. First a conference sum of all gain-scaled transmit samples is formed, a task requiring N multiply and accumulate operations. Then the receive samples are generated by subtracting out the receiving port's scaled transmit sample, an additional N operations, for a total of $2N$ operations per conference.

The chief advantage of the "$2N$" algorithm is its efficient execution, an important property in a centralized, intelligent TSI where processing throughput may be a constraint. A disadvantage is the inflexibility it imposes on conference call transmission gains, since all conferees must listen to a given port with the same gain. Often this results in loss in excess of that required for stability or optimum intelligibility.

The second method allows individually chosen interport gains between all conferees but requires considerably more processing effort for large conferences. It builds each receive sample R_k separately:

$$R_k = \sum_{\substack{i=1 \\ i \neq k}}^{N} g_{ik}T_i.$$

Samples from the other N-1 conferees are individually multiplied by the appropriate transmission gain coefficient and added to the partial sum as it is built up. This requires N-1 multiply operations. Since the sum must be formed separately for each receiving port (a total of N times), $N \times (N - 1) = N^2 - N$ operations are needed to form the conference. The advantage of this "N-squared" algorithm is the freedom it allows in choosing transmission gain coefficients. Each party can listen to the other conferees with an individually tailored gain.

System 75 achieves an N-squared conference algorithm since each of the N NPEs in a conference perform the required N-1 operations with individually chosen gain coefficients. The necessary processing throughput is obtained as a natural benefit of the parallelism inherent in the NPE-based network.

The importance of N-squared conferencing is illustrated by a three-way call involving two telephone line ports and a central office trunk port. PBX line ports optimally require about 6 dB of loss between them to simulate losses normally encountered in the loop plant, while trunk connections should have 0-dB transmission loss between them, as shown in Fig. 4. This combination of loss relationships can only be achieved with an N-squared algorithm since a $2N$ method would subject the trunk to the same 6-dB loss that is applied between lines. The NPE implementation of this three-party conference is illustrated in Fig. 5.

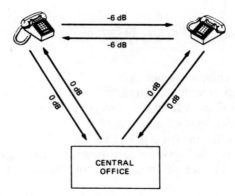

Fig. 4—Example of System 75 gain plan for three-party conference.

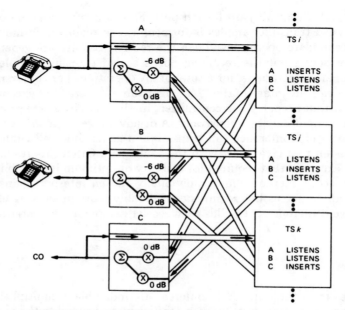

Fig. 5—System 75 implementation of three-party conference.

2.4 Intelligent port circuits

Figure 6 is a block diagram of a generic System 75 port circuit. The circuit pack interfaces to the TDM buses via the custom bus transceivers. Time-slot information, which may be either PCM voice samples or data, is handled by the NPEs (Section 2.3). The interface to the control channel is handled by another VLSI device, the SAKI (Section 2.2).

The BORSCHT circuitry contains whatever is necessary to interface to a particular type of line or trunk. (BORSCHT is an acronym for Battery feed, Overvoltage protection, Ringing, Supervision, Codec, Hybrid, and Testing.) In general, this block of circuitry is different for every type of port circuit.

The heart of the port circuit is the on-board microprocessor, or angel. Every port board in System 75 has an angel that controls the operation of the circuit pack. The angel is implemented as a single-chip microcomputer with up to 8K bytes of firmware. The firmware is divided into two sections: a common portion, which is essentially the same for all circuit packs, and an application portion.

Both the NPE and the SAKI are operated as peripherals to the angel. When setting up a network connection, for example, the following actions occur. The control complex formulates a down-link mes-

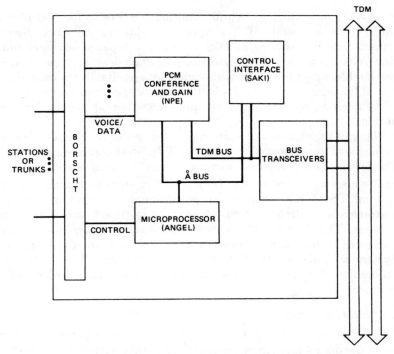

Fig. 6—Architecture of a System 75 generic port board.

sage and sends it over the control channel. The message is received by the SAKI and passed to the angel. The angel sends an acknowledgment (via the SAKI) and examines the message. It then loads the proper time slot and gain values into the NPE so that the desired connection is established.

The distributed intelligence of the angels is a key element which makes a common control channel message set possible. The angels also play an important role in the maintenance strategy. The angel's responsibilities include:

1. Scanning the station/trunk and reporting any state changes uplink to the control complex.

2. Interpreting received (down-link) control channel messages and taking the proper action. For example, when a 'ringer-on' message is received, an analog line-circuit angel must close a relay to provide 90-volt ringing, do ring-cycle timing, etc. A digital line circuit angel, however, would format a command to the station set and send it, using the DCP signaling channel.[2]

3. Handling all short-duration timing functions. Examples include ring cycle and LED cadence timing for stations, outpulsing for trunks, and interdigit timing for DTMF receivers.

4. Performing a variety of maintenance tests. In addition to tests which are run on command from the control complex, the angel does extensive in-line error testing during the normal operation of the circuit pack. Error pegs are kept and reported uplink.

In summary, the angel provides the intelligence necessary to isolate call processing software from port-specific differences and to off-load the control complex from having to do real-time intensive port scanning functions. These are reflected in the virtualization provided in the control channel message set.

This cleanly defined, message-based control interface, coupled with the universal slot concept provides yet another benefit. It is relatively straightforward to design new types of port circuits and integrate them into System 75. Our current family of port circuits includes five types of line circuits (analog, hybrid, Multibutton Electronic Telephone [MET], digital, and data); five types of trunks (central office, direct-inward-dial, tie, auxiliary, and DS-1[7]); and four types of service circuits (tone/clock generator, tone detector, pooled modem, and speech synthesis). Most port boards provide eight port circuits. In the case of the digital line circuit, this results in 16 network appearances, since each port supports two information channels in the digital communications protocol.

Many of the station types are supported across the product family. Analog sets are supported on all AT&T Information Systems PBXs. The hybrid set was adapted from the *Merlin*™ communications system and is also supported by the *Dimension*® System 85 communication system. The MET set provides an economical migration path for customers who already own *Dimension* or *Horizon*® communications systems. The digital stations provide advanced voice/data features using the digital communications protocol and are common with System 85. This variety of ports and stations allows the system to be tailored to the specific needs of each customer in a cost-effective manner.

2.5 Digital line circuit

To illustrate the concepts previously mentioned, a particular circuit pack, the digital line circuit, is discussed in this section in more detail. Like all System 75 port circuits, the digital line circuit makes extensive use of custom VLSI devices and performs many functions in firmware rather than hardware. The digital line circuit, which terminates eight DCP lines, is an evolutionary step towards an Integrated Services Digital Network (ISDN).[8] Like the proposed ISDN interface, each DCP line provides two information channels and a separate channel for signaling, thereby supporting simultaneous integrated voice/data communication. Thus, this circuit pack supports sixteen endpoints—a density unmatched by any of the other port boards.

2.5.1 Hardware configuration

Figure 7 is a photograph of the digital line circuit. The major functional blocks are indicated on the figure. The five integrated circuits at the upper left are the bus transceivers. The SAKI device, which provides hardware support for the control channel, is at the right of the bus transceivers. Note that the SAKI, like several other devices on the board, is packaged in a 68-pin surface-mount chip carrier. (Physical design considerations are explained in more detail in Ref. 9.) To the right of the SAKI are four NPEs, which provide access to the TDM buses for the 16 information channels that this circuit pack supports. The angel microprocessor that controls the operation of the circuit pack, and its associated RAM are at the right side of the circuit pack. All of the components mentioned above are common to all System 75 port circuits, and are also shown in the generic port board diagram (Fig. 6).

The remaining circuitry (called the BORSCHT in Fig. 6) interfaces directly to the DCP lines. The bottom half of the circuit pack contains eight identical blocks of circuitry. The Digital Line Interface (DLI) device contains a complete 160-kb/s modem packaged in a 40-pin DIP (Dual In-line Package). It provides full-duplex operation over up to 5000 feet of 26-gauge cable, and includes circuitry for framing, scrambling (to reduce radiated noise), clock recovery, and automatic equalization. The two SIPs (Single In-Line Packages) immediately below the DLIs contain the external resistors and capacitors needed by the DLIs. A pair of transformers provides the actual interface to the DCP

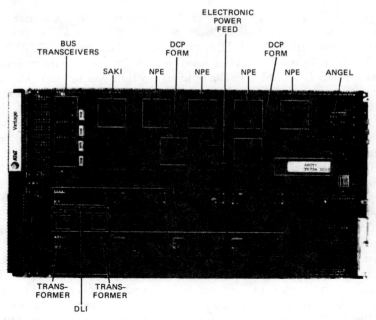

BUS TRANSCEIVERS

ELECTRONIC POWER FEED

DCP FORM

DCP FORM

SAKI NPE NPE NPE NPE ANGEL

TRANS-FORMER

TRANS-FORMER

DLI

Fig. 7—Photograph of System 75 digital line circuit with major functional blocks indicated.

lines. The use of a transformer-coupled interface has a couple of advantages: it protects the board against longitudinal surges, and it allows power to be supplied to the station over the same pairs of wire (via a technique known as "phantom powering").

The Electronic Power Feed (EPF) chips which are in the middle of the circuit pack control station power. An EPF is a microprocessor-controllable electronic circuit breaker. The EPFs automatically shut down when an overcurrent condition is detected and can also be turned on and off by the angel. In addition, the angel can read the status of each EPF and determine (1) whether it is supplying a normal amount of current to the station, (2) whether it is not supplying any current to the station (this normally means that the station is unplugged), or (3) whether it is in overcurrent mode, which indicates a fault condition in either the station or wiring.

The DCP formatters are custom integrated circuits that provide link-level hardware support for the DCP signaling channel. Since most of the DCP signaling channel protocol is implemented in firmware, further discussion of the DCP formatters will be deferred until the next section.

2.5.2 Firmware interactions

The digital line circuit angel firmware has three main functions:

1. It processes control channel messages to and from the archangel, as described in Section 2.2.

2. It translates between the control channel (CCMS) protocol and the DCP signaling protocol.

3. It performs a number of maintenance functions, such as logging and reporting transmission errors.

The digital line circuit angel firmware is built around a task dispenser known as APEX (Angel Processor Executive), while real-time I/O to the SAKI and DCP formatters is interrupt driven. Processing DCP messages will be described in more detail, since it is the most complex of the above functions.

The DCP provides an 8-kb/s signaling channel that uses a simplified

High-level Data Link Control (HDLC) protocol. In particular, the framing, bit stuffing, Frame Check Sequence (FCS), and link initialization commands (SABM, DM, and UA) are identical to HDLC.[10]

The DCP formatter devices each provide link-level hardware support for four DCP links. In the uplink direction (station to PBX), the formatters provide flag detection, bit de-stuffing, and message demarcation. They generate an angel interrupt when a message byte has been assembled (approximately every millisecond during a message transfer). The angel stores the received bytes in a buffer until the formatter indicates that the complete message has been received.

The completed message is processed at base level in the next APEX task cycle. The angel calculates and verifies the FCS, checks the sequence number, and transmits an acknowledgment—called a Receive Ready (RR)—to the station. The message is then converted to CCMS format, and moved to a different buffer to await uplink transmission over the control channel.

In the downlink direction, messages received over the control channel are converted to DCP format during an APEX task cycle. This includes prepending the correct header and sequence number, and calculating and appending the correct FCS. The message is delivered to the DCP formatters one byte at a time via an interrupt-driven routine. The formatters take care of flag generation and bit stuffing. The angel retains the message in its buffer until an acknowledgment is received from the station. As in HDLC, if none is received within a specified period, the message is retransmitted a maximum of two times. If no acknowledgment is received after the third try, the angel attempts to reinitialize the link.

The two information channels on a DCP link use different logical channels for signaling. Thus, the angel must maintain 16 HDLC-like protocols simultaneously. The angel has responsibility for all link-level functions, including link initialization, sequence numbering, FCS generation and checking, acknowledgments, and retransmissions. As mentioned previously, this allows the call-processing software to maintain a uniform message-based (CCMS) interface to all types of endpoints.

2.6 Digital signal processing technology

Digital signal processing technology is used extensively on all the System 75 service circuits. The AT&T Digital Signal Processor (DSP) integrated circuit[11] is used to implement the signal processing algorithms in System 75. Its advantages include small size, high reliability, low cost, low power consumption, and the availability of numerous development tools.

Some of the many uses of the DSP within System 75 are:

1. The tone/clock circuit pack uses two DSPs to digitally generate all the various tones used by the PBX (e.g., dial tone, ringback, busy tone, intercept tone).

2. The tone detector uses DSPs to implement both Dual-Tone Multifrequency (DTMF) receivers and general-purpose tone detectors (for detecting dial tone, modem answer tone, maintenance tones, etc.) on a single circuit pack.

3. The pooled modem circuit pack contains conversion resources to convert 212A modem signals into DCP format.[2] DSPs are used to implement two 212A-compatible modems on the circuit pack. The advantages of this circuit pack over conventional modem pools include lower cost, uniform administration, better maintenance, and reduction of PBX-room clutter.

4. The speech synthesis circuit pack uses DSPs both for DTMF receivers and for generating Multiple Pulse Linear Predictive Coding (MPLPC)[12] speech samples from stored coefficients.

III. CONTROL COMPLEX

The System 75 control complex is shown in Fig. 1. The control complex is often referred to as the Switch Processing Element (SPE). It consists of a processor, memory, and I/O connected by a single-master Memory Bus (MBus). This configuration meets the cost, performance, and reliability goals for basic service and it can be expanded to support optional services.

3.1 Processor

The processor consists of a commercial 16-bit *Intel** 8086 microprocessor and a Memory Management Unit (MMU) implemented in custom gate arrays. The microprocessor and the MMU functionality were chosen to provide good performance for the largest system configurations and minimum cost for the smallest configurations. Specific constraints are:

1. To minimize equipment cost, the processor and MMU are implemented on a single circuit pack.

2. For maximum performance, most memory accesses are accomplished with only two wait states, including memory management and error correction overhead.

3. Multiple contexts and fast context switching are supported to achieve maximum operating system performance.

4. A high degree of self-checking and protection is provided for call processing applications.

Design trade-offs were made between hardware and software to meet these constraints. The result is an MMU which supports 16-bit virtual to 24-bit physical address mapping, 15 segments of up to 64K bytes each, and the following protection features:

1. Two levels of execution privilege (system and user).

2. Bounds checking on any access, with an overflow stack to aid recovery from stack exceptions.

3. Illegal instruction detection (e.g., HALT instruction).

4. Segment write-protect capability.

5. Distinction between text and stack/data segments to prevent execution of data and to provide execute-only access of text.

3.2 Memory

Because System 75 is software-intensive, the memory can have a significant impact on system cost, reliability, and performance. To meet the system design objectives, the memory uses 256K Dynamic Random Access Memory (DRAM) devices and Error Detection and Correction (EDC) logic. Each memory circuit pack provides 2M bytes organized into 22-bit words (16 data bits + 6 check bits). The EDC circuitry provides single-bit error correction and double-bit error detection and therefore dramatically improves the system's mean time to critical failure. The memory uses VLSI devices to incorporate all refresh, control, and maintenance functions on each pack, thereby eliminating any external memory control function.

---

* Trademark of Intel Corporation.

3.3 Input/output

The I/O functions are implemented with intelligent interfaces which off-load the processor and shield call processing software from real-time-critical tasks. The processor communicates with the interfaces through dual-port memories on the MBus.

3.3.1 Network control

As previously discussed, the network control circuit pack provides the bridge between the control complex and the communication network. It is the master of the TDM bus control channel and, in addition, provides a time-of-day clock with battery holdover, a system clock failure detector, and four switched data channels used for dial-up maintenance/administration and printer output.

3.3.2 Tape interface

The tape interface circuit pack with associated tape drive provides 20M bytes of storage on a 1/4-inch cartridge tape for program load, patches, and translation. The tape drive provides an intelligent memory-mapped interface. It supports both streaming and edit modes and provides extensive error detection and correction capabilities, including the ability to correct very long burst errors.

3.3.3 Maintenance

The maintenance circuit pack uses microprocessors, VLSI, and digital signal processors to provide the following:

1. An RS-232-C interface to a hardwired maintenance/administration terminal (known as the system access terminal), and a low-level user interface in firmware that supplements the high-level interface in software.

2. A tip/ring interface to the remote maintenance center via the central office for automatic alarm reporting. It includes an autodialer, 212A modem emulation, and Level 2 X.25 protocol termination.

3. Cabinet environmental monitoring. If the temperature rises too high, the system is switched to power-fail transfer mode. When the temperature gradient across the cabinet increases to a predefined threshold, the user is reminded (via the system access terminal) to clean the air filters.

4. Power supply and battery holdover monitoring and control. The battery charger and power supplies are constantly monitored and controlled. On ac power failure, the entire system is powered from the batteries for ten seconds. Then the port carrier supplies are shut down and the control carrier is held over an additional ten minutes. Thus, most commercial power outages are bridged without any interruption in service.

5. Power fail transfer control. As explained above, after ten seconds of battery holdover, the port carriers are shut down. At this time, selected voice terminals are connected directly (via relays) to central office trunks to provide emergency phone service.

The maintenance architecture is described in more detail in Ref. 13.

3.4 Extensions

The control complex can be extended by adding an interface to a multimaster System Bus (SBus), as shown in Fig. 8. In System 75, the SBus supports an additional processor with I/O that terminates the

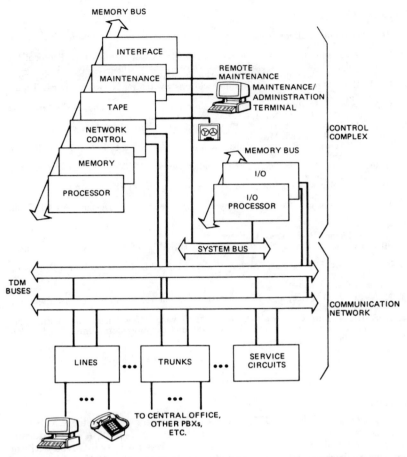

Fig. 8—System 75 communication and control architecture with I/O processor.

switched X.25 channels. These channels connect to adjunct systems such as the applications processor, or other nodes in a Distributed Communications Service (DCS) network.[14]

The I/O processor has its own 16-bit microprocessor plus 128K bytes of RAM and an SBus interface. It connects to the I/O interface through another short MBus. The I/O interface provides additional flexibility because it connects to the TDM buses and terminates the X.25 protocol and the underlying DCP protocol on all four channels. This permits the use of standard data switching, cabling, and termination features for a wide variety of system arrangements.

IV. SUMMARY

System 75 provides a digital communication network that serves the voice and data communication needs of medium-sized customers. Conferencing and a flexible gain plan are integral parts of the network. The control complex efficiently supports a modern operating-system-based software package, and can be expanded to support additional optional features. A wide range of station equipment is supported so that the system can be configured to fill the customer's needs in a cost-effective manner.

To protect the customer's investment, System 75 uses a flexible and highly modular architecture so that the system may be expanded to meet future needs. In addition, the modularity allows economical upgrading of the system as technology progresses.

REFERENCES

1. Bell Telephone Laboratories, *Transmission Systems for Communications*, Fifth edition, 1982, Chapter 28.
2. G. M. Anderson, J. F. Day, and L. A. Spindel, "A Communications Protocol for Integrated Digital Voice and Data Services in the Business Office," Proc. Sixth Int. Conf. on Computer Communication, London, September 1982, pp. 367–72.
3. H. G. Alles, "The Intelligent Communications Switching Network," IEEE Trans. Comm., *COM-27* (July 1979), pp. 1080–7.
4. L. A. Baxter, P. R. Berkowitz, and C. A. Buzzard, "Distributed Digital Conferencing System," U.S. Patent No. 4,389,720, June 21, 1983.
5. B. S. Moffitt and A. R. Ross, "Digital Conference Time Slot Interchanger," U.S. Patent No. 4,382,295, May 3, 1983.
6. B. S. Moffitt and A. R. Ross, "Multiport Memory Array," U.S. Patent No. 4,395,765, July 26, 1983.
7. "Digital Channel Bank Requirements and Objectives," AT&T Technical Reference PUB43801, December 1978.
8. "Integrated Services Digital Networks," Special Issue of IEEE Comm. Magazine, *22*, No. 1 (January 1984).
9. A. S. Loverde et al., "System 75: Physical Architecture and Design," AT&T Tech. J., this issue.
10. A. Meijer and P. Peeters, *Computer Network Architectures*, Rockville, MD: Computer Science Press, 1982, Chapter 4.
11. Special issue on "Digital Signal Processor," B.S.T.J., *60*, No. 7, Part 2 (September 1981).
12. B. S. Atal and J. R. Remde, "A New Model of Pulsed LPC Excitation for Producing Natural Sounding Speech at Low Bit Rates," Proc. ICASSP '82, Paris, France, May 1982, pp. 614–17.
13. K. S. Lu, J. D. Price, and T. L. Smith, "System 75: Maintenance Architecture," AT&T Tech. J., this issue.
14. R. S. Divakaruni, G. E. Saltus, and B. R. Savage, "New Directions in Enhanced Voice Networking," Proc. Sixth Int. Conf. on Computer Communication, London, September 1982, pp. 362–6.

AUTHORS

L. A. Baxter, B.S.E.E., 1975, Rochester Institute of Technology; M.S.E.E., 1977, University of Delaware; Bell Laboratories, 1977–1982; AT&T Information Systems Laboratories, 1983—. Mr. Baxter was initially involved with the exploratory design of office communication systems. Since 1980 he has worked on the design of the System 75 communications network. He currently is Supervisor of the System 75 Digital Switching Hardware group. Member, IEEE, Tau Beta Pi, Phi Kappa Phi.

P. R. Berkowitz, B.S.E.E., 1974, Columbia University; M.S.E.E., 1975, Columbia University; Bell Laboratories, 1975–1982; AT&T Information Systems Laboratories, 1983—. Mr. Berkowitz is Supervisor of the System 75 Circuit Design group. He was previously a member of the design team for the *Horizon®* communications system.

C. Alan Buzzard, B.S.E.E., 1964, M.S.E.E., 1965, Cornell University; Bell Laboratories, 1964–1982; AT&T Information Systems Laboratories, 1982–1983. Present affiliation Bell Communications Research, Inc. Mr. Buzzard's past responsibilities have included development of modems, data networks, and voice/data PBXs. His present interests are in speech coding, speech synthesis, and automatic speech recognition. Member, IEEE.

J. J. Horenkamp, B.S.E.E., 1964, St. Louis University; M.S.E.E., 1966, Columbia University; Doctor of Engineering Science, 1973, Columbia University; Bell Laboratories, 1964–1982; AT&T Information Systems Laboratories, 1983—. Mr. Horenkamp is Head of the System Design department responsible for hardware and firmware development and maintenance planning of System 75. He has been involved with exploratory development and final design of a variety of PBXs and customer premises telecommunication systems.

Frank E. Wyatt, B.S.E.E., 1969, M.S.E.E., 1970, University of Illinois; Bell Laboratories, 1971–1982; AT&T Information Systems Laboratories, 1983—. Mr. Wyatt has worked on a variety of business communication and management information systems. Since 1980 he has supervised the group that developed the control complex for System 75. Member, IEEE.

AT&T Technical Journal
Vol. 64, No. 1, January 1985
Printed in U.S.A.

System 75:

Switch Services Software

By W. DENSMORE, R. J. JAKUBEK, M. J. MIRACLE, and
J. H. SUN*

(Manuscript received July 11, 1984)

The switch services software of System 75 provides the basis for an extensible office communication system, supporting a wide variety of voice and
data-switching services. This paper presents the software architecture of the
System 75 switch services. The concepts of user, group, and process-per-call
form its foundation. We introduce the architecture by stepping through a
simple station-to-station phone call, and proceed to the derivation of a call
model based on the topology of a call. This call model is realized as a layered
set of cooperating processes that execute under the Oryx/Pecos Operating
System on the System 75 switch processor. The software layers and processes
are discussed and a call walk-through is used to illustrate the process interactions.

I. INTRODUCTION

The System 75 software supports a wide spectrum of terminals that
range from a single line station to a sophisticated digital station for
simultaneous voice and data communications. It also supports more
than 150 features for office and business communication needs. Among
them are the station features for handling multiple, simultaneous calls,
features used for covering unanswered calls, routing features to select
the least expensive network facility, capabilities that allow messages
to be left for the called party automatically, and terminal dialing and
modem pooling features for data communications services.

This paper describes the switch services software architecture of
System 75, a framework that supports a wide variety of features and
terminals. Section II states the design challenges and Section III
illustrates a basic call scenario leading to the derivation of an essential
call model. The concepts of user, group, and call are introduced, and
are mapped to a basic software structure. In Section IV, this structure
is generalized to a layered software architecture consisting of a set of
cooperating processes. A call walk-through is used to illustrate the
interactions among processes.

---

* All authors are members of AT&T Information Systems Laboratories, an entity of
AT&T Information Systems, Inc.

II. DESIGN CHALLENGES

The challenges in building switching software for an office communication system include:

- The vast number and variety of features to be supported
- The need to integrate different office services
- A wide variation in the capabilities of existing and future terminals
- Stringent real-time response criteria
- The asynchronous and concurrent nature of the external world.

Our design began with an analysis of the feature operations and resource management requirements of a switching system. Then, the essence of the feature and terminal operations were extracted into functional modules and the basic primitives of the system were defined. Next, we formulated a call model by analyzing the dynamic behavior of a call and the relationships between the functional modules. The functional modules were then layered into a set of cooperating processes, with the primitives of the system provided through message-based interfaces. Finally, information hiding and synchronization techniques were applied to simplify the software structure and to enforce stronger partitioning of functions.

A primary goal of this modular and disciplined architecture is to minimize the effort required to add new terminals and features. This architecture should also be easy to understand by software developers so that the architectural integrity is preserved over the product life. Moreover, the architecture should simplify the integration of more data-processing-like functions in the future.

Trade-offs exist between implementing the design goals and meeting the real-time requirements of a switching system. For example, a virtual terminal interface provides for uniform implementation of features across all terminal types at the expense of access time to the terminal. The primitives of the system are carefully defined to balance between generality and efficiency.

III. ARCHITECTURAL MODEL

To introduce the design of the switching software, we will step through a basic telephone call and develop a call model. Then the System 75 realization of this call model will be presented.

3.1 Basic call example

A user originates a call by going off-hook on his/her phone. This change is detected by an I/O peripheral and an off-hook signaling message is sent to the switch processor (see Fig. 1). On receiving this message, the resources for processing this call are allocated and messages are sent to the switch network for connecting a Dual-Tone Multi-Frequency (DTMF)* receiver and giving dial tone to the user. The call processing software interprets each digit dialed by the user and routes the call to the terminating station when all the digits have been dialed. The call is signaled to the terminating user with ringing, and the call progress is indicated to the originating user with ringback tone. When the terminating user answers the call, the switching network is instructed to remove the ringing signal and the ringback tone, and establish a talking path between the originator and the terminator. Finally, when either the originator or the terminator goes on-hook, the call processing software tears down the circuit connection and deallocates all the resources associated with the call.

---

*Acronyms and abbreviations used in the text are defined at the back of the *Journal*.

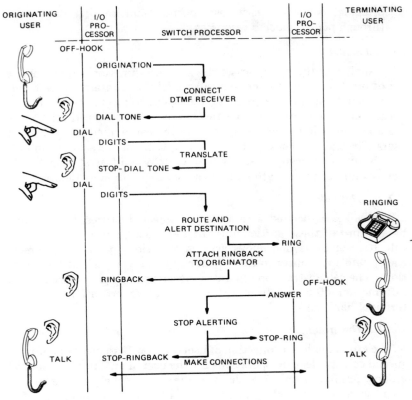

Fig. 1—Basic call example.

This example illustrates the major functions of System 75 in processing a basic call:

- Terminal Handling
  The voice terminals supported by System 75 range from a single-line analog station to a simultaneous voice/data station with display, multiple call appearances, feature buttons, and data module. A variety of trunks are used to interconnect with other switching systems or a central office switch.

- Resource Management
  In addition to terminals, there are other resources in the system that need to be managed. These include the DTMF receivers, the time slots for circuit connections, tone generators, and the internal software records for call processing, messaging, measurements, and call detail recording.

- Call Sequencing Control
  Of the more than 150 features suported, many involve complicated sequencing logic to bring a call from one state to another. For example, the call coverage feature specifies the selection of new call destinations (coverage users) if no answer occurs in a specified time interval at either the principal destination or the current coverage destination. A conference call is another example where, in response to a user's conferencing request, the internal records and the circuit connections of the two initially distinct calls are merged to establish a common talking connection for all the parties.

- Routing and Termination Selection
  Routing and termination refer to the selection of a terminating endpoint or set of endpoints for a call. There are a wide variety of algorithms for routing and termination selection, such as hunting,

bridging, coverage, least-cost routing, and routing data calls through pooled modem resources.

3.2 Call model

By analyzing the dynamics of the software functions in the above list of functions, we can derive a call model that contains three major components: the *call*, the *group*, and the *user*. Highest in the hierarchy is the call, which ties all the parties of a connection together. Next is the group, which appears as a party on the call and contains a set of users. The user is an entity that models a terminal or a set of terminals that belong to a system user. This hierarchy is illustrated in Fig. 2. Let's examine the call, group, and user concepts in more detail.

3.2.1 The call

The call is associated with a set of connected parties. It is defined by a record of these parties plus the sequencing control logic of the call. It resolves asynchronous actions of various parties and directs the system's responses to these actions for the various feature operations. The call abstraction separates the common sequencing logic and information of a call from the group feature operations and the terminal handling.

3.2.2 The group

The group models a collection of users who are associated to provide special call and feature operations for the customer. There are several group types in System 75; each one contains special algorithms for the operations of that group type. For example, the hunt group specifies how a user should be selected from a group to receive a call. The group abstraction hides the internal operation of group features from the call, and provides a uniform structure for handling such group-oriented features as hunting, bridging, multiple attendants, and trunk groups.

3.2.3 The user

The user models the end user in the system, who may have a single telephone instrument or possibly a collection of interacting terminals. The user abstraction is a terminal handler that provides a set of resources for which the different services compete: voice and data channels, status indicators, displays, ringers, etc. Trunks and data lines are also modeled as users. The intent of the user abstraction is to hide the terminal-specific operations from both the group and the call operations.

3.3 Realization of call model

The functional modules and dynamics of the call model are mapped into a software realization consisting of a layered set of cooperating processes. The processes execute under the Oryx/Pecos operating

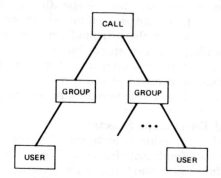

Fig. 2—Basic call model.

system[1] on the switch processor of the System 75 control complex.[2]

The call is realized by a call record and a single, transaction-oriented process that controls a call from origination to completion. A set of call processes exist to handle multiple calls in the system.

The group and user are realized by a group-manager and a user-manager process. Conceptually, there is a process per instance of each different group or user; in practice a single group-manager and a single user-manager process are multitasked to handle all instances of groups and users. The group-manager and user-manager processes provide a set of primitives that are independent of the group and user types of the system. These primitives serve as the building blocks for the upper-level software (e.g., the call process), where the feature sequencing is implemented.

This call model is a hybrid of both the process-per-call, or transactional structure,[3] and the process-per-terminal, or functional structure of switching systems.[4] The call process uses a process-per-call approach, whereas the group and user managers support the process-per-terminal approach. This hybrid design permits synchronizing the interactions of various terminals, as in the process-per-call approach, and permits isolating and distributing the terminal processing software, as in the process-per-terminal approach.

IV. SWITCH SERVICES SOFTWARE STRUCTURE

So far we have presented the call-process, and the group-manager and user-manager processes. There are additional processes that provide messaging and station services and the network and resource management functions of the switch. To organize the software, we define a service-control layer, a resource layer, and a driver layer of

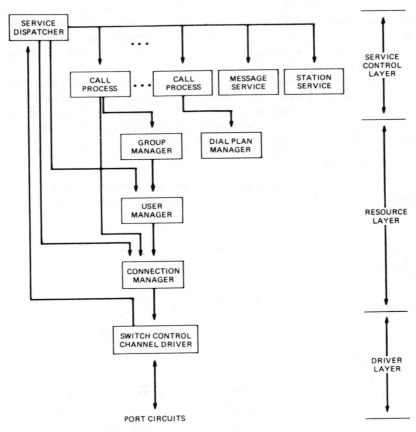

Fig. 3—Switch services software structure.

software (see Fig. 3). Within each layer, the various processes provide further information hiding and separation of functions.

4.1 Service control layer

The Service Control Layer contains a Service Dispatcher (SD) process and a process for each of the different services of System 75. The service processes execute concurrently under control of the service dispatcher. User actions are translated into service commands by the resource layer and these commands are executed by a service process through primitives supplied by the resource-layer processes. Attention is given to the proper synchronization of user actions to guarantee that out-of-sync situations and deadly embraces cannot occur between concurrent transactions.

4.1.1 Call service

The Call Process (CP) provides the control and sequencing logic for call set-up and take-down and for a variety of feature operations in the system. Since a process-per-call is costly in terms of system resources, only a limited number of call processes exist to serve all calls. A call process is allocated to each call in a transient (e.g., dialing) state and is deallocated from calls that have reached a stable (e.g., talking) state. The information is saved in a call record in the service dispatcher. The service dispatcher manages the pool of call processes, creating, allocating, suspending, and resuming them to/from calls.

There are many primitives supported by the resource-layer processes. The function of the call process is to invoke these primitives in the proper order and thereby produce the required response to the external user. It does this by analyzing the service-command message together with the current call state (e.g., idle or talking), and then invoking the appropriate sequencer code for that phase of the call. A call sequencer handles a special set of operations such as routing, answer, or drop (see Fig. 4). The operations are performed by invoking the primitives of the resource layer, which results in the driving of the hardware circuitry.

In addition to managing the call processes, the service-dispatcher process receives the signaling messages from the network and forwards them to the appropriate terminal handler process for interpretation into service commands. The appropriate service process, e.g., the call process, is then dispatched to execute the service command.

4.1.2 Message service

The control for the messaging services such as leave-word-calling and manual-message-waiting is provided by the Message-Service

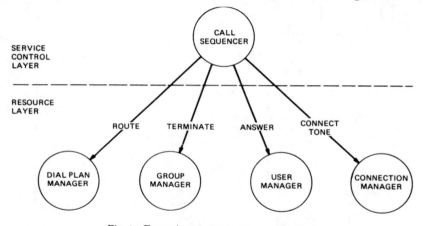

Fig. 4—Examples of resource layer primitives.

(MSG) process. The message service is a permanent server process, providing service for all users in the system. Message services can be invoked from a call by user terminal input or by other services. When message services are accessed via a call, the message-service process interacts with a call process. The message-service process accesses the resource-layer processes for terminal input/output and translation data access.

4.1.3 Station service

The Station-Service (SSV) process provides miscellaneous station services such as integrated directory service, time-of-day display service, and programming some translation data from the user's terminal. The station-service process is a permanent server, handling all users in the system, similar to the message-service process. The station-service process accesses the resource-layer processes for terminal I/O and data access. When station services are accessed via a call, the station-service process also interacts with a call process.

4.2 Resource layer

The resource layer provides general resource management for the services, service-specific functions, and the line/terminal signaling. The system resources managed include the switch network, DTMF receivers, tones, trunks, telephone terminals, data terminals, groups, and databases like the system dial plan and the name/number directory. Call routing, queueing, terminal administration and maintenance, and feature activation primitives are some of the service-specific functions provided.

The software at the resource layer is functionally organized around the user and group concepts, the switch network, and the routing and directory database. A process exists for each of these main functions. Primitives are provided to the services by these processes for resource access. These primitives use the synchronous message facilities of the operating system to support the processing of user actions.

4.2.1 Group management

The Group-Manager (GM) process has all the translation data for group membership and group properties, and it maintains the state of the group and its members. Service-specific functions are provided by the group manager to manipulate the groups for the different services and features of the system. The group executes the service command on the users or members in the group according to the type of group. For example, in response to a call service terminate command, the coverage group would sequence the termination to each idle member in the group (an alert-all algorithm), while the uniform-call-distribution hunt group would select the longest idle member to receive the call.

4.2.2 User management

The User-Manager (UM) process contains both the user and terminal management software and status information. It presents an abstract user or virtual terminal interface to the upper layers of software, while handling the signaling with terminals at the Driver Layer. Terminal access contention between the switch services, maintenance, and system administration is arbitrated by the user manager.

The user manager provides service-specific primitives to manipulate the users and terminals in the system. For example, the terminate primitive would, for a user with a multibutton telephone, select an idle call button, update the button-status lamp to flashing, and start ringing the telephone. The lamp and ringer operations are performed by sending signaling messages to the port circuit that interfaces the telephone.

Signaling-message interpretation is done by the terminal-handler functions in the user manager. Low-semantic-level messages, e.g., off-hook or button-push, are interpreted into service-level commands like originate and answer. A service-control-layer process then executes these commands by invoking the corresponding primitive provided by a resource-layer process.

4.2.3 Data management

The Dial-Plan-Manager (DPM) process provides access to and interpretation of translation databases in the system, including the system dial plan, the name/number directory, user permissions, least-cost routing patterns, and speed-calling numbers. Customized access to the data is provided for the call service. For example, digit analysis and the choosing of the initial routing destination are performed by the dial plan manager in response to a request from a call process.

4.2.4 Network management

The Connection-Manager (CM) process is responsible for the management of the network resources and for network control signaling. It abstracts the physical characteristics of the switch network and provides network connection primitives. Primitives to access the network resources like the DTMF generators and receivers are provided. Contention between the switch services, system maintenance, and system administration for the network resources is arbitrated here.

The communications network in System 75 is distributed onto each port board.[1] Network control messages to do time-slot assignment for listening and talking, gain adjustment, and six-party conferencing are sent by the connection manager to the port boards.

4.3 Driver layer

The driver layer encompasses the operating system drivers and the firmware in the intelligent port circuits of the System 75 communications network. The drivers include the Switch Control Channel Driver (SCD), asynchronous data channel drivers, a timer driver, and bus drivers. The SCD interfaces the switch processor to the network control channel of the Time Division Multiplexed (TDM) bus (see Fig. 5). The network control (or archangel) acts like a full-duplex message switch between the switch processor and the microprocessors in the port boards (commonly referred to as angels). Error correction is provided between the SCD and the angels and flow control is managed by the archangel.

A functional message set is used between the switch processor and the angels. Signaling and network control messages are sent from the user and connection manger processes to the port circuits. The signaling messages are common across all types of ports. This helps to isolate the user manager from terminal-specific differences, providing the first level of terminal abstraction.

The angels off-load the switch processor by executing the low-level port-scanning, driving, and timing functions. This includes the ringer, tone, and lamp cadences, button scanning, and the timing for trunk signaling.

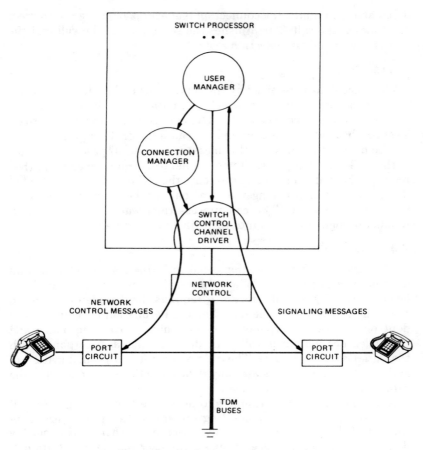

Fig. 5—Communications network control.

4.4 Call walk-through

A high-level walk-through of a call to aid in understanding the operation of the switch services software is presented here. The call is a station-to-station call between multifunction stations.

4.4.1 Origination

A station user goes off-hook with an idle call button selected, indicating the user wishes to "originate" a call. The off-hook signaling message is received by the service dispatcher, which forwards this signaling message to the user manager for interpretation. The user manager examines the terminal and user state and returns a call service originate command for the user back to the service dispatcher. The service dispatcher then allocates a call record, assigns an idle call process to this new call, and forwards the service command to the call process.

The call process enters the call setup mode, first requesting the connection manager to reserve network resources for the new call and then commanding the group manager and user manager to originate for the user. The user manager requests the connection manager to connect the originating user to the call and handles the lamp indications to the originating user. The origination message sequences are shown in Fig. 6.

Next, the call process requests the group and user managers to collect digits, or get a destination for the call. The user manager determines how to collect digits for this type of user and starts the collection, typically by requesting a DTMF receiver from the connection manager. Finally, the connection manager is requested to connect dial tone on the call. This finishes the setup part of the call and the user is now in a digit-collection state.

4.4.2 First digit

When the digit is received by the service dispatcher, it is forwarded to the connection manager, which has data about the call that the DTMF receiver is connected to. The connection manager returns a message that indicates a digit and the call number the digit is for. The service dispatcher forwards this message to the call process allocated to this call. The call process removes dial tone from the call and then requests the dial plan manager to route the call based on the collected digit. The dial-plan manager returns data on how many more digits need to be collected. The call process then waits for the required number of digits or for an interdigit time-out or a hang-up.

4.4.3 Last digit

The last digit of a call is the signal that triggers the selecting and alerting of a destination. When the required number of digits have been received by the call process, it requests the dial plan manager to route the call based on the dialed digits. Permission checking is also done on the routing attempt. If enough digits have been dialed and the dialed destination is valid, the dial plan manager responds with the initial destination user. The call process then stops digit collection and the user manager releases any digit-collection resources from the call, such as the DTMF receiver.

Next, the call process attempts to terminate (i.e., bring in) the call to a destination user by requesting the group manager to terminate the call. If the destination is simply a user, no further routing need be done, so the group manager requests the user manager to terminate the call. The user manager alerts the local user to the incoming call and replies to the call process that the user accepted the call. The call

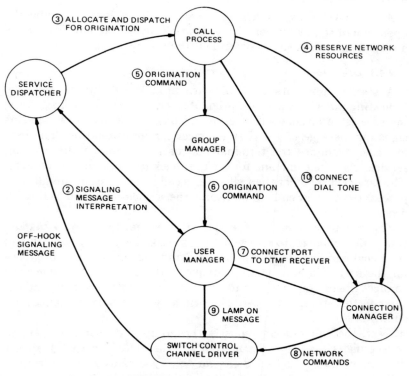

Fig. 6—Call origination sequence.

process then requests the connection manager to apply ringback tone to the originating user and "feeds back" the state of the termination and the identity of the destination to the group manager, which passes the information to the user manager—this results in the originating station receiving a display update if the user has a display.

The call is now in the terminating state, awaiting either an answer by a destination, an abandon by the originator, or a time-out to trigger the next stage of routing. Because the call is in a stable state, the call process is deallocated from the call.

4.4.4 Answer

The call is answered when a destination party goes off-hook. The off-hook is handled like the origination off-hook, except that the user manager determines that the user is now answering a ringing call. The service dispatcher allocates an idle call process and passes the answer command to it. The call process retrieves the call record from the service dispatcher and enters the call answer sequencer. First, any routing timers are canceled, ringback is removed from the connection, and the call process informs the originating user, via the group manager, that the call has been answered. Next, the call process commands the group manager to answer the call for the user. The group manager handles all users being alerted for the incoming call and commands the requesting user to answer the call. The user manager handles the answer command from the group manager by stopping the alerting indications at the user's station and commanding the connection manager to connect the answering port to the call.

Since the calling user was connected in the origination phase, the calling and called parties now have a two-way connection. The call is in a stable state so the call process is deallocated from the call.

4.4.5 Drop

The call is dropped when a user hangs up. The on-hook signal is processed in the same way as the off-hook, with the user manager interpreting the on-hook (or a button push) into a drop command. The call process receives the drop and commands the group manager to drop the user off the call. The gruop manager forwards this to the user manager, which then handles the lamp indications to the user and also commands the connection manager to remove the port from the connection.

Since there is only one party left on the call, the call process sequences the teardown of the call by sending drop commands to the group manager. The drop commands are forwarded to the user manager, which again changes the user's indications and commands the connection manager to remove the port from the connection. Since all parties have been dropped, the call process informs the connection manager that the call no longer exists. The connection manager idles its connection and resource records for that call. The call process then cleans the call record and tells the service dispatcher to free it and the call record from the call.

V. ACKNOWLEDGMENTS

The System 75 switch services software architecture reflects the efforts of many people within the AT&T Information Systems Laboratories. Many ideas were the fruits of exploratory work that preceded the development phase.

REFERENCES

1. G. R. Sager, J. A. Melber, and K. T. Fong, "System 75: The Oryx/Pecos Operating System," AT&T. Tech. J., this issue.
2. L. A. Baxter et al., "System 75: Communications and Control Architecture," AT&T Tech. J., this issue.
3. Wing H. Huen, "What Is Different About Operating Systems for Telephone Systems," IEEE Reprint CH1515-6/79/0000-0179.
4. L. E. McMahon, "An Experimental Software Organization for a Laboratory Data Switch," Proc. ICC '81 (June 1981).

AUTHORS

Wayne Densmore, B.S. (Electrical and Computer Engineering), 1975, Clarkson College of Technology, M.S. (Electrical and Computer Engineering), 1976, Clarkson College of Technology; AT&T Bell Laboratories, 1976–1983; AT&T Information Systems Laboratories, 1983—. Mr. Densmore has been involved with software development for the *Horizon*® communication system prior to working on System 75. He is currently involved with enhancements to the System 75 architecture. Member, IEEE.

Ray J. Jakubek, B.S. (Electrical Engineering), 1967, Manhattan College; M.S., 1968 and Ph.D. (Electrical Engineering), 1973, New York University; Bell Laboratories, 1968–1982; AT&T Information Systems Laboratories, 1983—. Mr Jakubek has been involved in the design of a wide range of customer premises support systems and products. He contributed to the software development of the CNCC, NCOSS and RMATS II support systems. Subsequently he has supervised the software development of many of the call processing features of System 75. Since December 1982, he has been Head of the Software Applications department. Member, Eta Kappa Nu, Tau Beta Pi.

Michael J. Miracle, B.S. (Electrical and Computer Engineering), 1979, University of Wisconsin; M.S. (Electrical Engineering), 1980, Stanford University; AT&T Bell Laboratories, 1979–1982; AT&T Information Systems Laboratories, 1983—. Mr Miracle worked on System 75 switch software development in the areas of call processing and data switching and is currently working on System 75 enhancements. Member, IEEE, Eta Kappa Nu, Phi Kappa Phi, and ACM.

John H. Sun, B.S. (Electrical Engineering), 1972, National Chiao-Tung University, Taiwan; M.S. (Computer Science), 1977, University of Connecticut; Taiwan Telecom Research Laboratories, 1974–1975; ACCO-Bristol Company, 1978; Bell-Northern Research, 1979; AT&T Bell Laboratories, 1980–1982; AT&T Information Systems Laboratories, 1983—. Mr. Sun has contributed to System 75 switch services software architecture, design and implementation both as a developer and in his current supervisory capacity.

Courtney A. Klinck, Courtney A. Klinck & Associates, Riverside, Calif.

Just when you thought it made sense to get rid of Centrex . . .

Instead of fading, Centrex and the newer central-office-based services have grown due to enhanced data options and a push from the BOCs.

A few years ago, most people knowledgeable about the communications industry predicted the rapid dismantling of Centrex, assuming that most users would convert from Centrex to third- or fourth-generation on-premises private branch exchanges (PBXs) within two or three years. One major underlying reason for this prediction was the assumption that AT&T Information Systems, the owner of the great majority of Centrex customer premises equipment (CPE) following divestiture, would replace Centrex with on-premises PBXs since AT&T no longer had any reason to perpetuate Centrex. Another was the assumption that users would be eager for the more sophisticated enhanced voice and data features that were available with on-premises PBXs but not with Centrex.

In 1983, the Bell operating companies (BOCs) were not generally optimistic about the future of Centrex and were endeavoring to develop product strategies to compete directly with AT&T and other independent PBX suppliers. The use of Centrex had taken a sharp decline in 1982 and did not seem to be making a very strong comeback (Fig. 1). Now, however, Centrex is projected to continue growing at a healthy pace, as it has done since 1983.

Rather than withering and dying on the vine, the service is showing such signs of vitality as add-on data options and third-party marketing arrangements. The BOCs and the companies that supply them with central-office (CO) gear are working to enhance Centrex's image in the marketplace and its data communications functionality. At present, most organizations using Centrex are generally satisfied with the basic service and support the moves made to sustain it (Figs. 2a and 2b). Such developments should make users think carefully before dismissing Centrex.

In the last two years or so the BOCs have adopted a much more positive, aggressive attitude about Centrex and have structured Centrex rates and services to be competitive with PBXs. The renewed emphasis on Centrex can be attributed to the following:
■ The BOCs have a need to protect their huge capital investment in CO equipment and circuits associated with Centrex.
■ There is also an interest on the part of the BOCs to retain the major customers currently using Centrex.
■ Many large Centrex users are reluctant to replace Centrex with on-premises PBX equipment requiring a multimillion-dollar capital investment and possibly a major disruption of service.
■ Since only a relatively small number (usually less than 10 percent) of all telephone station users have data and/or enhanced voice requirements, Centrex remains viable for a longer time.
■ New data-oriented technologies and products are emerging that upgrade Centrex to be more competitive with on-premises PBXs, as will be described below.

The Centrex service, while competing directly with PBXs, is distinct from these devices in that it:
■ Is a regulated service provided by the local telephone company under the authority of the state Public Utilities Commission and the Federal Communications Commission (FCC);
■ Performs switching functions in equipment located in the telephone company's central office as opposed to on the customer's premises;
■ Is leased or provided on a month-to-month basis with no option to purchase.

There are several user benefits associated with Centrex as opposed to PBXs (Fig. 2c). The major Centrex advantages are as follows:
■ *Reliability.* Centrex switching equipment, built primar-

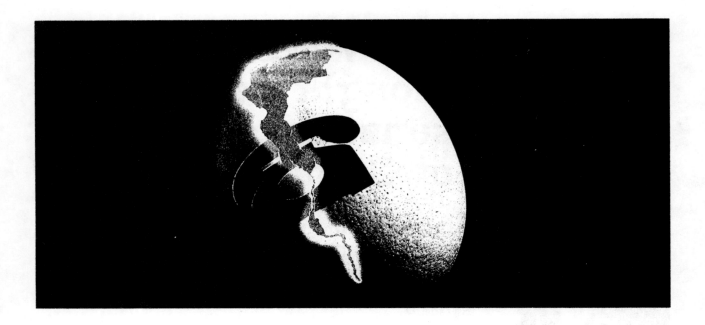

ily by Western Electric to handle public voice traffic, is engineered for reliability. Redundancy is built into the equipment and into the service, since standard voice equipment is available in the CO and can be brought to bear on any Centrex outages.

■ *Availability.* Since Centrex is located in the telephone company CO, there are maintenance and other technical support personnel available at all times.

■ *Unlimited capacity.* Centrex configurations can grow to as many lines as the customer desires. There is virtually no upper limit on size, as COs can be designed to handle as many as 50,000 lines or more.

■ *Multiple building coverage.* Centrex can serve many separate buildings, often miles apart, with no requirement for interbuilding trunking—every location is served in a star configuration from the CO.

■ *Avoidance of PBX-related expenses.* These include initial capital investment, insurance, switch repair and maintenance charges, and costs for space, environmental controls, and utilities.

■ *Continued updating of functionality.* Centrex switches are regularly updated with new software releases to provide additional features and capabilities without capital investment by the customer.

Thus, Centrex has a lot to commend it. However, with the advent of so-called third- and fourth-generation PBXs providing sophisticated voice and data capabilities and the parallel explosion of other communications products and services, users have found weaknesses in Centrex (Fig. 2d). These deficiencies, particularly in the data area, have caused customers to seriously consider replacing Centrex. A number of users have, in fact, replaced the service with on-premises PBXs and many more are seriously considering such a move.

In general, the major data communications short-comings of Centrex as compared to the new PBXs that handle voice and data are that Centrex is based upon analog, as opposed to digital, switching; it supports limited data speeds (typically 300 bit/s to 1.2 kbit/s with a practical maximum of 4.8 kbit/s); offers no significant integration of voice and data (typically, the new PBX station sets include data connections and the switches support application software); offers few of the management, reporting, and control enhancements now commonplace on PBXs, including traffic and accounting reports, networking capability (such as alternate routing and tandem switching), data security, and error control; and it provides no local (on-premises) distribution (everything is routed through the central office).

While Centrex CO equipment is very reliable, individual twisted-pair station links are more prone to failure and harder to reroute than those located within a building. In addition, customer-owned on-premises wiring may consist of coaxial cable or optical fiber, which offer higher bandwidth than that of twisted pairs. Also, from a cost standpoint, a one-time wiring charge seems more attractive than monthly leases of $16 to $22 per Centrex station line.

Fortunately for Centrex users, many of the above deficiencies are being mitigated by new products and services developed by AT&T, the regional Bell operating companies (RBOCs), and independent suppliers.

More attractive pricing
Ameritech has been one of the leaders in aggressive Centrex pricing. The company is restructuring Centrex rates in its five BOCs along similar lines to: absorb the monthly customer access line charges (CALC) imposed by the FCC subsequent to divestiture; provide long-term "rate stabilization" fixed-price contract op-

1. Projected growth. NATA figures show the market for Centrex, measured by the number of station instruments installed, taking off after a brief drop.

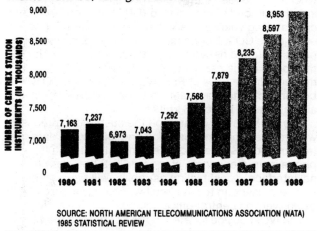

SOURCE: NORTH AMERICAN TELECOMMUNICATIONS ASSOCIATION (NATA) 1985 STATISTICAL REVIEW

2. Centrex survey. Most major users like Centrex's performance (a), rate-stabilization programs (b), and most features (c), though not all other aspects (d).

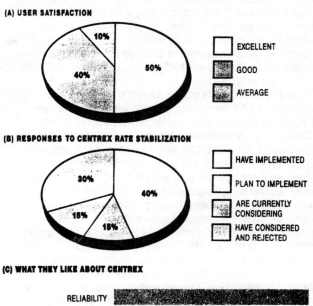

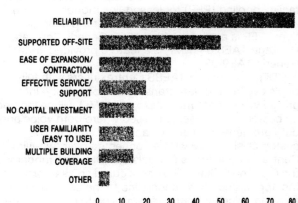

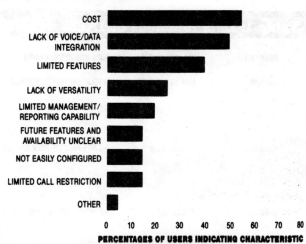

SOURCE: COURTNEY A. KLINCK & ASSOCIATES, MID-1985

tions; and provide a variable rate structure based on the number of lines involved and the distance from the central office.

Bell Atlantic has developed a Centrex rate structure similar to Ameritech's, including repackaging of features. The new rates range from 25 to 45 percent less than the old rates. US West (Northwestern Bell) has actually offered Centrex on an untariffed basis in a portion of its territory, while Nynex (New York Telephone) has filed new Centrex rates that give discounts to large users. The other RBOCs either have or are contemplating the same types of rate restructurings and reductions.

Several RBOCs are now beginning to make Centrex proposals to customers on a lower-cost, "special-case" basis as opposed to using standard tariffed rates. The RBOCs are arguing that if they don't lower Centrex rates to be competitive with PBXs, they will lose major customers and leave a huge amount of "stranded investment" in cable plant and switching equipment that still must be absorbed by the rate payers. The RBOCs claim that avoidance of this stranded investment justifies cutting Centrex station line rates to the $10-to-$12-per-line level from their current level of $16 to $22.

Enhanced services
The North American Telecommunications Association (NATA), a trade association of CPE suppliers, is strongly opposed to the reduction in Centrex rates and the addition of new and expanded features. NATA contends that the RBOCs are pricing Centrex services below cost and subsidizing Centrex with local exchange service revenue. NATA also argues that many of the new Centrex features fall under the category of "enhanced services" as defined by Computer Inquiry II and should, therefore, be provided by an unregulated subsidiary as opposed to a regulated operating company in order to eliminate the threat of cross-subsidies.

Centrex utilizes public central-office switching technology enhanced to form a virtual private switch. A large percentage of the Centrex feature upgrades originate with AT&T, the company that designed and installed the 1AESS (Electronic Switching System) analog switch currently used in most Centrex installations. AT&T's newer 5ESS (competing with the DMS-100 from Northern Telecom Inc., or NTI of Minneapolis, Minn.) is a digital CO switch. Both the 1AESS and 5ESS switches were designed to provide basic central-office functions with the capability also to be used as Centrex switching vehicles.

As part of AT&T's commitment to expand and enrich Centrex, AT&T Network Systems introduced a new Data Communications Package for Centrex in late 1984. A major portion of the new data capabilities are provided via 1AESS Centrex switch generic software packages (Table 1). The most significant elements of the new package are integrated voice/data access over leased or public-switched lines, alternate voice/data high-speed lines, packet-switching, and central-office-based local area networks (LANs).

Included in AT&T's new data communications offering is the Centrex Data Facility Pooling (CDFP), which integrates voice and data on the same circuit. CDFP will initially consist of two basic capabilities, private facility pooling (PFP) for leased-line modem pooling and network modem pooling (NMP) for dial-up connections. PFP is available with 1AESS Generic Software Package 1AE8.04. NMP is a feature provided under Generic 1AE9.02.

CDFP will provide 1AESS Centrex users with full-duplex voice and data transmission capability. Integrated voice/data multiplexers (IVDMs) will be located on customer premises and used where both voice and data are generated at the same user station. The customer-site IVDM will be connected over two-wire Centrex station loops to pools of IVDMs and modems in the CO. The IVDMs will be configured so as to provide the appearance of a second line to the user, thereby allowing voice and data transmission to occur independently and simultaneously.

One user can set up a voice/data connection with the telephone and data device of another connected to the same central-office switch by dialing directly through that switch (Fig. 3). Destinations located elsewhere can be reached by dialing out on an NMP port into the direct distance dialing (DDD) network, with one call handling voice and another carrying data. IVDM units in the CO will connect to computer ports over the public switched network up to 9.6 kbit/s (limited by the speed of the associated full-duplex dial-up modem). In the case of PFP, IVDM units in the CO will be connected to computer ports via dedicated circuits, also at speeds up to 9.6 kbit/s.

The IVDM device located on the customer's premises will provide simultaneous voice and full-duplex data transmission using a two-wire connection to the telephone instrument and a standard RS-232-C data terminal interface. Local testing capability will be available, so that a user can test a circuit without having to rely exclusively on central-office testing.

The IVDM device located in the CO passes the carrier signal from the remote device to the IVDM on the customer premises via the 1AESS switch. For the data portion of the signal, it provides a standard RS-232-C interface to a CO dial-up modem for NMP calls or a leased-line modem or direct-connect computer-port multiplexer for PFP traffic. (When data is transmitted without voice, the data device is connected directly to a multiplexer without an IVDM.) For the voice portion, the CO IVDM includes two two-wire connections. As in the case of the customer premises IVDM, local testing capability is provided.

Optional data features available with the new Centrex package provide access to public packet data networks and 56-kbit/s Circuit Switched Digital Capability (CSDC). In addition, where large volumes of data traffic are involved, the new data package can be configured with the AT&T Datakit Virtual Circuit Switch (VCS), which provides access to packet networks and functions as a powerful standalone switch. The Datakit VCS can also be connected to Datakit LANs and other data networks external to the premises.

CSDC is a sophisticated high-speed digital data network service provided with AT&T's 1AESS Generic

Table 1: Data functions of 1AESS generic software packages

RELEASE	AVAILABLE (TO THE BOCs)	DATA FUNCTION ADDED	DESCRIPTION
1AE7.00	MAY 1984	CIRCUIT SWITCHED DIGITAL CAPABILITY (CSDC)	HIGH-SPEED DIGITAL DATA NETWORK SEVICE INVOLVING END-TO-END ALTERNATE VOICE/DATA (56-KBIT/S CIRCUIT-SWITCHED DIGITAL) OVER THE PUBLIC TELEPHONE NETWORK (LONG-HAUL CIRCUITS VIA DATAPHONE DIGITAL SERVICE).
1AE8.04	MARCH 1985	PRIVATE FACILITY POOLING (PFP)	FULL-DUPLEX VOICE/DATA CAPABILITY USING DATA/OVER/VOICE TECHNOLOGY FROM CUSTOMER PREMISES TO TELEPHONE COMPANY CENTRAL OFFICE AND MODEM POOLING IN THE CENTRAL OFFICE, CONNECTING DATA USERS TO REMOTE COMPUTERS VIA DEDICATED CIRCUITS.
1AE9.02	EARLY 1986	NETWORK MODEM POOLING (NMP)	SAME AS PFP EXCEPT COMPUTERS ACCESSED VIA PUBLIC SWITCHED NETWORK AS OPPOSED TO DEDICATED CIRCUITS.

BOC = BELL OPERATING COMPANY   ESS = ELECTRONIC SWITCHING SYSTEM

3. Data communications package. *AT&T's new offering permits voice/data calls between users of a 1AESS, via dedicated or dial-up lines, or through data equipment, such as a packet-switching interface or a data switch. This scheme provides Centrex users with data capability over twisted pairs to the central office.*

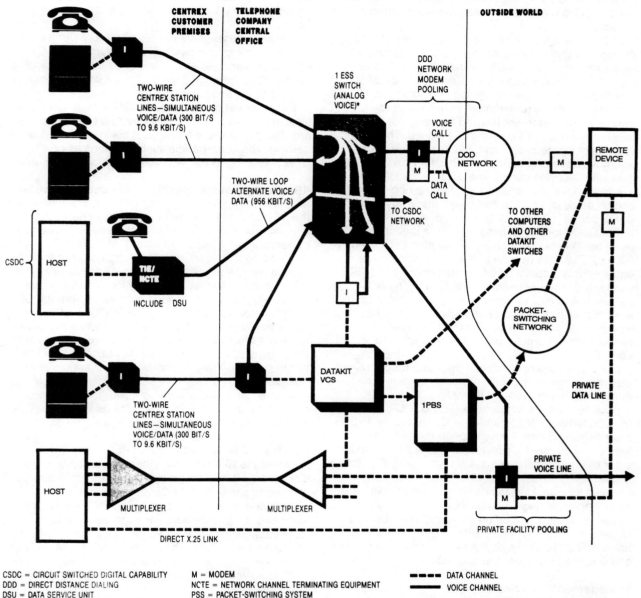

CSDC = CIRCUIT SWITCHED DIGITAL CAPABILITY
DDD = DIRECT DISTANCE DIALING
DSU = DATA SERVICE UNIT
ESS = ELECTRONIC SWITCHING SYSTEM
I = INTEGRATED VOICE/DATA MULTIPLEXER

M = MODEM
NCTE = NETWORK CHANNEL TERMINATING EQUIPMENT
PSS = PACKET-SWITCHING SYSTEM
TIE = TERMINAL INTERFACE EQUIPMENT
VCS = VIRTUAL CIRCUIT SWITCH

▬▬▬▬ DATA CHANNEL
─────── VOICE CHANNEL

*ALL VOICE AND DATA-OVER-VOICE CONNECTIONS
TO THE SWITCH CAN BE ROUTED TO ALL OTHERS

Software Package 1AE7.09 and is thus, like Centrex, a data service provided from the CO. CSDC involves end-to-end alternate voice/data capability over 56-kbit/s circuit-switched channels through portions of the public telephone network. Long-haul circuits are routed over Dataphone digital service (DDS). According to AT&T, the service has one error in 10^6 or better on at least 95 percent of the calls. The block-error rate is given as 99 percent error-free blocks on at least 95 percent of the calls. AT&T says that users of CSDC will be able to migrate to Integrated Services Digital Network (ISDN)

with minimal impact on their operations.

Applications well-suited for CSDC include such high-speed digital services as:
■ Distributed data processing,
■ Private-line backup and overflow,
■ Interactive computer graphics,
■ Intercompany interactive processing,
■ Communicating word processing,
■ High-speed facsimile,
■ Secure voice, and
■ Teleconferencing.

Customers of CSDC can select one of two types of service—a service allowing only CSDC calls or one providing calls over both CSDC and the public switched network. In addition, according to AT&T, the customer can select from a variety of features and custom-calling capabilities.

CSDC network applications can be implemented through a variety of loop configurations, the simplest arrangement being a two-wire metallic loop between the CO and the customer premises. AT&T specifies that this loop must be nonloaded and limited to 18,000 feet (though these requirements are subject to change). For customers beyond the range of the two-wire metallic loop, T1 carrier can be used. Those customers located in a non-CSDC serving area can also have access to central offices that provide CSDC through telephone company equipment, such as D4 channel banks. Customer access to the non-CSDC central office itself is the same as noted above for CSDC central offices.

The CSDC loop is terminated on the customer premises in network channel terminating equipment (NCTE). The data portion of the terminal interface equipment is actually a 504A data service unit (DSU) connected to the network via the NCTE. The NCTE buffers 144-kbit/s bursts coming in over the loop and passes data to the DSU at 56 kbit/s. The DSU buffers, stores, and passes the 56-kbit/s data to the user's terminal equipment either at the same speed or at 9.6 kbit/s.

CSDC calls are originated in the voice mode. Users can change back and forth between voice and data throughout the duration of the call.

Routine tests are implemented for CSDC trunks and lines, automatically scheduled and initiated by 1AESS diagnostic software. A loopback to the user's data terminal equipment is provided in the DSU. In addition, a parallel maintenance circuit provides overall digital and operational test capabilities, working in conjunction with loopbacks in various components. With the exception of testing the cable pairs when they are installed, all tests can be performed automatically by a single person at the central office. The 1AESS maintenance circuit performs digital tests of the entire end-to-end link, the local loop, and the NCTE.

Management modules

AT&T's new management software, called the Advanced Communications Package, allows Centrex users to integrate voice and data communications with local information processing. The features provided with this package are achieved through the addition of an AT&T 3B computer, including the 3B2, 3B5, or 3B20 models.

The Advanced Communications Package provides modules that can be made available in various combinations. These modules are grouped under the names "Communications Manager" and "Office Manager." Communications Manager modules include the following:

■ Centrex electronic key set. The set supports single-button feature access in addition to multiline capabilities over "skinny," rather than multipair, wires.

■ Advanced station message detail recording (SMDR). Provides custom-tailored call accounting.

■ Customer station rearrangement (CSR). Lets the user change Centrex features either one at a time or in batch.

■ Customer message center. Establishes a centralized message pickup and call-forwarding location.

■ Facilities management. Offers control over Electronic Tandem Switching (ETS) features and provides real-time traffic data reporting on facilities usage.

The other set of modules, the Office Manager modules, are not as well described by AT&T. However, they are meant to provide for general business data processing. They also handle such functions as providing reports relating to usage of automatic call distributors.

RBOCs: Aggressive development and promotion

Much recent activity related to Centrex, especially from the RBOCs, has to do with the nation's changing telecommunications environment (see "Regulatory, legal, and organizational impacts"). Most RBOCs have repriced, repackaged, and/or renamed Centrex, or are currently in the process of doing so. For example, ESSX is a "Centrex" service offered by BellSouth wherein station line rates vary based on distance from the CO to the customer premises. Those RBOCs that are most aggressively promoting Centrex generally serve older, more densely populated areas and have considerably more invested in Centrex equipment and station lines, not to mention a large installed base of major Centrex customers.

In an effort to provide competitive Centrex services, the RBOCs are upgrading the software in their 1AESS analog switches and are also, on a selective basis, replacing the analog switches with digital devices, such as NTI's DMS-100 and AT&T's 5ESS. The RBOCs may then use the AT&T products described above, those supplied by third parties, or perhaps a combination of the two to enhance their Centrex offerings.

Software upgrades to the analog Centrex switches, such as the 1AE8 generic package, provide added data capabilities, including data over voice (DOV) and modem pooling. Upgrades to the 5ESS or DMS-100 digital switches theoretically allow for virtually all the enhanced voice and data features offered by an on-premises PBX.

The RBOCs are faced with myriad alternatives as to specifically how they can incorporate enhanced data features into Centrex. RBOCs, such as BellSouth and Nynex, are pursuing services centered around the installation of digital central-office switches. AT&T's Centrex features for the 5ESS were released in 1985; NTI's advanced feature set for the DMS-100 switch will be available this year.

Pacific Bell has developed a seven-channel, single-line voice and data capability called "Project Victoria." This new digital service provides two digitized voice channels and five data channels (four at 1.2 kbit/s and one at 9.6 kbit/s) over a single circuit connecting the central office to the customer premises (either residential or business). A six-month field test of Project

Regulatory, legal, and organizational impacts

The Computer Inquiry II decision by the Federal Communications Commission (FCC) creating a separate unregulated subsidiary of AT&T (AT&T Information Systems) and the divestiture of the Bell operating companies (BOCs) by AT&T on January 1, 1984 both had a major impact on the status of Centrex. Computer Inquiry II allowed AT&T to compete head-on with unregulated suppliers in the area of customer premises equipment (CPE) and market integrated voice and data products competing directly with Centrex.

Divestiture left the BOCs without an installed base of on-premises private branch exchange (PBX) equipment, which was transferred to AT&T. Thus, the BOCs initially depended upon Centrex alone for business telephone revenue (although they have subsequently marketed PBX products obtained from third-party suppliers as well).

Divestiture also separated the CPE portion of Centrex, which consists of the telephone station equipment, from the station-line and central-office (CO) switching portions. Under this arrangement, AT&T and various independent suppliers provide the station equipment on an unregulated basis and the BOCs offer the balance of Centrex as a regulated, tariffed service.

This split in Centrex service responsibility has created coordination problems between the BOCs and the station equipment suppliers. It has made the user's job more difficult as well, since the voice-network manager has to deal with at least two different companies instead of just one for orders, maintenance, administration, and the installation of wiring. (The voice environment today is more like the one familiar to the data processing professional, who has always had to deal with multiple vendors.) On the positive side, separating CPE from the rest of Centrex has opened a whole new market for independent suppliers providing "Centrex-attached" devices that give the Centrex user a much broader range of station equipment options.

The FCC has recently initiated another comprehensive proceeding—Computer Inquiry III—aimed at clarifying the competitive interrelationships within the telecommunications industry and simplifying the regulatory process. CI III will explore the implications of competition and regulation on such emerging services as protocol conversion, Integrated Services Digital Networks (ISDN), and voice messaging. The results of CI III could remove many of the present barriers that stand in the way of the BOCs offering a wide variety of enhanced data services, both from the CO and on the customer premises.

CI III is addressing and reviewing issues of separate subsidiaries and the relationship between the regulated and unregulated parts of the RBOCs. If new regulations give the operating companies more flexibility, there is no technical reason why they could not provide a variety of enhanced data services to the user. Many if not all of the RBOCs are working behind the scenes to strengthen their data positions. Within two or three years, assuming favorable CI III rulings, users will have even more RBOC-provided data options to choose from.

Since divestiture, the seven RBOCs have set up their own unregulated subsidiaries to compete with ATTIS and other independent equipment suppliers. The establishment of these unregulated companies has further complicated the Centrex issue since now the RBOCs, as well as AT&T and others, have on-premises PBX products competing with Centrex.

A good example of the organizational and marketing issues surrounding Centrex can be seen from a review of the situation in Pacific Telesis, one of the seven RBOCs. The standard tariffed Centrex offering is provided by Pacific Bell, the regulated BOC. On-premises PBX equipment, such as Northern Telecom's SL-1, is provided by PacTel Communications, the unregulated subsidiary of Pacific Telesis. A separate group within PacTel Communications, PacTel Connections, provides CPE associated with Centrex as well as various networking services, such as custom network design and installation.

In this case, there is competition between PacTel Communications and Pacific Bell (on-premises PBXs versus the service portion of Centrex) and between PacTel Communications and PacTel Connections (again, on-premises PBXs versus the Centrex station equipment). Thus, various groups within the same organization are competing for the same business.

AT&T-IS and other independent suppliers are also involved in supplying CPE associated with Centrex. In addition, companies like the San Francisco start-up Centex are selling Centrex service along with other services (such as low-cost long-distance and detailed traffic and accounting reports). All in all, the Centrex picture in the Pacific Telesis territory, as elsewhere, is quite diverse and complex.

Another example of the organizational changes surrounding Centrex is the use by Nynex and other RBOCs of third-party companies (representatives and distributors) to sell Centrex service. In the case of Nynex, a good portion of Centrex revenues is generated by third-party sales.

It seems inevitable that the regulated and unregulated organizations within the BOCs will eventually be allowed through new regulations and court rulings to put aside their competitive differences (which they would prefer to do) and join forces. In this way, they will be able to develop a coherent strategy for providing users with a choice between CO-based and customer-premises data solutions and a migration path, if required, from one type of service to the other. The FCC's recent ruling that allows ATTIS to be recombined organizationally offers the prospect that the RBOCs will also be permitted to integrate regulated and unregulated operations and to deregulate offerings, like Centrex, are in competitive markets.

Victoria is scheduled to begin in the first quarter of 1986, with about 200 Pacific Bell customers participating. Currently, Project Victoria is being tested operationally within Pacific Bell.

New network services, such as Project Victoria and ISDN, when it becomes available, will no doubt be used in the future to more effectively connect customer premises voice and data terminal equipment to CO-based configurations, such as LANs, protocol converters, data switches, and packet networks. In addition to Pacific Bell's Project Victoria approach, other RBOCs are developing similar Centrex-based data services as enhancements to 1AESS analog central-office switches. Digital CO switches are also being installed to provide enhanced voice and data Centrex features including the implementation of ISDN beginning in 1986.

Bell Atlantic: A CO-based LAN
In May 1985, Bell Atlantic made a major move in offering data capability as part of Centrex with the introduction of what it termed a central-office-based local area network (LAN). Aiming to compete with on-premises LANs, the new service supports simultaneous voice and data transmission over existing Centrex circuits. The heart of the offering is a data switch at the central office.

Asynchronous speeds up to 19.2 kbit/s can be handled, as can 1.544-Mbit s T1 connections, to host computers. The Datakit LAN from AT&T Technologies has been employed in initial installations of the new service. However, other LAN products are also being considered.

The Bell Atlantic setup employs DOV technology involving a voice/data multiplexer on the customer premises and a comparable unit in the telephone company CO. At the customer location, the data terminal device and a telephone are connected to the voice/data multiplexer. Both voice and data are transmitted from the customer premises to the CO over standard Centrex station-line twisted pair. At the CO, the voice transmission is routed to the Centrex switch and the data traffic to the LAN where it is connected to another Centrex station or to a statistical multiplexer for T1 transmission to a host computer. Modem pooling access to the public telephone network is also provided.

Bell Atlantic believes that installing the LAN in the CO provides economy-of-scale advantages. Since multiple customers with multiple locations can be served by the same network, the associated cost savings can be passed along to the user.

Bell Atlantic plans to upgrade its new service to include 56-kbit/s support for such protocols as binary synchronous communications and synchronous data link control, as well as protocol conversion. It should be noted that adding protocol conversion to Centrex requires an FCC waiver of the Computer Inquiry II restrictions dealing with the separation of regulated and unregulated services. A 1985 FCC order has authorized asynchronous-to-X.25 conversions but under conditions that the RBOCs have generally found

too restrictive and are currently attempting to modify and expand.

Also in May of 1985, Nynex announced Pathways, a new family of advanced network services that allow users to exchange voice, data, and images on a single, high-speed digital line. With the exception of Infopath Packet-Switching Service, as described below, all services are currently available. The Pathways services emphasize digital technology and provide a link in Nynex's evolution to ISDN. At the same time, the new services maintain Nynex's commitment to Centrex.

Nynex: New and renamed services
Software installed in digital CO switches will offer a variety of new features and services, as noted below, to Centrex users served from these offices. (Currently, more than 500,000 Nynex customers are connected to digital switches.) Nynex is also upgrading analog Centrex service with packages of new features, rate stabilization plans, and restructured pricing.

Pathways services use several wideband (high-speed) arrangements that provide value-added services in the Nynex network. These services include switched 56-kbit/s services as well as more flexible and higher-capacity links to the interexchange network and information databases.

The following five specific Pathways services were announced:
- *Infopath Packet-Switching Service.* Provides users with X.25 access to other computer users, remote interactive databases, and interexchange carriers without costly and time-consuming conversions from ASCII to X.25. Customers pay only for the information transmitted. Voice is also supported via DOV technology.

Infopath, like similar packet-switching services announced by other RBOCs, is an outgrowth of the Local Area Data Transport (LADT) service developed by AT&T and, prior to divestiture, field-trialed by BellSouth with Knight-Ridder. Packet switching involves complex regulatory issues and approval to provide service that has not yet been obtained.
- *Flexpath Digital PBX Service.* Provides users of digital PBXs with T1 (1.544 Mbit/s) connectivity to the central office and the ability to configure the number and type of individual incoming and outgoing circuits.
- *Switchway Switched 56-kbit/s Service.* Provides end-to-end switched digital data transmission service at 56 kbit/s. Customers pay based on usage as opposed to a flat private-line rate. The service uses standard two-wire, nonloaded subscriber loop switched through the network on an intra-LATA or (in conjunction with another carrier) an inter-LATA basis.
- *Intellipath Digital Centrex Service.* Consists of a software package, installed in digital central-office switches, that provides such features as direct inward dialing, automatic route selection, trunk queuing, electronic tandem networking, customer station rearrangement, and traffic information.
- *Quickway Digital Service.* Provides private-line, point-to-point and multipoint, full-duplex data-only service at speeds of 2.4, 4.8, 9.6, and 56 kbit/s. Quickway will extend Nynex's digital data circuit capability into the

states of Maine, New Hampshire, and Vermont.

Nynex folded two existing services into the Pathways family:

■ *Digipath Digital Service (formerly Dataphone digital service)*. Provides tariffed, full-duplex, data-only service over private lines at speeds of 2.4, 4.8, 9.6, and 56 kbit/s. This service is available in both point-to-point and point-to-multipoint arrangements.

■ *Superpath 1.5-Mbit/s Service (formerly High-Capacity Digital Transport Service)*. Provides tariffed, high-capacity point-to-point transmission at 1.544 Mbit/s for voice, data, facsimile, and video.

Nynex is also planning several additional Pathways services:

■ *Virtual Network Services*. Will utilize the public switched network but provide capabilities similar to those of private lines, such as conditioning, error testing, and higher-speed, full-duplex, four-wire transmission with a line quality adequate for data. They will also offer features useful to multilocation business customers, such as uniform numbering and LATA-wide Centrex.

■ *Integrated Digital Services*. Involve a wide variety of services currently under evaluation. These include letting PBX customers dynamically control the mix of trunk features based on individual needs and letting Centrex customers accurately record call-completion and billing information (similar to PBX call accounting).

As the Pathways family of services grows, Nynex plans to offer its Centrex customers access to the Integrated Digital Services through Intellipath Digital Centrex Service. This will provide Nynex customers with ISDN capabilities without having to replace Centrex.

As a result of Nynex's moves in Centrex pricing, repackaging, and feature-enrichment and rate-stabilization plans, the company's customer base has increased during the past two years to the point where Nynex now has more Centrex customers than ever before. Nynex is evaluating additional Centrex capabilities to support this large customer base including:

■ Modemless data transmission at 9.6 kbit/s.
■ Switched modem pooling at 9.6 kbit/s.
■ A message-desk interface.
■ Real-time management, administrative, and control interfaces.

Nynex expects the Pathways family of services to include services defined by ISDN standards development, exchanging voice, data, and image information without concern for different interfaces, standards, and bandwidth limitations. The goal of Pathways Integrated Digital Services, like that of ISDN, is to give all customers a variety of voice and data services on a single circuit. Lines and terminals will utilize a limited number of standard interfaces, allowing customers to choose from a variety of terminal equipment. Nynex plans to offer 1.5-Mbit/s primary access to ISDN for large business customers and 64-kbit/s ISDN access to small business and residential customers.

Another RBOC, Ameritech, has announced two CO-based data transport services to enhance the data communications capabilities of Centrex—Public

Switched Digital Service (PSDS) and Packet Switching. The packet-switching service allows users to efficiently share high-speed data communications lines. Variable-length messages are received by a local packet-switching node and arranged into packets. Packets are buffered at each network switching node and rapidly routed from node to node through the network. If a link is congested or inoperative, packets are routed over alternate paths.

Under the Ameritech's packet-switching service, voice and data can be simultaneously transported over existing Centrex stations using DOV. Data speeds up to 9.6 kbit/s are possible. At the telephone company central office, dial-up modems connected to packet assemblers/disassemblers (PADs) accept incoming data over Centrex station lines and route it through the packet-switching network to the final destination.

Ameritech's other CO-based offering, PSDS, is equivalent to AT&T's CSDC. It will provide for circuit-switched voice or, alternately, 56-kbit/s data transmission. Under this service, a Centrex station can be used for regular voice communications or to send and receive high-speed data. An additional dial access code preceding the dialed telephone number is used to signal the CO to establish digital facilities to handle the high-speed data.

PSDS will be aimed at a number of business communications requirements including bulk data transport, electronic document distribution (such as high-speed facsimile, electronic mail, and communicating word processing), DDS overflow and backup, and teleconferencing (alternately or simultaneously transmitting voice and graphics information).

Tariffs have been and are being filed for the packet-switching service with approval expected sometime this year, assuming resolution of current regulatory issues. PSDS is currently available.

Fancy FN/SI

In addition to the two Centrex-based services described above, Ameritech is also pursuing a new network concept called Feature Node/Service Interface (FN/SI). This concept involves a novel network architecture and common interface standards for interconnecting users with a variety of voice and data telecommunications services via the public switched network (Fig. 4).

Through the FN/SI approach, Ameritech hopes to create a standard that will be used by the providers of voice and data services to achieve more efficient accessibility by the user. The concept provides "open" (that is, publicly specified) interfaces to the telephone network, allowing customer access to service vendors and offering those vendors the ability to provide their services through the network to virtually all telephone subscribers.

Customer access to third-party service capabilities will occur through an interface called a routing control point (RCP). The third-party services will be accessed through a service vendor-owned device called a feature node. Some services that could be provided through the feature node include automated answering ser-

4. FN/SI footwork. *Ameritech's Feature Node/Service Interface plan is touted as a way to rearrange voice-and-data building blocks to construct new services. This goal, if realized, could get users and vendors cooperating on new services and deploying them more rapidly than would otherwise be possible.*

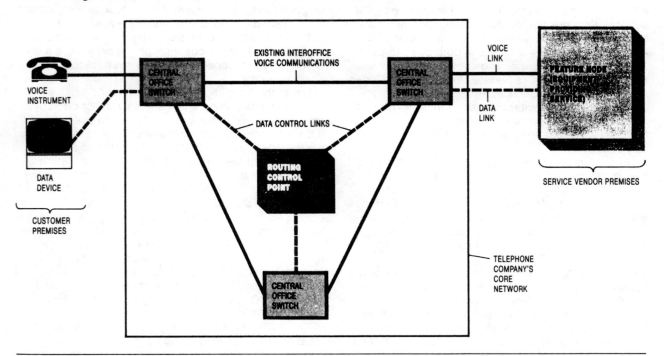

vices, voice store-and-forward, and database access. As an example, a voice-mail provider may wish to use the telephone company as a core network for a voice store-and-forward service.

The FN/SI concept will allow Ameritech's operating telephone companies to define and configure new services through the selection and combination of various call-processing building blocks residing in a central network location. The building blocks will be automatically activated in all COs within the core network. Rearranging the building blocks to provide various services could be done by the local telephone company in a relatively short period of time, without CO redesigns or costly additional equipment.

Ameritech believes that FN/SI represents a new approach by telecommunications service companies. Getting suppliers and users to work together may encourage them to create and customize a variety of applications and services to satisfy different market needs. Ameritech has made the FN/SI specifications available and is not seeking a patent for the concept.

In 1985, Ameritech issued a request for information, soliciting suggestions from almost a hundred communications equipment manufacturers on the development of a working FN/SI. In reaction to the positive responses, FN/SI is currently under development in Bell Communications Research (Bellcore, the research and development group supporting all the RBOCs). This development has the backing of Ameritech as well as other RBOCs. Ameritech hopes to have a pilot customer trial operational in late 1986 or early 1987.

FN/SI, if pervasively implemented, would create an information utility where the telephone company would define standard interfaces and provide the network transport over which various new services would be connected. It is, to be certain, quite an ambitious program, the viability of which will be tested in the face of current and planned competing services.

CPE Centrex enhancements

New customer-premises products have recently been introduced that enhance the functionality of Centrex. They are offered by David Systems, based in Sunnyvale, Calif., Northern Telecom, and Telegence Corp. of Westlake Village, Calif.

David Systems offers a unique approach for providing sophisticated data capability to Centrex users. The company's initial product, the David Manager (DM), provides data communications features, such as Ethernet LAN and RS-232-C switching, as well as access to the voice capabilities of Centrex over the same pair of wires (Fig. 5).

David's strategy assumes that no more than 10 percent of Centrex users currently need data capability. The company's approach is to provide only those users with the kind of data communications features available with on-premises voice/data PBXs while the remaining users retain their basic Centrex service. The fact that data users can be satisfied without replacing Centrex lets organizations that basically like Centrex but need more data functionality have their cake and eat it, too.

Among its most significant features, the David Manager provides:

5. CPE for Centrex. *A number of data terminals and hosts can be linked through David Systems' product over standard twisted-pair wiring. This arrangement al-* *lows the Centrex user to provide, to stations that need them, the advanced data features available on other customer-premises equipment.*

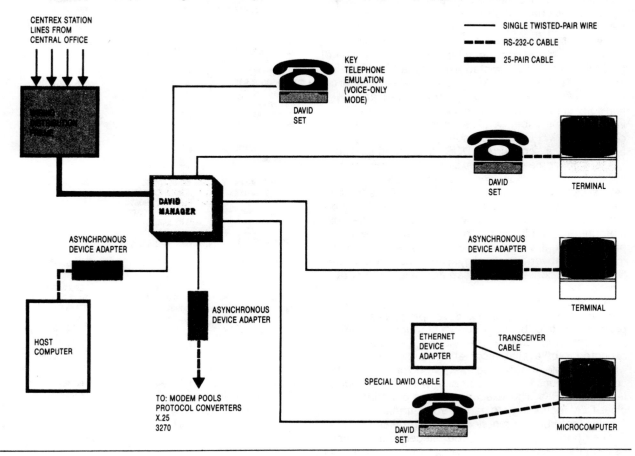

■ Simultaneous voice and data transmission over a single twisted pair at 2 Mbit/s to the user station by means of a proprietary protocol.

■ End-to-end error detection and retransmission for on-premises connections.

■ Speed matching (conversion and buffering).

■ Support for up to 120 voice-and-data users per DM or 384 users per DM in a data-only configuration (although note that multiple DMs can be networked together).

■ A digital telephone set including up to 16 programmable function keys that can be used for single-button access to Centrex features and/or multiple line appearances, a liquid-crystal display, a built-in speakerphone, and separate connectors on the telephone set for RS-232-C and IEEE 802.3 LAN terminal interfaces. (The RS-232-C link can support up to 19.2-kbit/s traffic, while the 802 connection buffers 10-Mbit/s traffic from the terminal to 1 Mbit/s over the pair of wires used for local distribution.)

David Systems has developed its DM using standard interfaces so that users can achieve network connectivity with a wide variety of equipment. The company has focused on providing an efficient network transport mechanism with users selecting whatever types of

terminals, computers, and application software packages they require.

One of David's strengths against the widespread popularity of LANs is its use of standard twisted-pair telephone wiring for network distribution. By avoiding the installation of the costly coaxial wiring often found in LANs, users can realize considerable savings in terminal installations and moves. David has signed a distribution agreement with Ameritech to market and service the DM. Other RBOC agreements are expected. In addition, David has signed six other distributors worldwide. David is also selling directly to users and has successfully installed its devices in a number of major corporations.

Northern Telecom's Meridian DV-1 Data Voice System is another new product that can be installed "behind" Centrex to add substantial data communications capability. The DV-1 is capable of supporting up to 100 users, as many as 35 of whom can have access to a broad range of business software, such as word processing, electronic-mail, and spreadsheet packages in addition to the standard voice capability.

The DV-1 was developed to serve the needs of individual departments and work groups having a significant "community of interest." Through the combination of voice and data communications as well as data processing (including application software), the DV-1 gives Centrex users a cost-efficient alternative to replacing the entire Centrex service with a voice/data PBX.

The DV-1 highlights a proprietary terminal, the M4020, which is capable of interfacing with the DV-1 at 2.56 Mbit/s over two twisted pairs of standard telephone wires. The DV-1 also accommodates asynchronous terminals and microcomputers. Several application software packages are available on the DV-1, initially running under Xenix. An added multitasking capability of MS-DOS (Microsoft's disk operating system) known as ''concurrent DOS'' is planned for 1986.

There are some conceptual similarities between the DM and the DV-1. However, as of late 1984, there were several significant differences as noted in Table 2. The David approach emphasizes a standard interface and basic transport architecture while the DV-1 focuses on a proprietary, applications-oriented architecture. Depending upon the user's preferences, existing environment, and networking strategy, either the David or the Northern Telecom approach can add significant data functionality to Centrex.

Telegence Corp. (formerly Reichert Research) is a start-up company that is also developing a customer premises product based around a standard twisted-pair distribution scheme. The Telegence product is expected to be available sometime in mid-1986 and will be capable of providing integrated voice and data communications behind Centrex. It will also work with a number of PBXs. The Telegence design concept employs DOV technology to handle voice traffic.

Voice features and functions will be provided by the Centrex (or PBX) switch with the Telegence unit simply passing the voice traffic from telephone set to switch in a transparent manner. The Telegence product will support RS-232-C communications up to 19.2 kbit/s through an interface module. Also, a plug-in card for the IBM Personal Computer (PC) will connect the twisted-pair wiring directly to the PC's bus at 1 Mbit/s.

Since the Telegence architecture will be highly distributed and modular, it will allow for terminal connectivity of only a few terminals or expansion to support hundreds of data devices. This design means that small or large configurations can be cost-efficient. Data features and functions that will be provided by the Telegence product include:
- Fixed and selectable data connections.
- Shared resources.
- Pooled resources.
- Contention and queuing.
- Dissimilar device type and speed connection.
- Error control.
- File transfer.
- Local area networking.

Given the wide variety of recent and pending improvements to Centrex, particularly in the data area, users can now seriously consider Centrex and related CO-based data services as viable alternatives to on-premises PBXs and other privately owned voice and data equipment. Careful consideration should be given

Table 2: Comparing two Centrex add-ons

FEATURE/FUNCTION/ CAPABILITY	DAVID MANAGER	NORTHERN TELECOM DV-1
ANALOG TELEPHONE SET	NO (AVAILABLE 1986)	YES
ELECTRONIC TELEPHONE SET	YES	NO (AVAILABLE 1986)
MAXIMUM TRANSMISSION SPEED TO DESK	2 MBIT/S (1 PAIR)	2.56 MBIT/S (2 PAIR)
MAXIMUM NUMBER OF VOICE USERS	120	100
MAXIMUM NUMBER OF VOICE/DATA USERS	120 (96 WHERE A SIGNIFICANT NUMBER OF HOST COMPUTER PORTS IS REQUIRED)	35
NETWORKING OF MULTIPLE SWITCHING DEVICES	YES	NO
RS-232-C ONLY (DATASWITCH) CONFIGURATION	YES, UP TO 384 DATA PORTS (NONBLOCKING)	YES, LIMITED TO 44 DATA PORTS (35 CONCURRENTLY). IMPROVED CAPABILITY UNDER DEVELOPMENT
LOCAL AREA NETWORK INTERFACE	STANDARD IEEE 802.3 ETHERNET	PROPRIETARY LANSTAR
TERMINAL	NONE	PROPRIETARY M4020
APPLICATIONS SOFTWARE	NONE	SEVERAL PACKAGES SUPPORTED
3270 EMULATION	YES, THROUGH ATTACHED THIRD-PARTY DEVICE	YES, M4020 TERMINAL CAN EMULATE 3270
STANDALONE CAPABILITY (WITHOUT CENTREX)	NO	YES

before replacing Centrex, and users should work closely with their local BOCs to determine what specific Centrex functionality, prices, and delivery schedules will apply in their area. ■

Courtney A. Klinck, president of Courtney A. Klinck and Associates, has more than 20 years of professional experience in voice and data communications with such firms as Bell Laboratories, Intel Corp., and Rockwell International. In recent years, Klinck has been involved in a variety of marketing consulting assignments, including market research, strategic planning, and product development. Klinck is a graduate of the the Bell System Data Communications Training Program and holds a B. S. degree in English.

Chapter 2: PBXs as Office Service Centers

The change in perception regarding the private branch exchange (PBX), from a switching device providing solely voice services to a computerized office service center enabling a multitude of functions, has evolved over the last 10 years from several sources. First, the increasingly powerful and modern digital PBXs have been able to add software and hardware that enable them to perform new functions. New applications and more powerful processors and ever greater memories act synergistically. Other applications, some provided by third parties, are often piggybacked via attached processors. Second, PBX architectures have incorporated extensive measures for fault tolerance and thus provide a reliable platform for office services. Large PBXs, those over 400 lines, usually incorporate duplexed main processors, battery backup to bridge power failures, solid state devices to keep the batteries charged, and external AC generating plants to operate through longer power outages. Third, the ubiquity of twisted pair telephone wiring, and a newly found respect for its capacity, has often given PBX-extended services price and convenience advantages over networks, such as local area networks (LANs), that would require various kinds of labor-intensive rewiring.

The deregulation of the long-distance market has probably provided the biggest push for enhanced administrative features. Long-distance costs typically represent a major and visible cost item. Optimizing long-distance costs yields relatively quick and very visible rewards. Not surprisingly, most PBX manufacturers offer various combinations of software and hardware to: (1) prevent access by the unauthorized; (2) optimally distribute in-WATs (wide area telephone network) calls; (3) optimally route outgoing calls by time of day, day of week, long distance resource and carrier; (4) capture, store, and analyze call data, and (5) process, format, and manipulate data for chargeback.

More recently, the demand for ever improved features has expanded to include data switching services. Providing data switching for office workstations to access electronic mail, departmental management information system (MIS) processors, or reference data bases has become popular. Typically, this workstation support involves asynchronous protocols at speeds of 19.2 Kbps or less by using all or part of a 64 Kbps voice channel. But a number of PBX manufacturers have, by employing innovative technologies, implemented LANs whose performance and features are comparable to cable-based LANs (Chapter 4). In some instances, the PBX can interconnect otherwise incompatible hosts and terminals by converting protocols or network gatewaying. Where outbound modem pools or facsimile transmission are used there is the added advantage of being able to employ the same cost optimization and accounting techniques used to control voice traffic. Finally, in the data services area, the popularity of desktop personal computers (PCs) and the convenience of twisted pair media have led two of the major manufacturers, AT&T and Northern Telecom, to introduce LANs that employ a high-speed packet switch in an attached processor role to the PBX.

The use of twisted pair wiring for data transmission, particularly at rates in excess of 64 Kbps, has brought increased attention to wiring and wire plants. AT&T, Northern Telecom, and IBM have all announced premises distribution systems for those wiring a new building or rewiring an old one. Overall, several tendencies are apparent. All wiring schemes are characterized by multiple kinds of physical media, usually copper and fiber optic cable.

Copper is usually employed from the intermediate distribution frames (small wiring centers typically on each hall of each floor) to the individual office. Either copper or fiber optic cable is employed on the runs from the intermediate distribution frames to the main distribution frame (the building's wiring center, usually in the basement and often adjacent to where the central office trunks enter the building). Fiber optic cable is becoming the cable of choice for inter-building and/or inter-switch links, though often less for its high capacity than for its small size and ability to be pulled through existing cable ducts.

Depending on the use of the wiring, gauges from range from the heavier 19 (AWG) to the lighter 26, and the addition of foil wrapping to shield all or selected pairs from interference may be recommended. A number of the wiring plans allow the elimination of present coaxial cabling, such as that used with IBM 3270 family of terminals or Wang word processors, via baluns or resistance balancers, with or without switching the terminal session connection through the PBX.

Copper twisted pairs are being employed to supplant other forms of cable, reduce capacity problems in overhead cable trays, minimize missing or damaged ceiling tiles, and reduce the management problems inherent with several kinds of coaxial cables. Nonetheless, copper twisted pairs offer no panecea. One of the advantages of coaxial cables is the relative unconcern with which it could be employed; if it could be physically installed, the signal attenuation characteristics were such that one could be reasonably sure that it would work. With longer runs employing copper twisted pairs, even when part of the run uses a fiber optic bridging segment, the same uncritical assumption cannot be made, particularly at the high speeds used on some of the PBX LANs. Some of the older cables, for instance, employed a lead/copper alloy in the conductors. The tradeoffs involved with using twisted pairs for high speed data transmission link the low cost and convenience—particularly in the case of preexistant wiring—with the added costs associated with careful verification of wire quality and other site engineering activities.

Recent additions to PBX functionality include voice mail, monitoring of building temperature, lights or fire alarms, and various kinds of security systems, including compressed video. Further additions seem likely. Compressed video will likely be combined with existing voice conferencing. High resolution video display work stations will increasingly input, display, and store complex facsimile images, with a PBX-implemented "fax pool" with laser printing devices one possible outcome. About the only major impediment to additional office-oriented service offerings is the proprietary nature of the protocols employed by current PBXs in the station-to-switch area.

In the papers in this chapter, the Sharma paper, "Some Architectural Considerations for Local Area Networks," aside from arguing that PBXs are the natural mechanism to provide office services—contains useful summations of various work breakdown studies for office workers. Horwitt's short paper, "Surveying the PBX Path," summarizes why, for most offices, a PBX approach to office automation combining voice and data services is preferable to separate systems. Coover, in "Voice-Data Integration in the Office: A PBX Approach," starts with assumptions about a typical office and argue that a PBX approach to voice/data integration offers the best path to eliminating equipment duplication, implementing a uniform wiring plan, and achieving device interconnection. In the Enomota and Kadota paper, "The Role of Information Management System in Business Communication," the authors detail the very wide range of computer and communications services offered by NEC's 2400 Information Managment System (IMS) PBX and how these services can be provided intra-office, inter-office, and intra-corporationally. Thomson's paper, "Helping Data Managers Find Their Voice," though appearing in *Data Communications*, is oriented toward voice management and PBX administration, and its emphasis is cost minimization. Coover and Kane, in "Analog vs. Digital," focus on the ongoing costs of PBXs and argue that the costs of moves and changes in a dynamic environment justify the higher purchase costs of digital station sets. Sagaser, in "Use Integrated PBXs and X.25 in Today's Networks; Don't Wait for ISDN," makes the case that the combination of data-capable PBXs with X.25 interfaces and public data networks (PDNs) can deliver many of the services promised by ISDN today. Last, Creadon, in "Voice-Data PABX Makes Data Analysis Easy for E-Tech," provides a case study involving a Siemens Saturn II PBX for E-Tech, a Florida systems-and-training analysis house. Note that for E-Tech the salient requirements were voice and data over twisted pair wiring, cost optimization for long distance calls, and, in the Florida location, the ability of a PBX to operate normally over a broad temperature range.

Data Communications

Reprinted from Business Computer Systems, pages 56-65.
Copyright © 1985 by Cahners Publishing Company.

Surveying the PBX Path

Telephone technology has been an office staple for nearly a hundred years. Now, the modern PBX offers business users the most cost-effective solution for voice/data networking.

By Elisabeth Horwitt

Two managers in separate offices at First Interstate Capital Inc., a Los Angeles capital investment firm, are holding a phone conference with a Boston client. While they talk, one manager (manager A) peruses the client's file, displayed on the screen of his IBM PC. He asks his colleague, manager B, to take a look at a relevant entry in the file. Since the office computers are linked on a digital PBX system that simultaneously handles voice and data, this is easy to do.

While dialogue with the client continues, manager B puts his PC in terminal mode with manager A's system acting as host. By hitting the Shift/Print keys, manager A sends the contents of his screen over to manager B's display. If manager A wants his colleague to see another part of the document, he can scroll until the desired segment is on his screen, and then hit the Shift/Print key again to send it. By this method the two managers can exchange relevant infor-

mation about the account, without interrupting the flow of their conference call with the client.

Simple, useful applications like this at First Interstate Capital are the trademark of voice/data PBXs: network systems that deliver both voice and data transmissions to users' workstations. Sometimes called "digital PBXs," these products handle both voice and data transmissions in the digital form characteristic of computer communications. Their forebears, voice-only PBXs such as Centrex and AT&T's Dimension, transmit analog signals — the wavelike frequencies that carry voice transmissions between most types of telephone equipment. Analog systems can also link computers, but only if a modem first converts the computer's digital transmissions into analog frequencies. (For an overview of succeeding generations of the PBX, see The Story So Far, page 62.)

While the first digital PBXs were de-

signed by vendors as more efficient voice networks, the newer systems were designed specifically to handle both voice and data. And as such, they are becoming an increasingly viable alternative to local area networks (LANs).

The one area where PBXs cannot compete with LANs is speed. A typical PBX data transfer rate is 19.2 kilobits per second. According to an August 1984 integrated voice/data PBX report published by Minneapolis, Minn., consulting firm Architecture Technologies Inc., even the maximum digital PBX data-transfer rate of 64 kilobits-per-second is too slow to handle intensive peer-to-peer computer communications. For example, users are likely to grow impatient waiting for the network to deliver a 50-page document, or a series of CAD/CAM designs. LANs can transfer up to 10 megabits per second.

However, this shortcoming does not stop PBXs from being extremely effective switching devices that link a group

Sound investment: Investment firm comptroller Keith Larson brought in a PBX and networking computers became a matter of plugging in a phone jack.

Jim Caccavo

of low-speed ASCII terminals to two or three hosts. That is their traditional function. But more recently they have come into their own as office automation networks carrying users' voice and data messages, as well as short documents, spreadsheets, simple graphics, and so on. In this role digital PBXs offer real advantages over their high-speed competitors. In terms of hard cost benefits, they are inexpensive and easy to install and maintain.

But PBXs' greatest potential lies in their ability to carry both voice and data transmissions. For some installations, this simply means that one centralized switching system and one set of wires connect both computers and telephones. But digital PBX vendors such as Inte-Com Inc., Allen, Texas, IBM/Rolm Corp., Santa Clara, Calif., and Northern Telecom, Richardson, Texas, have recognized a growing consumer demand

for a communications system that not only delivers voice and data on the same hardware, but also offers applications that combine the two.

For example: A flashing red light or message on his display notifies a user of either voice or data messages waiting. Voicemail and electronic mail would be sent via the same addressing and routing system, and users could append voice comments to electronic documents (for instance: "I have doubts about paragraph two"). An auto-dial feature automatically keys in not only frequently called telephone numbers, but also the baud rates, parities, and sign-on procedures that link the workstation with a mainframe host or an information service such as The Source.

All of the features described above are available on existing digital PBX systems, which will be discussed shortly. According to Yankee Group consul-

tant Amy Smith, such leading-edge PBX products are still very expensive: "But vicious price wars are going on right now, and the user will be the only victor."

Basic digital PBX systems ungarnished by the new, sophisticated add-ons and upgrades still make ideal voice/data office networks for companies with less exotic communications needs. For example, First Interstate Capital Inc., a capital investment subsidiary of Los Angeles-based financial holding company Interstate Bank Corp., found that Telenova 1, a digital PBX from Telenova Inc., Los Gatos, Calif., filled its needs very nicely.

First Interstate wanted a communications system for a new branch office. According to controller Keith Larson, who participated in the selection process, voice criteria included telephone features such as call waiting and conferencing ("We're a phone-intensive business.") On the data side, the firm was interested in a cost-effective way to enable two IBM XTs, two Compaq Pluses and a Dictaphone word processing system to communicate, and all of them to share resources such as modems and printers.

First Interstate's Telenova 1 installation includes eight voice-only telephone sets and five voice/data stations. The latter consist of a Telenova phoneset linked, via twisted-pair wiring, to a microcomputer's RS232 port. "Data ports" connect shared devices such as modems and printers to the central PBX.

The subsidiary had seriously considered installing a LAN to handle data transmissions, and a voice PBX to handle voice. But Telenova's solution seemed

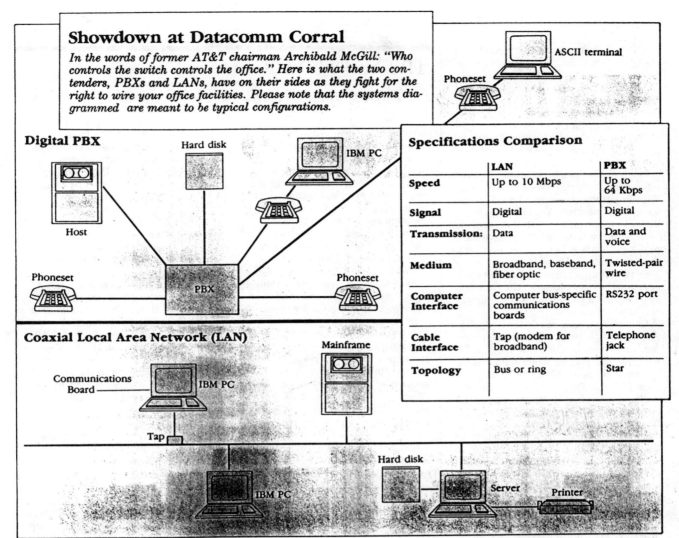

Showdown at Datacomm Corral

In the words of former AT&T chairman Archibald McGill: "Who controls the switch controls the office." Here is what the two contenders, PBXs and LANs, have on their sides as they fight for the right to wire your office facilities. Please note that the systems diagrammed are meant to be typical configurations.

Digital PBX

Hard disk
IBM PC
Host
Phoneset
PBX
Phoneset
ASCII terminal
Phoneset

Coaxial Local Area Network (LAN)

Communications Board
IBM PC
Mainframe
Tap
IBM PC
Hard disk
Server
Printer

Specifications Comparison

	LAN	PBX
Speed	Up to 10 Mbps	Up to 64 Kbps
Signal	Digital	Digital
Transmission:	Data	Data and voice
Medium	Broadband, baseband, fiber optic	Twisted-pair wire
Computer Interface	Computer bus-specific communications boards	RS232 port
Cable Interface	Tap (modem for broadband)	Telephone jack
Topology	Bus or ring	Star

superior on several counts. In terms of up-front savings, differences between the two types of network were negligible. Larson estimates that the $40,000 price tag of his company's Telenova installation is approximately the same as the initial cost of installing a LAN, even though a LAN requires specially installed coaxial cable, while Telenova 1 uses ordinary telephone wire.

But First Interstate expects Telenova 1 to be significantly less costly and troublesome than a LAN would be, in terms of maintenance and management. Says Larson: "We felt that having voice and data on one centralized system would make administration far simpler. We didn't want to hire a person just to be administrator." Digital PBXs also are much more economical than coaxial networks when it comes to moving workstations around. Even IBM reportedly admits that when terminals are "hardwired" (linked directly) to a host with 3270 coaxial cable, moving one to another office costs between $1,200 and $1,500. In a PBX configuration it becomes a matter of unplugging the phoneset, to which the terminal is at-

tached through an RS232 port, and moving the whole contraption to another telephone jack.

Telenova also saves the subsidiary money by permitting workstations to share expensive peripherals, such as printers, modems and Winchester disks. A manager finding a printer busy can leave the file to be printed in his IBM PC's buffer. The PBX lets him know as soon as the printer is free; meanwhile, he can be working on something else.

First Interstate chose Telenova over other digital PBXs because the system offers enough, but not too much power and functionality for the subsidiary's needs. Telenova 1 is designed to support up to 100 users, so that the initial 18-node set-up can grow considerably before it overburdens the system. At the same time, Telenova does not need to be supporting 100 workstations in order to be cost-effective.

And when it comes to applications, branch managers have the functions that help them do their jobs, but the subsidiary need not pay for sophisticated features it can do without. On the voice side, Telenova 1 offers all the

standard telephone features, plus a two-line liquid crystal display that allows a user who is unable to answer an in-house call to send the caller a short message such as "call back later."

On the data side, First Interstate can hook an IBM or DEC system into the network anytime it chooses. Gateways on the PBX device enable users to put their workstations in terminal mode and then pick up electronic mail, run software programs, or access a printer or modem connection to the outside world through the host.

First Interstate is considering this latter application for some future time, but does not now see the need. Admits Larson: "We are on the low end of Telenova's user base." Even the comparatively simple capability of one user transferring the contents of his screen to another user's display, Larson feels, would be "more useful in a bigger office than ours."

•

SATISFYING THE COMMUNICATIONS needs of bigger offices in bigger firms is the business of Telenova's high-end competitors: industry giants like Northern Telecom, IBM/Rolm, and AT&T Information Systems. Perhaps the most significant difference between high-end and low-end digital PBX vendors is that the former offer a proprietary host system which, in conjunction with the PBX switching device, delivers "personal productivity" applications to users' workstations.

One of the most common services offered by such PBX hosts is voicemail, the computer equivalent of a telephone answering machine. Voicemail systems, like electronic mail systems, provide "mailboxes" where users can leave messages for each other. Some products, such as Rolm's PhoneMail, include a remote dial-up feature that allows managers on the road to contact the voicemail system, listen to messages, and dictate and send off replies.

Considering how much time employees waste playing in-house telephone tag, voicemail is almost a bargain. A four-channel unit of Rolm's PhoneMail system, said to handle up to 80 users, is priced at $45,000. But as the market heats up and digital PBX vendors seek out ways to differentiate their products from competitors, features keep getting more exotic. It will be up to customers to decide which features to purchase immediately, which to hold off on until the price goes down, and which to ignore indefinitely. The following section summarizes some of the major vendors' most recent offerings.

Earlier this year, Northern Telecom introduced the Meridian line. According to the Yankee Group's Smith, Meridian is meant to upgrade the vendor's old SL1 and SL100 PBX products — "dinosaurs," in Smith's opinion, that originally were designed to handle voice only. However, it probably will take a year for Northern Telecom to start shipping add-ons to bring Meridian features to its old systems. Right now, Meridian is a standalone system composed of 4020 voice/data workstations, and DV1, which acts both as voice/data PBX and office automation host.

Meridian functions both as a communications processor and office automation system. As terminals connected to the host processor, workstations can run UNIX applications like database management, word processing and electronic mail. Plug-in 8088 and 8086 microprocessor boards provide MSDOS applications as well. Communications applications integrate voice and data on an unusually high level. For example, users can store, display and automatically dial both frequently called telephone numbers and the sign-on codes, baud rates, and so on of computer hosts and database services like Dow Jones.

The Story So Far

Evolution of the PBX and computer networks

While voice PBXs are technically capable of carrying computer transmissions, they are far from ideal. First of all each computer must be equipped with a modem to convert its digital (alternating high and low frequencies) transmissions to analog (unbroken, wave-like frequencies characteristic of telephone voice transmissions). As well, voice and data networks have very different requirements.

The total number of telephone conversations is more predictable than data transmissions. And so voice PBXs provide far fewer circuits than phonesets, the ratio determined by an estimate of what percentage of users will be using the phone at peak periods. As soon as the number of callers equals the number of circuits, everyone else is "blocked" — and hears a busy signal. When voice PBXs network computers as well as telephones, the likelihood of transmissions not getting through rises dramatically.

Thus the need for digital PBXs incorporating a "non-blocking" architecture that denies access to no one. First came data PBXs from vendors like Gandalf Data Inc., Wheeling, Ill., and Micom Systems Inc., Chatsworth, Calif. Such systems handle computer transmissions and leave the voice to the voice PBX. An optional data-over-voice (DOV) box takes transmissions from both PBX types and sends them over the same twisted-pair wires to another DOV, which separates them out again. It sends voice to the phoneset and data to the computer or terminal. A Gandalf data PBX costs $125 to $250 per line, depending on the number of lines. A pair of DOVs costs $590.

The latest addition to the PBX family is voice/data PBXs that handle both voice and data as digital signals. Some incorporate two switching devices, one for each type of transmission, since voice and data needs are so different.

They also can send appended voice comments to data text messages.

For $65,000 a customer can purchase a DV1 configuration that includes communications and personal productivity software, 20 Touchphones, ten 4020s and an 80-megabyte disk. Touchphones can be equipped with RS232 ports so that they become network interfaces for standard ASCII terminals or IBM PCs.

Thomas Mercer of Wang Laboratories points out the major problem (or advantage, depending on the point of view) of Northern Telecom's system: "Some PBX vendors are trying to do it all." Meridian's pricing is reasonable if the customer uses the system as Northern Telecom suggests, for both communications and office automation applications. But many companies prefer to buy their office automation systems from vendors like IBM (or Wang, naturally), who have solid reputations in those areas, not from a vendor whose main expertise is in PBXs. What these customers have been waiting for is a PBX system that works in concert with, and adds valuable communications functions to, hosts and workstations that have already been installed. Fortunately, IBM's acquisition of Rolm and Wang's alliance with Intecom promise to produce just such hybrids.

The IBM/Rolm partnership has blossomed forth with the "tree" series:

three voice/data workstation products designed around the PC. Cypress, $1,995, is a telephony device that emulates a Digital VT100 or IBM 3270 terminal. Cedar, $4,995, is an IBM PC-compatible voice/data workstation. Juniper, $1,495, is a communications board that converts an IBM PC into a voice workstation with access to all functions offered by Rolm's PBX.

All three products can use the vendor's voice communications features such as voicemail; as terminals they can display electronic directories of frequently used phone numbers. Users can also store "profiles" including the baud rate, parity, and other parameters necessary to sign on to a host or service such as The Source.

Still to come (and probably a year or two off) are IBM/Rolm offerings that integrate voice and data on the application level — that mesh, for instance, IBM's electronic mail system with Rolm's PhoneMail. Not the least of the problems to be overcome is deciding on one electronic mail system (IBM has several, although PROFS may be the preferred choice) and one phonemail system. An IBM spokesperson cautiously admitted that the vendor probably will withdraw its own phonemail product, Audio Distribution System (ADS), in favor of Rolm's clearly superior PhoneMail.

The Wang-Intecom alliance, cemented by Wang ownership of 20 percent of Intecom's equity, is committed to producing integrated voice/data products very similar to the present and future offerings of the IBM/Rolm contingent. Right now, Wang Professional microcomputers can hook up to a Wang VS through the PBX system.

More significant is a product which the duo hopes to start marketing in a month or two: a Professional-based workstation with its own built-in telephony features. When the two vendors consummate their plans for an integrated voice/data messaging system (Mercer of Wang claims that his company's own voicemail system, Data Voice Exchange or DVX, is likely to emerge as the standard), there will be a workstation to deliver it to the user's desk already in place.

For many communications managers, the acronyms PBX and AT&T are practically synonymous. Certainly it seems puzzling that the creator of Dimension, the Old Faithful of voice PBXs, as well as digital PBX Systems 75 and 85 PBX series, has failed to come out with a voicemail system or a voice/data workstation.

In the last few months, however, AT&T has begun to get its voice/data act together. The new UNIX PC voice/data workstation is one important step. Costing $6,095 (hard disk included), the micro features an electronic directory of phone numbers and host computer log-on procedures, and electronic mail. Another recent AT&T introduction of note is the phonemail system, Audix, which should be available in June. And a third AT&T move toward integration is a fiber-optic link between the System 75 and 85 digital PBXs and Information Systems Network (ISN), a data network that links hosts, terminals and microcomputers in a star configuration with a high-speed (8.64 megabits per second) bus in the center. The link allows workstations on either the PBX or ISN to access hosts on either system.

Slow as it has been to draw level with its competitors, AT&T has assigned itself the task of standardizing the PBX industry. The vendor has already proposed Audix as a universal voicemail standard. Says AT&T spokesperson Barry Campbell, "Audix is an attempt to develop unified messaging standards so that voicemail can be exchanged in a multi-vendor environment. One key concept is that anyone's message will activate the mail notification device."

Industry response to the Audix proposal has not yet gelled: As the two alliances described above indicate, vendors are finicky about whose voicemail system they adopt. A surer hit is AT&T's proposal of Digital Multiplexed Interface (DMI) as a standard for linking minicomputer and mainframe hosts to PBXs. As of March 1, 1985, 36 companies have answered AT&T's call, including Prime, Harris, NCR, Hewlett-Packard, Wang and Data General; PBX vendors Intecom, NEC, Ztel, Gandalf and CXC; local area network vendor Ungermann-Bass; and T-1 vendor Timeplex Inc.

This impressive display of enthusiasm reflects computer vendors' growing conviction that they cannot afford to ignore PBXs, and communications vendors' decision not to ignore AT&T. Ungermann-Bass is playing it safe by negotiating with Northern Telecom and Rolm as well. Vice president James F. Jordan says, "We think it is important to interface with PBXs. LANs won't link everything." Quite an admission for a LAN vendor to make.

On paper, DMI does not seem to do anything impressive: All it does is specify how one T-1 link (T-1 is an industry standard for a link that transfers 1.544 megabits per second) should be divided up into 23 terminal-to-host channels transferring 64 kilobits per second, and one channel for control signals. But DMI could become a rallying point for industry support of an emerging communications standard: Integrated Services Digital Network (ISDN).

ISDN still has a long way to go before it is completed, and it will be a long time before the industry accepts it. But eventually it will standardize long-distance digital transmissions, many hope. Even ordinary telephone lines will support it (the phone company will have to renovate some of its equipment before it can reliably transmit 1.544 megabits per second). When all is in place, all brands of computers and all communications devices will send voice, data and text as digitized signals between facilities, over distance, via ISDN links. And within a facility via the flexible, cost-effective, versatile, digital PBX. ∎

Voice/Data Connections

The following vendors currently offer voice/data PBX systems. Business Computer Systems welcomes comments and additions.

AT&T Information Systems
1 Speedwell Ave.
Morristown, N.J. 07690
Information Service No. 382

Anderson Jacobson
521 Charcot Ave.
San Jose, Calif. 95131
Information Service No. 383

CXC Corp.
2852 Alton Ave.
Irvine, Calif. 92714
Information Service No. 384

Ericsson Inc.
7465 Lampson Ave.
Garden Grove, Calif. 92642
Information Service No. 385

GTE
P.O. Box 2392
North Lake, Ill. 60164
Information Service No. 386

InteCom Inc.
601 InteCom Drive
Allen, Texas 75002
Information Service No. 387

Mitel Inc.
54 Broken Sound Blvd., N.W.
Boca Raton, Fla. 33431
Information Service No. 388

Nippon Electric Co. (NEC)
8 Old Sod Farm Road
Melville, N.Y. 11747
Information Service No. 389

Northern Telecom
2100 Lakeside Blvd.
Richardson, Texas 75081
Information Service No. 390

Rolm Corp.
4900 Old Ironsides Drive
Santa Clara, Calif. 95050
Information Service No. 391

Siemens Communications Systems Inc.
Telephone Division
5500 Broken Sound Blvd., N.W.
Boca Raton, Fla. 33431
Information Service No. 392

Telenova Inc.
102-B Cooper Court
Los Gatos, Calif. 95030
Information Service No. 393

United Technologies Communications Co.
Lexar Division
31829 W. La Tienda Drive
Westlake Village, Calif. 91362
Information Service No. 394

Ztel Inc.
181 Ballardville St.
Wilmington, Mass. 01887
Information Service No. 395

RICHARD D. CREADON

Voice-data PABX Makes Data Analysis Easy For E-Tech

EFFICIENT communications is essential to collecting and analyzing data for the preparation of government and industry analyses. The need becomes even greater when the effort requires heavy daily data transfer and telephone communications. Optimizing communications was a major concern of Eagle Technology Inc., Winter Park, Fla., during a major facilities expansion last year.

From its Florida directorate, E-Tech performs system and training analyses for government and industry organizations. It also designs various automated management training and delivery systems. To develop a typical training course, for example, E-Tech first defines the problem and outlines a solution approach. It then sends field teams to various operational sites, where the teams develop and send data to a central location. There the data are collected and sorted to create a database. The data arrive via written and verbal messages, drawings and sketches, as well as digital communications. Then the data are fed into E-Tech's mainframe for cataloging, retrieval and analysis.

This clearly illustrates E-Tech's need for total information management and control. Linking diverse data types and sources to a central database requires many tools. Among them are the ability to process a variety of inputs, the capability of a large, on-line and expandable system, the availability of around-the-clock access with occasional unattended operation, and a flexibility in functional access and operating modes. Also important are features such as conference calling, message receipt and forwarding, anywhere-page-pickup, voice mail, speakerphones, easy design configurability and others.

The hope for voice-data integration was equally important. It would provide E-Tech with a variety of enhanced overall communications capabilities and qualities inherent in digital technology. A special advantage digital technology would provide is the ability to quickly connect office PCs and other devices into the user network without cabling requirements. This was important to E-Tech when one of the company's terminal-

Richard D. Creadon is District Manager, Tel Plus Communications Inc., Orlando, Fla.

equipped employees changed locations. Finding a system that would provide the necessary digital port capacities without using up the ports allocated for system changes and expansions was a problem.

Michael Hart, an E-Tech training analyst, was assigned to find a solution when the company moved 70 workers last year from Orlando to Winter Park. Hart might not have considered the PBX solution had he had more data training. His experience taught him that most data people customarily think in terms of coaxial cable when they consider data communications, and at the old location, the company was constantly re-

wiring the coax and drilling holes in the walls, ceilings and floors to lay cable. The mess, Hart said, was costly because E-Tech made changes to accommodate new clients.

Because of such expenses, E-Tech did not consider a non-PBX solution. The company contacted the leading PBX manufacturers and their sales organizations for evaluation purposes. Comparative shopping for price and switching capabilities nar-

Author Richard D. Creadon, district manager, Tel Plus Communications (l), and Michael Hart, an Eagle Technology Inc. training analyst, discuss Siemens Information Systems' Saturn II System.

Photo courtesy of Siemens Information Systems Inc.

rowed the choices to switching systems made by Northern Telecom, Rolm and Siemens. Hart knew each offered workable configurations to meet E-Tech's specific needs.

The search concluded with the Saturn II System manufactured by Siemens Information Systems Inc., Boca Raton, Fla., which was recommended by Tel Plus Communications Inc. of Orlando, Siemens' sales, maintenance and servicing organization.

Recommended by Tel Plus in response to E-Tech's expressed specifications, the system was favored for several reasons. Foremost among them were the enhanced and proven digital capabilities of the system. The digital PBX case histories and actual customer references supplied by Tel Plus proved valuable. In addition, E-Tech pointed out to Tel Plus that going to digital systems usually required the use of many peripherals, which often caused clutter at system stations. This was not the case with the Saturn system. Spatial requirements of the switch and housing also proved desirable for E-Tech's headquarters facilities. It afforded more space in the computer room for other computers, facsimile machines and additional equipment. It also provided more room for facilities expansions or

fuller use of computer room space for other purposes.

Moreover, the system offered the widest optimal operating temperature ranges, an important consideration for a Florida location in the event of air conditioning or dehumidification outages. The temperature operating parameters were viewed favorably from the standpoints of year-round, daily operations and the elimination of maintenance costs. The benefit was considered equally attractive to Tel Plus, which would provide all continuing diagnostics, service and maintenance.

Its least cost routing scheme was another savings benefit. Users call long distance via direct dial, or WATS, or other carrier services over the least costly medium to the next station by the least expensive medium selected by the system.

The specific system Tel Plus proposed, and the one E-Tech finally chose, is a stored program controlled, digital voice-data PBX configured on a port basis (offering from 40 to 224 ports). The system uses a microprocessor-based controller to perform all call processing. And, because of the combined use of pulse code modulation and time division multiplexing, the system can handle a high volume

of call traffic over standard, single twisted-pair wiring. Compared to other systems that require the use of high-frequency coaxial cable, and specifically compared to the other systems evaluated by E-Tech, it offered a much lower cost.

Another reason for E-Tech's selection was a feature recommended by Tel Plus in view of E-Tech's desire for voice-data integration capabilities. Called Office Communications II, or OC II, it provides special software to integrate voice and data. With the growth of word processing, batch and real-time data processing, plus electronic mail, OC II provides for efficient, low-cost voice-data integration compared to other systems.

The system's ability to connect office PCs into the communications network was also an important factor. Using transparent, digital-switched communications among multiple data information systems, OC II transforms E-Tech's existing telephone wiring—which is standard twisted pair—into a powerful communications network. Word processing, data batch processing, electronic mail services and MIS inquiries are now all available.

OC II allows E-Tech to pool facilities and existing installation wiring;

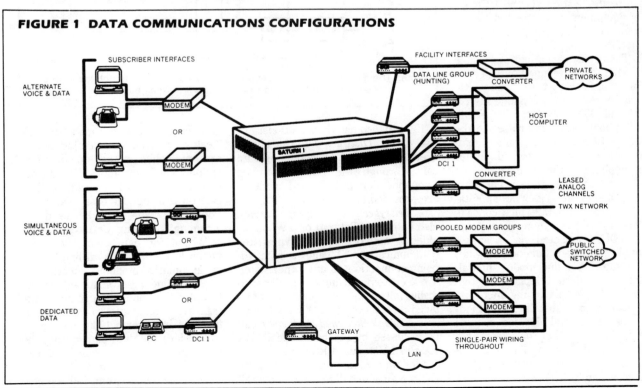

FIGURE 1 DATA COMMUNICATIONS CONFIGURATIONS

improve control of network configuration through the use of administration and management switching capabilities; access information storage facilities via the system; and use both terminal and conventional dialing.

For handling digital and modal data, modems pooled by the system allow interconnection of the data types. The data call origination capabilities (Figure 1) include:

• analog call setup via conventional telephone with switchover to data via a modem;

• data hot line calling via a data terminal wired to the system via a modem;

• automatic calling unit dialing via a modem/ACU combination;

• voice call setup via a conventional telephone set and add-on (or switchover) to digital data via an associated data terminal wired to the system via a data communications interface; and

• terminal dialing via the DCI. Data calls can be terminated either by manual answer or by automatic answer operation. Flexible data trunking is provided by leasing analog channels interfaced to the system via a modem and DCI combination, or leased digital channels (or digital switched networks), which are interfaced to the system via the DCI. All user equipment is connected to the system via subscriber line modules (digital or analog) and DCIs.

E-Tech enjoys a number of possible subscriber data arrangements. These include alternate voice and data, which consists of a telephone instrument and DTE connected via a data set and appropriate interfaces to the system. (Because only a single communication path is provided, either the telephone or the terminal may be on-line at any one time.) The data arrangements also include simultaneous voice and data, which consists in this case of a telephone instrument and a DCI connected via an appropriate line interface (analog or digital) and a subscriber line module digital. When a standard telephone instrument is used in this arrangement, its interface is the subscriber line module analog. Digital devices require the SLMD. Another data arrangement is a dedicated data configuration, which consists of a DCI connected via an SLMD.

FIGURE 2 POOLED MODEM GROUPS WHICH CAN BE SHARED AS NEEDED

1 DIGITAL LINE TO ANALOG TRUNK

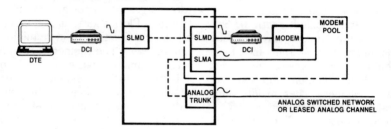

2 ANALOG DATA LINE TO DIGITAL TRUNK

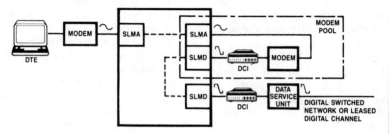

3 DIGITAL DATA LINE TO ANALOG DATA LINE

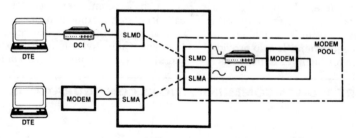

Note: The dashed lines represent a temporary connection through the MTS established by the Common Controller on a per call basis.

E-Tech also has these options:

• *Direct Distance Dialing* allows user origination and reception of data calls through the public switched telephone network. Pooled modems are provided for these calls.

• *Teletypewriter Exchange Service,* also known as Telex II or TWX, provides a network for teletypewriter communications. It allows point-to-point connection to any other listed TWX customer using direct-distance dialing. The system may be interfaced directly to the TWX network.

• *Digital Switched Networks* can be accessed by means of the PABX via an EIA standard RS232C.

Modem pooling in the system gave E-Tech an added dimension in overall data communication capabilities. E-Tech presently uses a combination of modem-pooling and nail-up communications configurations. Pooled modem groups can be shared for data calls on an as-needed basis (Figure 2). Pooled modems may be used for incoming, outgoing or internal data calls and for data connections between analog and digital lines. Also, in automatic modem selection, a pooled modem automatically is selected if the connection requires the transition from digital to analog interface, or vice versa. The modem is placed into the data path during the call setup phase and requires no special call setup procedures.

With the new system, E-Tech now

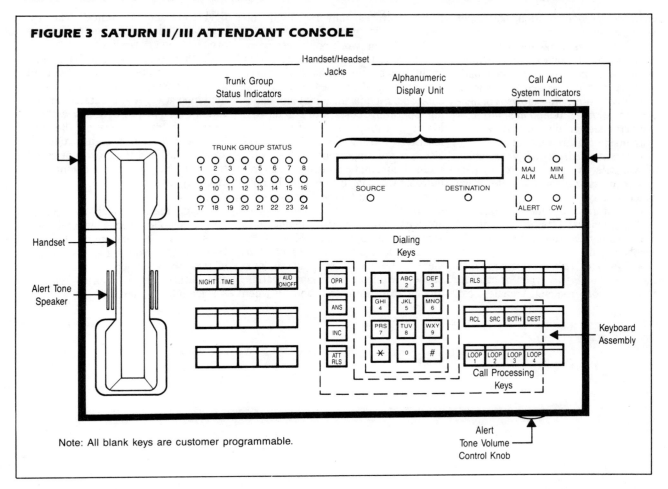

FIGURE 3 SATURN II/III ATTENDANT CONSOLE

Note: All blank keys are customer programmable.

uses one attendant console and can use as many as 12. A basic console layout is shown in Figure 3. To aid the operator, the console has a 40-character call information alphanumeric display plus 30 status-indicating LEDs. The call information display provides visual information on calls and system conditions. Between calls, the display provides the number of calls waiting in real time to be processed in queues, such as operator, recall, incoming and centralized attendant service. During call processing, the display provides the call identification (extension number or eight-character alphanumeric trunk designation), class of service, condition and destination identity.

Indicators used to monitor system and console conditions include trunk group status, MAJ ALM (major alarm) and MIN ALM (minor alarm) LED signals, ALERT LED to indicate a call received on one or more of the answer keys, and a call waiting LED indicator. The keyboard assembly contains a 12-button digital key pad

for dialing, 34 non-locking function keys for call processing, and speaker and volume controls.

Although E-Tech does not prioritize calls, the company can do so. For that function, the call waiting the longest, or having the highest priority, may be taken by depressing the answer key. Call answer keys include incoming, operator, recall, answer and centralized attendant service. The system has six types of call processing keys and four loop keys. The loop keys offer the attendant several features, including the ability to put four calls on hold simultaneously, to monitor the length of a call through serial calling, and to supervise a trunk-to-trunk connection. There are 18 optional feature keys, which are left blank so E-Tech may assign them to perform a wide variety of functions. These features of the attendant console illustrate the system's flexibility. Not only can E-Tech operate the system with multiple attendant consoles, but it can move a console with minimal rewiring, since the con-

sole controls the system via the software connection.

Three digital telephones come with the system, either an 18- or 26-button DYAD or a JR-DYAD. As many as 184 DYADs can be used in any combination within a single system. These phones offer several noteworthy features, including:

• *on-hook dialing* by means of one programmable button, which permits the user to dial without lifting the handset until the call is answered;

• *hands-free operation,* which, in addition to the on-hook dialing, permits the user to talk with the 18- or 26-button DYAD to a called party without lifting the DYAD handset;

• *an automatic display,* which, with the 18- or 26-button DYAD, provides an attendant identification indicator when communicating with attendant, a forwarded destination number during call forwarding, location numbers during call park, call back numbers and conference mode displays, and other displays, such as time of day, call duration, account code, recall identification, and input verification;

• *full-feature intercom operation,* either automatic or executive. Automatic intercom provides a talk path between two designated telephones. Executive intercom allows connection to another telephone within a pre-arranged group through a programmable feature button;

• *message waiting displays,* which allow a selective response to any particular message, including selective cancellation;

• *speed dialing,* which provides up to 64 individual lists with up to 10 numbers per list (up to 18 digits per number); and

• *advanced call processing,* which includes redial, five types of call forwarding, conferencing, transfer, call pickup features, universal night answer and voice paging access.

E-Tech's network objectives were well-achieved, especially in terms of facilitating efficient linkage of diverse data types and sources to a central database. Significant initial and ongoing operating costs were achieved. Major operating cost benefits include least-cost-routing. Soft dollar savings also proved important to E-Tech's intense information management requirements. The urgency and precision required in taking data in a variety of types and from a variety of sources can create considerable stress among personnel. Most of the features described provide more efficient and easier access among personnel within each main location, among main locations, and between field teams and their offices. The system has all but eliminated equipment problems and downtimes.

Further E-Tech savings have been achieved in terms of advanced switching capabilities, which allow simultaneous voice and data communication over twisted wire pair without increasing traffic loads on the system's switch. The system's modular architecture and its growth-oriented software can quickly accommodate changing needs.

Design features such as modular power supplies and universal card slots permit expansion of the system's digital voice and data capabilities to keep up with virtually any future requirements—without service interruption. □

Voice-Data Integration in the Office: A PBX Approach

Edwin R. Coover

This article posits the role of a PBX in achieving voice-data integraton in the office

This article posits the role of a PBX in achieving voice-data integration in the office. For realism, it makes a number of unglamorous assumptions about the typical office:

- The office already possesses personal computers, shared logic word processors, a facsimile machine, and some kind of departmental processor, and must communicate with a remotely located large-scale IBM mainframe.
- The office knowingly or unknowingly, currently supports at least three levels of file storage: individual, on the PC's; group, at the office level on the shared logic word processors and/or departmental processors; and corporate, on the remotely located IBM mainframe.
- The office manager does not have the luxury of replacing all present office equipment at one time.
- The office manager has neither the budget nor the in-house expertise to integrate all equipment immediately even if s/he could, so an incremental approach is essential.
- Lastly, the office has or is in the process of getting a modern PBX.

Such a scenario raises the question of why, regardless of method (PBX, cable LAN, host mainframe), integrate voice-data at all? The answers here are several, and, fortunately for one who must sell the idea to management, eminently quantifiable:

Eliminate equipment duplication. If one device can perform two or more tasks, multiple physical devices do not have to be bought or installed, consume space, or be serviced. To cite a typical example, if a PC can access a time-sharing service by emulating a terminal, and neither the PC nor time-sharing service is used continuously, then a PC can replace a terminal.

Implement a uniform wiring plan. Data devices that require coaxial cabling impose special pain on the office manager. They are expensive to install and expensive to move. Something must be done to prevent employees from tripping over the cabling, much less to hide it for aesthetic reasons. Yet retrofitting coax connection networks (Ethernet, broadband) into existing offices is expensive and disruptive in the extreme. By contrast, employing extant telephone wiring for data use has an irresistible appeal: ubiquity and low cost. Furthermore, as typically everyone in the office has a telephone, and nearly everyone some kind of video workstation, and personnel rearrangements are common, being able to implement most moves and changes in software at a PBX administrative console can be a substantial money-saver.

Device interconnection. Here, the problem is the necessity of transferring information files (electronic mail, word processing test, data base subsets) across physical space (countries, states, cities, blocks, floors) and between different types of devices (often with different equipment vendors). This often involves specialized equipment and a significant amount of technical understanding. Nonetheless, once the necessary equipment is in place and the requisite training accomplished, the process can become routine and the

Reprinted from IEEE Communications Magazine, July 1986, pages 24-29. Copyright © 1986 by The Institute of Electrical and Electronics Engineers, Inc. All rights reserved.

costs of reentering "foreign" information can be eliminated.

To further clarify the assumptions regarding the archetypical office environment described earlier, Fig. 1 summarizes the five categories of devices and their typical uses, and presents some representatives. Figure 2 attempts to summarize their communications capabilities in terms of the ISO seven-level typology. Note that in Fig. 2, "Media," technically a sublayer of the Physical Layer or Level 1, is listed separately in order to emphasize the wiring aspect of the overall connectivity problem. Finally, the term "modern PBX" used to refer to a digital circuit switch of current manufacture (AT&T's Systems 75 and 85, NEC's IMS 2400, Rolm's CBX II, Northern Telecom's SL-1, InteCom's S/80, and the like) whose primary mission is voice switching, but which can be augmented to provide a variety of data services as well.

Having set the scenario and limitations, how might such a PBX-implemented, incremental voice-data integration strategy for the office proceed? Specifically, what roles might a PBX play?

Port selector. Perhaps the simplest cure is where the PBX functions as a port selector. Should there be multiple hosts accepting ASCII/asynchronous terminal traffic, a user with an ASCII/asynchronous terminal (or PC emulating the same) reaches either host by dialing or keying a four-digit number. The PBX equipment needed to effect this integration can take several forms. One consists of an electronic phone with asynchronous data interface. The terminal or PC is hooked RS-232C to RS-232C to the voice-data set at the user station. Another form is a plug-in card for the IBM PC, which may (Northern Telecom) or may not (Rolm) allow a 2500 analog phone to be attached as well. On the port side of the PBX several arrangements are also possible. On some models, a dedicated data pair is terminated directly on a data card; on others, a multiplexed voice-data stream is demuxed and then split. On the DCE side of the PBX, the PBX and host

front-end processor (FEP) are similarly connected. Commonly, a port card with RS-232C connects with a RS-232C port on the host FEP. If distances over 50 feet are involved, an additional RS-232C connecting "line driver" (signal amplifier) may be employed. The direct advantage to this scheme over the traditional, dial-up modems is that transmission speeds up to 19.2 Kbps are possible, and, where phones are present, the station-to-switch connection can be used simultaneously for voice.

Port selector with PBX-provided coax elimination. A slightly more complicated case involves the port selector function in conjunction with high speed devices and the elimination of coaxial cable. Some PBX's support coax elimination directly. AT&T, Northern Telecom, and InteCom currently support 3270 terminals in this manner. InteCom expects to add support of Wang shared-logic word processing workstations soon. Other vendors are likely to follow. In this case, the PBX directly time-slots the data between the workstations and host. This involves a special balun (resistance balancer) from the workstation to the electronic voice-data station and high-speed data cards on both the port and DCE sides of the PBX connecting to the terminal controller. With this scheme, coaxial cable is eliminated and terminal moves and changes are facilitated.

Coax elimination using PBX wiring. A less elegant alternative solution is to use a spare twisted pair of the telephone wiring, bypass the PBX, and hook directly to a cluster controller or host FEP. In this scheme, for which IBM and Lee Data provide equipment, a special cable connects the workstation, a balun (which varies by workstation type), and a data connector, which attaches to two twisted pairs. Then, at either a wiring closet or the main distribution frame cross connects, the pairs are terminated by another balun and connected to the computer equipment. One should bear in mind that permissible device distances between terminals and controllers are altered under this scheme, as

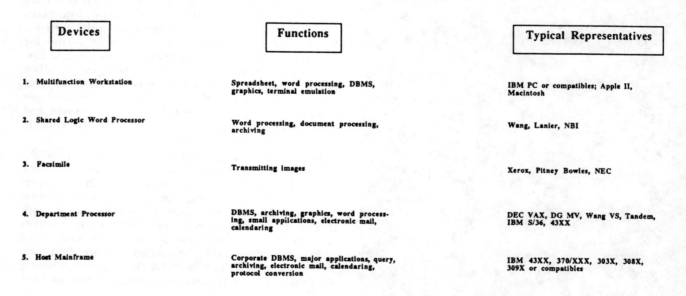

Devices	Functions	Typical Representatives
1. Multifunction Workstation	Spreadsheet, word processing, DBMS, graphics, terminal emulation	IBM PC or compatibles; Apple II, Macintosh
2. Shared Logic Word Processor	Word processing, document processing, archiving	Wang, Lanier, NBI
3. Facsimile	Transmitting images	Xerox, Pitney Bowles, NEC
4. Department Processor	DBMS, archiving, graphics, word processing, small applications, electronic mail, calendaring	DEC VAX, DG MV, Wang VS, Tandem, IBM S/36, 43XX
5. Host Mainframe	Corporate DBMS, major applications, query, archiving, electronic mail, calendaring, protocol conversion	IBM 43XX, 370/XXX, 303X, 308X, 309X or compatibles

Fig. 1. Devices, functions, and typical representatives.

	-Media-	Level 1 -Physical-	Level 2 -Data Link-	Level 3 -Network-	Level 4 -Transport-	Level 5 -Session-	Level 6 -Presentation-	Level 7 -Application-
Multifunction Workstation	Twisted pair, various coax	RS-232C; with cards, numerous options	Asynchronous; with co-processors, HDLC, SDLC, BSC	Can be used on circuit & packet-switched networks & LANs	In terminal or cluster controller emulation modes, ASCII/asynchronous, 3278, 3274; upper layers host & application-dependent			
Shared Logic Word Processor	Various coax	Various proprietary	Proprietary; optionally, HDLC & BSC	Proprietary LAN with external 3278/4 or asynchronous bridges	In terminal or cluster controller emulation modes, ASCII/asynchronous, 3278, 3274; upper layers host & application-dependent			
Facsimile (Categories I-IV)	Twisted pair, usually shielded	RS-232C, V.35	Proprietary & HDLC	Can be used on circuit or packet-switched networks	Conformity to one or more CCITT (Category I,II,III,IV) capabilities under proprietary upper layers			
Department Processor	Twisted pair, usually shielded	RS-232C, RS-449, Ethernet, proprietary channel	Asynchronous, HDLC, BSC, SDLC & Ethernet	Can be used on circuit or packet-switched networks or LANs	Limited SNA (PU 2, 2.1, LU 2, 6.2) capabilities, or pre-SNA device emulation (BSY 3274, 3276, 3777, 3780, HASP) with upper layers host determined; if 370 architecture, full SNA or pre-SNA			
Host Mainframe	Twisted pair, usually shielded, various coax	RS-232C, RS-449, Ethernet, proprietary channel	Asynchronous, HDLC, BSC, SDLC & Ethernet	Can be used on circuit or packet-switched networks or LANs	Various implementations of SNA or pre-SNA functional layers			

Fig. 2. Device communication capabilities.

the twisted pair run can only be about half the maximum distance of a coax run. To overcome this limitation, IBM offers a fiber optic channel extension kit (the IBM 3044) which allows the 3274/76 controllers to be moved closer to the terminals they support. Although the voice/data co-existence occurs only in the station-to-closet or station-to-main distribution frame wiring, and supported distances become less, this less elegant scheme both gets rid of coaxial cable and greatly eases future terminal moves and changes, although manual wiring reconnects are necessary at the closet or main distribution frame.

Port selector with out- or in-bound modem pooling. Another extension to the PBX-implemented port selector, resource sharing, has a number of advantages, providing users with a hierarchy of modem speeds, saving on long distance accesses, and centralizing the devices for maintenance. Administratively, the port selection can be implemented in several ways. In a "hardware" approach, the user can dial different numbers for different computer/speed combinations. Alternately, in a "class of service" or software approach, the user can dial a unique number for each computer, and the PBX will employ the fastest modem supportable by the connection, terminal, and host computer. For those accessing distant hosts, the PBX offers the additional advantage of economically routing dial-out data calls via its Advanced Route Selection (ARS) software. Physically, on the station side, the setup is the same as with other port selector roles: the electronic voice/data set connects the terminal with the twisted pairs via RJ-11 to the electronic voice/data set by RS-232C. A transformer that powers the line driver plugs into an electrical wall outlet. Call set-up varies by manufacturer with several sequences popular. On some, the user turns on the terminal, hits a data button on the voice-data set, and dials the station number. On others, the user turns on the terminal and types in the called

number on the keyboard. Some allow both. On the network side, the modems are connected between the trunk side data ports and the analog trunk groups.

Port selector with X.25 PAD. X.25 packet assemblers/disassemblers (PAD's) are accommodated in a manner very similar to the in- and out-bound modem pool—and are potentially modem pool users themselves. Depending on the number of simultaneous connections supported, port selection for the PAD can be a single "hardwired" number, multiple numbers, or a single number associated with a "hunt group" of lines assigned to the PAD. As with other PBX port selector services, an array of voice management features is available: camp-on, busy line call back, automatic call queuing, and recorded messages, depending on the contention/queuing philosophy.

Port selector/device emulation/protocol conversion. Incorporating protocol conversion capabilities into the PBX switch is consistent with accessing multiple hosts and reducing the number of physical devices on a user's desk. Most typically, this involves converting line-oriented asynchronous ASCII terminal streams into an IBM EBCDIC coded, 3270 full screen mode, 3270 bisynchronous, or 3270 synchronous data link control (SDLC) format.

The usual approach is for the PBX to perform the three conversions (ASCII to EBCDIC, line to screen, and stream to block) and to emulate a 3274/76 controller (Physical Unit 2, Logical Unit 2) to the FEP. (With the increasing popularity of IBM host-based Distributed Office Support System (DISOSS) services, such as electronic mail, revisable format document interchange, and peer-to-peer links, carrying multiple sessions via Logical Unit 6.2/Physical Unit 2.1 is likely to be added to the PBX protocol conversion repertory.) In sum, via the PBX's emulation/protocol conversion, an ASCII/asynchronous workstation may connect to various remote IBM hosts and interact as a monochrome 3278, or, in some cases, as a color 3279 terminal where the physical connection is dedicated or dial-up.

Bandwidth allocation via multiplexing. The employment of digital T1 (1.544Mbps) trunks has become increasingly popular in the larger private networks and can be expected to become more so with multiplexers that allow direct—and possibly dynamic—user bandwidth allocation. Conventional techniques place the muxes on the DCE side of the PBX and divide the bandwidth on a preallocated basis, with voice channels going into the PBX and the data into FEP's. This is likely to change for three reasons. First, modern PBX's have considerable data switching capability and may, along with their voice station wiring, furnish the connection medium to the workstations. Second, multiplexers are available that allow scheduled allocation (for instance, voice during the day and channel-speed data transfers at night) or dynamic allocation (for example, a dial-up 2 to 3 p.m. video conference, where preempted voice traffic is routed through the common carrier network). In any of these integrated voice-data contexts, the bandwidth request will begin with a voice-data station call, requiring close integration of PBX and multiplexer functions. Third, given the cost of T1 links, one-time costs in multiplexer intelligence that offer substantial and recurring utilization benefits become extremely attractive. All modern PBX's today offer T1 interfaces, and scheduled bandwidth allocation schemes are commonplace, though most often implemented by a network control group.

High-speed data services. High-speed data services with data rates over 64Kbps are, as mentioned earlier in 3270 cable replacement schemes, still the exception among modern PBX capabilities. Nor is it certain how much demand exists, beyond the 3270 cable replacement market. Relatively few present-day offices employ LAN's where the devices themselves (workstations, facsimile machines, printers)—as opposed to the signaling used on the LAN—are capable of transmitting at speeds exceeding 64 Kbps. The requirements situation may, however, change. Should the PBX makers decide to provide high-speed support, several approaches are possible. Some, such as InteCom, may have sufficient bandwidth in the main switch to support high-speed data within their present switch design. Others, such as AT&T's System 85 and Northern Telecom's Meridian, have already announced data LAN services via an attached processor concept, where data is split off at the switch backplane and passed to the extra processor. In the AT&T scheme, where data devices are remote and clustered, a fiber optic link directly connects the 327X type controllers to the attached LAN processor.

More probable is that the impetus for high speed support (3270 polling, Wang shared-logic word processors, Ethernet, IBM token ring) may come for more mundane reasons. Data devices, taken as a class, have been very inferior to voice devices in several respects. They have been expensive to install and hard to move, with IBM 3270 coax the prevalent *bête noire*. Also, the data gathering regarding their use has been extremely deficient; the emphasis has been placed, rather, on diagnosis and service routines to keep them working. Few cable-based LAN's produce easily available statistics on usage. Strongest in this area are those data PBX's that provide Station Message Detail Reporting (SMDR) records on all calls. Most voice-data PBX's

	Multifunction Workstation	Shared Logic Word Processor	Facsimile	Departmental Processor	Host Mainframe
Multifunction Workstation	Terminal-to-terminal: PBX as port selector	Terminal-to-terminal: PBX as coax eliminator, protocol converter		Terminal-to-host: PBX as port selector, possible protocol converter, coax eliminator	Terminal-to-host: PBX as port selector, protocol converter, device emulator, poss. PAD, A/D, coax eliminator
Shared Logic Word Processor		Possible terminal-to-terminal: PBX as port selector, coax eliminator		Terminal-to-host: PBX as port selector, possible protocol converter	Terminal-to-host: PBX as port selector, protocol converter, device emulator, poss. PAD, A/D, coax eliminator
Facsimile			Device-to-device: PBX as port selector, possible PAD, A/D conversion		
Departmental Processor				Host-to-host: PBX as port selector, protocol converter, possible mux, device emulator, PAD, A/D conversion	Host-to-host: PBX as port selector, protocol converter, possible mux, device emulator, PAD, A/D conversion
Host Mainframe					Host-to-host: PBX as port selector, protocol converter, possible mux, device emulator, PAD, A/D conversion

Fig. 3. PBX-provided device connection matrix.

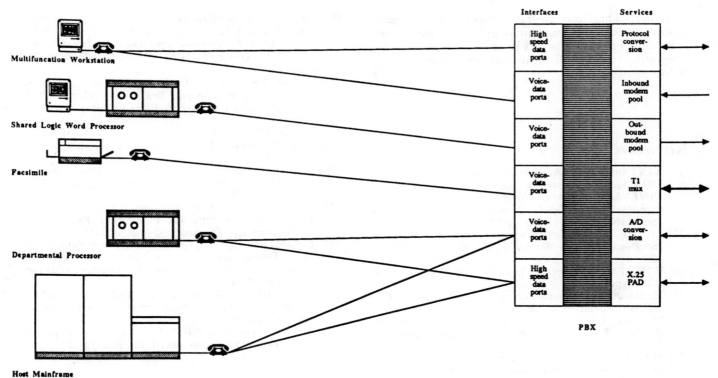

Fig. 4. PBX as voice-data integrator.

produce peg counts on feature use (such as data calls) on inside (4-digit) calls, and full SMDR records on outside (7 or more digit) calls. In sum, the mundane advantages of PBX-provided voice-data integration—the multiple use of extant wiring and the ability to collect limited usage data, impose classes of service, enforce security, and provide other low-tech, low-glamour administrative features essential to management control—may tilt decision makers in the direction of PBX's for voice-data integration in the office.

Figure 3 provides a device connection matrix wherein the PBX role is summarized. Figure 4 diagrammatically depicts the PBX functions to supported devices. As the figures illustrate, the initial predicament posed—that of interconnecting disparate devices (Wang, IBM PC's, office minicomputers) within the office, and between the office and the remote IBM mainframe—can be done today on a modern PBX with modest incremental costs. In most cases, the salient cost item is upgrading the individual's voice-data station. Although prices for integrated voice-data stations are falling, many are still in the $300–500 range, and, unlike data items located at the switch (line cards, port cards, modems, PAD's, protocol converters), they are neither shared resources nor do they service multiple users. Nonetheless, when compared with the annual costs of moves and changes for data devices, redundant terminal devices, redundant data entry, and the accompanying lack of usage statistics and management control, PBX upgrade costs can seem modest.

Despite the advantages of employing the PBX as an office voice-data integrator, several limitations need to

be stressed relative to this level of voice-data integration. Unlike voice, connectivity does not imply transparency, and the PBX data connection functionality is at a relatively low level (see Fig. 2). For example, if the user is employing an IBM PC to run Lotus 1-2-3 and wishes to access a department DEC VAX for electronic mail, and then to access the latest corporate data on a remote IBM 3081, the physical connectivity—dialing the right numbers for the DEC and IBM processors—is the easy part. He or she must be knowledgeable enough in PC DOS to exit Lotus and load the ASCII communications package, familiar enough with DEC VMS to access the mail package, and sufficiently savvy in CICS/MVS to get to the IBM host's DBMS, enter the DBMS, and get what is needed. As noted in Fig. 2, the top three protocol layers are visible, vendor-unique, and, for security reasons, often unforgiving. It is a speculative statistic, but it is unlikely that the number of computer users who regularly grapple with three different operating systems on a daily basis exceeds one percent. The return argument is that exposure to alien operating systems can be minimized via macros, particularly for such trivial functions as sending and receiving electronic mail, posting or checking the latest data base numbers, and the like. Despite the bridging strategies, highly dissimilar, upper-level protocol user interfaces ensure that training costs will remain high and usage low.

In sum, such device connectivity and primitive levels of voice-data integration represent early points of progress down the path of information integration and presentation, not to speak of cognition and understanding. They nonetheless represent a promising beginning offering clear and quantifiable benefits, and a modern PBX enables this progress to proceed incrementally, economically, and at an understandable velocity.

References

[1] AT&T Information Systems, *Reference Manual DIMENSION System 85 Data Management, Reference Manual DIMENSION System 85 Voice Management,* and *Reference Manual DIMENSION System 85 Hardware* (Indianapolis, IN: Western Electric, 1983).

[2] Ericsson Information Switching Systems, *MD110 General Description* (Garden Grove, CA: Ericsson, February 1983).

[3] GTE, *The GTD-4600-E Digital PBX* (McLean, VA: GTE, April 1984).

[4] Harris, *20-20 Integrated Network Switch* (Novato, CA: Harris Corporation, undated).

[5] InteCom, *General Description IBX S/80 Integrated Business Exchange* (Allen, TX: InteCom, February 1983).

[6] Mitel, *SX2000 General Description* and *SX2000 Features and Services Guide* (Boca Raton, FL: Mitel, 1984).

[7] NEC, *NEAX 2400 IMS General Description* (Melville, NY: December 1983).

[8] Northern Telecom, *SL-1 Business Communications System, Generic X11 Feature Document Business Features* (Santa Clara, CA: Northern Telecom, March, 1983), *Electronic Switched Network Feature Document* (Santa Clara, CA: Northern Telecom, March, 1984), and *SL-1XN System Description* (Santa Clara, CA: Northern Telecom, April 1984).

[9] ROLM, *CBX II Business Communications System* (Santa Clara, CA: ROLM, October, 1983) and *9000 System Data Communications Feature, System Administrator's Guide* (Santa Clara, CA: ROLM, September 1983).

[10] United Technologies Lexar, *UTX-5000 Voice/Data Switching System General Description* (Westlake Village, CA: Lexar Corporation, February 1984).

[11] Western Union, *Vega System General Description Manual* (McLean, VA: Western Union, March 1984).

[12] Ztel, *PNX System Description* (Wilmington, MA: Ztel, May 1984).

[13] S. L. Junker and W. E. Noller, "Digital Private Branch Exchanges," *IEEE Communications Magazine* (May 1983), pp. 11-17.

[14] E. R. Coover and M. J. Kane, "Notes from Mid-Revolution: Searching for the Perfect PBX," *Data Communications* (August 1985), pp. 141-150.

[15] N. Janakiraman, "An Overview of Recent Developments in the Designs and Applications of Customer Premises Switches," *IEEE Communications Magazine* (October 1985), pp. 32-45.

[16] E. E. Mier, "PBX Trends and Technology Update: Following the Leaders," *Data Communications* (September 1985), pp. 82-96.

[17] A. Sikes, "The Data PBX Solution Local for Networks [sic]," *Teleconnect* (December 1985), pp. 112-125.

Edwin R. Coover has taught at Indiana University, Bloomington, and is currently with the MITRE Corporation in McLean, Virginia. In the last several years he has been heavily involved in the specification of PBX and office automation systems. He holds Masters and Doctorate degrees from the University of Virginia and University of Minnesota, respectively.

SOME ARCHITECTURAL CONSIDERATIONS FOR LOCAL AREA NETWORKS

Roshan L. Sharma

Rockwell International
P. O. Box 10462
Dallas, Texas 75207

Abstract

In order to meet the diverse needs of emerging office workers, the local area network (LAN) architecture must be flexible. It cannot be overly simple to meet only the voice or data needs. The intelligent work stations are constantly being harnessed to meet the needs of office workers and hence increase their productivity. Since approximately 75 percent of the office worker's time is spent on communications, it is but natural to link the user terminals such as telephones, data terminals, graphic terminals and the emerging intelligent work stations. This paper describes the basis for a composite LAN architecture consisting of coaxial cables that use the packet switching techniques and digital Private Branch Exchanges (PBXs) that use the circuit switching techniques.

INTRODUCTION

Just a few years ago, the population of office/information workers suddenly exceeded that of the blue collar workers in America. Unfortunately, this event was also followed by lower productivity of the office workers and the rising costs of labor resulting from inflation, etc. This gave rise to a sudden interest in the drive to achieve higher productivity through the use of office automation technology. Fortunately, during this period, the technological advances had been bringing down the costs of data processing, word processing, information storing/retrieving and communication processing functions. One is no longer limited by large centralized DP centers, expensive storage devices and complex communication interfaces. Instead, distributed information processing (data, text and communication) technology can finally equip each office worker with sufficient information processing power to increase workers' productivity and at the same time achieve some degree of job satisfaction.

In order to fully harness this emerging office automation technology, we still need to consider several systems engineering and network architectural aspects. Recent studies (Ref. 1) of user requirements and habits show us that at least 75 percent of office workers' time is spent in communication with other office workers. This

clearly shows why we must link workstations into not only LANs but also link the LANs to one another and to public data/voice networks. This also clearly shows why earlier attempts to equip secretaries with stand-alone electronic typewriters (ETWs) and word processors (WPs) failed. A later attempt to integrate ETWs, WPs and intelligent WSs into a stand-alone LAN is also turning into an ineffective approach to increase productivity.

Our study of user requirements shows that the need for data communication as compared to voice communication is expected to grow dramatically in the future. This presents a great challenge to system scientists and network architects for an integrated office system (IOS) that can enhance productivity through unfettered communication and office automation for the least expenditure of money. It is the purpose of this paper to present an IOS architecture involving a digital PBX. The manner in which a digital PBX is applied to achieve an overall integration of the IOS within the corporation will also be discussed.

User Requirements and Technology Considerations

American industry is witnessing a crisis in productivity on the one hand and the need to take advantage of the cost effectiveness of information processing (word processing, data processing, and telecommunication) technology on the other. This situation is presenting a unique challenge to all U. S. corporations to plan IOSs that will increase the productivity of their office workers. This challenge is no different from the ones that brought about a significant increase in the productivity of the industrial and farm workers of America through mechanical automation only a short period ago.

In order to appreciate the suggested IOS architecture, one must first study the demographics of the U. S. office workers, then their daily activities and finally the composition of their inputs and outputs.

Figure 1 illustrates the makings of the U. S. office workers during the years 1981 to 1986 (Ref. 2). The distribution of office workers in Japan is quite different. The number of technical personnel (e.g. engineers) per 10,000 office workers is almost 6 times that in the U.S.

Figure 2 represents our compromise models of the activities of a typical office worker in the years 1982, 1987 and 1992, using the sources of Ref. 1.

According to these charts, the present time spent on telephone conversations and desk work (thinking/planning, and organizing) do not change over the years. The time spent on video conferencing assumes a special room and therefore it is similar to a scheduled meeting. Time spent on information gathering/processing is identical to teleprocessing and involves data terminal equipment (DTE), hosts, H.S. graphics terminals and data bases. The times spent on video desk work assumes a video terminal that employs compressed full motion video at 1.5 Mbps for in-office training, data gathering/dissemination, one-way controlled interviews and meetings. Assuming some duty cycles in the form of think times and amounts of bits transmitted using typical data transmission devices, one can obtain the average number of data bits transmitted per second per 100 U. S. workers. The results are shown in TABLE I. It should be emphasized that the results of TABLE I may vary from application to application and from corporation to corporation. But we consider it as a good starting point for planning.

It may be interesting to relate the results of TABLE I to current perceptions of voice-data mixes in terms of either transmission costs or pure bandwidth utilizations. Our models for computing such voice-data mixes assumes the following:

1. Each voice conversation involves 2 long-haul trunks, each rated at 4 KHz, one for each direction of flow.

2. Each data call assumes a 1-way flow of data with bandwidth multiplication factors of 11 and 1 for circuit switching (CS) and packet switching (PS) respectively. Misuse of bandwidth in CS systems is caused by the duty cycles assumed in TABLE I.

Using the results of Figure 2, TABLE I for data requirements (with no video) and the above assumptions, one obtains the ratios of data and voice bandwidths as shown in TABLE II.

The TABLE II results for 1982 agree with the current perception that data accounts for 10-20% of total corporation communication costs that are based upon the predominant CS environments. In order to handle the tremendous growth in the user data requirements during coming years without

TABLE I: DATA RATES GENERATED BY OFFICE WORKERS

PARAMETER		1982	1987	1992
		FRACTION OF 8-HRS SPENT		
MEETINGS		.66	.44	.39
TRAVEL		.03	.02	.01
DESK WORK		.22	.21	.21
TELEPHONE		.08	.08	.08
VIDEOCONF.		0	.04	.08
INF. GATHERING/PROC		.01	.14	.09
WP/EM		0	.05	.09
VIDEO DESK WORK		0	.02	.05
TOTALS		1	1	1
	MINUTES	BITS	BITS	BITS
MEETINGS	60	9600	9600	9600
TRAVEL	1440	9600	9600	9600
DESK WORK	60	9600	9600	9600
TELEPHONE	5	1000	1000	1000
VIDEO CONF.	30	9600	9600	9600
INF. GATH/PROC.	.17	9600	9600	64000
WP/EM	10	0	9600	9600
VIDEO D/WORK	.17	0	6400	1540000
		BPS	BPS	BPS
TERMINAL		9600	9600	9600
		9600	9600	9600
SPEEDS		9600	9600	9600
		300	300	300
IN		9600	9600	9600
		9600	9600	9600
BPS		9600	9600	9600
		64000	64000	1540000
Kbps PER	(NO VIDEO)	1.1	12.28	51.8
100 OFFICE WORKERS	(WITH VIDEO)	1.1	23.7	739.3

significantly increasing the transmission costs, one should employ the PS technology which is ideal for bursty data traffic. Of course exceptions could be made for cases where majority of the data traffic is between intra-plant work stations or shared DP resources.

TABLE II: (DATA BW)/(VOICE BW) IN EITHER PS OR CS SYSTEMS FOR 1982, 1987 and 1992 (FOR NO VIDEO CONTENTS)

1982		1987		1992	
CS	PS	CS	PS	CS	PS
19%	2%	244%	20%	938%	81%

Another interesting aspect of user requirements is the geographical ranges of various communities-of-interest as shown in Figure 3 (See Ref. 3). These results show that up to 60 percent of all communication traffic takes place within a building or local complex. Only 8 percent of the user traffic goes beyond the 500 miles distance. These facts pinpoint the need for local area networks (LANs) within the corporate network architecture. Both the digital PBXs and conventional data-oriented LANs should play a significant role in private networks.

The factors dealing with the office worker requirements should create a tremendous increase in the demands for intelligent work stations (WSs), LANs and digital PBXs (influenced by increasing office worker population and replacement of older PBXs). Figure 4 attempts to focus on the dynamics of the new market place. The phenomenal growth in the WSs and LANs especially, and integrated office systems (IOSs) in general can finally be visualized as the natural extension of the VLSI/Distributed DP technologies.

The Solution

A great deal of consensus is emerging among the major vendors such as American Bell Inc., IBM/Rolm, Northern Telecom, Harris etc. (See Ref. 4, 5, 6, 7) that an Integrated Office System (IOS) architecture as shown in Figure 5 is the most appropriate for handling the user requirements discussed earlier. According to this architecture, a digital PBX is destined to be the integrating element of a multi-node local area network (LAN) and also a multi-node wide area network (WAN) belonging to a corporation. There are two main reasons for this architecture to merge: cost effectiveness in handling both voice and data, and easy interfacing to public voice/data networks. Furthermore, a digital PBX forms an excellent starting point around which all of the integrated office systems of the future will evolve during the next few years.

It should be instructive to define some key concepts before describing the applications of a digital PBX in local area networks (LANS) or corporate wide area networks (WANS).

A digital PBX in this discussion implies a fourth generation development which allows a full integration of voice and data through a stored program controlled (SPC) wide-band (>64 Kbps per port) digital switch. The preceding generations being (1) hardwired step-by-step electromechanical switch-based, (2) SPC controlled cross-bar switch based and (3) distributed microprocessor controlled digital switch-based.

A local area network (LAN) may consist of one or more individual networks based on either PS technology (e.g. baseband bus with either CSMA/CS or token architecture), or CS technology (e.g. a digital PBX). A wide area network (WAN) may consist of 2 or more LANs as defined above. A LAN usually spans a building or campus. A WAN generally spans several cities. Implementation of a WAN will require (1) leasing of transmission facilities from a common carrier (e.g. AT&T) and/or specialized common carriers (e.g. MCI or SPCC) and/or (2) buying services from a value added network (VAN) such as Telenet.

The digital PBX is assumed to possess sufficient intelligence to at least perform the following network-oriented tasks:

1) Subscriber-to-subscriber switching with multitudes of features such as 1-5 digit dialing, authorization codes, queuing, call transferring, conferencing, etc. An integrated work station may provide a more friendly access to these features.

2) Subscriber-to-off-net locations calling via public central office (CO) trunks.

3) Tandem Overlay Option that allows uniform dialing among subscribers located in different cities, head-end-hop-off (HEHO) or tail-end-hop-off (TEHO) routing to off-net locations, automatic route selection (ARS) to choose alternate routes when the primary route is busy, trunk-to-trunk routing, and common channel interoffice signalling (CCIS) preferably based upon the emerging CCITT #7 standard. The CCITT #7 signalling will also enable interfacing with the emerging Integrated Services Digital Network (ISDN) which will be a near-future enhancement of all public voice networks to handle both voice and data. This tandem overlay option should also allow off-net users to access on-net and other off-net users of the private WAN. For some large

corporations, one or more stand alone tandem nodes (sometimes called Super Controllers) may be necessary for implementing an economical WAN. The need for a CCITT #7 derived signalling scheme becomes even more obvious.

4) Integrated data and voice switching capability which allow the sharing of a switch port by voice and low-to-medium speed ($\leq$ 9600 bps) asynchronous data or a full port utilization for high speed ($\leq$ 64Kbps) synchronous data. Since data terminals usually transmit information in a bursty fashion on local CS facilities, bandwidth misutilization (by a factor of 11) and long holding times ($\geq$ 1 hour), the PBX switch should have a non-blocking architecture. Otherwise a good deal of traffic engineering will be needed to maximize the utilization of the switch bandwidth. The availability of HS synchronous data switching on some 3rd generation and all 4th generation PBX enables exchange of information between local terminals without the need of modems.

5) Voice messaging capability that allows users to forward their incoming calls automatically to a voice message storage node in the LAN. Although this capability appears quite exciting to many, its long term effectiveness within the office-of-the-future remains to be proved.

6) Traffic logging and resource utilization reports for all WAN resources provide the controls necessary for a corporate network. Station Message Detail Recording (SMDR) with either automatic number identification (ANI) or authorization codes at a centralized location and timely LAN/WAN resource utilization reports help ensure smooth WAN operation and cut communication costs. The 100% traffic logging on tapes not only helps assign costs to network users but also obtain global WAN optimizations at proper intervals.

7) Protocol conversion capability will ultimately enable the digital PBX to act as a bridge between divergent trends in distributed DP. But it will take many years before such a capability will be realized. In the meantime, special outboard devices will be employed to achieve protocol conversion and format translation required for exchanging information between myriad of existing and emerging data terminals and intelligent work stations. The emerging International Standard Organization's (ISO) open system interconnect (OSI) architecture will accelerate the development of the above-mentioned universal PBX.

Application of Digital PBX to a Corporate LAN/WAN

To justify a digital PBX as the controlling hub of an Integrated Office System (IOS), one must first show how all the activities of an office worker will be improved through such an IOS. Following set of applications are considered:

1) Telephone Calls: Since about 90 percent of the total telecommunication costs of a typical corporation are voice related, an IOS must enable office workers to place telephone calls in a much more friendly manner than possible today. To achieve this goal, an intelligent work station (WS) and a digital PBX are essential. An intelligent WS can help in directing hook-ups, repeat dial-up and using any special PBX feature. A well-planned WAN consisting of many LANs can also provide economical inter-city telephone calling. This will require the tandem overlay, alternate routing (including day-time routing) off-net access, 100 percent call logging, etc. In some cases, 1 or more stand-alone tandem switches may be involved in the WAN. Such an environment can't help but stimulate productivity and provide job satisfaction to any office worker.

2) Data Calls: A digital PBX can enable information exchange between a local data terminal equipment (DTE) and a local host processor (HP) on a CS basis. Applications such as teleprocessing and data base entry/retrieval can thus be accomplished. If the HP is a large system and tied to other HPs and expensive storage/printer devices via a Bus based LAN one will need a PS interface between the PBX and the LAN. If the HP is located in another city, one will need an interface to either ISDN or a public data network (e.g. Telenet, Tymnet). One could employ the existing PBX interface to the public voice network to establish a CS data call but the misuse of transmission facilities should discourage such a solution. Data calls between local WSs and DTEs should also enable EM. If intelligent WSs are employed, word processing (WP) capability can be combined with EM capability. Combination of WP, EM and data calls can really bring about significant savings and faster communication between office workers. Information gathering and processing requirements of office workers can also be accommodated via data calls between WSs, DTEs and HPs. The H.S. graphics and database entries and retrievals also come into this area.

Data calls between large host processors may require data rates approaching 1.54 Mbps (or even higher rates). An ideal solution will be PS-LAN with wide-band bus architecture (e.g. Xerox's Ethernet$^{(TM)}$ and Network Systems Hyperchannel$^{(TM)}$). For file transfer applications (implying longer call durations) and local HPs, one may be able to transfer data at 1.54 Mbps via digital PBXs if port-strapping capability exists.

3) Video Desk Work: Even today, an office worker spends a good deal of time in reading, learning and attending special courses to improve his/her skills. The special training courses demand strict schedule adherences. A work station equipped with a multi-purpose screen can receive slow-scan full motion video at 1.54 Mbps from a HP-based system which is capable of sending any number of courses to any one or more office worker. A freeze-frame video-call can

also be established at 64Kbps. Using the arguments of the preceding discussion on data calls, a digital PBX can be employed for this application that is bound to grow in popularity with time. Any office worker that cannot keep up with the advancing technology is bound to join the unemployed segment of our population.

This capability can also be used to set up a 1-way video call between a prospective client or job application or any office worker who wants to communicate with an employee with slides or any visual material. This 1-way video call set up requires only one audio-visual conference room. Such a capability can really save travel expenses. Of course, the 2-way video-conferencing capability will also help reduce the travel expenses. However, in that case, two or more very expensive video-conference rooms equipped with wide-band video transmission and high fidelity will be required by the corporation. Furthermore, the utility of the digital PBX becomes questionable due to very high bandwidths necessary.

The preceding discussion should illustrate the manner in which a digital PBX can be applied to increase the productivity of an average office worker and increase his/her job satisfaction. As the technology explodes, additional applications are bound to arise within the above mentioned three categories of voice, data and narrow-band video calls. But in any case, a digital PBX is destined to remain as an integrating element to any IOS since it provides easy connectivity and control for all the geographically scattered IOSs of a corporation.

CONCLUSIONS

An attempt was made to characterize the office workers during the period 1982-1992 in terms of their daily activities, and requirements of voice, data and video communication and information/textual processing. Based on this model, an architecture of an integrated office system (IOS) was discussed. It was shown a digital PBX forms a central hub that provides controlled interfaces to data oriented LANs, public voice/data networks, private communication networks, and the emerging Integrated Services Digital Network (ISDN). Within the constraints of this architecture, several applications of a digital PBX were shown to help office workers to improve their productivity and achieve work satisfaction.

References

1) Bell Northern Research Data from J. H. Blair covering 2.5 years, 691 people and 7 organizations.

Bolt Beranek and Newman Inc. Cambridge, Mass., data from the 1980 Report "Electronic Mail: The Messaging System Report".

Booze Allen and Hamilton figures from G.Tellefsen based on a 1-year study of 300 people in each of 15 organizations.

IBM/SRI data from A. Martin represents SRI's synthesis of an IBM study of time usage in a single large manufacturing organization, 1979.

2) Intl. Data Corporation Report and U.S. Dept. of Commerce Statistics.

3) J. Boyd, "Communication Cost-Containment Through High Technology", The Jrnl. of Telecommunication Networks, Date Unknown, 1982, p. 95.

4) R. Violono "Baby Bell's 1st Step: Digital PBX, LAN", Information Systems News, Monday, Jan. 24, 1982 p. 1, 14

5) R. Violono "IBM Seen Aiming at Big Communication Role", Information Systems News, Exact Date Unknown, 1982.

6) A. Negrin, "Today's PBX Trends and the Switch of the Future: A Slightly Reactionary View", Communication News, February 1983, p. 28.

7) D. Hudson, "PBX Will be the Integrating Element of Multi-Node Information Systems", Communication News, August 1981, p. 22.

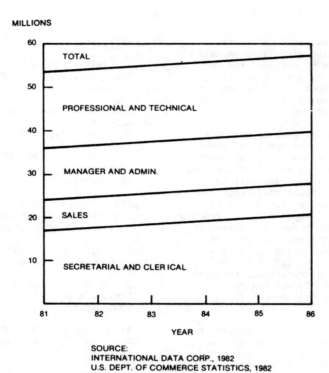

MILLIONS

TOTAL

PROFESSIONAL AND TECHNICAL

MANAGER AND ADMIN.

SALES

SECRETARIAL AND CLERICAL

YEAR

SOURCE:
INTERNATIONAL DATA CORP., 1982
U.S. DEPT. OF COMMERCE STATISTICS, 1982

FIG. 1: CATEGORIZATION OF U.S. OFFICE WORKERS

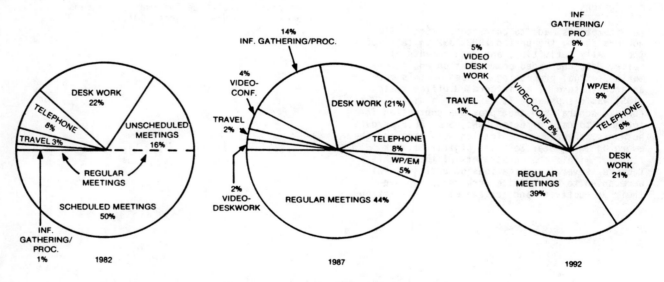

FIG. 2 DAILY ACTIVITY PROFILES OF OFFICE WORKERS

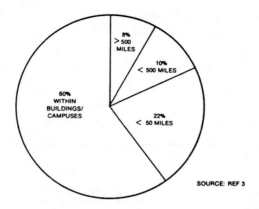

FIG. 3 GEOGRAPHICAL RANGES OF BUSINESS COMMUNICATION COMMUNITIES-OF-INTEREST.

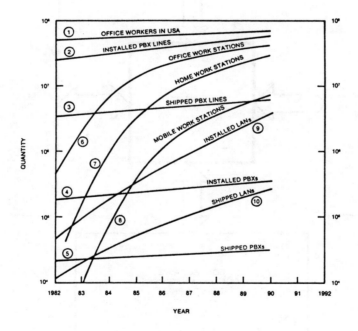

SOURCES:
①,⑥,⑦,⑧ . Scientific American, Sept. 1982, p.149
②,③,④,⑤ . Eastern Mngmnt. Gp.:PBX MKTS/Competitors:A 10 Year Strategy Analysis,Nov. 1982
⑨,⑩ . Inf. Systems News, Jan. 10, 1983, p.37

FIG. 4 DYNAMICS OF THE IOS MARKETPLACE

127

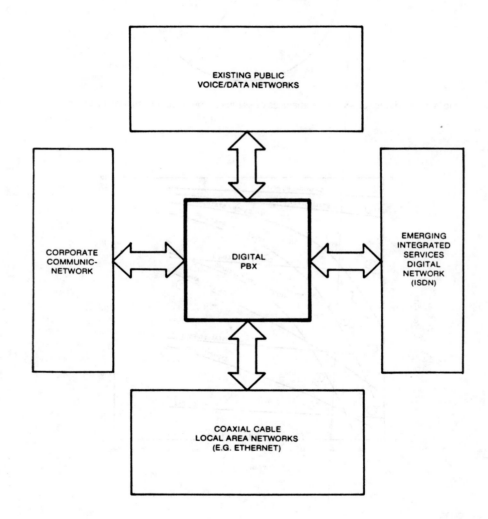

FIG. 5: INTEGRATED OFFICE SYSTEM ARCHITECTURE

Don Thomson, Rolm Canada Inc., Willowdale, Ont.

Helping data managers find their voice

In the wake of divestiture, versatile PBXs are flooding the market. What functions should a communications manager look for?

Many data communications managers, thrown into supervising PBX operations, now have to bone up fast on what voice network officials have known for years—how to keep telephone costs low. While familiar digital transmission capabilities of modern PBXs can start data communications managers hopping, various telephony features can set heads spinning. And the ongoing merger of voice and data networks requires an understanding of perhaps alien equipment and features.

Since the breakup of AT&T, businesses might think they need a full-time accountant to figure out their telephone bills. First there is the depreciation, lease payments, or capital cost allowance on the newly acquired telephone gear. Then there is AT&T's regular long-distance bill, not to mention bills from WATS (Wide-Area Telecommunications Service), MCI, Sprint, or other common carriers. Last but not least is the local Bell bill for local trunks.

Just around the corner is yet another twist—equal access. With this feature, the local Bell operating companies must make available full access to AT&T and the other common carriers on regular local Bell lines. By dialing a long-distance prefix, the business user can be connected to AT&T or one of the other common carriers. No longer will the business user require special trunks or complicated dial-up access codes and authorization sequences.

How can communications managers manage the new PBX, this supermachine with all its potential, before it becomes a monster? That depends on what type of business wants it. For example, any consideration of PBXs will eventually take in the question of voice and data integration. If the office is like most, requests for computer or display terminals are depleting computer resources.

Many of these new requests are from employees who are not typical users, in that they are only occasionally on the network. They need to get into a multitude of different applications, which reside in several different mainframes, both inside and outside of the building. These users, who are neither programmers nor data-entry clerks, spend from 10 to 40 minutes per hour at the terminal. Response time is not critical to them. They can continue to use terminals hardwired with expensive coaxial cable, the same as used for the other users, and pay for moving them throughout the complex. Or the new PBX can act as a central hub or controller for these occasional-use devices. Speeds of 19.2 kbit/s asynchronous and 56 kbit/s synchronous are quite common using the normal twisted-pair wiring that forms the backbone of the telephone network. In essence, the PBX acts like a port selector.

Used in conjunction with modem pools, protocol converters, and gateways, these voice/data PBXs can allow more flexibility and control than a full coax network. Many of the vendors have X.25 packet-switching interfaces for either private or public data network access. Ask to see a working demonstration. Many vendors can provide intelligent terminals or microcomputers that are fully integrated with the telephones.

As it does with any major purchase that a company would consider, it makes sense to put together a business plan for the new PBX and its related communications equipment. The first stage in this business plan is to determine what is now on hand, what it costs, and how much growth remains in the present equipment. Current telephone bills for equipment should help. And if there isn't one already at your company, ask the local telephone supplier for an equipment record and have someone in your organization do a

Add to this a minute to tell the joke, and the company has lost an employee's productivity for a year.

physical inventory of all the sets, bells, whistles, and the like. Make sure that the inventory matches the equipment record because the company may be paying for equipment that was removed years ago and has been long forgotten.

The next step is to determine how much money is being spent on long-distance calls. In most organizations, long-distance charges can usually exceed the monthly equipment cost. Most of the older PBXs still in place in North America cannot track calls. In many states, message units for local calls can mount up quickly. An analysis of the telephone bill may lead to a big surprise.

Analyzing the telephone bill

How do communications managers analyze their telephone bills? The first step in this complicated procedure is to be certain that all the information is available: Long-distance, WATS, Sprint, MCI, and other common-carrier bills, as well as any other foreign exchange or tie-trunk bills, form an essential part of the process. Have a WATS service area map at hand—available from an AT&T or local Bell office—to help determine the categories.

Next, break down the calls into the various areas. This can be done in a variety of ways. Many firms use a microcomputer, while others do it by hand. If the analysis is being done manually, use colored highlighter pens, one for each different area. The five areas that need to be analyzed are the date, duration, cost, type of call, and area code of each call. Have someone mark each call to a single service area in a separate color. Do one complete area first. When finished, add up the number of calls, their duration, and the cost by area. Break this down for each day.

If your company already has WATS service, use the WATS bills to determine the length of an average call. If the firm doesn't have WATS, it might help to conduct a time study on calls. Have the participants of the study track the length of long-distance calls to the nearest 10 seconds. This should be close enough to get started. A random sample of 5 to 10 percent for two to three days should be sufficient. Remember, Bell will round up the call duration to the next minute. If the average long-distance call is, say, two minutes and 20 seconds, Bell rounds this up to three minutes. Depending on the calling volume, WATS service may save up to 25 percent.

If the telecommunications manager is using a microcomputer, use a spreadsheet program that is capable of sorting data. If your telephone bills are too large, it may be more cost-effective to use a service bureau of a manufacturer or distributor who maintains this type of

center for this purpose. The charges may be nominal, and often the payback is more than justified as savings to the company.

With WATS tariffs in hand, simple calculations can help determine whether WATS service will save more or less money than other common carriers. This will depend on several factors, including the calling patterns and the quality of service required. Depending on the distances between calls, WATS may save money, but it will not save money if many calls span short distances or are brief in duration. Calling patterns should be examined in terms of how geographically spread out they are: If one-sixth of the calls fall into each zone, then you may never reach the reduced calling prices that result from having the necessary number of calls on each trunk.

With "900" service, which takes the form of calls to area code 900 in some areas, and to a 9XX exchange in others, a new wrinkle comes into play. These 9XX calls are the service and information lines used for dial-a-prayer, dial-a-joke, dial-the-weather, and so on. Areas where area code 900 is used are easy to locate on the telephone bill. However, in other areas, the number may look like a local call with the prefix 9XX. How do they work? Callers are billed a flat amount, usually 50 cents for each minute, plus any regular applicable long-distance charges.

A recent analysis of a company in California that had 300 employees showed that on an average day, there were over 150 calls to these numbers. With an average 22 working days a month, this amounted to nearly $20,000 in a year. Adding in the average 30 seconds it takes to dial the number and make the connection, time lost to the company amounts to half of a work year. Add to this the roughly minute and a half that it takes the caller to tell someone else the joke, the prayer, or the weather report, and the company has lost an employee's productivity for a full year. Combined with the cost of the calls and the fact that a trunk line is tied up for three to four hours each day, the cost to the company added up to almost $45,000. How to explain this expenditure to stockholders? (This California company didn't have any call-detail recording on its equipment, so the calls made to the 9XX number on the WATS, FX [foreign exchange], and tie trunks were not shown.)

The reason to check off the type of call is to distinguish regular calls from credit-card calls. In the California company cited above, over 5 percent of its calls were made with credit cards from outside. And more than 70 percent of those calls were made from the local calling area of its office from pay telephones. Almost 30 percent were local calls for which the callers didn't have the correct change. The use of a credit card invokes a rather stiff premium, especially on local calls. Two solutions could have eliminated almost $400 of misuse for this company. First of all, install a DISA trunk. The DISA (direct inward-system access) is usually available as a software option on most modern PBXs. DISA allows employees to call into the PBX without the intervention of attendants and enter a security or account code. When these are entered, the callers' Touch-Tone telephone can use the least-cost routing features of the PBX. Low-cost WATS, or other common-carrier service, can be reached.

Long distance

Having a mechanism in place that automatically tracks all outgoing calls and, in some cases, incoming calls, will allow costs to be controlled. But even if these calls can be tracked, there is no guarantee that the calls are being placed properly, taking into account the lower-cost routes, the discount hours, and many other factors. Busy trunk routes in a telecommunications network can lead to people in the organization using more costly facilities when less-costly routes are available. For instance, say your company has an FX line from New York to your office in Miami. When employees are calling the New York area, they should be using the FX line, which is charged at a flat monthly fee regardless of usage. When calling within the New York City area, the calls are made as if the company had an office in New York. When calling from New York to New Jersey, a nominal long-distance charge applies, a charge that is usually considerably less than the long-distance charge from Miami to New Jersey. Depending on calling volume, one or more of these FX lines should be installed. When they are in use, however, many people, upon receiving a busy tone, will immediately use the more expensive WATS or long-distance facilities. Had they been forced to queue for a few minutes, they probably would have gotten through on less-expensive lines without complaint.

Keeping track of which area codes can be called from which WATS band, which cities can be called from foreign exchange lines, and which cities service MCI, Sprint, or other common carriers would take Einstein's memory and the patience of Job. Each and every month Rolm receives the revised telephone company tariffs, including updates on the V&H coordinates and the changes in the NNXs. The V&H coordinates are the way the North American telephone operating companies and airlines keep track of where everything is located. The entire North American continent is divided into a grid with vertical and horizonal lines. The distance between these lines is two to three miles. Each and every city, town, or village that has a telephone exchange is assigned a V&H coordinate. Using a complex formula, the distance between any two centers can be determined without a map. The NNXs are the central office prefixes commonly known as the first three digits of a telephone number. Each area code can have up to over 700 different NNXs, and each NNX can support up to 9,999 different lines. In addition to the Bell tariff information, Rolm also receives regular up-

dates on the area codes and NNXs that MCI and Sprint are serving.

Most of this is inserted into a database program which keeps track of this information. The program takes data from the tariffs, town names, and associated NNXs and puts it into a file where it can be read and sorted by various criteria, including distance from a particular area. On request, least-cost routing schemes, local calling areas, and much more can be determined quickly.

Most of this information is downloaded to a new or existing telephone switch and forms the routing scheme for the route optimization software and the call-costing equipment. Whereas it used to take one company two-and-a-half days to change rate structures manually, it now takes that same company 53 minutes to download new information.

In many cases, businesses putting in a new PBX that has least-cost routing or route optimization have realized savings of 5 to 50 percent in their long-distance bills. Others have found that their actual network costs have gone up. Why? The best guess in these cases is that before the arrival of the new PBX, excessive blockages were encountered. Callers just stopped trying. However, once the new equipment was installed, they were able to get through on a more regular basis. This doesn't mean that costs are going to skyrocket. Initially, higher costs do mean that more calls are getting through, which may result in greater productivity and higher overall profits.

Tracking calls

Consider some means of tracking calls. Most of the new PBXs on the market today can provide some means of call-detail recording. Expect this feature to track outgoing trunk calls and, optionally, incoming trunks calls. Do not expect it to track internal calls. Such a feature may be available, but don't expect it to be cheap. Most businesses do not have the time to do extensive studies into intracompany calling patterns. Leave that job to the individual department managers to control. The amount of real-time that internal call-detail recording would eat up in certain cases could seriously affect operation of the equipment, thereby reducing its efficiency.

Once the company is keeping track of its calls, it must be determined which tracking method to use. Most manufacturers will offer a variety of options in this area. Some will use a standard CDR (call-detail recording) port. Generally, this will take the form of an RS-232-C asynchronous connection, which gives a simple chronological listing of all the tracked calls. This port may be connected to a standard printer, a collection device, or a modem, for downstream processing by a service bureau. Downstream processing means that the reports are collected, sorted, and priced at a location outside of the actual business.

There are several ways to accomplish this, and the size of your business may determine which method you choose. For smaller PBXs, some sort of memory buffer can be located next to the PBX equipment. At regular intervals, the service bureau will place a call through the normal telephone lines and remotely poll the collection device. Another way is to lease a data trade channel from the common carrier that hooks up the PBX to other service bureau computers. In this way, any interruption in the flow of calls can be responded to in a timely manner. Others may have a software package that does some of the sorting and tabulating internally, with no other external processing required. Again, on smaller machines, this may be more than sufficient, but as your PBX grows, memory or real-time limitations may cause a degradation of service below an acceptable level. Many other methods, ranging from storage on floppy disks to huge magnetic tape devices, are available for storing CDR records. Once the raw data is reformatted into a report from external sources to the PBX, even bigger problems await.

It sometimes seems that for every new type of PBX at least one new machine is introduced to report on long-distance calls. But many of them suffer from a variety of ailments, the most common of which is inaccuracy. With rates frequently changing across North America, the basic accuracy of most of these machines is poor.

Don't blame the machine for everything. In many cases the PBX is partially to blame, because of the way it connects with the Bell switching offices. The long-distance charges incurred on regular calls are calculated from the time a call is actually answered until one of the two parties disconnects. For billing purposes on regular long distance, this time is rounded up to the nearest minute. The actual timing is done in six-second increments (one-tenth of a minute). WATS calls are actually billed in these six-second increments. Even the most modern digital PBXs cannot determine when a call is answered, if the Bell central office does not normally give the PBX what is called answer supervision.

Answer supervision would tell the PBX if and when a call is actually answered. Rumor has it that future enhancements to the Bell switching offices will allow them to provide such supervision to the PBXs that can recognize it. Although you are not urged to wait for this enhancement, look for the capability of recognizing the reverse battery supervision when shopping for a new PBX. It may not need it or be capable of using it now, but it probably will eventually.

In lieu of answer supervision, most modern PBXs will estimate the time of a call based on the time the call occupied a trunk. Some machines will allow the user to set a minimum time, such as 30 to 45 seconds, under which a call will be deemed to have been abandoned, busy, or otherwise not completed. These calls will not be recorded as legitimate. Some of the off-line processing devices for CDR will allow an adjustment to the call length to offset the call setup time. If the gear being considered can do this, make sure that the adjustment can be tailored to individual trunk groups. Remember, a call between New York and California takes more time to connect than does a call down the road. This variance will range from 5 to 30 seconds in most cases. Even more important is to be sure that the PBX will include in call record the time the call was originated and the duration, which should be in the form of hrs:min:ànd 1/10th min. This last field is especially important in tracking WATS calls. If the PBX rounds up a 2-minute, 20-second call to 3 minutes, no amount of downstream adjustment will help.

If your company intends to provide some form of emergency backup power for the PBX, don't forget to size any CDR device and other peripheral devices that should stay working with the setup. Thus, when installing battery backup, it is important to take into consideration the amps that backup printers, CDR devices, 48-volt to 110-volt inverters, and the like draw on power supply. So make sure to include all draws on equipment. Don't forget to include all these extra boxes in air-conditioning estimates if special air conditioning is required. Remember, computers will always work more efficiently in a temperature/humidity-controlled environment. Many vendors or distributors say that normal room temperatures are fine for their gear. Beware! With all that new equipment the normal room temperature will rise significantly, and long-term damage can be done to any electronic equipment at excessive temperatures.

So now there is a way of tracking all those calls. But are they being placed by the best route? What is the best route? In most cases, it is the least-expensive route, placed on a facility that will provide the required level of service for the call. Some calls, such as data calls, will not work as well on some of the lower-cost facilities. Some long-haul voice calls, using certain facilities of the other common carriers, can be of such poor quality that the cost saving is lost by the need to repeat every second sentence, making the call longer.

Most manufacturers now offer some method of least-cost routing, alternate-route selection, basic alternate-route selection, or route optimization. There are almost as many terms for this as there are vendors. But be careful. The cost of the software usually reflects its complexity.

Some of these products will only allow the user to connect to the least-cost route. Others allow overflow to alternate routes somewhere between the least costly and regular long distance. Some are simple, while others are extremely sophisticated. If your firm has

Beware! With all the new equipment, the normal room temperature can rise and damage be done.

simple needs, don't look for or demand a sophisticated solution. On the other hand, if your company has so many different facilities that it is impossible to keep track of them, don't settle for a simple solution. Be prepared to make adjustments to the software after it is installed. Ask for examples of analyses done before and after the new PBXs were installed, what differences they made, and what savings were realized through adjustments to the software.

Make certain that the route optimization scheme is easy to use. Look for a PBX that users can handle without a two-week training course. If they currently are dialing 9+(number) for long distance, don't change it. If a series of access and authorization codes are being used for WATS and FX, get rid of them. Back in the early days of WATS, there was a minimum billing for up to 160 hours. Companies that were not using the minimum each month would tell their employees to feel free to use the WATS line after normal working hours, because it didn't cost anything extra. WATS became what many people considered toll-free service. Now most areas bill for WATS based on actual usage with a reduction in the cost per minute depending on total usage.

Even with the recent change in WATS, all calls are charged for. Also be aware of the minimum billing requirements for most WATS trunks. If your firm is in a business where most of the long-distance calls are short, it may get caught in the minimum one-minute syndrome. For example, if WATS calling amounted to a total of 400 minutes per month, but there was a total of 600 calls, Bell would charge for 600 minutes, or a minimum of one minute per call.

Many companies currently use a numbering scheme that varies depending on the trunk route being used. On some, WATS will be accessed by dialing 7X + area code + 7 digits with the X being the band or zone in which the call is to be placed. For tie trunk access, the digits 8X seems to be popular. Many employees seem to think that if the numerals 9+1 are not dialed, the call is toll free.

By eliminating the access code for WATS and foreign exchange lines and implementing a 9+(1) +(area code)+ 7-digit number, the concept of toll-free calling will eventually disappear. If the company uses MCI, Sprint, or some other regional common carrier, these services can be handled on a dedicated or dial-up basis by several of the more sophisticated route-optimization or automatic-route selection schemes on the market today. Many will allow a bad trunk to be bypassed for another route when poor quality transmission is encountered. Some won't even make the call be redialed. Others will alert the user who is being forced into a more expensive trunk route. Some will even allow the PBX administrator to determine who can use the more expensive route and who will have to queue.

A commonly underused feature that most new PBXs have is called toll restriction. With toll restriction, the communications manager can control who calls where and when they can call. Many schemes will allow only simple restrictions. For instance, anyone with a certain class of service can only make local calls or calls within their own area code. Others are more flexible and allow the caller to place calls to certain predetermined area codes. Still others can filter out calls to specific area codes and NNX prefix combinations. Managing these toll tables can become a problem. With the advent of the 9XX-service numbers, it will soon be a requirement that full 7- and 10-digit numbers be allowed in the restriction tables.

Some means should be available for the exception to the rule. Perhaps your company has one customer in a particular area, but for all intents and purposes, no calls should go to any number but that one within that area. Some PBXs allow specific numbers to be placed in special speed lists that when called using the speed index, will override the normal restrictions. If your PBX does this, make sure that it doesn't override the least-cost routing scheme as well.

Making the most of equal access

Equal access is about to hit the industry in a big way. How does this work, and can your telecommunications network take advantage of it?

In most areas, firms will have to presubscribe to one of the common carriers. This may be AT&T, MCI, Sprint, or one of the other local common carriers. If users dial 1+(area code)+7 digits, their calls will be routed to their presubscribed carrier. They will have the choice of dialing 1XXXX (+area code) — + 7 digits (the XXXX will be the common-carrier code), and the call will be placed over that carrier's facilities. Changing your presubscription will take the normal lead time for most telephone company transactions (one week to six months). This doesn't help when other common-carrier sales representatives drop in and say that they are running a special next month with 30 percent off all calls placed using their service. When you call your local telephone company with this information, they may say that it will take three to four weeks to implement that change, and, by the way, it will cost $XX per trunk connection to do it. In the meantime, your employees must be told that until then, they will have to dial 1XXXX instead of just 1 when placing long-distance calls. Naturally, at least 15 to 20 percent of them will remember, but the person who does your call costing may not have been told, and since your calls might be costed at the time they were placed, any resemblance of the CDR costing and your actual costs could be purely coincidental. Try doing a cost justification for any carrier using those sets of limits.

Ask your PBX vendors how their equipment will handle this nightmare. Some will handle it well, others will tend to leave your office quickly and quietly. It should be possible to make simple changes to your PBX's routing table that will automatically insert the required codes, so that it is completely transparent to the users.

Remember, before buying any route optimization or automatic route selection software package, find out exactly what it can and cannot do. Can it handle equal access, least-cost routing overflow to various other routing schemes, full toll restriction? Simply providing

If users are required to wait during peak periods, a fair savings can be realized.

least-cost routing or a busy signal when calls don't go through doesn't really help.

Some method of determining traffic patterns is a must. The call-detail recording reports may give some information, but in most cases, separate traffic data on all types of transactions within the PBX may help more than just call reports. In order to determine the proper balance between good service and good economics, it is normal that a reasonable amount of blockage on the trunks be maintained. Completely nonblocking trunking is not economically sound. If users are required to wait a reasonable time during peak calling periods, a fair savings can be realized.

Most companies do not have the same calling patterns all day long, day after day. Traffic statistics can certainly help in determining how many trunks of a particular type can be justified. If a company still leases or rents Bell equipment, most operating companies will conduct a traffic study for it on request. In most cases there will be a charge for this. In virtually all cases, there will be a wait. Most traffic studies are done months after they are requested. Know your firm's busy time of the year and try to arrange a study three to six months in advance of it. Don't worry about having to analyze the results right away. Most operating companies won't get them back for 30 to 90 days. Some take even longer. If a business has unusual calling patterns, the operating company may wish to do another study to confirm the first one. Traffic studies should not be done when external events are acting on a company. For example, if the price of its stock soars or dips or splits or its employees are on their annual two-week layoff during the summer slow period, the traffic study will not accurately represent how the company functions.

Traffic studies from Bell will generally indicate how many trunks are required. If your telecommunications manager does obtain a Bell traffic study that says you can reduce your trunks overall, do yourself a favor and keep it in a safe place. Someday it may be valuable, as most rare documents usually are. If the local BOC will not do a traffic study to your liking, find another vendor who will.

Once the new PBX is installed, a company should be able to run its own traffic studies. Find out at the beginning of the sales cycle how much traffic information is available. Some traffic statistics will require downstream processing before they are useful. Check to see if the PBX will assist in analyzing these reports. Find out what type of information is available. Does it just track trunk information, or can it track attendant consoles, common equipment, and feature usage? All these items are important in managing a new PBX.

Another feature to consider for new equipment is called queuing. If users are going to be allowed to queue for facilities during peak periods, equipment must have this capability. Generally, queuing should only be permitted on one-way trunks. Queuing on busy two-way local trunks or on tie trunks is just asking for problems. Tie trunks in particular are notorious for a condition known as glare. When a tie trunk group is busy, and users at both ends are allowed to queue, the normal result is that when a trunk becomes available, both ends grab it at the same time, resulting in two users connected to each other, both really wanting to talk to someone else.

Two different types of queuing are possible: ring-back or call-back queuing, and off-hook or camp-on queuing. Many PBXs will allow both. In some PBXs, especially those that have automatic route selection or route optimization, the users will dial entire numbers, the software will determine that no least-cost facilities are available and will signal the user with a fast busy tone. They then choose to remain off-hook or to request a call-back. When the facility is available, users are notified that their call can now be placed and they can either allow the PBX to place the call or refuse the call-back. If they had requested call-backs, and they are on the telephone, the PBX should keep track of the call-back requests and attempt the call when the telephones are free. Without the queuing feature, it is not possible to properly fine-tune the entire operation. The PBX should also remember the dialed number, for in most cases the user will have forgotten it. Some switches will return a fast busy tone to the caller as soon as the trunk group access code is dialed.

Now that all these new, cost-saving features are installed, what to do with them? For the most part, if they are properly maintained and used, they can enable the communications manager to determine which facilities the company should have, and how many of each type are economically justified.

Traffic patterns

To receive the best possible at the lowest price, a business should have just enough low-cost trunks to handle the average traffic load and allow overflow to more expensive routes for peak traffic times. Expect some blockage on trunks during peak periods. Something between 2 and 5 percent is usually quite acceptable during the busiest periods.

Don't use the actual figures from WATS or regular long-distance bills to determine the amount of traffic trunk groups must carry. Trunks are in use for longer periods than just that which is billed. The call setup time and the time involved in busy, toll-free, local, and uncompleted calls must all be taken into account. The traffic studies obtained from a new PBX should give sufficient information.

There always has been controversy over which traffic model should be used under the many circumstances one may encounter. There are three popular theories currently used in sizing facilities. These are Erlang B Erlang C, and Poisson. The first two are based on classic models developed by Agner Erlang, a mathematician who worked for the Copenhagen Telephone Co. Poisson is a compromise between the first two models. It is generally considered an outdated and unrealistic model, usually oversizing the trunk requirements.

Erlang B assumes that a blocked incoming or outgoing caller is cleared from either the entire PBX or from one trunk group. If the PBX does not have route optimization or automatic route selection (ARS), the caller would receive a "fast busy." If it does have ARS, then the caller is overflowed to another trunk group. Without the overflow, the caller simply hangs up. With the overflow, the caller gets through and the blockage is transparent. In both cases, the PBX is assumed to not have any form of queuing. Erlang C assumes that the PBX has queuing, and the caller queues and waits as long as necessary to seize a facility.

In 1956, James Jewett of Vanderbilt University argued that if overflow is not allowed, and there is no queuing, some fraction of callers will retry their call, while others will abandon the effort. This is known as extended Erlang B. As blockage increases, and hence recall traffic is larger, this model becomes unreliable. If callers are blocked and allowed to queue, but are allowed to overflow after a maximum time period in queue, the extended Erlang C model may by used to analyze what is known as the finite time queue.

The Poisson model assumes that a blocked caller will wait for one holding time and, if not connected, will go away. If the caller is connected after a delay, this call will only last the remainder of the holding time. Simply put, if an average call lasts 2.5 minutes, then a caller who is delayed for 1 minute will engage in a call lasting only 1.5 minutes. This is why it is considered outdated.

The accompanying diagrams (see figure) give a sample comparison of the Erlang B, Erlang C, and Poisson models. The diagrams illustrate the actual traffic that is carried for 1 to 10 and from 10 to 50 trunks. For the Erlang C, a typical 2.5-minute call was used. It was assumed that 90 percent of the queued callers would receive service within 60 seconds. Note how much additional traffic can be handled when the queuing is properly used.

Contrary to popular belief, with automatic route selection either with or without queuing, a PBX should not be using the Poisson theory for blockage. Most traffic engineering uses either the Erlang B or Extended Erlang C tables. From actual experience, Rolm has found that in hundreds of PBXs analyzed, the Erlang solution is generally closer to reality than the Poisson solution. The telephone companies will normally suggest the Poisson. Most companies with a communications department can usually determine a proper bal-

Traffic models. *Whether with one to ten trunks (a) or ten to 40 trunks (b), the Erlang and Poisson traffic models describe drastically different patterns.*

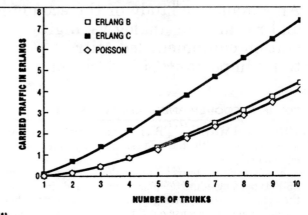

(A)

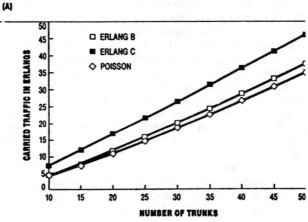

(B)

ance for their particular setup. If your company is not yet large enough to have a full-time staff just to manage your communications network, your telephone equipment supplier should be able to assist in determining just what that balance is.

The best way to be sure of getting help in fine-tuning new equipment once it is installed is to make sure any final adjustments are provided before signing that contract. Ask for examples, talk to the sales or traffic engineering department and determine the level of expertise available. Ask to see live examples of their work. Some suppliers may charge for special analysis. Depending on the size and complexity, simple traffic analysis can cost $50 to $500. It can be made part of a regular maintenance agreement. Even if there is a charge, this is one time when it will be worth the expenditure. With a little care, the super PBX can be kept from becoming a monster. ∎

Don Thomson studied engineering at the University of Western Ontario. He spent ten years with Bell Canada doing PBX maintenance, where he specialized in transmission, traffic, and special assemblies. In 1981, he joined Rolm Canada, where he is manager of sales engineering.

Analog Vs. Digital

An exhaustive study of the economic risks
and rewards of analog vs. digital
station equipment led Mitre
to its important CPE decision

EDWIN R. COOVER AND MICHAEL J. KANE

WHEN CHANGING PBXs, what does one do with the old station equipment?

For standard items, like analog 2500 type, single-line telephone sets, the choices seem straightforward enough. One can use the old sets with the new system. Or one can replace them with digital sets with additional capabilities for about $125 each. A number of studies have shown that few users take full advantage of the additional features. Single-line phones are often present in great numbers. So why replace the reliable 2500 sets with more expensive digital sets? It would appear to be one of the easier decisions for a PBX procurement team.

For many environments, the easy answer is the wrong answer.

Few PBX environments have a

Edwin R. Coover is Lead Engineer and Michael J. Kane is Systems Engineer at The Mitre Corp., McLean, Va.

tight rein on their operational and maintenance costs, and on the annual cost of moves and changes. Most companies employ a mix of multiline and single-line station equipment. Often hidden behind these kinds of stations are different requirements for station wiring, line cards and controllers or ac power sources. When a company has a lot of personnel movement, many moves and changes cannot be performed exclusively in software but require one or more types of rewiring. The annual operational and maintenance costs can involve substantial sums, particularly when rewiring services are performed by vendor staff.

Most modern digital station equipment allows moves and changes to be performed completely in software. Does the cost savings from moves and changes justify the higher costs of all new digital station equipment? The answer is yes—maybe.

The Mitre Corp., a non-profit organization that provides a variety of systems engineering services, was housing its Washington-area offices in nine buildings in McLean, Va. At the time, 2100 employees were on-site, with another 200 off-site accessing the PBX. The centrally-located AT&T Dimension PBX serviced remote buildings via a combination of Mitre-owned and leased twisted-pair tielines. One distant building employed a Rolm CBX II in a voice-data role. The Rolm switch acted as a remote to the Dimension for voice traffic. For data traffic the Rolm switch bridged to a Sytek Localnet 20 dual cable broadband local area network.

The station equipment consisted of roughly two-thirds single-line sets, and one-third multiline sets, with the latter requiring a controller within 1500 ft. and special station-side wir-

Continued on page 80

FIGURE 1   COSTS OF MOVES AND CHANGES WITH DIMENSION TECHNOLOGY

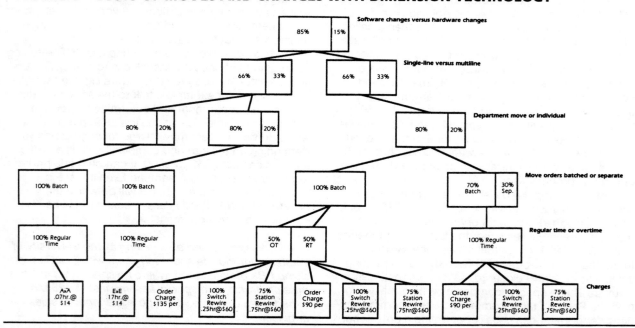

ing. The requirements for a Department of Defense secure facility (that is, no tenants) and the shifting nature of projects combined to render space allocation volatile, resulting in numerous Mitre employee moves. Eighty percent of these moves were by department, 20% by individual. Quantifying the costs of the frequent moves and changes turned out to be surprisingly complex.

Because of its situation, Mitre considered replacing its leased AT&T Dimension PBX. Previously, it took advantage of an AT&T offer to purchase its in-place 2500-type analog telephone sets at $20 per set, which was roughly equivalent to one year's lease. There was considerable uncertainty regarding the operational costs of the switch, since divestiture bills came from several sources and delays in AT&T billings often resulted in lags of months. Mitre performed some of the operations and maintenance work. Other work was done by AT&T. It was difficult to pinpoint the origin of the costs.

Whether or not to keep the 2500 analog station equipment with the new PBX, rather than purchase new digital single-line sets, seemed to be an easy decision. It involved comparing a sunk cost vs. lease or purchase of a set that cost about $125.

After considerable study, Mitre found the cost of its moves and changes with Dimension-era analog station equipment to be substantial. Further, the company found that digital single-line sets with a new PBX represented a cost-effective alternative for the Mitre environment as well as offering important qualitative benefits. Figure 1 provides a schematic view of the costs of moves and changes with the Dimension technology. Additionally, to ascertain annual costs, it is important not only to calculate charges connected with moves and changes but also the costs of growth and attrition.

Table 1 presents in tabular form the cost components by environment (a mix of analog types vs. all digital) and by cost (equipment vs. labor). In the analysis of retaining the owned analog sets vs. procuring new digital sets, Rolm equipment was employed for costing.

Several observations can be made. First, although these cost compo-

TABLE 1   COST COMPONENTS BY TYPE

MITRE Values Appear In Brackets [ ]

Mixed Analog & Digital Equipment	All Digital Equipment

Equipment Costs • Single-line sets
  –Number of single-line sets to be purchased [initially, 280]
  ––Cost [$24]
  ––Annual attrition of analog sets [7%]
  ––Number due to growth of staff population [100/yr.]
• Controllers
  –Number of controllers [89]
  ––Number of single-line sets to be supported by all controllers [1412]
  ––Number of single-line sets which could be supported by one controller [16]
  ––Lease cost of controllers [$734/yr.]
• Cabinets
  –Number of cabinets [12]
  –Lease cost of cabinets [$1090/yr.]
	• Digital single-line sets
	  –Cost [$125]
	  –Number required [2140]
	  ––Initial provisioning [ ]
	  ––Annual growth [100]
	
	• Cabinets
	  –Lease cost [$1090/yr.]
	  –Number [8]

Labor Costs • Database changes
  –Homogeneous moves (analog to analog, electronic to electronic)
  ––Number of moves and changes annually [2140]
  ––Percent of moves which are homogeneous [85%]
  ––Time to make single-line homogeneous moves [0.07 hr.]
  ––Time to make multi-line homogeneous moves [0.17 hr.]
  ––Labor hour rates [$14]
• Rewiring
  –Heterogeneous moves (analog to electronic)
  ––Number of jobs [321]
  ––Time to rewire only at the main distribution frame [0.25 hr.]
  ––Time to rewire only at the station [0.75 hr.]
  ––Time to rewire both at the main distribution frame & station [1 hr.]
  ––Ratio of batched jobs to individual jobs [70:30]
  ––Labor hour rates, regular vs. overtime [$60 hr. vs. $90 hr.]
  ––Ratio of departmental moves vs. individual [80:20]
  ––Ratio of move orders, batched vs. separate [70:30]
  ––Ratio of department moves performed during regular time vs. overtime [50:50]
  ––Percent of individual moves performed during regular time [100%]
• Installing cabinets
  –Number to be installed [12]
  –Cabinet cost (installed) [$12,000]
• Testing existing single-line analog sets
  –Number to be tested [1412]
  –Time required [0.10 hrs./set]
  –Labor hour rates [$68.75 hr.]
	• Database changes
	  –Number of moves and changes annually [2140]
	  –Time to make moves & changes [0.02 hr.]
	  –Labor hour rates [$14]
	
	• Installing cabinets
	  –Number to be installed [8]
	  –Labor cost [$12,000]

nents must be understood and measured to perform a quantitative analysis, the values are unique to a particular environment at a particular time. Second, collection of these values involves both expert knowledge and considerable time and effort.

Mitre considered the following qualitative variables in analyzing the analog vs. digital station equipment question:
• the greater reliability of new, digital station equipment;
• the value of user-friendly fea-

tures on new, single-line digital sets, such as hold button, call transfer button, call repeat button and call volume controls;
• the easier and faster moves with all digital station equipment via a 100% software move;
• the analog "penalty" in a digital switch—that is, less efficient port cards (costed) and possibly more space for the switch (uncosted); and
• the capability to integrate data communications interfaces with the digital sets.

The analysis took into account three important questions. First, what were the current annual costs of moves and changes in a changing environment? Second, if a new digital switch were installed, how would these annual costs be impacted by retaining the single-line analog sets in a mixed station equipment environment with multiline digital sets? Third, what would the annual costs be should a decision be made to go to all new digital station equipment?

The analysis employed the Lotus 1-2-3 spreadsheet program and the values shown in Table 1.

The calculations revealed some unanticipated factors. First, there appeared to be a subtle analog penalty to employing analog station equipment in a digital switch. This analog penalty extends beyond the analog/digital conversion at the line card. It extends to the number and space/efficiency of the line cards, which can affect, in a large system, the number of cabinets. In looking at switch architectures other than Rolm, it seemed as if each manufacturer decided to accommodate the analog penalty differently, though frequently a manufacturer's digital line cards will handle twice the station instruments as the analog cards.

Second, and paradoxically, the analog penalty may not appear in switch pricing. Usually an analog line card with codecs should cost more than a digital line card without. Nonetheless, with Rolm, the digital line card cost more, the analog less. The rationale seemed to stem from marketing factors rather than manufacturing costs: Systems with a large number of analog phones are aimed at the low end of the market where costs are determinate. Conversely, predominately digital systems are features-driven, and costs are perceived as secondary. Finally, although the system with analog line cards may consume more space because of less efficient cards and more cabinets, it may not matter on a small system or systems with partially filled cabinets.

Our calculations revealed results that were not initially obvious. Given Mitre's size and the numerous employee moves due to the shifting nature of contracts, station equipment-related operational and maintenance costs with a heterogeneous station equipment environment cost about $80,000 each year. The digital set environment, where moves and changes can be performed completely in software, is cost competitive with the analog 2500 equipment immediately. In Mitre's case, the all digital set option was calculated to save about $126,000 over five years.

Viewed from some distance, there appear to be two larger issues. The first is familiar—the frequent economic advantages of trading off what are largely one-time charges for equipment vs. recurring charges for labor.

The second is more elusive and concerns the relative dynamics of a particular work environment. In Mitre's case, the combination of labor costs and high corporate horizontal (moves) and vertical mobility (changes) were the driving forces, and made the all digital station equipment the appropriate technology. □

The Role of Information Management System in Business Communication

Osamu Enomoto
Shinsuke Kadota

This article examines the role of information management system in business communication and what system is best suited for a particular busines

The term business communication is opposite in meaning to the term public communication. A public communication system is planned and constructed as an infrastructure for the purpose of providing general communication services for the best interests of the society as a whole, rather than providing some specific or convenient services only for some specific groups. In other words, a public communication system is planned, constructed, and utilized so that ordinary telephone services may be offered to as many people as possible in a fair and economical manner.

A business communication system is planned, constructed, and utilized for the purpose of a specific business, so that the system may provide the maximum services to the business concerned. Therefore, the services considered to be offered are those that are best suited to the specific business organization that uses the communication system. The system is designed so that these services may be executed with ease.

The PBX has provided a variety of services, mainly by telephone services, as the hub of business communication. However, not only telephone sets but data terminal equipment, image terminal equipment, and facsimile equipment are connected with the communication system, and the service features to be provided are more and more diversified according to the types of business involved. Table I shows some typical examples of services (strictly service packages which are assemblies of services) for different types of business.

Services for General Business: These are basic services of business communication, and all others can be considered as additions. In most cases, a nationwide business organization is either required to be equipped with its own tie lines, besides ordinary telephone service facilities, so that it can set up the most economical connections like Least Cost Routing, or it is required to have facilities for data communication, mail service, and so on. Furthermore, key telephones are sometimes used as the terminal equipment in addition to ordinary telephones.

Services for Hotels/Motels: There are services such as billing on external calls originating from each guest-room, a unique numbering system in which room numbers and guest services are interrelated, room cut-off which restricts a guest telephone from outgoing calls after the guest checks out, and which includes guestroom status indication to the maid room and other services.

Services for Hospitals: There are a variety of unique services, including: a service that ensures that if a patient's telephone is left off-hook for a long time, the nurse's station receives a signal indicating that a patient's phone was not hung up properly.

Business organizations including railway services, electric power supplies, and banks are often required to employ some special specification and to be equipped with exclusive terminal equipment. In the example of an electric power supply system, there is a case where image signals are used for the purpose of supervising instruments at remote power plants for their external environments.

Reprinted from IEEE Communications Magazine, July 1986, pages 37-44. Copyright © by The Institute of Electrical and Electronics Engineers, Inc. All rights reserved.

From the viewpoint of communication networks, a large business organization is often required to be constructed so that its geographically scattered offices may seem to be one body of offices and to have the function of a centralized attendant console at which calls to and from plural areas can be handled. In addition, a tie-line network system within one business organization is required from the economical and operating points of view. In the case of a rental building, it is required to have these functions so different tenants appear as if each had its own PBX.

In the case of lawyers, accountants, and so forth, a service is required so telephone billing on a client can be administrated by adding a specific identification code when dialing a specific client's number.

For a business communication system, there are requirements that do not exist in the case of public communication systems, and the line size requirements range widely from tens of lines up to several ten thousand lines. It may be more appropriate to describe a PBX as being used in a business communication system such as an IMS (Information Management System) rather than to describe it as a telephone exchange.

When considering the conditions for an IMS which is capable of meeting diversified requirements, they can be summarized as:

TABLE I
EXAMPLES OF SERVICES IN BUSINESS COMMUNICATIONS

Type of Business	Network Configuration	Services	Typical Terminals Other Than Tel.	Remarks
General Business	Tie Line + Public Network PKT Network, LCR OCC	Voice/Data Mail Function CA Voice Mail	Key Telephone Facsimile	LCR: Least Cost Routing CA: Computer Access OCC: Other Common Carrier PKT Network: Packet Switching Network
Hotel Motel	Tie Line (If necessary) OCC	Message Registration Message Waiting Room Status	Telephone equipped with Message Waiting Lamp Dedicated Console	
Hospital	Public Network Image Network PKT Network LAN	Nurse Call Paging	Data Terminal	LAN: Local Area Network
Electric Power Supply (Example in Japan)	Tie Line Network	Talkie Function Toll Function (Private) Voice/Data	Facsimile Data Terminal	
Railway Service (Example in Japan)	Tie Line Network	Train Operation Function High-fidelity Line	Dispatcher Telephone	
Banking/Stock Market (Example in Japan)	Tie Line + Public Network PKT Network	Broadcast Message (FAX.) Dealing System	Facsimile Key Telephone Data Terminal	
Business Distribution System	Tie Line Network + Public Network	ACD Centralized Attendant Console	Facsimile Key Telephone POS	ACD: Automatic Call Distribution POS: Point of Sales
University	Public Network	LAN Voice/Data LAN INT	Data Terminal	LAN INT: LAN Interface
Intelligent Building		Tenant, Building Management LAN INT, DATA	Facsimile Key Telephone	
Lawyers/ Accountant	Public Network	ACCOUNTING CODE	Key Telephone	

1) For setting up the most efficient system for the required line size, the building block configuration is employed.
2) For system flexibility, the port-free system is employed so it can accommodate either terminal interface or trunk interface as required.
3) For allowing new terminals or value-added modules to be added to the system, each of the interfaces is standardized.
4) Information transmission, information storage, and information processing are possible.

As an IMS having these itemized conditions, we developed the NEAX2400IMS, offering various kinds of services shown in Table I. By introducing the system configuration of the IMS and its example of applications, this article describes the role of IMS in business communication systems.

System Configuration of IMS

The IMS was developed to serve as the node of a business communication. Figure 1 shows the total system concept of the IMS, and Fig. 2 shows the block diagram of the IMS. The IMS has digital circuit switch modules as its hub, with such peripheral equipment as value-added modules, network service, and terminals.

Digital Circuit Switch Modules

Digital circuit switch modules (DCS modules) are provided as the hub of IMS. According to the line size, the system is available as SIM (up to 384 lines), IMG (up to 560 lines), MMG (up to 4000 lines), and UMG (up to 16,000 lines). Since the systems from IMG to UMG are employing the complete building block configuration, the system can be expanded from IMG to UMG by adding port interface modules (PIM's) which accommodate lines/trunks and some dedicated control modules. Distributed control system and distributed speech path system are being employed for the DCSM to create the building block system.

Figure 3 shows the configuration of the distributed control system in the MMG. In this distributed control system, each kind of processor has the following functions:

1) *The Local Processor* (LP)—upon receiving the standardized information from the PM, analyzes the content of the status change of the speech path system equipment and sends necessary processing requests to the MP. One LP is capable of controling line circuits and trunk circuits of a maximum of 768 channels. As the LP, a 16-bit μ-processor μPD8086 is used. Where necessary,

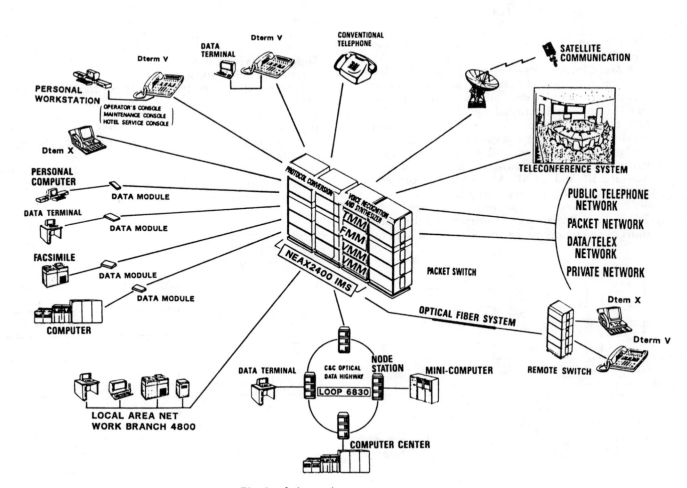

Fig. 1. Information management system.

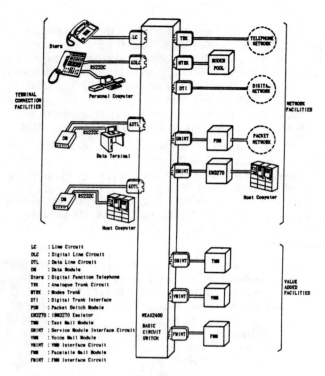

LC   : Line Circuit
DLC   : Digital Line Circuit
DTI   : Data Line Circuit
DM   : Data Module
Dters : Digital Function Telephone
TRK   : Analogue Trunk Circuit
MTRK   : Modem Trunk
DTI   : Digital Trunk Interface
PSM   : Packet Switch Module
EMX3270 : IMX3270 Emulator
TMM   : Text Mail Module
SMINT : Service Module Interface Circuit
VMM   : Voice Mail Module
VMINT : VMM Interface Circuit
FMM   : Facsimile Mail Module
FMINT : FMM Interface Circuit

Fig. 2.   IMS block diagram.

the LP can be provided in a redundant configuration.

2) *The Main Processor* (MP)—is a processor of both function sharing and load sharing. The required quantity of MP's is determined according to the line size and the functional size of the system. The MP performs logical processing of call status transition such as management of call status, analysis of call status, and so on, according to various kinds of processing requests which arrive from the LP and the SP. As the MP, a μ-processor μPD8086 is used. For searching reliability, an "N + 1" redundancy arrangement is employed.

3) *The System Processor* (SP)—supervises the operations of the whole processor system and manages

the normality of system operations. It also manages restart processing functions and fault information indication at the time of a system fault occurrence, and system maintenance through the man-machine interface. In an event of a fault occurrence, the SP performs system backup by controlling the external memory device (Bubble Memory).

In the SIM/IMG, the centralized control system is being employed in which the LP, MP, and SP have been logically concentrated on one processor to improve the system cost/performance. Figure 4 shows the method of expanding the system from IMG to MMG with respect to the control system.

For a port-free system in the case of the NEAX2400-IMS, another processor Port Microprocessor (PM) is used at the interface port of each line/trunk circuit. The PM is connected with the LP by the interface, which is physically and logically standardized. Also, by absorbing the controls related to the attributes unique to each line/trunk circuit by means of the processings on the PM, the requirements on each kind of line/trunk can be met with ease simply by making available the standard connecting services on the LP. By improving the processing capability of the PM, highly sophisticated terminals or computer interfaces which require higher-level controlling can be accommodated into the system without any large-scale modifications of the main software, and therefore, the system can be easily adapted to ISDN (Integrated Digital Service Network), DMI (Digital Multiplex Interface), and so forth.

Figure 5 shows the speech path configuration of the MMG. In this speech path configuration, the time division switches are provided by the three-stage configuration of Time-Space-Time (T-S-T). T-switches are distributed for each LP. In the IMG, the time division switches are provided by a two-stage configuration of T-switches only (T-T). Thus, IMG can be easily expanded to MMG by simply installing a space switch between the two time switches.

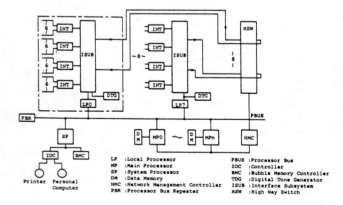

LP : Local Processor   PBUS : Processor Bus
MP : Main Processor   IOC : Controller
SP : System Processor   BMC : Bubble Memory Controller
DM : Data Memory   TDG : Digital Tone Generator
NMC : Network Management Controller   ISUB : Interface Subsystem
PBR : Processor Bus Repeater   HSW : High Way Switch

Fig. 3.   Distributed control method by multiprocessors.

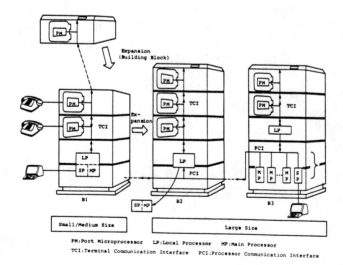

PM:Port Microprocessor   LP:Local Processor   MP:Main Processor
TCI:Terminal Communication Interface   PCI:Processor Communication Interface

Fig. 4.   Method of control-block expansion.

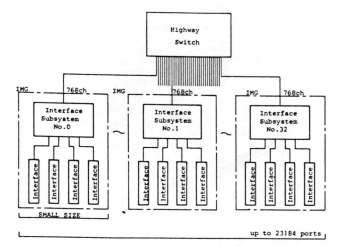

Fig. 5. *Time-division switch configuration.*

To meet ever-diversifying needs, the service features of the DCS consist not only of the conventional voice switching services but also of data switching services of the digital base. In addition, more sophisticated and more integrated services, for example, voice and data simultaneous switching service, can be made available.

Network Service

In order to support business activities of a wide range, a variety of network services are available by the IMS. One is related to digitalization of the conventional telephone network and enables more sophisticated internetwork services, including digital trunk interface (DTI), common channel interoffice signaling system (CCIS), electric private network (EPN), and so on. The other is a network service which allows access to packet switch network and computer network including packet switch module (PSM), 3270 Emulator (EM3270), and so forth.

1) DTI/CCIS/EPN

DTI is an interface that accommodates channel banks directly into the IMS to meet the digitalization of interoffice networks like T1 Carrier. At present, the DTI can be connected with digital lines of either 1.544M or 2.048M. Not only is DTI a cost-effective channel bank, but it also makes possible, by directly connecting digital speech paths to digital lines, such high-speed data communications with a data speed up to 56 Kb/s, image communications, and so on, which have not been available by the conventional analog network.

The CCIS enhances the effect of utilization of CCITT No. 7 Signaling System:

- Abundant kinds of signals
- Signal transfer during a call in progress
- Transmission of audible signals from the calling side
- High-speed signal transfer

- Automatized administrative control on signal network
- Independency of signal network

By utilizing these features, services appear as if they were handled by a large EPBX, even though the terminals distributed within the network are connected to different EPBX's. A feature transparency network can be realized through use of the above-mentioned services (functions). When utilizing DTI and CCIS in a T1 Carrier, one of the 64 Kb/s 24 channels can be used for common channel signaling, and the remainder can be used as the speech channel. Figure 6 shows the signal framing in this case.

EPN is the software service assembly which is provided by use of the equipment associated with these networks. The following services can be provided as EPN:

① Authorization Code
② Least Cost Routing—Deluxe
③ LCR—Automatic Overflow to DDD
④ LCR—Time of Day Routing
⑤ LCR—Controlled Alternate PRSC
⑥ LCR—Clocked Manual Override
⑦ LCR—Attendant Manual Override
⑧ LCR—Special Line Warning Tone
⑨ Primary Call Restriction
⑩ Priority Restriction Class
⑪ Outgoing Trunk Queuing—Deluxe
⑫ Traveling Class Mark
⑬ Uniform Numbering Plan
⑭ Indialing Through Main
⑮ Automatic Circuit Assurance

These devices and software enable the following:

- sophisticated network services
- sophisticated network management
- construction of a more economical network

2) PSM/EM3270

As previously stated, PSM and EM3270 are used to connect the IMS to the packet switch network and to the IBM Computer Network. These modules not only have the interfaces with packet switch network and SNA, but also have the functions for protocol conversions between the ASCII Dumb Terminal and X.25/X.75, SNA/SLDC. Therefore, with these modules added to the DCS of the IMS, an access can be made to a Packet Switch Network like TELENET, TYMENET, or to the IBM Computer Network from an ASCII Dumb Terminal accommodated in the DCS.

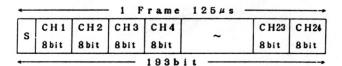

Fig. 6. *Frame configuration of DTI.*

Value-Added Services

As value-added services, there are mail service and automatic call distribution (ACD).

To provide mail service, modules such as voice mail module (VMM), text mail module (TMM), and facsimile mail module (FMM) are used. Each of these mail modules is independent of DCS and is capable of providing mail functions including information store and edition, mail delivery, message broadcast, and system answering. These mail modules are connected with DCS by means of the standard interface—service module interface (SMI)—and access can be made to these mail modules from a terminal accommodated in DCS.

The ACD system is used in:

—Acceptance of seat reservation requests at airline companies and travel agencies
—Acceptance of mail order business and TV shopping
—Verification of bank accounts (including checking and savings)
—Acceptance of orders from taxi services, etc.

The functions of the ACD system are to distribute ACD incoming calls efficiently to ACD agent consoles and reduce waiting time for connection of incoming calls. In addition, to manage agents' work performance at ACD agent consoles and system traffic, the Management Information System is being used. ACD is realized by means of the software on DCS and of ACD consoles specifically developed for the ACD system, and MIS operations are processed by the dedicated management control processor (MCP) connected to DCS.

Terminal Equipment

In the IMS, conventional telephone sets and Dterm digital electronic multi-function terminals are used in voice communications. The Dterm is available in two forms. One is represented by Dterm III, IV, and V, to be used in common by Digital Key System ELECTRA—IMS, and the other is represented by Dterm X equipped with a CRT Display.

Fig. 7. Dterm V.

Fig. 8. Dterm X (digital multi-function display telephone).

As seen from Fig. 7, a Dterm V is a terminal telephone set equipped with a 16-digit liquid crystal display panel. It is connected with the IMS by means of one 2-pair cable, using one pair for sending and one pair for receiving.

A Dterm can send and receive data signals along with voice signals if it is equipped with a digital adapter (DA) as an optional circuit. DA is available in two types: synchronous data and asynchronous data. The DA for synchronous data is capable of handling data rates up to 56 Kb/s, and the DA for asynchronous data is capable of accommodating data rates up to 9.6 Kb/s.

Dterm V is presently connected to the IMS by means of one 2-pair cable and employing a transmission speed of 256 Kb/s. It has a design consideration so that if the connection between the terminals and PBX is standardized by CCITT, accommodation of the proper kind of terminal into the IMS can be done with ease by designing a new line circuit card with which to replace the present one.

A data module is available for data handling only. The data module is connected directly to the data terminal. For example, if it is connected to the display terminal, a connection path can be established by keyboard dialing from the display terminal in the same way as in dial access. Dterm X shown in Figure 8 is a highly intelligent terminal having telephone functions, data terminal functions, and stand-alone functions, and is used as a voice terminal (telephone), online/offline terminal. Dterm X is connected with the IMS by means of one 2-pair cable, the same as in the case of Dterm IV and V.

As a telephone service feature, Dterm X has a direct recall service feature plus a variety of service features of Dterm V. By this direct recall service feature, a Dterm X user intending to make a call dials by designating with the cursor the telephone directory (listing of work section names, persons' names, and telephone numbers) on the display of Dterm X.

By means of data terminal functions of Dterm X, a user can perform data communications with other data

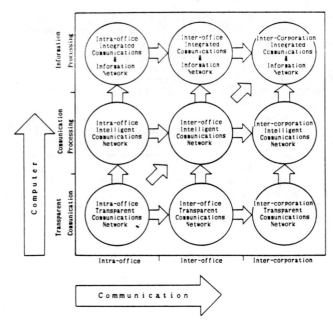

Fig. 9. *Application of business communication systems.*

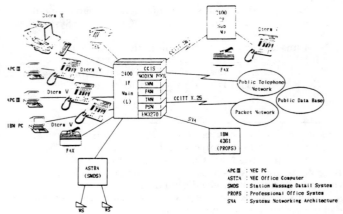

Fig. 10. *System configuration of NEC America OA system.*

terminals such as Dterm X, host computer, value added modules, and so on via the IMS, using the CRT Display (9 inches, Amber) and ASCII-type keyboard of Dterm X. Furthermore, with such emulation functions as VAX VT100 added, Dterm X can also be used as a terminal against the Host Computer.

Applications

It has been previously stated that the applications of business communication systems are diversifying at various types of business. If a business communication system is considered where computer and communication functions are integrated, it is possible that applications of a business communication system can be divided into several levels, as seen in Fig. 9. If applications are considered to the areas which the network covers, there are networks within a communication network in an office, a communication network among offices within a business organization, a communication network among business organizations, and others. Also, if applications are considered with respect to the intelligence level of a network, there is a wide range of levels, from transparent communication level to communication processing level through interactions with computer systems, up to an information processing level. Enhancement of applications of business communication systems means to materially find out the needs at each network hierarchy and each communication level. The IMS plays the most effective role in its realization. As an integrated OA System constructed by connecting the IMS to computer systems, the OA System at the headquarters of NEC America is shown in Fig. 10.

Conclusion

This article has described, by way of actual examples in the case of the Information Management System (IMS), that the characteristics of a PBX as the integrater of business communication systems are becoming much clearer than before. It is anticipated that business environments will become more complicated in the future, and the role of the IMS is expected to become more significant and to function as the hub of business communication systems. When considering possible development and growth of business communications in the future, the items listed below are especially important for the IMS.

1) Integrated multi-media services to meet requirements along with development of multi-media.
2) Wide-band/high-speed communication services including image information service, animation service, and so forth.
3) Standardized but flexible network access services which may become necessary, along with various kinds of work stations and terminals expected to advent in the future.
4) Wide area business communication network services.
5) Network management techniques for providing highly-reliable network services.

Acknowledgments

The authors would like to express their thanks for the encouragement and support extended by H. Koseki, the Vice President of NEC; Dr. N. Shimasaki, General Manager of C&C Systems Development Division; and T. Aomori, General Manager of Business Communications Division.

References

[1] O. Enomoto, et al., "Digital EPBX for future office communication," ISS '84, 1984.
[2] T. Aomori, et al., "Distributed microprocessor control architecture for versatile business communications," *IEEE Jour. on Sel. Areas in Commun.*, July 1985.

[3] K. Fujita, et al., "NEAX2400 information management system," NEC R&D, Special Issue on C&C Office System, 1985.

[4] N. Shimasaki, et al., "A study on multiple mode business information service networks," ICC '84, 1984.

[5] H. Goto, et al., "Office communication technology," NEC R&D, Special Issue on C&C Office System, 1985.

Osamu Enomoto received the B.E. degree in Electrical Engineering from Keio University, Tokyo, Japan, in 1958.

Since he joined the NEC Corporation in 1958, he has been engaged in the research and development of electronic switching systems, integrated digital transmission and switching systems, and electronic key systems, first at the Central Research Laboratories and then at the Business Communications Division. He is currently an Assistant General Manager of the switching planning office and the Integrated Switching Development Division.

Mr. Enomoto is a member of the Institute of Electrical Engineers of Japan and the Institute of Electronics and Communications Engineers of Japan.

Shinsuke Kadota received the B.E. degree in Electrical Engineering from Tokyo Metropolitan University, Tokyo, Japan, in 1966.

Since he joined the NEC Corporation in 1966, he has been engaged in the development and design of electronic PBX system and information management systems. He is currently a Manager of the Mass-Merchandise Products Development Department, and the Basic Technology Development Department, Business Communications Division.

Mr. Kadota is a member of the Institute of Electronics and Communication Engineers of Japan.

B. Chester Sagaser, GTE Communication Systems Corp., Phoenix, Ariz.

Use integrated PBXs and X.25 in today's networks; don't wait for ISDN

There are benefits to applying existing data communications technologies now. Do so until the standards are sorted out.

As the demand for digital services continues to increase, so too does the need for more efficient, hence more sophisticated, data communications equipment. Particularly, there is a growing demand for networking the devices that handle both data and digitized voice.

The response to the resultant growing device complexity has been the continuing integration of computers and communications. As today's office becomes ever more integrated, there is an increasing emphasis toward office equipment that can itself integrate with, and function in, an environment designed to meet evolving communications needs. This has created a corresponding demand for strong, formal links between various office communications products in order to prevent duplication and maximize integration.

Where PBXs are concerned, this developing interdependence of applications and products presents both a challenge and an opportunity. PBXs can now integrate voice and data (see below). In addition, they have more efficient switching capabilities than ever before. To function most effectively, however, PBXs must be able to work with other types of business information products, supporting everything from local area networks (LANs) and terminal-to-host/terminal-to-terminal connections, to direct access to public and private voice and data networks.

To establish a solid niche in the integrated communications environment, therefore, PBXs must serve as a common interface for integration of communications amidst overlapping technologies. By facilitating the physical transmission between devices, PBXs provide the connections to combine applications, thereby enhancing information exchange.

While voice switching will probably continue to remain a mainstay of PBX applications, it will be to the data side of communications that the PBX will make its most important contribution. Foremost among these contributions is the PBX role of integrating voice and data into a single, multifaceted communications network. The PBX will act as a communications hub for both voice and data interchanges. Moreover, when combined with packet-switching capabilities, the PBX can function as an integrated business communications hub. As such, it is able to handle all types of voice, data, and text information generated by a wide variety of terminals.

A multivendor environment
These new integrated capabilities make the PBX ideally suited to serve as a central data switch in addition to its traditional role as the central voice switch. Growing user acceptance of this fact can be tied to increased awareness as to just where and how the integrated PBX can support multivendor data-device applications, such as:
- Relatively low-speed (operator-keyed) interactive communications.
- Communications between formerly standalone devices (increasingly, these will be microcomputers).
- Terminal-to-host communications.
- The ability to mix and match the above combinations, subject to user needs .
- Elimination of costly multiplexer/port-contention devices by providing multiple data interchanges via a single PBX output port — be it to a host or to a public data network (this in turn provides significant cost reductions in multiplexing applications involving PBX-to-computer interfaces).

On a local basis, the PBX can support a LAN as well as direct communications from the local loop to local-communications or long-haul networks. What's more, in supporting a LAN, the PBX may serve as a parallel or interconnecting network that handles lower-speed (lower than megabit-per-second rates) interactive communications like electronic mail. Or it may be used as an efficient interconnecting device to two or more widely separated LANs.

By using packet-switching technology on the local loop, the integrated PBX can unify many widespread microcomputers into a single network—without the cost of installing coaxial cable or some form of broadband medium between them. In addition, packet switching, with its inherent protocol conversion capabilities, enables the PBX to connect a variety of incompatible devices.

But those local packet networks do much more than provide the connections. Because packet switching simultaneously supports a number of devices, user costs are kept down. Interactive packet switching means that data moves accurately (with error detection and correction) and quickly (in real time) over a packet network to and from the remote device. Thus, local packet networks can provide the most economical transmission medium for a variety of communications applications.

By providing simultaneous virtual connections, packet-switching networks eliminate the need for multiple leased lines and increase the flexibility of connecting a wide range of host computers. Even batch communications could be placed on a dedicated LAN: By reserving higher-speed batch and file transfer traffic for the LAN and lower-speed interactive communications for the PBX, the switch can provide the LAN with even greater capacity and value. And the two can be designed to allow a transfer of traffic between them.

Simplifying network access

Note that packet switching establishes a virtual connection between end points. This means that packet-switching PBXs are capable of putting their data streams directly onto public or private packet-switching networks without requiring any protocol conversion beyond that of the packet assembler/disassembler (PAD) at the user-device location. Also avoided is the data handling required with individual-network PADs or with conversions to circuit-switched transmissions.

The end-to-end use of packet switching eliminates many of the interfaces inherent in transferring data communications from the local loop (with its protocol) to an interconnect link (with *its* protocol) and, finally, to the long-haul network (using still a third protocol, X.25). If the connections are through a standard PBX, this entire protocol sequence must be reversed at the receiving end, resulting in a significantly increased risk of induced bit errors.

Use of X.25-compatible packet-switching protocols on the local loop eliminates much of this redundant processing. For instance, key X.25 functions include establishment of a permanent virtual circuit, setup and supervision of virtual calls, packet sequencing, and flow control of individual virtual calls and permanent virtual circuits. When site networks are X.25 compatible, traditional gateways are no longer necessary.

1. Conversion. *In its support of asynchronous terminals, the packet assembler/disassembler provides a form of asynchronous-to-synchronous conversion.*

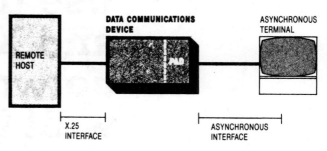

PAD = PACKET ASSEMBLER/DISASSEMBLER

Note in Figure 1 that the X.25 companion specification X.3 defines the operation and functions of the PAD in support of an asynchronous (start-stop) terminal. The functions performed by the PAD are determined by a set of variables, called parameters (discussed below), internal to the PAD protocol.

An added benefit of packet-switching local loops resides in the ability of a complete X.25 network to provide asynchronous-to-synchronous signal conversion, also shown in Figure 1. That set of rules defined in CCITT Recommendations X.3, X.28, and X.29 ensure efficient call-message interchanges between asynchro-

PARAMETER #	NAME
1	ESCAPE FROM DATA TRANSFER
2	ECHO
3	DATA FORWARDING
4	IDLE TIMER
5	PAD-TO-TERMINAL FLOW CONTROL
6	CONTROL OF SERVICE SIGNALS
7	BREAK HANDLING
8	DISCARD OUTPUT
9	CARRIAGE RETURN PADDING
10	LINE FOLDING
11	COMMUNICATIONS SPEED
12	TERMINAL-TO-PAD FLOW CONTROL
13	LINE-FEED INSERTION
14	LINE-FEED PADDING
15	EDITING (IN DATA TRANSFER)
16	CHARACTER DELETE
17	LINE DELETE
18	LINE DISPLAY

PAD = PACKET ASSEMBLER/DISASSEMBLER

nous devices and synchronous hosts, for example.

The 18 X.3 control parameters listed in the table provide terminal users with a command and control structure that less sophisticated protocols cannot. For example, a conventional circuit-switched approach would be hard pressed to provide these control abilities — the 18 parameters — without extensive interference with its data-transfer capacity.

In local-to-long-distance data communications, therefore, the use of a packet-switching PBX serves to eliminate additional equipment and processing between the LAN and long-haul networks, thereby improving data throughput and reducing communications costs. In addition, packet switching permits existing telephone lines to be used for access to the long-haul networks, reducing the number and costs of dedicated lines between the local-loop interface points and the packet-switching network. In this application, a PBX that switches data via internal packet-switching mechanisms has a distinct advantage over those that depend on circuit switching. Public data network and host ports can support multiple virtual connections simultaneously, thus eliminating the need for most contention mechanisms external to the ports.

Hubbing with the local loop

Many major manufacturers of office automation equipment have begun to view the integrated PBX as a communications hub. Much of this acceptance is a direct result of the unit's ability to move data in the local loop on existing twisted-pair telephone wires. Also, more than 200 X.25-certified devices from more than 100 manufacturers are currently registered by the major public packet networks.

The benefits of twisted-pair wire and X.25 compatibility are many. The most obvious is the ability to use the existing installed base of telephone wiring for both data communications and the normal flow of voice traffic. As user acceptance of the PBX as a data switch develops further, PBX manufacturers will begin to implement interfaces for a large portion of the existing coaxial, twinaxial, and shielded twisted-pair installations currently serving terminal-to-host data users.

Coaxial- and fiber-based LANs alone will not meet total short- and long-haul communications needs and can actually generate their own set of interconnection problems. For instance, without X.25, each interconnection would require a conversion in the transition from short- to long-haul transmission. Thus, PBXs that can offer capabilities such as simultaneously connecting hundreds of terminal users over a single 56- or 64-kbit/s channel will become accepted communications adjuncts to existing broadband networks. In fact, the entire spectrum of X.25-compatible technology will become a cost-saving and speed-enhancing service that will be increasingly valuable to the user over time.

Use of the X.25 protocol is significant in that it enables office automation equipment to gain transparent access to national and international packet-switching networks. At the same time, it improves the quality — through inherent error control — of terminal-to-host communications, while allowing users with X.25

mainframe links to reduce their long-distance communications costs. This reduction is brought about because of the economy of local X.25 access.

A growing number of equipment manufacturers are recognizing the utility of the X.25 packet protocol for data switching. As a protocol that allows access to a public packet data network, X.25 is structured to provide users with data communications equipment's full spectrum of capabilities. That is, X.25 serves as the interface between the user device and the network. Additionally, X.25 has been more developed and improved than any de facto standard to date, and it is the most widely accepted protocol with both domestic and international compatibility.

Compelling reasons

The CCITT adopted X.25 as *the* domestic and international standard for connecting user equipment to public packet-switching networks. Besides that fact, there are compelling technical reasons for the use of this protocol as the PBX data standard, such as:

■ There are significant cost reductions in multiplexing applications involving PBX-to-computer interfaces. For example, X.25 interfaces exist today that can replace up to 255 ports on a multiplexer — that is, these interfaces can handle 255 simultaneous asynchronous connections to a single host or front-end processor (FEP) port. In this manner, utilizing the PBX as a controlling hub, with its X.25 port in place of a port contender, very real savings may be realized in the use of high-priced FEP ports — which cost anywhere from four to 100 times that of X.25 ports.

■ Ready connection may be made to outside packet-switching networks via a single, standardized protocol for all communications on the PBX link.

■ This *workable* protocol provides immediate hub benefits. Thus the risks involved in proving a new protocol are avoided — not to mention the expense to manufacturers of providing new protocol conversion equipment for older, current, and future products. This is indeed a critical area since manufacturers are seeking the most workable and economical solution to integrating their LAN and terminal equipment with PBXs. The more difficult it becomes to forge those integrated connections, the greater the cost. As a known and proven standard, X.25 makes it easier to plan, design, and implement economical protocol conversion equipment — such as PADs — for manufacturers' LAN and terminal equipment.

■ A standard X.25 packet-switching interface provides direct PBX access to any LAN and provides such access at no additional equipment cost. This results from the ability of PBX manufacturers to implement the interface during the assembly operation.

■ By packetizing data, X.25 allows the dynamic allocation of bandwidth on standard twisted-pair lines, enabling data to share the same lines with voice. In addition, this dynamic bandwidth allocation, inherent in X.25 packet switching, is responsible for lower PBX equipment costs. (Fewer ports are needed than with circuit switching for the same grade of service. The per-port cost is about $600.) Dynamic bandwidth allocation also substantially reduces leased-line costs—one needs fewer leased lines for the same or better service—as well as the need for more twisted-pair wiring.

■ X.25 gives users tremendous flexibility and resilience. For instance, terminals connected to the network can access remote mainframes via X.25. Moreover, users can "converse" with multiple destinations.

■ A public X.25 network is clearly defined in its interfaces to both the service provided and the organizations requiring that service—the destination and source of a transmission.

■ By eliminating unnecessary links, X.25 allows timesharing, not only of multiplexer channels, but also of computer ports, modems, and the transmission medium itself—enhancing communications efficiency.

Using the X.25 protocol ensures compatibility in the interconnection of computing devices. Moreover, for those PBXs that employ some form of internal packet switching, this compatibility and interconnectability can be combined with multiple simultaneous connections to single-host or public-data-network ports (Fig. 2).

Maximizing the benefits

Integrated PBXs provide totally digital communications for both voice and data. To eliminate the need for costly rewiring every time functions or features are added, deleted, or modified, these PBXs utilize a building's existing twisted-pair wiring scheme. In addition, if the PBX is a switch with a dual-bus architecture, the voice processing can use pulse code modulation. This allows the data throughput to be greatly expanded—typically to three to four times the previous amount—by using proven packet-switching technology, with no effect on the voice-switching capacity.

A packet-switching capability allows dissimilar terminals to communicate with each other or with central computers. It also allows tandem (nodal) switching of data circuits, use of different types of various protocols, and more efficient—because of packet switching—use of available bandwidth. Data may be switched at speeds up to 64 kbit/s synchronous and 19.2 kbit/s asynchronous via a packet-switching PBX.

In creating a parallel local communications network for data, integrated PBXs offer the potential for unnecessarily creating confusion in the data communications industry. The crux of the matter lies in how well the communications industry handles the issue of compatibility, or the standardization of protocols.

Most major PBX and LAN manufacturers have fully accepted the viability of the PBX as a data communications hub. But even though X.25 has received widespread acceptance, it is not the only protocol available. The longer that the marketplace waits for a "final" PBX protocol standard, however, the greater the likelihood that numerous de facto standards will be established— a different one from each PBX manufacturer.

There is currently a controversy over an acceptable standard for PBX-to-computer interfaces, with two opposing factions (see DATA COMMUNICATIONS, "Comparing the two PBX-to-computer specifications," May 1984, p. 215). The eventual merging of the two into the ultimate controlling standard, the Integrated Services

2. Application. *Besides compatibility between connected devices, the packet-switching PBX provides multiple simultaneous connections to X.25 ports.*

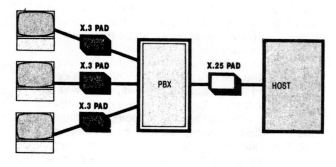

Digital Network (ISDN), has focused attention on requirements that will apply later in this decade (see "X.25, ISDN, and the office").

However, users must have assurances that the PBX office-equipment hub can enable communications between existing DTE (data terminal equipment) devices and today's host computers, LANs, and public data networks. Such a PBX, supporting internal packet switching, offers today's data communicator an immediate interface between existing asynchronous terminal equipment and those destination devices required by the user's business needs. Moreover, by providing direct T-carrier connections—which have been offered on modern digital PBXs since 1976—a digital PBX can offer, today, the utilization of multiple simultaneous X.25 connections over a single channel on that T-carrier.

For that future world when both the ISDN network standards and the ISDN terminal standards are agreed upon, the PBX will be able to continue supporting existing devices as "R" (non-ISDN standard) interfaces. With the installation of relatively simple program and hardware modifications, the PBX will provide an ISDN connection and local support of the ISDN standard, as well as of R terminal devices.

What ISDN may mean

Although the production of numerous incompatible protocols may benefit some manufacturers in the short term, in the long run such a dispersed marketplace will probably produce more confusion than advantage. Next-generation standards such as ISDN will parallel and may eventually absorb X.25. But a plethora of manufacturer-based protocols may create extensive and long-lasting confusion in the crucial area of standards. Helping to forestall this eventuality are the most recent CCITT/ISDN-committee recommendations, which have suggested X.25 for the first three layers of ISDN's seven-layer protocol.

When it comes to integrated PBXs and their coexistence in a multidevice environment, the issue of compatibility through a single communications standard is a critical one. Acceptance of a single standard should not curtail development of further improvements to current communications protocols. But the search

X.25, ISDN, and the office

Probably one of the least controversial of the forthcoming data communications offerings insofar as switched network services is concerned is the proposed Integrated Services Digital Network (ISDN). However, this offering has caused, and will continue to cause, a considerable amount of controversy in the area of terminal planning.

ISDN will become a reality. Interested potential suppliers include the nationally owned public telephone administrations in Europe and many other parts of the world, AT&T, the RBOCs (regional Bell operating companies), and the large independent telephone companies in the United States. They see a very real need to establish ISDN to protect against bypass and its erosion of their revenue base. Since these organizations are also determined to cooperate—ultimately—in establishing interface standards, this too will occur.

The controversy, however, arises in the marketplace, that common meeting ground where potential suppliers solicit the cooperation of potential purchasers. By the time ISDN becomes a viable, widespread offering, the North American market will surely contain a considerably greater quantity of terminal equipment than today's sizable numbers. Users of mainframes and minicomputers, smart and not-so-smart terminals, microcomputers, word processors, file servers, graphics devices, and communicating copiers will be encouraged to resist embracing a standard that might require extensive replacement or modification of their existing devices. On the other hand, a large segment of the industry with considerable resources is going to be pushing for the development and introduction of "true-ISDN" terminals.

Until such time as the dictates of the marketplace are more clearly understood, existing, well-defined, protocols and interface rules such as the CCITT Recommendation X.25 would seem to offer the most economically viable solution.

for such enhancements should not be allowed to interfere with the acceptance of a standard *now*, along with the substantial benefits such a standard can provide.

A single standard permits the users *and* the marketplace to continue with their business in an environment of stability and compatibility. Furthermore, it promotes future stability by setting a precedent for agreement among competitors. But most of all, it enables users to gain the greatest benefits from their existing and future communications investments, simply by *not* forcing them to choose among many incompatible options. ∎

Chet Sagaser has been with GTE since 1974, holding positions in marketing, sales, and sales support. He earned a B. S. degree in business administration from Charter Oak College, Hartford, Conn.

Chapter 3: A Modern PBX in Detail: NEC's NEAX2400 IMS as Example

One of the principal impediments to evaluating private branch exchanges (PBXs) is their sheer complexity. Were one to compare them to the world of computers, a PBX selection is similar to buying a processor (the PBX), a front-end processor (the line cards), a local network (the various house cables, tie lines, and twisted pairs), terminals (the station equipment), a network control center (the operator console and administrative/maintenance terminal), UPS and/or emergency generator, site preparation and installation, applications software (the various call processing options), and possibly one or more dedicated processors (for electronic mail, station message detail record processing), all in one purchase decision. As was previously argued, the greatest obstacle to effective PBX use is assembling the talent to unequivocally state the organization's requirements, describe its operating environment, and define its future directions.

Once assembled, a PBX evaluation committee must expect to encounter a wide range of various manufacturer's technical literature and specification documents. The large variety in these documents belies their ambitious and often ambiguous objectives. Variously conceived for readerships as diverse as technical support/marketing, maintenance, PBX administration, they are often monuments to the fact that it is harder to write less than more, more difficult to say no than yes. Invariably they are in the form of huge, undated, three-ring binders with numerous tabs, along with the inevitable loose, and usually undated, supplements and updates.

In the following section appears a complete, technical specification for the NEAX2400 IMS. NEC's was selected because it was (1) far more compact than most, (2) reasonably well written, and (3) contained a relatively high number of illustrative photos and diagrams. For a reader interested in contrasting PBX designs, another specification that meets these criteria is International Business Machines Corporation's (formerly Rolm) IBM 9751 CBX, Voice/Data Controller (145 pages).

In reading the document, the reader should pay special attention to terminology. Note the large number of proprietary terms and abbreviations introduced in "Description of Equipment—Voice and Data" (Section I, Part 2) and "Description of Operation" (Section I, Part 3). Although most modern PBXs are grossly capable of the same functions, the terminology often varies significantly between manufacturers. In an effort to mitigate the terminology problem, a generic glossary is provided at the rear of this volume.

The best way to surmount the array of proprietary terminologies in the PBX domain is to know what to expect in generic terms so that the reader can mentally construct a "terminology translation table" while ploughing through technical manuals. For instance, within PBX hardware, the reader should be looking for voice terminals (or stations), data modules, operator and maintenance consoles, line cards for stations, ties (for connections with other switching elements), and several kinds of central office trunks, some variety of synchronous, time-division multiplexed high speed bus, one or more main processors and the possibility of various specialized applications processors for electronic mail, gateways, protocol conversion, and the like.

A similar situation exists for software. There will be an operating system for a single PBX node and perhaps an optional network operating system. Depending on the functional role of the PBX (business, hotel, perhaps tandem switching), a large, generalized application package with many modules—which the user may or may not choose to activate—will handle the higher level call processing. Additional software, usually embedded in the line cards, will be present for the lower level call processing functions. If there are attached applications processors, they, too, will have software as well as hooks to the main processor's operating system.

Finally, there will be the various components of the PBX physical plant environment: the main distribution frame and a number of intermediate distribution frames employing some kind of punch down block or connector technology, batteries for UPS, charging equipment, power protection equipment, supplementary air conditioning, and, oftentimes, one or more diesel- or gas-powered generators capable of maintaining the batteries and the air conditioning over an extended outage.

Finally, bear in mind how NEC integrates voice and data features in its NEAX2400 IMS. This reflects a conscious attempt to add value (often billed as "enhanced functionality" and/or "user friendliness") at the primary user interface. In later papers where the products of other well-known PBX manufacturers such as AT&T, Northern Telcom, and IBM/Rolm are decribed, note the wide variation in designs and in the way vendor's multifunction (voice and data) stations communicate with the PBX. Depending on the manufacturer, the physical interfaces, number of pairs required, bit rate, multiplexing scheme, extent of integration of voice and data, etc., all appear to be different. Some manufacturers, for compatibility with older equipment, offer several schemes.

The effect of all this innovation—or proprietary fence building—at the user station is that one cannot connect an integrated voice/data terminal from one vendor to another's PBX. Although these integrated terminals have of late fallen drastically in price owing to offshore manufacture, the selection of a PBX using such a proprietary scheme will "lock in" an institution to a particular PBX vendor. Happily, this situation will likely change with the gradual adoption of integrated services digital network (ISDN) standards. Each ISDN voice/data set will manage two 64 Kbps and one 16 Kbps channels (the 2B+D "basic service") in a uniform manner. Interchangeable ISDN terminal devices, with software to adapt to the differences in manufacturers' feature sets, as well as competing sources of supply, will be the likely result.

SECTION I.   GENERAL DESCRIPTION

1.   Introduction

1.1   Scope

1.1.1   The NEAX2400 Information Management System (IMS) is an intelligent Digital Communications controller that offers all the functions and services of an Advanced Private Branch Exchange and can provide various integrated information services.

1.1.2   NEAX2400 IMS station users have access to over 200 service features which enhance user productivity, reduce operating costs, and improve communications efficiency.

1.1.3   Innovative modular hardware and software design allows the NEAX2400 IMS to serve efficiently over its entire size spectrum, from less than 34 to an excess of 20,000 ports.

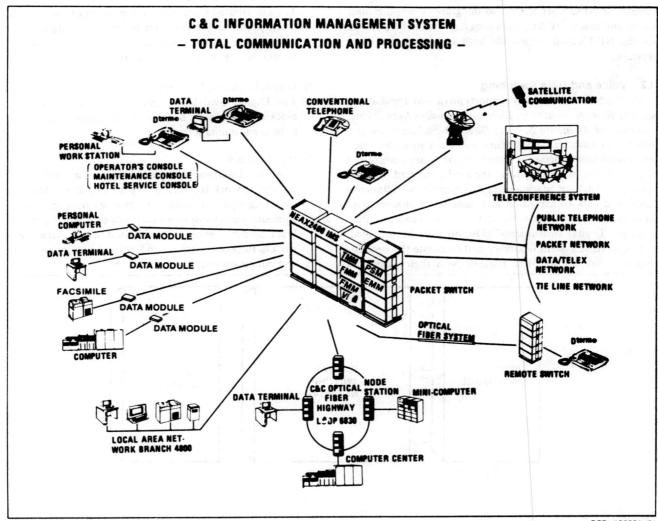

C & C INFORMATION MANAGEMENT SYSTEM
– TOTAL COMMUNICATION AND PROCESSING –

BCD-100001-01

Figure 1-1   C&C Information Management System

1.1.4 The NEAX2400 Information Management System has the ability to be expanded from its minimum configuration to its maximum capacity with virtually no loss of existing hardware. This unique expansion capability allows the system to grow in a cost effective manner as the user's requirements expand.

1.1.5 By serving as the controller of an integrated information network, the NEAX2400 IMS is able to integrate and enhance all forms of advanced information processing and management services. The NEAX2400 combines voice and data, store and forward, packet switching and other advanced features. This reflects the philosophy of NEC Corporation to integrate "Computer and Communications" (C&C) technology, and makes it possible for the NEAX2400 to provide both voice and non-voice services.

1.2 Voice and Data Switching

The NEAX2400 IMS supports advanced applications such as *Remote Switching, Electronic Tandem Networking, Centralized Attendant Service, Direct Digital Interface* and other voice features. The system's non-blocking architecture and distributed processor control hierarchy are designed to support the traffic and control the load generated by voice and data switching. Station users may perform simultaneous voice and data transmission at speeds up to 56Kbps over universal 2-pair wiring without the use of modems. Proprietary "Digital Instruments" (Dterm) may be provided, which increase system flexibility and eliminate the need for conventional multiple line stations with their associated

control equipment and cable plant requirements. The control, network and interface portions of the NEAX2400 IMS can accommodate features, services and subsystems as required by the specific application.

1.3 System Architecture

1.3.1 The NEAX2400 IMS architecture consists of 3 major functional components: Distributed Controller, Digital Switching Network, and Port Interface as shown in Figure 1-2.

(1) Distributed Controller
The Distributed Controller is composed of distributed multi-processing units, generic memory, data base instructions, system interface, and interface ports for system maintenance and administration.

(2) Digital Switching Network
The Digital Switching Network consists of a non-blocking digital time division switch, allowing all ports to be used simultaneously.

(3) Port Interface
The Port Interface provides access to the public and private network for various types of terminal devices, including digital and analog telephones, data terminals, computers and subsystems such as Voice Mail Systems, Data Switch Networks, wide band Local Area Networks, Packet Switches, and related communication and information services.

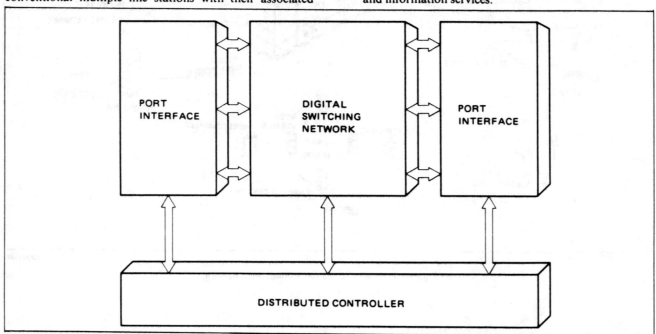

Figure 1-2 NEAX2400 IMS Architecture

BCD-100003-01

1.4 ADVANTAGES

1.4.1 The NEAX 2400 IMS provides a unique set of advantages to users who quest for an advanced information system that is both flexible and dependable. Through the use of state-of-the-art computer controlled telecommunications technology, NEC is able to provide the following advantages:

a)   Full-Featured System

NEAX2400 IMS station users have access to more than 200 service features which enhance user productivity, reduce operating costs, and improve communication efficiency. In addition, the control, network and interface positions of the NEAX2400 IMS can accommodate features, services and subsystems as required by specific applications.

b)   Network Integration

The NEAX2400 IMS offers business, industries, hospitals and hotel/motels the ability to access an extensive array of information processing and management services by serving as the central controller of an integrated information network.

The NEAX2400 IMS reflects the philosophy of NEC Corporation to integrate "Computer and Communications" (C&C) technology, and can provide many voice and non-voice services.

c)   Flexible Line Size

Innovative modular hardware and software design allows the NEAX2400 IMS to efficiently serve from under 64 ports up to 23,184 ports. Modularity gives the system the ability to be expanded from its minimum configuration to its maximum capacity as the need arises. This unique expansion capability allows the system to grow in a cost effective manner as the user requirements expand. Your initial system investment is protected through growth capability.

d)   Energy Savings and Space Savings

Through employment of state-of-the-art technology in the system circuity design, NEC has been able to reduce power consumption. As a result the current consumption of the system has been reduced to 60% of that of any conventional electronic PABX system. This energy saving oriented system design allows for the use of much smaller capacity main power equipment and air conditioning equipment. In parallel with the energy savings, the space requirement for the system has been reduced to one third when compared with that for a conventional electronic PABX system.

e)   Building Block Configuration

In a conventional switching system, various kinds of equipment are mounted in a cabinet group and are connected to each other by use of connecting cables. The NEAX2400 IMS, however, utilizes a building block modular design. When installing the system, the required blocks are placed on top of each other in a building block formation and interconnected by flat cables.

f)   Flexible Interface Port

The NEAX2400 IMS employs a "Universal Port" architecture that has the flexibility to accommodate station terminal equipment, trunks and adjunct processors. This universality allows the NEAX 2400 IMS to maximize slot space utilization and lower expansion costs.

g)   Redundancy

The basic design philosophy of the NEAX2400 IMS is to provide redundant capacity for all critical circuity such as processors, control circuity, memory, switching network and power, By providing this redundancy, NEC is able to offer a highly reliable system with an extremely high grade of maintainability and sophisticated trouble diagnoses. The NEAX2400 IMS has all the inherent reliability required to control a total network integration service. It is important to note that system redundancy is not available on the smallest NEAX2400 IMS configuration.

h)   High Reliability

The NEAX2400 IMS is designed and manufactured to provide the highest level of system reliability. The NEAX2400 IMS is designed with such features as: remote maintenance, distributed call processing, error-correcting memory, equipment redundancy (except on the smallest configuration), battery backup, self-testing and automatic system alarm indications to insure unsurpassed reliability. In addition, only the finest components have been used. In addition, through the employment of LSI and custom LSI technology the number of component parts has been greatly reduced, thus lessening possible failures and insuring continuous operation.

i)   Intelligent Attendant Console

The NEAX2400 IMS attendant console is a compact, desk top unit equipped with non-locking keys and Light Emitting Diodes. The LED's provide continuous information relative to the status of calls in progress. The display will provide station and trunk identification, class of service and the number of calls waiting. The attendant console also provides a flexible busy lamp field and digital clock.

j)   Intelligent Digital Multi-Function Terminal

In addition to supporting conventional station equipment, the NEAX2400 IMS can be equipped with the Dterm series digital electronic multi-function terminals. The Dterm terminals are intelligent microprocessor controlled terminals, which enhance the feature capabilities offered by the system and provide the service of conventional "key" telephones over 2-pair wiring. The Dterm instrument may be equipped with an interface adapter to allow simultaneous voice and data switching, without compromising the voice communication system.

k)   Ease of Installation

Because the NEAX2400 IMS utilizes preassembled modules and plug-in type circuit packs, it is easy to install. Wiring connections, both internal and external, are made through simple to use standard plug-ended cables. In addition, with each unit and system having been fully factory tested prior to shipment, potential obstacles to easy installation have been held to a minimum. Furthermore, in the case of the Single Interface Module (SIM) configuration, Default Data architecture has been employed to simplify installation.

l)   Ease of Maintenance

Because the system is constructed with first quality components, reliability is high and operation is trouble free. However, should a minor fault occur, the self diagnosing programs will detect the fault, and automatically make the needed corrections. If the problem is beyond the internal correction capabilities of the system, the self diagnosing programs will automatically print the nature of the fault and the involved unit is identified on the man-machine interface equipment. The faulty plug-in unit can then be quickly replaced with little or no interruption of service.

m)   Flexible Numbering Plan

The NEAX2400 IMS provides the ability of flexible numbering assignment to meet all forms of network integration service.

n)   Future Capabilities

Because the NEAX2400 IMS utilizes a stored program control, performance enhancements and new features can be easily incorporated by simple changes in software. The NEAX2400 IMS is upgradeable by utilizing the most current software release and, when necessary, additional hardware. This means that the system will not become obsolete.

o)   Cost Controls

With telecommunications costs growing, it is becoming increasingly important to control them. The NEAX 2400 IMS makes it possible for you to get a firm grasp on telecommunication costs. Through the use of such features as least cost routing, class of service and detail call recording, cost reduction and control are possible.

2. Description of Equipment — Voice and Data

2.1 General

2.1.1 The NEAX2400 IMS hardware is designed to meet the following parameters:

a) flexibility and capacity

b) reliability

c) optimum space utilization

d) minimal environmental requirements

e) ease of installation and maintenance

2.1.2 A unique modular design is employed throughout the NEAX2400 IMS, from the smallest Single Interface Module (SIM) configuration to the largest Ultra Modular Group (UMG) configuration. Figure 2-1 illustrates the modular growth of the NEAX2400 IMS. As additional equipment modules are required, they are stacked on top of each other, versus mounting the equipment in conventional frames or cabinets. This innovative method reduces installation time, avoids the use of bulky frames and provides for manageable future expansion, virtually eliminating the possibility of outgrowing your NEAX2400 Information Management System.

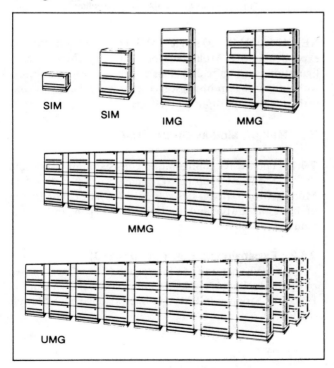

Figure 2-1 NEAX2400 IMS

2.1.3 The structure and physical appearance of each module of the NEAX2400 IMS is identical. The overall view, appearance and dimensions of each module is the same.

2.2 Single Interface Module (SIM)

2.2.1 The minimum equipment configuration for the NEAX2400 IMS is the SIM configuration. The SIM configuration is designed to efficiently meet the needs of installations requiring from 60 to 100 station lines. Its basic configuration consists of one Basic Module (BSCM), as shown in Figure 2-2. This configuration provides up to 160 interface ports for stations/trunks and other terminal devices.

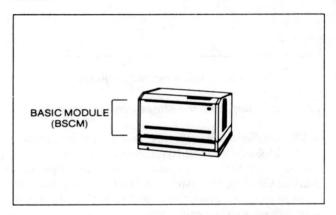

Figure 2-2 Minimum SIM Configuration

2.2.2 The SIM can be expanded to accomodate up to 472 ports in its maximum configuration. This configuration consists of three modules: BSCM, PIM0 and PIM1, which are described below.

a)   Basic Module (BSCM)
       The Basic Module houses the SIM system program and up to 160 ports for lines/trunks and other devices.

b)   Port Interface Module 0 (PIM0)
       PIM0 can accomodate a total of 184 ports for lines/trunks and other devices.

c)   Port Interface Module 1 (PIM1)
       PIM1 is identical to PIM0, but due to limitations of the unique SIM system processor, only 120 of the 184 ports are accessed.

Figure 2-3 Maximum SIM Configuration

2.3 Expansion from the Single Interface Module

2.3.1 Configurations over 464 ports require an Interface Module Group (IMG) configuration, which accomodates up to 736 ports. Because of the modular design of the NEAX2400 IMS, this expansion can be accomplished by using most of the equipment from the SIM configuration and some additional components.

2.4 Interface Module Group (IMG)

2.4.1 An Interface Module Group (IMG) consists of one Miscellaneous Module (MISCM) and up to four PIM's (see Figure 2-4), depending on the system requirements. The basic modules are described below:

a)   Port Interface Module (PIM)
The PIM provides access to 184 ports for stations/trunks and other terminal devices.

b)   Miscellaneous Module (MISCM)
The MISCM provides locations necessary for Alarm, Maintenance and other hardware controllers.

2.5 Expansion from the Interface Module Group

2.5.1 For configurations greater than 736 and less than 5520 ports, a Multiple Modular group (MMG) configuration is used. Due to the modular design of the

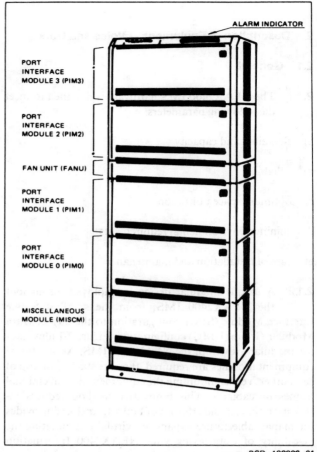

BCD-100006-01

Figure 2-4 Maximum IMG Configuration

NEAX2400 IMS, a single IMG configuration can be expanded into a Multiple Module Group (MMG) configuration (see Figure 2-5). This expansion is accomplished without causing an obsolescence of existing hardware and with minimal interruption of service.

2.6 Multiple Module Group (MMG)

2.6.1 The MMG configuration consists of an Expansion Module Group (EMG) and up to seven Interface Module Groups (IMG's), see Figure 2-6. The configuration of both the EMG and the IMG's is dependant on the required capacity.

2.6.2 Expansion Module Group (EMG)
The Expansion Module Group (EMG) is comprised of up to two Port Interface Modules (PIM's), two Control Processor Modules (CPM) and one Miscellaneous Module (MISCM).

2.6.3 Control Processor Module (CPM)

The Control Processor Module (CPM) consists of three sections; Main Processor (MP), System Processor (SP) and Highway Switch (HSW) section as shown in Figure 2-7.

(1) Main Processor Section

The Main Processor Section is responsible for overall system control of all switching operations and operates in a load sharing mode. This section also provides a location for storage of data base information.

(2) System Processor Section

The System Processor Section performs supervision of switching operations, faults analysis, backup memory control and overall man-machine interface.

(3) Highway Switch Section

The Highway Switch Section serves as a portion of the multistage Time Division Switch.

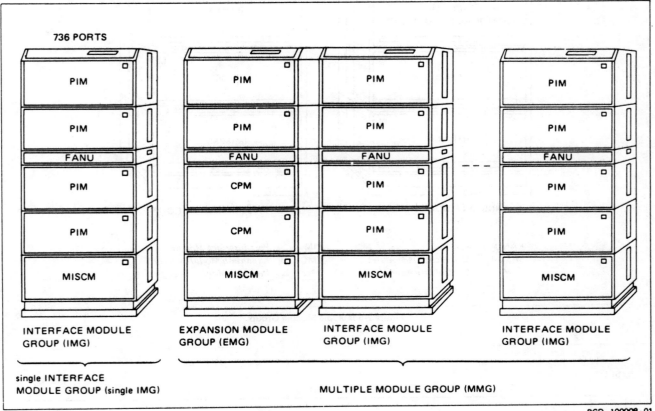

BCD-100008-01

Figure 2-5 Expansion from single IMG to MMG

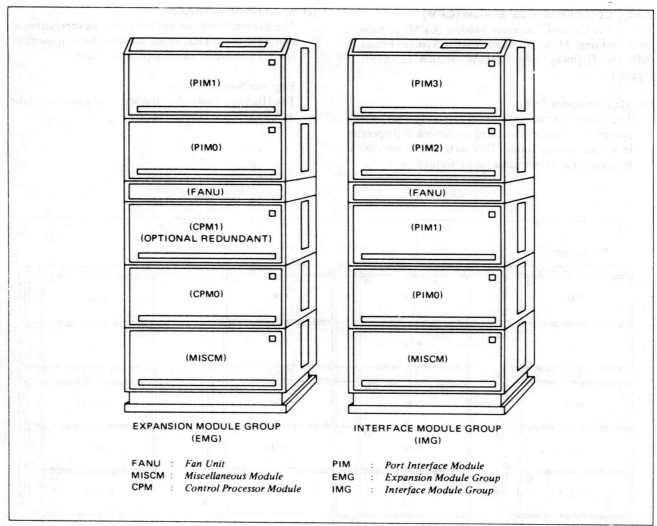

EXPANSION MODULE GROUP
(EMG)

INTERFACE MODULE GROUP
(IMG)

FANU   :   *Fan Unit*
MISCM :   *Miscellaneous Module*
CPM   :   *Control Processor Module*

PIM   :   *Port Interface Module*
EMG   :   *Expansion Module Group*
IMG   :   *Interface Module Group*

Figure 2-6 Multiple Module Group   BCD - 100009 - 01

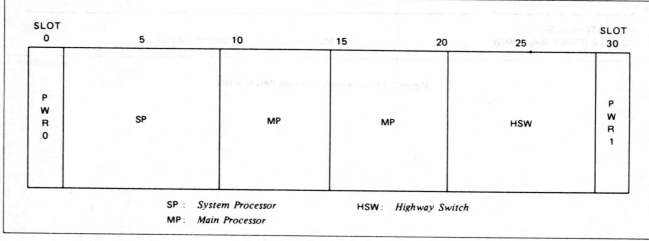

SP :   *System Processor*   HSW: *Highway Switch*
MP :   *Main Processor*

Figure 2-7 Control Processor Module (CPM)   BCD - 100010 - 01

2.7 Expansion from the Multiple Module Group (MMG)

A system requiring more than 5,520 ports requires an Ultra Module Group (UMG) configuration. Because of the modular design of the NEAX2400 IMS, an MMG can be expanded to a UMG configuration with no obsolescence of hardware, and minimal interruption of service. Expansion of the MMG to a UMG is accomplished by the installation of the Highway Module Group 0 (HWMG0). With the installation of the HWMG0, up to 15 Interface Module Groups may be assembled together. Beyond 15 Interface Module Groups, a HWMG1 must be installed, allowing the system to grow to its maximum capacity of 23,184 ports. This expansion is illustrated in Figure 2-8.

2.8 Ultra Module Group (UMG)

The UMG is the ultimate system configuration for the NEAX2400 IMS. The UMG is composed of one EMG, one or two Highway Module Group(s) (HWMG) and up to 31 IMG's. The number of HWMG's and IMG's is determined by the required system capacity. The basic groups comprising the UMG are described below:

a) Highway Module Group (HWMG)

Provides the highway portion of a 3-stage Time Division Switch.

b) Expansion Module Group (EMG)

Provides space for system control and some ports for lines/trunks and other terminal equipment.

c) Interface Module Group (IMG)

Provides ports for lines/trunks and other terminal equipment.

Figure 2-8 Expansion from MMG to UMG

3. Description of Operation

3.1 Configuration

3.1.1 Single Interface Module (SIM)

The SIM provides access of up to 464 ports for stations/trunks and other terminal devices. It's smallest configuration consists of one Basic Module (BSCM) for 160 ports, and can grow to its maximum configuration of one BSCM and two Port Interface Modules (PIM's).

(a)　Basic Module (BSCM)

The BSCM provides access to 160 ports for station/trunks, and the necessary locations for Alarm, MAT interface, Common Controller and other devices.

(b)　Port Interface Module 0 (PIM0)

PIM0 provides access to 184 ports for stations/trunks and other devices.

(c)　Port Interface Module 1 (PIM1)

PIM1 provides access to 120 ports for station/trunks and other devices.

3.1.2 Interface Module Group (IMG)

An IMG provides access to 736 ports for stations/trunks and other terminal devices. An IMG can support up to 4 Port Interface Modules (PIMs) and 1 Miscellaneous Module (MISCM).

(1) Port Interface Module (PIM)
Each PIM provides access to 184 ports for stations/trunks and other devices.

(2) Miscellaneous Module (MISCM)
The MISCM provides locations necessary for Alarm Supervisory Circuitry, interface to the Maintenance Administration Terminal (MAT) and other hardware controllers.

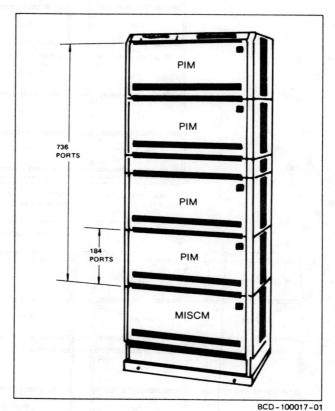

BCD - 100017 - 01

Figure 3-2 Configuration of Interface Module Group (IMG)

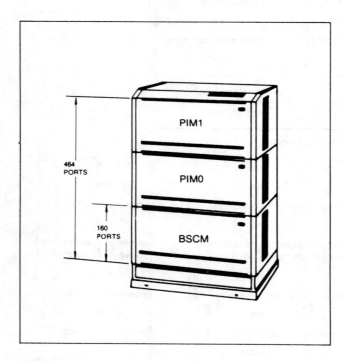

Figure 3-1 Configuration of Single Interface Module (SIM)

3.1.3 Multiple Module Group (MMG)

An MMG consists of an Expansion Module Group (EMG) and a number of Interface Module Groups (IMG), depending on required capacity. An MMG is used for configurations greater than 736 ports and less than 5521 ports.

(1) Expansion Module Group (EMG)
An Expansion Module Group (EMG) is comprised of two or three PIM's, one or two Control Processor Modules (CPM's) and one Miscellaneous Module (MISCM).

(2) Control Processor Module (CPM)
A Control Processor Module (CPM) houses the distributed Multiprocessor control circuitry.

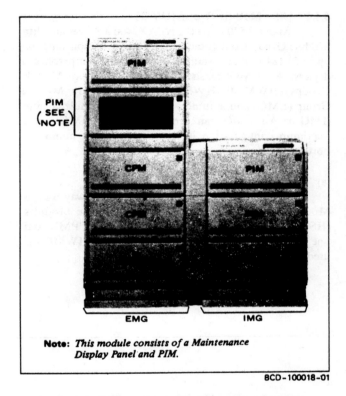

Note: *This module consists of a Maintenance Display Panel and PIM.*

8CD-100018-01

Figure 3-3 Configuration of Multiple Module Group (MMG)

3.1.4 Ultra Module Group (UMG)

Above 5,520 ports the NEAX2400 assumes an Ultra Module Group Configuration. This configuration supports up to 23,184 ports for stations/trunks and other perpherial devices. An UMG consists of 1 or 2 Highway Module Groups (HWMG0, HWMG1), an Expansion Module Group (EMG) and a number of Interface Module Groups (IMG's). An UMG can provide up to 31 IMG's in a maximum configuration, in accordance with the number of ports required.

(a)   Highway Module Group 0 (HWMG0)

A HWMG0 is composed of two Highway Switch Modules (HSWM's), two Control Processor Modules (HSWM's), two Control Processor Modules (CPM's) and one Miscellaneous Module (MISCM). This HWMG0 is used in all UMG configurations.

(b)   Highway Module Group 1 (HWMG1)

A HMG1 is comprised of two Highway Switch Modules (HSWM's), two Control Processor Modules (CPM's) and one Miscellaneous Module (MISCM). The HWMG1 is required for systems with more than 15 IMG's.

(c)   Expansion Module Group (EMG)

An EMG is comprised of two PIM's, two Control Processor Modules (CPM's) and one Miscellaneous Module (MISCM).

(d)   Interface Module Group (IMG)

An IMG is comprised of one Miscellaneous Module (MISCM) and up to four Port Interface Modules (PIM's), depending on the number of required ports.

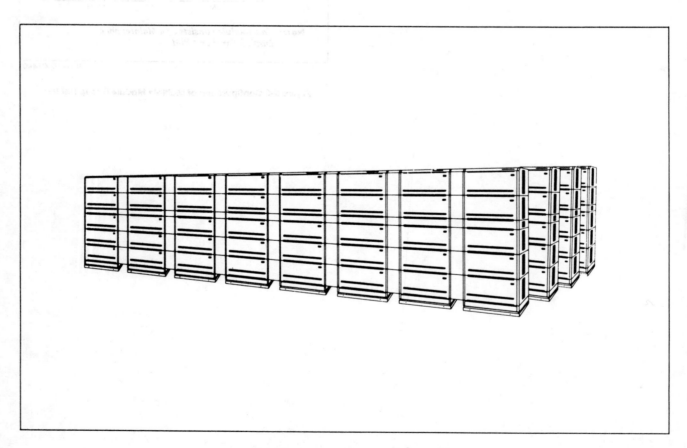

Figure 3-4 Configuration of Ultra Module Group (UMG)

3.2 Port Interface

3.2.1 Port Interface consists of interface hardware and supervisory components for connecting various types of stations, terminals and other devices to the networks. Each individual station/trunk or terminal circuit card within the PIM is equipped with a Port Microprocessor (PM) containing a unique set of software instructions. Under the direction of, and in communication with the Central Processor Unit (CPU), the Port Microprocessor performs the "real time" functions required by the specific type of station/trunk or terminal devices.

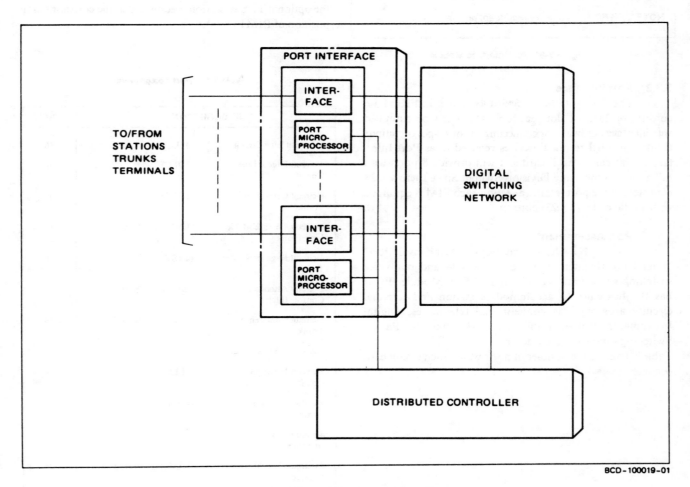

BCD-100019-01

Figure 3-5. Block Diagram of Port Interface

3.3 Port Interface Module (PIM)

3.3.1 The PIM consists of three sections, Port Interface, Time Division Switch (TDSW) and Processor. Figure 3-6 shows Port Interface Module (PIM).

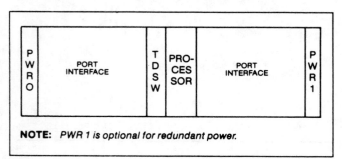

NOTE: *PWR 1 is optional for redundant power.*

Figure 3-6   Port Interface Module

3.3.2   Port Interface

The Port Interface Section as shown in Figure 3-6 consists of 23 card slot locations which can be equipped with interface hardware for connecting various types of station/ trunk and other terminal devices referred to as Port Interface circuit cards. Each card slot will provide 8 ports with full access to the Time Division Switch. Since there are 23 card slots in the port interface section each PIM is equipped with a total of 184(8 x 23) ports.

Port Assignment

As the NEAX2400 employs a "Universal Port" architecture to interface voice, voice/data and non-voice switching subsystems, each Port Interface Module (PIM) has the flexibility to accommodate station and terminal circuit cards such as conventional telephones, Digital Instruments, Private/Public Network Interface, Packet Switching subsystem Interface, etc.
Table 3-1 shows the number of ports which each circuit card occupies.

3.3.3 The Processor Section mounted in PIM0 (Active Processor) and in PIM1 (Optional Standby Processor) is capable of controlling four (4) PIM modules for an IMG configuration or larger. Each port interface card contains its own Port Microprocessor (PM) which communicates with an active processor.

3.3.4 In the IMG configuration, the TDSW Section contains a non-blocking PCM Time Division Switch and all related circuitry which are located on a single printed circuit card mounted in PIM0. If full redundancy is desired the optional TDSW section is equipped in the corresponding position of PIM1.

Table 3-1   Port Assignment

DESCRIPTION			PORTS
16-circuit Line Circuit	(16LC)	Note 1	16
4-circuit Digital Line Circuit	(4DLC)		8
8-circuit Central Office Trunk	(8COT)		8
4-circuit 4W E & M Tie Line Trunk	(4ODT)	Note 1	16
4-circuit Register Sender Trunk	(4RST)		8
2-circuit Attendant Console Interface	(2ATI)	Note 2	8
8-circuit Conference Trunk	(8CFT)	Note 2	16
4-circuit Tie Line Trunk	(4TLT)		8
8-circuit Electronic circuit	(8ELC)		8
4-circuit Data Line circuit	(4DTL)		8
4-circuit Modem Interface	(4MDMI)		16
24-circuit Digital Interface	(24DTI)	Note 3	24

Note 1: *16LC and 40DT take up 2 card slots.*
Note 2: *Location of 2ATI and 8CFT are fixed.*
Note 3: *24DTI utilizes up to 4 card slots.*

3.4 Digital Switching Network

3.4.1 Time Division Switch (TDSW) for Single Interface Module Group (SIM) and Interface Module Group (IMG)

The Time Division switch is a full availability, nonblocking device. The TDSW employs solid-state memory to exchange digital PCM formatted information between terminal devices. Operating under the control of the CPU, the TDSW establishes a communication path between the desired instruments, terminals, or other devices. The TDSW and all related digital tone synthesizer and network timing circuits are contained on a single printed circuit card. The TDSW can be equipped with redundancy to provide complete network reliability.

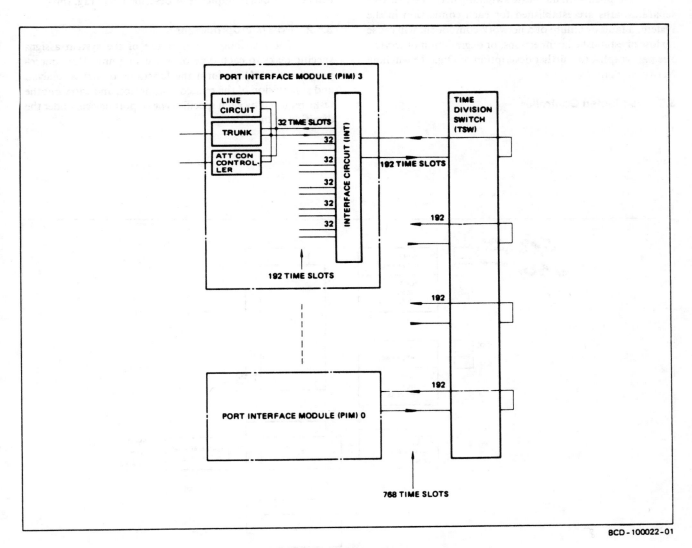

Figure 3-7 Time Division Switch - single IMG

BCD - 100022 - 01

3.4.2 Time Division Switch (TDSW) for Multiple Module Group (MMG) and Ultra Module Group (UMG)

For applications in excess of 736 ports, the TDSW functions as a multistage network. In all configurations and all sizes, the switching network remains a "Virtually Non-Blocking" Device.

3.4.3 To provide maximum reliability, the system can be equipped with duplicate switching networks. As two separate paths are established for each connection in the system, a fault or failure of a network component will cause no loss of established connections, or degradation of service. See paragraph 4 for further description of Digital Switching Network reliability.

3.5 Distributed Controller

3.5.1 Central Processing Unit (CPU) for single IMG Configuration

The CPU is comprised of a single 16-bit microprocessor (PD 8086), associated memory and interface components. Under the direction of the program data stored within the memory section, the CPU maintains overall control and supervision of the Digital Switching Network (TDSW) and Port Interface Module (PIM) sections of the system. The CPU and all related memory can be provided in duplicate as described in paragraph 4.

3.5.2 Port Microprocessor

The distributed architecture of the system assigns specific tasks to each type of processing unit. Port microprocessors (PM) perform the functions of status analysis and supervision of the related port device, and carry out the tasks required by each specific type of port device, under the direction of the CPU.

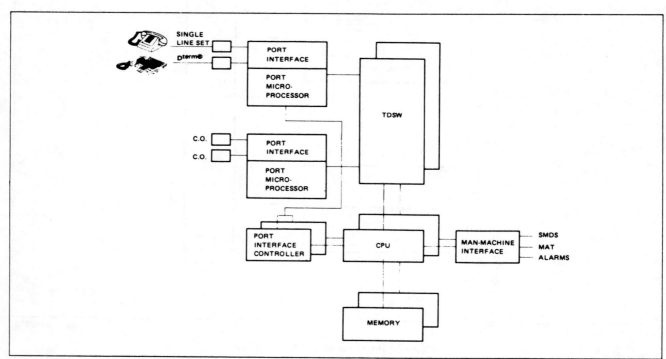

BCD - 100023 - 01

Figure 3-8 Block Diagram of single Interface Module Group (IMG)

3.5.3 CPU Memory

The memory system of the SIM and IMG is divided into 3 major sections; generic memory, operating memory, and system data (data base) memory. The following is a brief description of the three types of memory.

(1) Generic Program Memory

The generic memory is composed of solid state Programmable Read Only Memory (PROM) circuits.

The generic program consists of instructions relating to overall control of the system and provides the CPU with the necessary intelligence to execute the tasks required by the system, and carry out continuous diagnostic and fault isolation routines.

To ensure the integrity of the generic program, the memory devices used within the system are non-volatile and noise-immune. This program storage technique insures that all critical data is retained by the system without reliance on commercial power, or an external mechanical medium such as a cassette tape, or disk system. Program maintenance using this type of storage is not susceptible to faults caused by commercial power line fluctuations, or failures common to mechanical devices.

(2) Operating Memory

The operating memory of a single IMG system consists of solid-state Random Access Memory (RAM) which is used for temporary storage of data related to tasks required by port interface circuits. This information is subject to continuous updating by the CPU. Information such as circuit status, status of features and related data is inserted into the operating memory and recalled or altered by the CPU as required.

(3) Data Memory

The Data Memory of a SIM and IMG consists of Random Access Memory (RAM) containing information relating to service feature assignments, port class of service and terminal device type. System configuration and other site-specific data is stored in the data memory. To preserve this data without reliance upon external power sources or storage devices, the data memory is equipped with Ni-Cad batteries. These batteries will preserve all information stored within the data memory for a minimum of 1 week without any external power source. Further security is provided by a floppy disk containing the data memory contents. If power is not restored before the batteries run down, the floppy disk can be used to restore the lost memory.

3.6 Multi-Processor Architecture for Multiple Module Group (MMG)

3.6.1 Applications requiring in excess of 736 ports are equipped with a Multi-Processor configuration. The use of a Multi-Processor architecture increases system real time processor capacity by distributing the functions typically assigned to a Central Processor Unit over a network of interactive controllers.

3.6.2 The distributed/load sharing technique employed in the control hierarchy allows the NEAX2400 IMS to add services, features, and functional modules without exceeding a predefined control capacity. All processors are identical in hardware, and vary only in their resident programs.

3.6.3 The following is a functional description of the components and subsystems within the Multi-Processor Configuration.

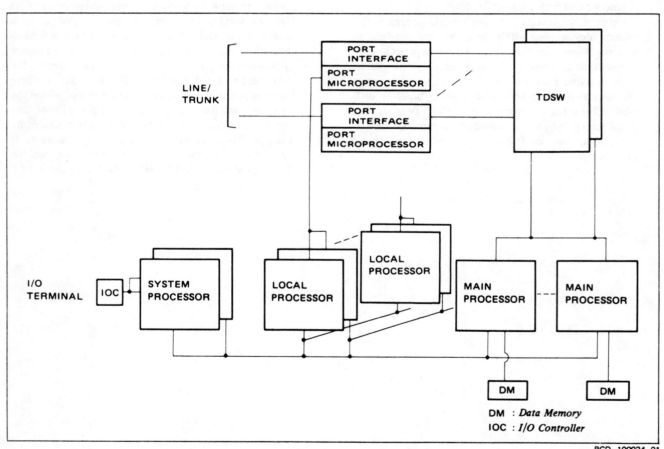

BCD-100024-01

Figure 3-9. Block Diagram of Multiple Module Group (MMG)

(1) Local Processor (LP)

Local Processors analyze, control and supervise the status of terminal devices accommodated in the Port Interface Modules (PIM). LP's perform their tasks based on the stored program resident on each LP. By interacting with the Port Microprocessors (PM) located on each port interface circuit card, and the other components of the multi-Processor configuration, Local Processors assist in establishing and maintaining the required connections within the switching network. Each Local Processor can control up to 4 PIM's (736 ports). LP's and their associated memory and interface may be duplicated for redundancy.

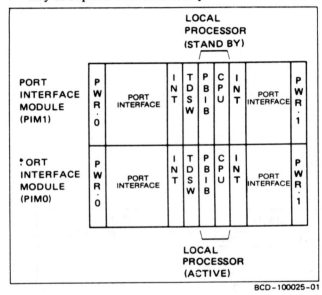

Figure 3-10 Local Processor - Redundancy

(2) System Processor (SP)

The System Processor performs management and supervisory functions for the multi-Processor configuration. Resident program instructions allow the SP to perform continuous system diagnostic routines, and isolate faults or failures to the smallest replaceable unit. Data relating to traffic, message detail information, and other related management data are collected by the SP and stored within its Data Memory. System Data examination and administration are provided by the SP in conjunction with the Maintenance Administration Terminal (MAT) which is described in paragraph 5.

As with all control components of the NEAX2400 IMS, the SP may be provided in duplicate for maximum system reliability.

(3) Main Processor (MP)

The MP interacts with the LP(s) and SP to execute tasks relating to system status transition, such as call status management, input data analysis, and control of the Time Division Switches. These functions are carried out in accordance with the instructions and information stored within the generic program and data base memory.

The Multiple Module Group Configuration is always equipped with a minimum of 2 MPs. All MPs operate in a load sharing/function duplicating mode, providing full redundancy. The number of MPs may be increased in a modular fashion as the system is expanded.

3.7 System Memory (Multiple Module Group)

3.7.1 Each processor within the system is equipped with memory for generic instructions (PROM), storage of call processing data (RAM) and system data base information (RAM). For detail operation of system see paragraph 3.5.3(1) - (3).

3.7.2 To provide for additional backup of operating memory and data base information, copies of the SP(s), MP(s), and data memory are contained on Bubble Memory devices. Bubble Memory will retain all contents without reliance on external power sources.

3.8 Processor Architecture for Ultra Module Group (UMG)

The processors used in the UMG are theoretically the same as the MMG. However, due to the total system traffic capacity, two additional types of processors, which were originally a part of the MMG's Main Processor, are required.

3.8.1 The following is a brief description of each of the processor's function:

a)   Local Processor (LP)

Provides local control for up to 736 ports or four PIM's.

b)   Main Processor (MP)

Provides overall system functional control.

c)   System Processor (SP)

Provides man/machine interface, and system maintenance control.

d)   Time Division Processor (TP)

Provides hardware control of the time division switch.

e)   Administration Processor (AP)

Performs SMDS data collection and reporting functions for the NEAX2400 IMS UMG configuration.

f)   Processor Bus

Provides local processor with information regarding the status of its associated ports.

g)   Data Memory

Two sets of Data Memory are provided in the UMG. These two data Memories are accessible from any MP through the data path in the UMG.

4. System Reliability

4.1 General

4.1.1 The NEAX2400 Information Management System has all the inherent reliability required to control a total Integrated Office System. Through employment of VLSI and custom LSI technology, the total number of parts has been greatly reduced from that of a conventionally designed switching system, resulting in overall reliability. The basic design philosophy of the NEAX2400 IMS is to provide in duplicate all critical circuitry such as processors, control circuitry, memory (Data Base and Generic), Switching Network and Power. The modular system design and sophisticated Diagnostics Routines enable maintenance personnel to isolate and replace faulty modules in a minimum amount of time.

4.2 System Redundancy

4.2.1 The NEAX2400 IMS single IMG, MMG and UMG can be configured as a dual, or redundant system (see Table 4-1). This provides a high grade of maintainability and continuity of service. The dual system employs high-level fault diagnosis programs to control the redundant components should they become necessary. All critical circuity, including processors, control, generic memory, switching network and power circuits can be provided in duplicate.

system	IMG		MMG		UMG	
grade	Dual-1	Dual-2	Dual-2	Dual-3	Dual-2	Dual-3
block						
CPU (LP)		X	X	X	X	X
MP	—	—	X	X	X	X
SP	—	—	X	X	X	X
TP	—	—	—	—	X	X
SMDS (AP)	—	—	—	—	X	X
INT		X	X	X	X	X
TDSW		X	X	X	X	X
HSW	—	—	X	X	X	X
HSWC	—	—	X	X	—	—
PLO		X	X	X	X	X
NMC	—	—	X	X	X	X
DLKC	—	—	X	X	X	X
GT		X	X	X	X	X
BUS	—	—	X	X	X	X
DSPC	—	—	X	X	X	X
BM	—	—	X	X	X	X
PWR		X		X		X

Table 4-1 NEAX2400 IMS Redundancy

4.3 Single Interface Module Group (IMG)

4.3.1 Redundancy

A single Interface Module Group can provide the following sub-systems in duplication, as shown in Figure 4-1 and Table 4-2.

Processors
Memory
Switching Network
Control Circuitry
Power Supplies
Ringing Generator Circuitry
Digital Tone Synthesizer

4.3.2 Central Processors (single IMG)

The Central Processor Unit (CPU) and all its related memory and circuitry for a single Interface Module Group can be provided as a duplicate system operating in an active/standby arrangement. The CPU maintains overall control and supervision of the Network (TDSW) and Interface (PIM) sections and is constantly providing automatic diagnostic information to the Maintenance Administration Terminal (MAT). The CPU is constantly performing diagnostics on itself and all related operating memory. If a fault is detected, it will automatically changeover to the standby system without any interruption or degradation of service. Therefore, all existing connections will be maintained during switchover. The active CPU is continuously checking the standby CPU and its associate memory and control circuitry to determine if any problem exists before a need arises to changeover. On a daily basis the active and standby CPU and all related circuitry change their roles automatically, manually, or via a command from the MAT.

4.3.3 Port Microprocessor (PM)

The Port Interface Module houses stations/trunks and terminal interface cards, each equipped with its own Port Microprocessor. The Port Microprocessor operates in a distributed mode, constantly in communication with the CPU. If the CPU detects a faulty Port Microprocessor or its related interface circuitry, it will automatically send the status of the fault, the Processor Number and the Time and Date of occurrence to the Maintenance Administration Terminal (MAT) so that corrective action can be taken.

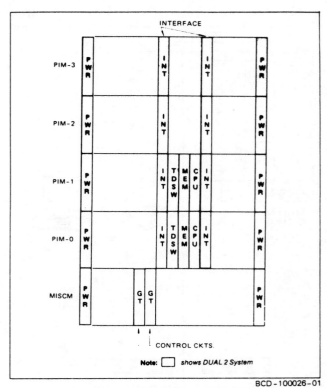

BCD-100026-01

Figure 4-1 single IMG - Redundancy

Table 4-2 single IMG - Redundancy

SYSTEM	TYPE
CPU	ACT/ST-BY
GT	ACT/ST-BY
INT	ACT/ST-BY
TDSW	ACT/ST-BY
MEM	ACT/ST-BY
RG	ACT/ST-BY
PWR	PARA-OP

ABBREVIATION	DESCRIPTION
ACT/ST-BY	Active/Standby Type of Redundancy
CPU	Central Processor Unit
GT	Gate Circuit
INT	Interface between TDSW and Port Card
PARA-OP	Parallel Operation
MEM	Memory Circuit
RG	Ringing Generator
PWR	Power
TDSW	Time Division Switch

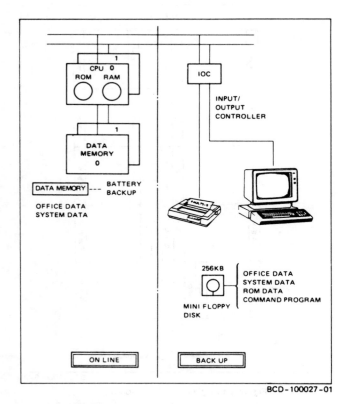

BCD-100027-01

Figure 4-2 Backup System for single Interface Module Group

4.3.4 System Memory

The NEAX2400 IMS uses the following three types of memory: Generic Memory, Operating Memory and System Data (Data Base) Memory, as shown in Figure 4-2.

(1) Generic Program

The Generic Program contains the main instruction set for each processor in the system. By employing PROM, the information resident in the generic is non-volatile and noise immune. This type of storage medium can be used because of the inherent stability of the NEAX2400 software, which elimintes the need for constant reprogramming of the generic operating system. As new features and service enhancements are offered they can be implemented without any interruption of system operation.

(2) Operating Memory (single Interface Module Group)

The operating memory of a single IMG system uses Random Access Memory (RAM) for temporary storage of call processing data. If a fault occurs in the active operating memory section, it will automatically switch-over to the standby section with no effect upon existing connectings.

(3) Data Memory (single Interface Module Group)

The Data Memory, RAM, employs Ni-Cad batteries to retain stored contents, thus eliminating the requirement of an external power source for memory retention. Further safeguard of the Data Memory is provided by duplicated backup information contained in the standby memory section.

"Off-line" storage is provided through the use of a floppy disk accessible through the Maintenance Administration Terminal (MAT).

4.3.5 Switching Network (single Interface Module Group)

The switching network consists of a nonblocking digital Time Division Switch (TDSW) and related components contained on one circuit card to provide maximum reliability. A redundant TDSW can be equipped operating in a parallel switching network mode as shown in Figure 4-3.

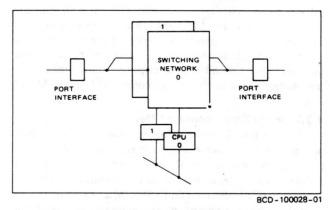

BCD-100028-01

Figure 4-3 Switching Network for IMG

Unlike conventional switching systems, if a failure occurs in the Switching Network, there will be no interruption of service.

4.3.6 Power Distribution - single Interface Module Group

In a single IMG, each individual module (MISCM, PIM) has its own power module circuit card which is used as a DC-DC Power Supply, Ringing Generator Supply, and Howler Tone Supply as shown in Figure 4-4. Each module can be equipped with Dual Power circuit cards which work in a parallel operation mode of operation.

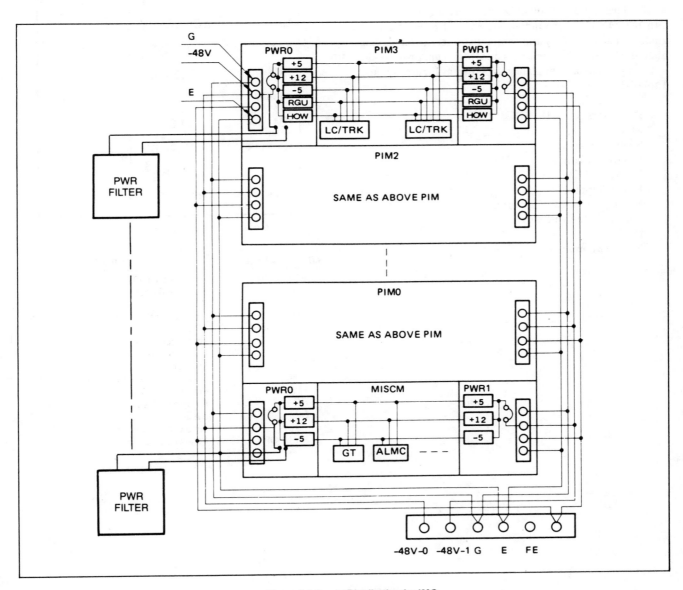

Figure 4-4 Power Distribution for IMG

4.4 Multiple Module Group (MMG)

4.4.1 A Multiple Module Group (MMG) (over 736 ports) has all the reliable characteristics of a single IMG. The MMG possesses increased processor capacity, as the work load offered by the system is distributed over a number of interactive processing units

4.4.2 Redundancy

4.4.2 This section will describe the inherent reliability built into the MMG. A redundant system will provide a high grade of maintainability and continuity of service employing high level fault diagnostics. An MMG can provide the following grade of redundancy as shown in Figure 4-5, and Table 4-3.

Processors (LP, MP, SP)
Memory
Switching Network
Control Circuitry
Processor Data Bus
Power Supplies
Ringing Generator Circuitry
Digital Tone Synthesizer
IO Bus
Bubble Memory

Table 4-3 MMG Redundancy

SYSTEM	TYPE	GRADE	
		DUAL-2	DUAL-3
BUS	LS	X	X
MP	LS	X	X
SP	ACT/ST-BY	X	X
LP	ACT/ST-BY	X	X
NMC	ACT/ST-BY	X	X
INT	ACT/ST-BY	X	X
TDSW	ACT/ST-BY	X	X
BM	LS	X	X
IO BUS	ACT/ST-BY	X	X
RG	ACT/ST-BY		X
PWR	PARA-OP		X

Note: X *shows redundancy provided.*

ABBREVIATION	DESCRIPTION
ACT/ST-BY	Active/Standby Type of Redundancy
BM	Bubble Memory
BUS	Bus
INT	Interface between TDSW and Port Card
IO BUS	I/O Bus
LP	Local Processor
LS	Load Sharing
MP	Main Processor
NMC	Network Management Controller
PWR	Power
RG	Ringing Generator
SP	System Processor
TDSW	Time Division Switch
PARA-OP	Parallel Operation

4.4.3 Multi Processor System

A Multiple Module Group (MMG) is equipped with a Multi-Processor system, where each Processor independently shares a function or load and communicates with each other through a common data BUS. Each Processor in a Multi-Processor system has its own CPU, Generic Memory (PROM), Operating Memory (RAM), and Processor BUS Interface circuits controlled by the switching program stored in the individual Processor Program (PROM). The type of redundant processor system provided (Active/Standby, Load Sharing, Distributed) depends on the type of Processor (LP, MP, SP, etc.) and functions as shown in Figure 4-5.

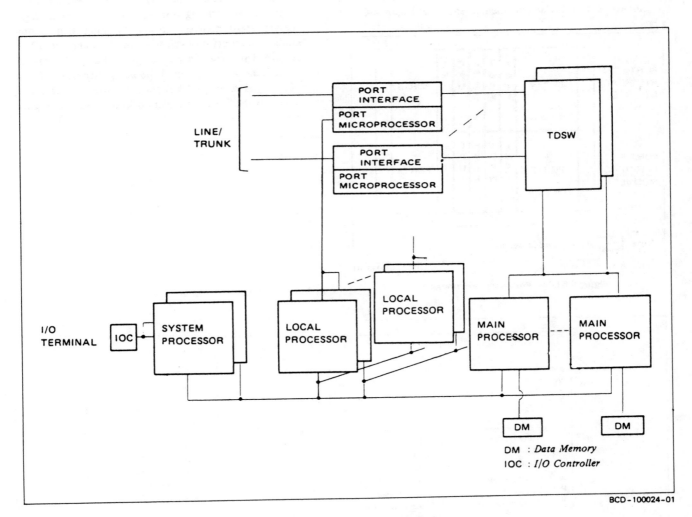

BCD-100024-01

Figure 4-5 Block Diagram of MMG

(1) Local Processor (LP)

A Local Processor (LP) is a distributed processor capable of controlling up to 4 Port Interface Modules (736 ports) as described in detail in paragraph 3. Each Local Processor has its own Generic Program, Working Memory and control circuitry to interface the Port Microprocessors and Data BUS. A Local Processor can be equipped with a redundant system operating in an active/standby arrangement as shown in Figure 4-6.

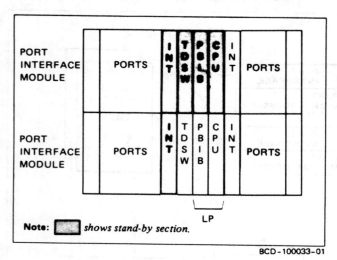

Note: ▓▓ shows stand-by section.

BCD-100033-01

Figure 4-6 Local Processor - Redundancy

(2) The Main Processors (MP) are equipped according to Processor occupancy, and operate in a load sharing mode. mode. The Multiple Module Group uses the "N+1" redundancy concept when determining the actual number of MP required. This means that the actual Processor load capacity is calculated and an additional MP is equipped so that if one fails no degradation of service will occur. The MP is the overall system controller performing port status management, digit analysis, status analysis, etc. in accordance with various kinds of requests received from the LPs and the SP as described in detail in paragraph 3. Each MP has its own Generic Program, Operating Memory, Data Memory, Processor BUS Interface and interface circuitry to the Switching Network as shown in Figure 4-7.

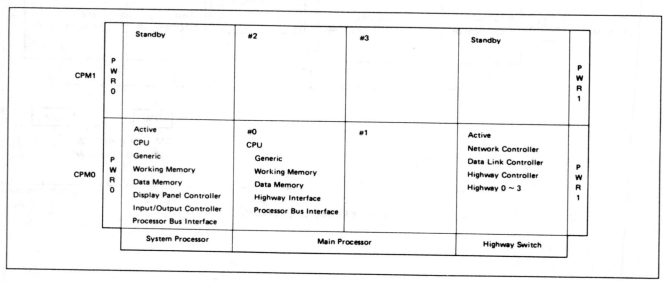

BCD-100034-01

Figure 4-7 Control Processor Module - Redundancy

(3) System Processor (SP)

The System Processor (SP) is actively engaged in supervision and control of the overall Multi-Processor system, constantly providing management information to the Maintenance Administration Terminal. The System Processor can be equipped with a redundant system operating in an active/standby arrangement.

Each SP is comprised of a Central Processor Unit with its related memory, Processor BUS Interface circuitry, interface to Input/Output device (Man-machine Interface) and interface to Backup Memory controller (Bubble Memory), as shown in Figure 4-8.

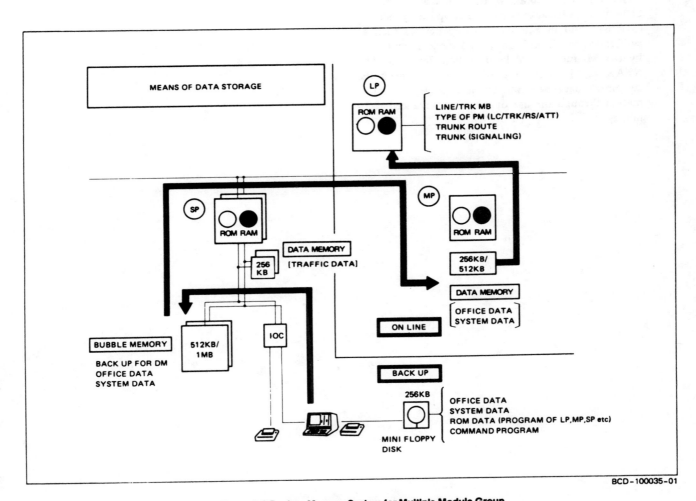

BCD-100035-01

Figure 4-8 Backup Memory System for Multiple Module Group

(4) System Memory (Multiple Module Group)

The Multi-Processor system uses non-volatile Programmable Read Only Memory (PROM) as Generic program instructions. The operating memory (RAM) is used for temporary storage of data related to tasks required by Port Interface Circuits for each processor. The Main Processor uses RAM as a storage medium for data memory for the office and system related information, while the System Processor has a data memory area for storage of traffic management. The Data memory is backed up in each duplicate processor (MP, SP) and in external memory devices such as Bubble Memory and Floppy Disk devices, controlled by the Maintenance Administration Terminal. The NEAX2400 IMS is designed to always protect the customer's data base and traffic management information through the use of the most advanced technology.

(5) Switching Network (Multiple Module Group)

The Switching Network for a system greater than 736 ports consists of a multistage, virtually non-blocking digital Time Division Switch and related components. To provide maximum reliability, the TDSW can be equipped in a redundant configuration. The standby TDSW is provided to ensure continuity of service.

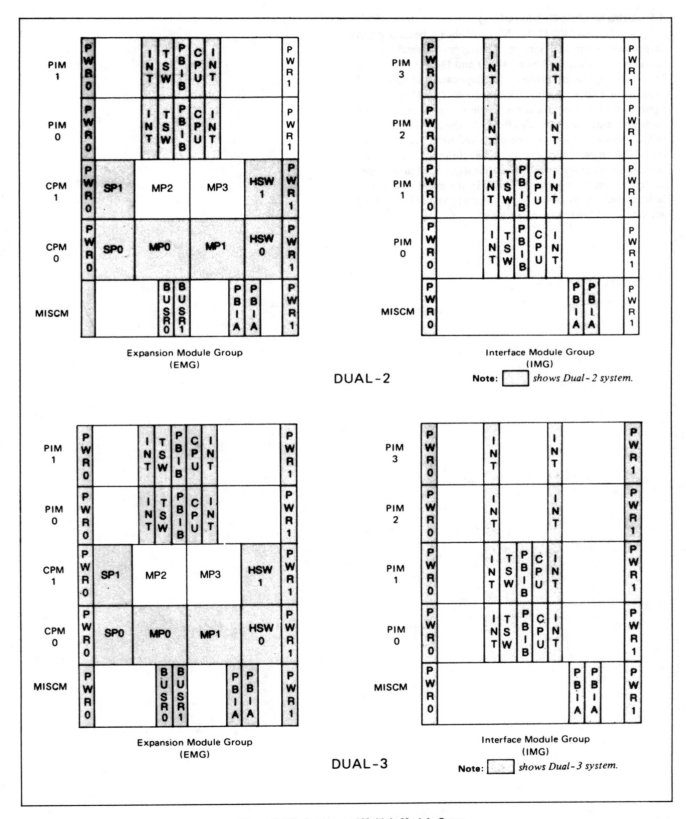

Figure 4-9 Redundancy of Multiple Module Group

183

4.5 Ultra Module Group (UMG)

The NEAX2400 IMS UMG will always be configured to provide redundant service. Two types of dual configurations are available, Dual-2 (D-2) and Dual-3 (D-3). The Dual-2 configuration provides for duplication of all critical processing and switching components. The Dual-2 configuration also provides for a single power supply within each Port Interface Module (PIM). In the event of a power failure, service would only be disrupted to the 184 ports of the PIM in which the power has failed. The Dual-3 configuration provides for the duplication of all processing, switching and power components. In the event of a power failure, the second power supply within the PIM would be activated and full service maintained.

DUAL-2 Redundancy

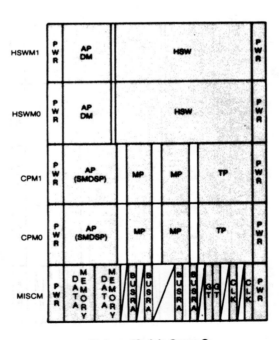

Figure 4-10 Redundancy of Ultra Module Group (1/2)

DUAL-3 Redundancy

Expansion Module Group (EMG)

Interface Module Group (IMG)

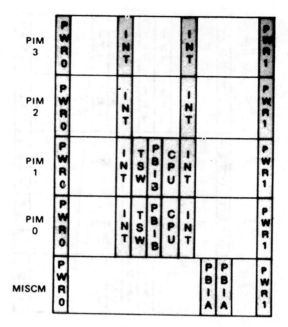

Highway Module Group O (HWMG 0)

Highway Module Group 1 (HWMG)

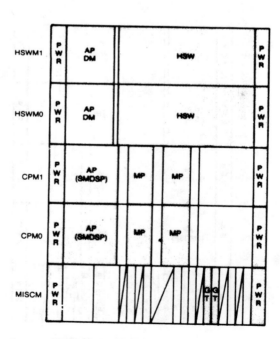

Note: ☐ shows Dual-3 system

Figure 4-10 Redundancy of Ultra Module Group (2/2)

5. Maintenance Administration Terminal

5.1 General

5.1.1 Man-machine interface to the NEAX2400 IMS is performed by a sophisticated Maintenance Administration Terminal (MAT), which is shown in Figure 5-1. This terminal provides access to, and management of, the system's Diagnostic and Data Base software. To facilitate system management, all supervisory software operates in an interactive, user friendly format.

BCD-100037-01

Figure 5-1 Maintenance Administration Terminal (MAT)

5.2 Menu Driven Software

The illustration below demonstrates the menu driven administration software.

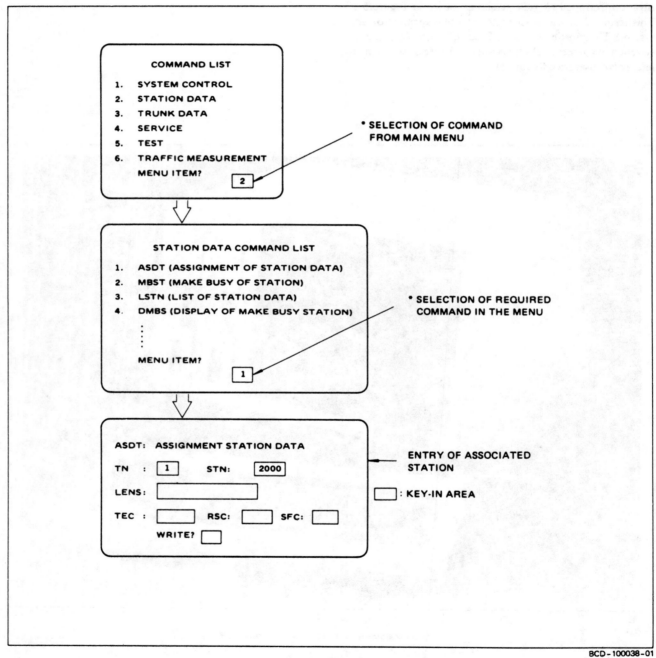

Figure 5-2 Menu Driven Software

BCD-100038-01

5.3 Maintenance

5.3.1 The design philosophy of the NEAX2400 IMS is to provide for duplication of all critical common control circuitry and power functions. The modular system design and sophisticated software diagnostics enables maintenance personnel to isolate and correct problems in a minimum amount of time. The maintenance personnel can interface with the system using the MAT from on site and/or a remote diagnostic center to determine overall system status as shown in Figure 5-3.

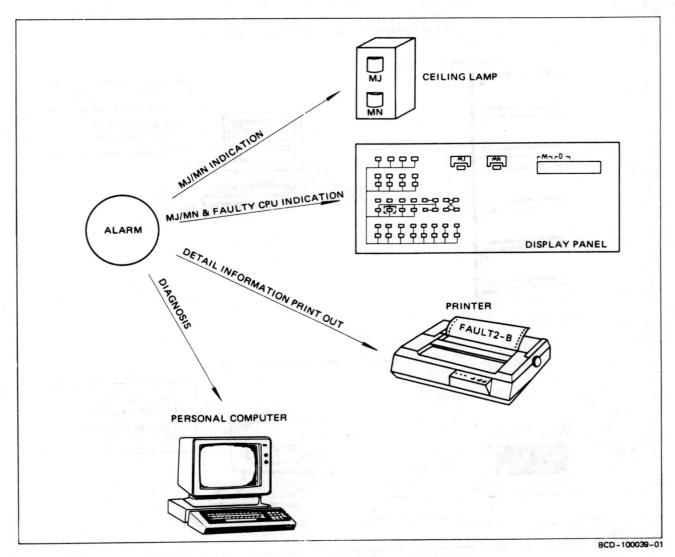

BCD-100039-01

Figure 5-3. Maintenance and Administration System

5.3.2 The NEAX2400 IMS diagnostic software is constantly managing the operating system and automatically generates supervision and control fault information on the overall system performance to the MAT and related supporting hardware. The maintenance management system is interactive with the system processors, switching network interface section and miscellaneous section (Power, Fuse, Temp., etc.) for fault indications. If a fault should occur, information is automatically supplied to the MAT, alarm panel, and an indicator panel to assist the maintenance personnel in analyzing and correcting the fault.

5.3.3 A typical Fault printout will supply the following status information for system processor failure.

Processor Identification Numbers
Processor status (Active, Standby, Make busy)
Time and Date fault occurred

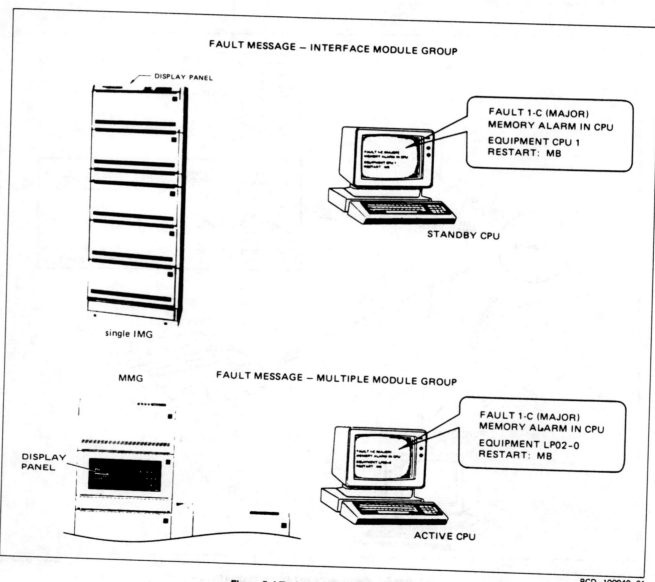

Figure 5-4 Typical Fault Message Display

BCD-100040-01

5.4 Remote Maintenance and System Administration

5.4.1 Access to the NEAX2400, for the purpose of system diagnosis, status reporting, and data-base reconfiguration can be performed from remote locations, (i.e. Maintenance Test Facilities, Technical Assistance Service Centers — TASC, etc.). The system is polled by a remotely located MAT via a data modem over DDD or other facility.

5.4.2 This capability allows station/trunk changes or reassignments to be performed without a site visit by technical service personnel, and can be used to detect fault tendencies before they affect service.

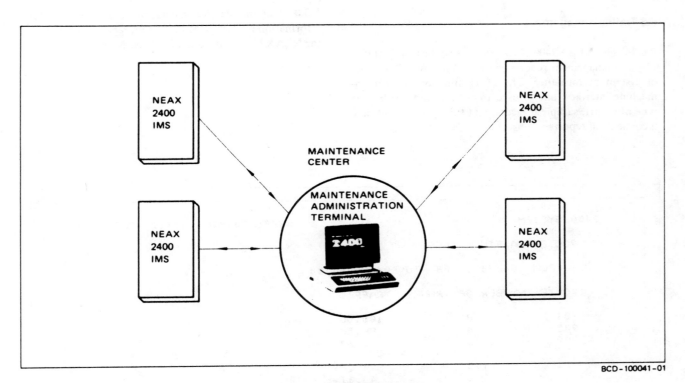

BCD-100041-01

Figure 5-5 Remote Maintenance

5.5 Automated Facilities Testing

5.5.1 The NEAX2400 IMS can be equipped to provide automatic polling and testing of station and trunk facilities as an optional feature. Test equipment and software within the system can isolate faults related to transmission and continuity, and overall connection status, such as answer/release detection and circuit supervision.

5.5.2 This service can assist in locating and identifying problems related to trunks and other network facilities before problems occur.

5.6 Traffic Management

5.6.1 The NEAX2400 IMS provides sophisticated traffic management reports to be used for overall analysis of system performance. The MAT functions as a man-machine interface, and is used to request and display the type of report, sample measurement time period, and time increments of reports.

5.6.2 The NEAX2400 IMS will provide the following typical types of Traffic Management Information.

Individual Attendant Console activity for all call processing.
Port Traffic Usage in CCS.
Terminal Peg Count for port access, incoming and outgoing trunk route selection.

The following illustration provides a typical MAT display for Port Traffic usage (CCS).

5.6.3 Communication management can obtain detailed information on Traffic Management reports by referring to the NEAX2400 Application Practices Manual.

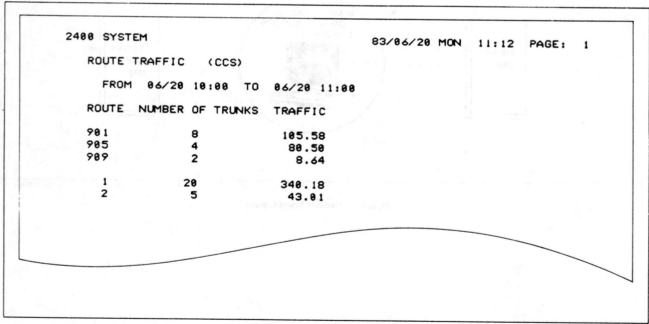

BCD-100042-01

Figure 5-6 Typical Traffic Management Report

192

6. Technical Specifications

6.1 Transmission Characteristics

6.1.1 The transmission characteristics of the NEAX2400 switching network and interface components are as follows:

a) Crosstalk
   Attenuation: More than 65 dB at 1 kHz

b) Idle Circuit Noise: Less than -65 dBmp

c) Insertion Loss (Relation to 1 kHz ~ -10 dBm)

   Station to Station: 5 dB (±0.8dB) with 16LC

   Station to Trunk: 0 dB (±0.8 dB)

   Trunk to Trunk
   (Analog 2 wire): 0 dB (±0.8 dB)

   Trunk to Trunk
   (Analog 4 wire): 0 dB (±0.8 dB)

d) Longitudinal
   Balance: More than 40 dB
   (300 — 600 Hz)
   More than 46 dB
   (600 — 3400 Hz)

e) PCM Characteristics:

   * Line Rate: 2.048/1.544 Mbps

   * Meets CCITT and North American T1-D3 standards.

f) Return Loss: More than 20 dB (300 — 3400 Hz, 600 ohms termination)

g) Loop Resistance: 1200 ohms with 16LC (including instrument)

h) Line Impedance: Station (Analog) 600 ohms Trunk (Analog 600/900 ohms selectable)

j) Leakage Resistance: More than 20,000 ohms

6.2 Rotary Dial Pulse and DTMF Signalling

a) Rotary Dial Pulse

   1. Receiving Conditions

      — Speed: 8 ~ 25 pps

      — Break ratio: 55 ~ 77% (10pps)

      — Minimum
      inter-digit
      pause: 180 ms

      — Switch hook
      flash
      detection: 360 ms to 2040 ms. (programmable)

   2. Sending Conditions

      — Speed: 10 pps (±0.8 pps)

      — Break ratio: 67 ±3% or 61 ±3% (10 pps)

      — Minimum

         inter digit
         pause: 800 ms

b) DTMF Signalling

   Frequency Combinations

	1209 Hz	1336 Hz	1477 Hz
697 Hz	1	2	3
770 Hz	4	5	6
852 Hz	7	8	9
941 Hz	✳	0	#

1. Receiving Conditions
   (Measured at Receiver Input)

   Signal duration: More than 40 ms

   Inter-digit
   pause: More than 30 ms

   Signal level: More than 0 ~ -30 dBm
                     (S/N: More than 14 dB)

2. Sending Conditions

   Signal duration: More than 50 ms

   Interdigit
   pause: More than 30 ms

   Signal level: -6.5 ~ -10 dBm

6.3 Ringing Signal

a) Conventional Instrument

   Signal voltage: 90 Vrms (±10 Vrms)

   Frequency: 20 Hz (±1 Hz)

b) Interruption (Internal/External Calls)

   Software selectable in any one second increment
   combination.

   1 ~ 15 sec. ON . . . . 1 ~ 15 sec. OFF

c) Feature Ringing

   0.4 sec. ON . . . . 0.2 sec. OFF
   0.4 sec. ON . . . . 1 sec. OFF

6.4 Audible Tones

Table 6-1 Audible Tones

TONE	DEFINITION	FREQ. (Hz)	INTERRUPTION
Dial Tone (DT)	Originate call	350 + 440	Continuous
Special Dial Tone (SPDT)	Require any service with switch hook flash	350 + 440	240 IPM
Ring Back Tone (RBT)	Calling the destination	440 + 480	1 sec. — ON 3 sec. — OFF
Call Waiting Tone (CWT)	Inform waiting call to called party	440	80 ms ON — 80 ms OFF — 80 ms ON
Busy Tone (BT)	Called party busy	480 + 620	60 IPM
Reorder Tone (ROT)	For restricted call	480 + 620	120 IPM
Service Set Tone (SST)	Confirmation of Service Set	440	Continuous
Warning Tone (WT)	For Executive Right of Way/Attendant Override	440	80 ms ON — 80 ms OFF — 80 ms ON
Camp ON Tone (CPT)	Confirmation of Camp On set	440	200 ms ON
Second Dial Tone (SDT)	Incoming call from distant PBX	350 + 440	Continuous

7. Environmental Requirements

7.1 Operating Conditions

Table 7-1 Operating Conditions

		TEMPERATURE	REL. HUMIDITY
Operational Limits	Normal Steady	41°F → 86°F (5°C → 30°C)	15% → 65%
	Short* Term Interval	32°F → 104°F (0°C → 40°C)	15% → 90%
Storage Only		0°F → 122°F (−18°C → 50°C)	8% → 90%
Rate of Temp. Change		41°F/30 min (5°C/30 min)	——

*Note: Short term intervals are to persist no longer than 72 hrs. with a total accumulation no longer than 360 hrs. per year or system performance may be impaired. All memory sections during this Time Interval must be air cooled.

7.2 Thermal Output

Figure 7-1 Thermal Output for SIM and IMG

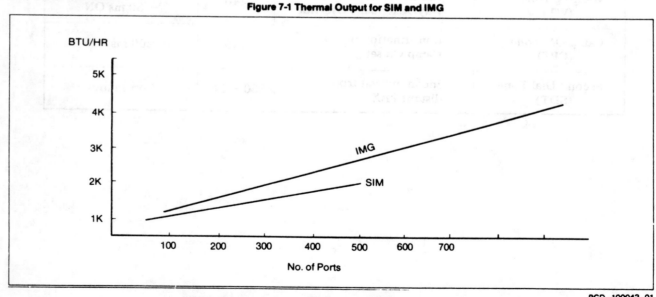

NOTE: The thermal outputs represented are based on the following assumptions:

per line traffic:	7.2 ccs
extensions:	20% digital instruments
	80% single line with PB
traffic distribution:	internal 40%
	external 60%

Figure 7-2 Thermal Output for MMG and UMG

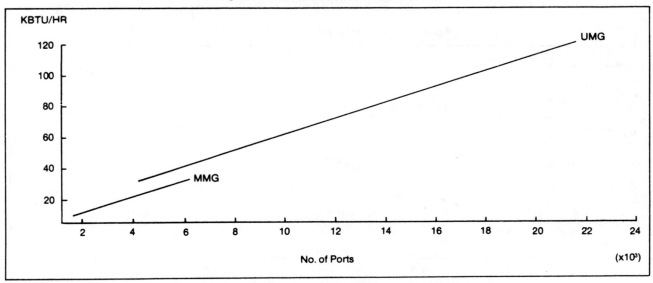

NOTE: *The thermal outputs represented are based on the following assumptions:*

per line traffic:	*7.2 ccs*
extensions:	*20% digital instruments*
	80% single line with PB
traffic distribution:	*internal 40%*
	external 60%

7.3 Power Consumption

7.3.1 The NEAX2400 IMS operates from a primary power supply of -48V ± 5V DC. This primary operating supply is provided by rectifiers which isolate the system from commercial power line fluctuations, and enhance the overall stability of the system.

7.3.2 The use of technologically advanced components such as C-MOS, VLSI and bubble memory circuits results in lower specific power consumption and heat dissipation, minimizing the system's environmental requirements.

Figure 7-3 Power Consumption for SIM and IMG

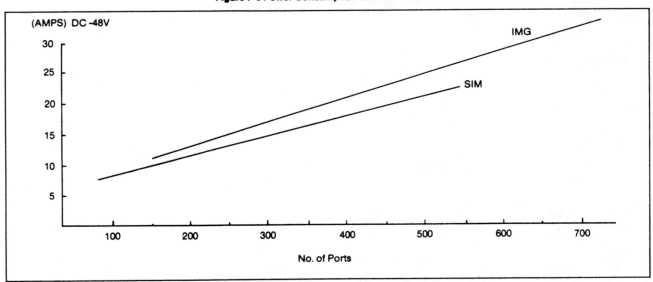

NOTE: *The power consumptions represented are based on the following assumptions:*

per line traffic:	*7.2 ccs*
extensions:	*20% digital instruments*
	80% single line with PB
traffic distribution:	*internal 40%*
	external 60%

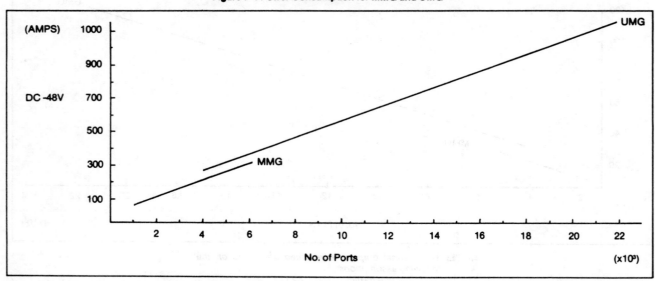

Figure 7-4 Power Consumption for MMG and UMG

NOTE: The power consumptions represented are based on
the following assumptions:

per line traffic:	7.2 ccs
extensions:	20% digital instruments
	80% single line with PB
traffic distribution:	internal 40%
	external 60%

7.4 Floor Plan Considerations

7.4.1 Equipment Dimensions and Weights

Table 7-2 Equipment Dimensions and Weights

EQUIPMENT	DIMENSION mm/Ft.	WEIGHT kg/lbs.
1. Single Interface Module (SIM) (See Note 3)	730(W) x 550(D) x 1170(H) 2'5" 1'10" 3'10"	160 kg 352 lbs.
2. Interface Module Group (IMG) (See Note 2)	730(W) x 550(D) x 1910(H) 2'5" 1'10" 6'3"	280 kg 617 lbs.
3. Expansion Module Group (EMG) (See Note 1, 2)	730(W) x 550(D) x 1910(H) 2'5" 1'10" 6'3"	280 kg 617 lbs.
4. Highway Module Group (HMG)	730(W) x 550(D) x 1910(H) 2'5" 1'10" 6'3"	280 kg 617 lbs.
5. Basic Module (BSCM)	730(W) x 550(D) x 320(H) 2'5" 1'10" 1'1"	70 kg 154 lbs.
6. Port Interface Module (PIM)	730(W) x 550(D) x 320(H) 2'5" 1'10" 1'1"	46 kg 101 lbs.
7. Control Processor Module (CPM)(See Note 1)	730(W) x 550(D) x 320(H) 2'5" 1'10" 1'1"	40 kg 88 lbs.
8. Miscellaneous Module (MISCM)	730(W) x 550(D) x 320(H) 2'5" 1'10" 1'1"	35 kg 77 lbs.
9. Fan Unit (FANU)	730(W) x 550(D) x 100(H) 2'5" 1'10" 4"	12 kg 26 lbs.
10. Attendant Console	440(W) x 320(D) x 165(H) 1'6" 1'1" 7"	4 kg 9 lbs.
11. Maintenance Administration Terminal (MAT)	——————N/A——————	37 kg 82 lbs.

Note 1: *Not required for application less than 736 ports.*

Note 2: *The module group shown includes 5 module blocks.*

Note 3: *The module group shown includes 3 module blocks.*

7.4.2 Typical Floor Plans

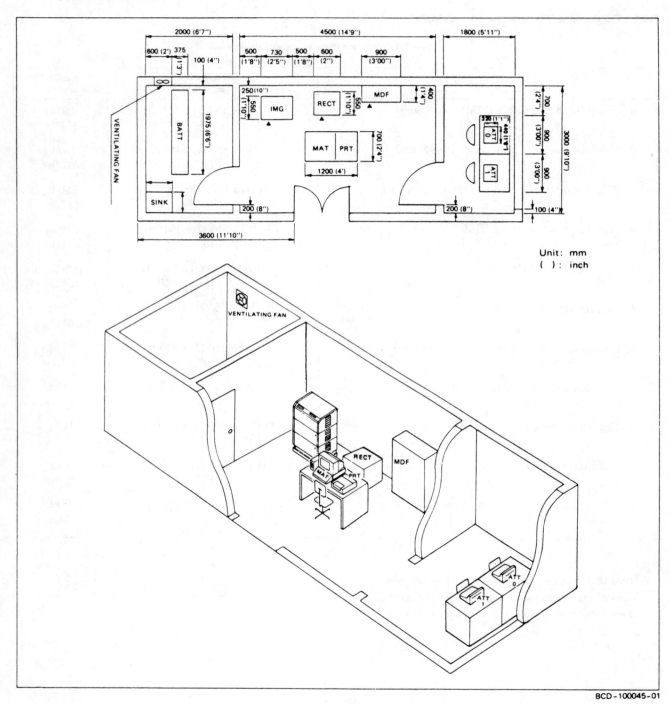

Unit: mm
( ): inch

BCD-100045-01

Figure 7-5 Typical Layout of 464 Port System

7.4.2 Typical Floor Plans

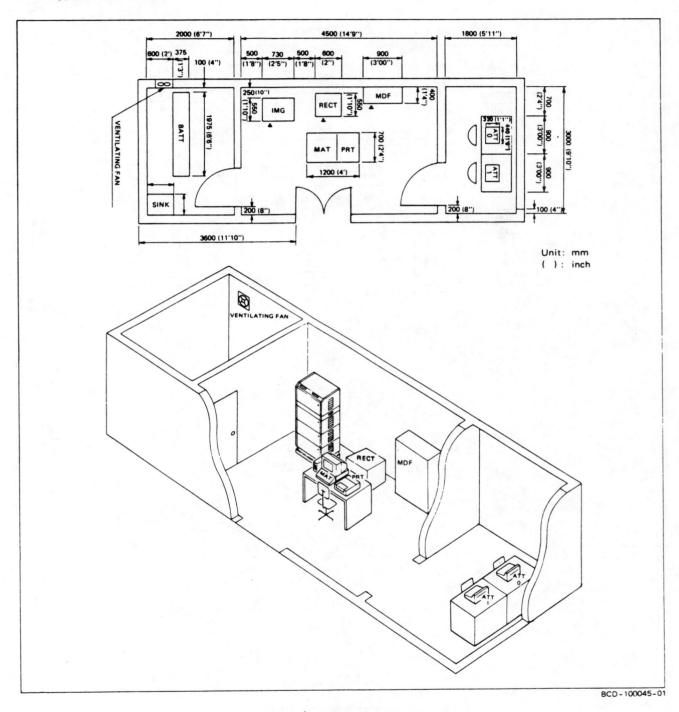

Unit: mm
( ) : inch

BCD-100045-01

Figure 7-6 Typical Layout of 522 Port System

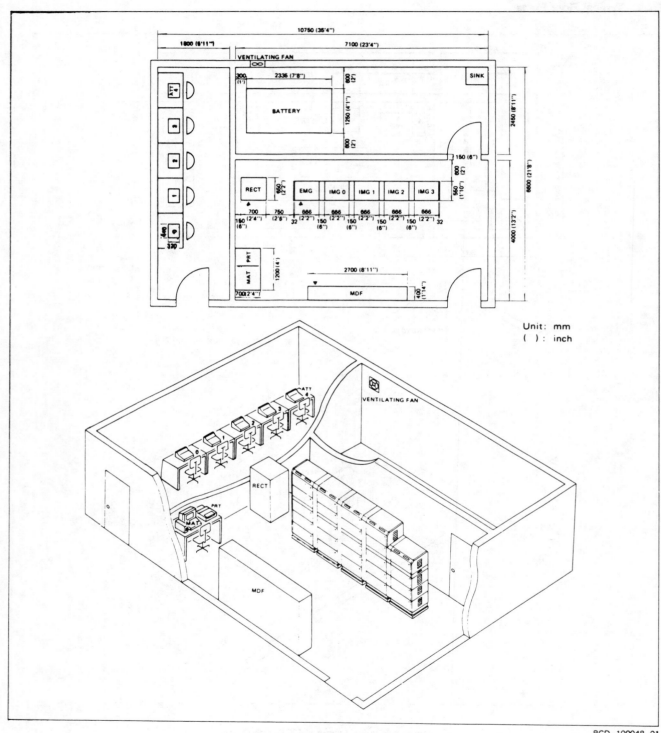

Unit : mm
( ) : inch

BCD - 100048 - 01

Figure 7-7 Typical Layout of 3128 Port System

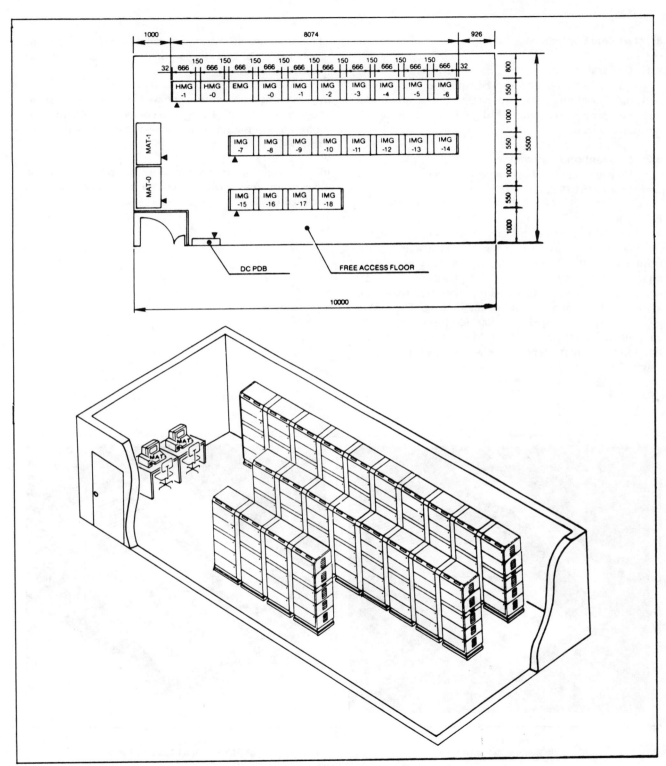

Figure 7.8

FLOOR LAYOUT NEAX2400 UMG SYSTEM (~ 14,352 PORTS)

8. Terminal Equipment

8.1 General

The NEAX2400 IMS can be equipped with a combination of conventional telephone sets and/or any of the series of proprietary Dterm® digital electronic multi-function terminals.

8.2 Conventional Telephone

Conventional telephones access all system features via simple one or two digit access codes.

8.3 Dterm®

The NEC Dterm® instruments are intelligent microprocessor controlled terminals, which enhance the feature capabilities offered by the system. They provide the service of conventional "KEY" telephones over 2 pair wiring, and may be equipped with an interface adapter to allow simultaneous voice and data switching. The Dterm instrument can be located up to 4000 feet (using AWG24 or equivalent) from the NEAX2400 IMS and, with the exception of the interface adapter, does not require the use of an external power source.

The Dterm line of instruments ranges from the single line telephone to the executive work station. The Dterm line includes the following:

a) A single-line instrument with a flash hook button and Message Waiting Lamp. It is compatible with the Type 2500 and is installed using one-pair analog wiring (see Figure 8-1).

b) A eight-line button digital instrument set that can be adapted for simultaneous voice and data communications (see Figure 8-2).

Figure 8-1
Single-Line Instrument

Figure 8-2
Eight-Line Button Digital Instrument

c)   A sixteen-line/feature button digital instrument set equipped with 20 speed call buttons. In addition, dedicated function keys are provided for Do Not Disturb, Three-Way Calling, Call Back (Outgoing Trunk Queuing), Call transfer - All Calls, and Line Reconnect - same Line features. Dedicated function keys are also provided for speaker and microphone control, Call Waiting Answer, Exclusive Hold, and Non-Exclusive Hold operations (see Figure 8-4). This set is also able to provide simultaneous voice and data communications through the use of a Data Adapter that is connected to the instrument. There is also a 24-line adapter (see Figure 8-5) to expand the number of lines available at the set.

d)   An advanced Integrated Voice/Data Workstation (IVDW) incorporating the advantages of a conventional "KEY" telephone with an ASCII terminal. This workstation features: three station line appearances, 10 function keys that can be shifted to provide 20 functions, personal directory service, 10-year calendar, world clock and map, voice memo recording, advanced telephone features, a 9" amber CRT screen, cordless computer keyboard that can be used up to 6 feet away from the terminal, and an RS232C access for printer hookup (see Figure 8-3).

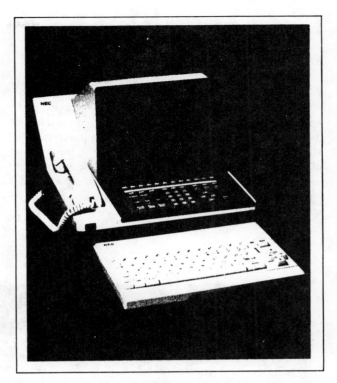

Figure 8-3
Advanced Integrated Voice/Data Workstation

Figure 8-4
Sixteen-Line / Feature Button Digital Instrument

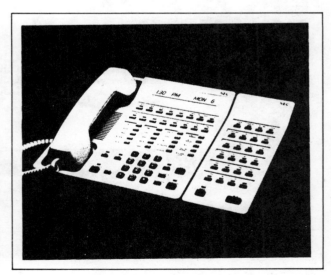

Figure 8-5
Sixteen-Line/ Feature Button Digital Instrument with 24-Line Adapter

9. Attendant Console Description

9.1 General

9.1.1 The NEAX2400 IMS attendant console is a compact, desk top unit equipped with non-locking keys and light emitting diodes. The LED's provide continuous information relative to the status of calls in progress. The alphanumeric display will provide station and trunk identification, class of service, time and date, and number of calls waiting. The attendant console also provides a flexible busy lamp field accommodating both three and four digit numbering plans.

9.1.2 All calls placed by the attendant to trunks and stations are made using a centrally located push-button dial pad.

9.1.3 The NEAX2400 IMS attendant consoles are designed to operate on a switched loop basis. Six independent loop circuits terminate at each console through its associated position circuit. The attendant can originate, answer, hold, transfer, and re-enter call back calls via an attendant loop key on a switched loop basis. The location of the console keys is designed to maximize operation efficiency. NEAX2400 IMS software allows the attendant to perform call processing functions by depressing a minimum amount of keys.

BCD-100056-01

Figure 9-1 Attendant Console

10. Digital Data Communication (DDC)

10.1 General

The Digital Data Communication (DDC) feature is an integral part of the NEAX2400 IMS. It makes it possible for a four or sixteen-line digital telephone to be equipped to accommodate digital equipment such as data terminals, printers and computers without the use of modems or other dedicated facilities. Each four and sixteen-line digital instrument can be equipped with a Data Adapter, and does not require any additional port interface hardware. The DDC option can be provided as asynchronous or syn- chronous transmission through the use of two types of Data Adapters. Figures 10-1 and 10-2 illustrate the two types of Data Adapters.

The most common application is to provide data terminal equipment with an RS232C interface, together with a Dterm digital instrument with a data adapter. This appli- cation allows the station user to use the telephone set for conventional voice communication while using a data terminal.

Applications not requiring simultaneous voice and data communications, but only data, can be met through the use of a Data Module.

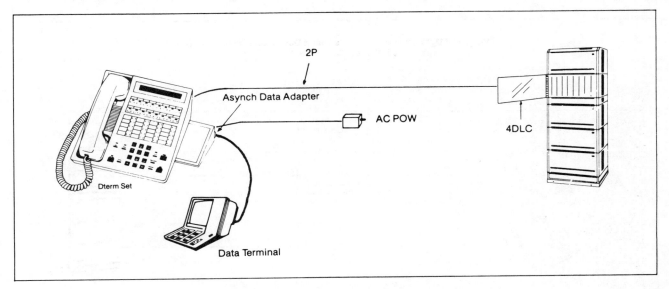

Figure 10-1 Asynchronous DDI

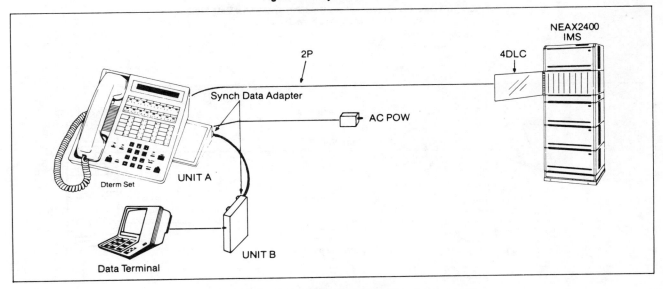

Figure 10-2 Synchronous DDI

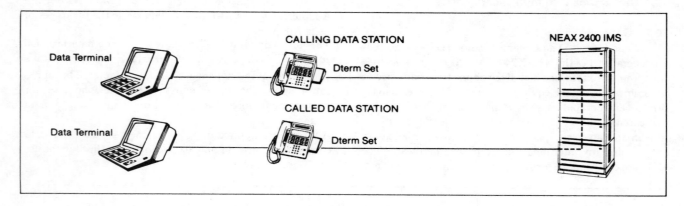

Figure 10.3 Data Connection from a Dterm Data Station to Another Dterm Data Station (Dial up Connection)

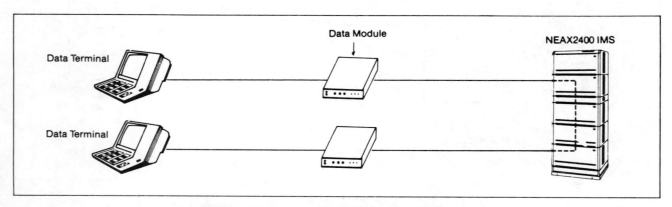

Figure 10.4 Data Connection from a Data Terminal with a Data Module to Another Data Station (Character Dial)

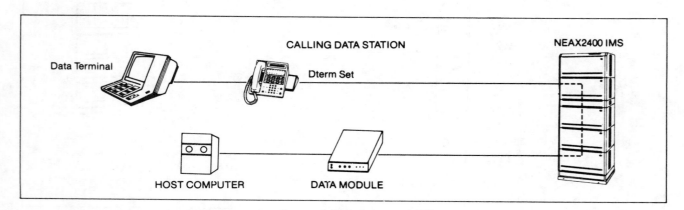

Figure 10.5 Data Connection from a Dterm Data Station to an Internal Computer Port

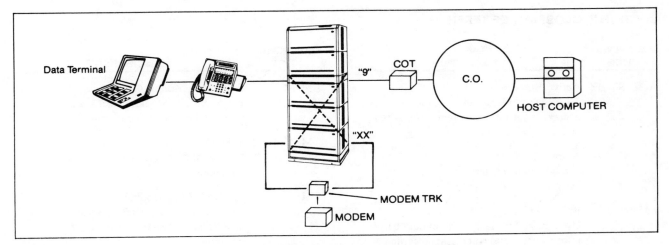

Figure 10.6 Data Connection from a Dterm Data Station to an External Computer Port

10.2 Dterm® Features

The Dterm® instrument can be equipped with either of 2 types of Data Adapter units. Both units are identical in physical appearance and dimensions, and vary only in their electrical and performance characteristics.

	ASYNC DATA ADAPTER	SYNC DATA ADAPTER
Physical Interface	RS232C	RS232C/V35
Data Speed	Up to 19.2 Kbps Async. (Note 1)	Up to 56 kbps Sync.
Error Rate (Note 2)	$\leq 10^{-6}$	$\leq 10^{-6}$
Mode	Full/Duplex Half Duplex	Full/Duplex

a) Modem Pooling

This feature allows modems to be arranged in groups for both incoming and outgoing data communications over analog network facilities.

b) Data Transparency

As all data communications paths established within the system are transparent to terminal and computer equipment, the user may select the most efficient protocol for their requirements.

c) Data Terminal Traffic Measurements

Allows periodic traffic studies to be performed to determine the optimum data communications configuration for the user.

d) Data Interface Auto Answer

Dterm® DDI's may be programmed by the station user to automatically answer all incoming data transmissions while unattended. This feature facilitates the Text Mail and other related message switching services offered by the NEAX2400 IMS.

e) Half/Full Duplex Switchover

Station users may select the mode required based on the type of device or network to be accessed.

f) Simultaneous Voice and Data

Ability to transmit simultaneous voice/data or use Dterm® for other applications during data transmission.

g) Alphanumeric Display

The Dterm® 16-character display provides the following services:

— CALLED STATION STATUS DISPLAY

The status or condition of the called station is indicated on the calling party's display. For example, station users calling a party who has set "call-forwarding," or encountering a "hunting" condition will be informed of the feature set and of the actual terminating station by their alphanumeric display.

— CALLING NUMBER DISPLAY

Dterm® station users will know the identity of the calling party before answering. Call forwarding and call pickup situations will also be accommodated by displaying both the originating and original destination's station number.

Notes: 1) A terminal which can accept up to 45% signal distortion at 19.2 Kbps, must be used
2) 16-point sampling method.

SECTION II GLOSSARY OF TERMS

ABBREVIATION	DESCRIPTION
ACT/ST-BY	Active/Stand-by Type of redundancy where standby processor automatically takes over when active processor fails.
ALMC	Alarm Controller Provides control circuitry for all alarm indications.
BM	Bubble Memory Used for additional backup of data base information, and will retain all contents without reliance on external power sources.
BSCM	Basic Module Provides 160 ports for station/trunk and other devices for the Single Interface Module configuration.
BUS	BUS Communication link between processors.
C&C	Computers and Communications Provides total communication and processing within the system.
C.O.	Central Office
CPM	Control Processor Module Houses the distributed multi-processor control circuitry in a Multiple Module Group.
CPU	Central Processor Unit Houses the microprocessor control circuitry in a single-Interface Module Group and is located in the Port Interface Module.
DDI	Digital Data Interface Unit which can be added to a Dterm® when using any of the data features.
Dterm®	Digital Terminal An intelligent microprocessor-controlled device which will enhance feature capabilities offered by the system.

ABBREVIATION	DESCRIPTION
EMG	Expansion Module Group Is requried for MMG, over 736 ports to provide the necessary common control.
TMM	Text Mail Module Storage and retrieval system for text information.
FMM	Facsimile Mail Module Used for storage, retrieval, retransmission or broadcast of facsimile information.
HOW	Howler Circuit Provides the NEAX2400 with a Howler Tone.
HWMG	Highway Module Group Provides the highway portion of a 3-stage Time Division Switch for the Ultra Module Group configuration.
IMG	Interface Module Group Consists of up to 4 Port Interface Modules and one (1) Miscellaneous Module.
IMS	Information Management System
I/O BUS	Input/Output Bus Communication link between processors, SMDS and IOC.
IOC	Input/Output Controller Used to control external devices such as Maintenance Administration Terminal. Interface Circuit Interface between TDSW and Port Card.
LAN	Local Area Network Controller of an intelligent wide band transmission system.

ABBREVIA-TION	DESCRIPTION
LP	**Local Processor** Analyze, Control and supervise the status of terminal devices accommodated within its Port Interface Module.
LS	**Load Sharing** Type of redundancy where multiple processors share the processing load.
LSI	**Large Scale Integration**
MAT	**Maintenance Administration Terminal** Used to perform system diagnostics and provide access to Data Base software.
MISCM	**Miscellaneous Module** Provides locations necessary for Alarm, Maintenance and other hardware controllers.
MMG	**Multiple Module Group** Consists of an Expansion Module Group and a number of Interface Module Groups used for configuration greater than 736 ports.
MP	**Main Processor** Interacts with Local Processors and System Processor to execute tasks relating to system status.
NMC	**Network Management Controller** Overall controller of Time Division Switch.
PCM	**Pulse Code Modulation** Method of transforming an analog signal into a digital signal using a specified sampling ratio.
PIM	**Port Interface Module** Provides access to 184 ports for station/trunk and other devices.
PLO	**Phase Lock Oscillator** Used to keep two switches in sync while using Direct Digital Interface.
PM	**Port Microprocessor** Performs status analysis and supervision of its related port devices.

ABBREVIA-TION	DESCRIPTION
PWR	**Power Unit** Supplies the NEAX2400 with various operating voltages.
PROM	**Programmable Read Only Memory** IC chips which can be reprogrammed according to system design and are not affected by power loss.
PSM	**Packet Switching Module** Acts as a switching network for dissimilar devices. All data is converted to a uniform protocol and format.
RAM	**Random Access Memory** IC chips which store information that is constantly changing (port status and system data).
RG	**Ringing Generator** Supplies the NEAX2400 with ringing voltage.
RS232C	**RS232C** A method of transmitting serial data information using half or full duplex. RS232C can transmit data at the rate of 19.2 kbps, async. or sync. with a maximum cable length of 50 feet using a 25 pin connector.
RS449	**RS449** A method of transmitting serial data information using half or full duplex. RS449 can transmit data at the rate of 2 Mbps, async. or sync. with a maximum cable length of 2000 feet using a 34 pin connector.
SIM	**Single Interface Module** Consists of one Basic Module and up to two Port Interface Modules - used for configurations up to 464 ports.
SMDS	**Station Message Detail Recording System** Provides a detailed record of all outgoing station to trunk calls and incoming trunk to station calls.

ABBREVIA-TION	DESCRIPTION	ABBREVIA-TION	DESCRIPTION
SP	System Processor Performs management and supervisory functions in a Multi-Processor configuration.	UMG	Ultra Module Group Consists of one or two Highway Module Groups, one Expansion Module Group and up to 31 Interface Module Groups - used for configurations up to 23,184 ports.
TDSW	Time Division Swtich A full availability, non-blocking communication path between ports.	VSLI	Very Large Scale Integration
TP	Time Division Processor Provides hardware control for the time division switch of an Ultra Module Group.	VMM	Voice Mail Module Stores digitized voice information on a mass storage media for retrieval by the directed user.

Chapter 4: PBXs as LANs

At least taxonomically speaking, anyone who accesses a local computer via a private branch exchange (PBX) is using his or her PBX to implement a local area network (LAN). Virtually all modern digital PBXs offer facilities to connect, by means of circuit switching and time-division multiplexing, terminal devices to hosts. Most support asynchronous data transmission to 19.2 Kbps by using IEEE RS-232C interfaces and synchronous devices to 56 or 64 Kbps by using CCITT V.35 interfaces. Several PBXs go considerably further, supporting Ethernet (IEEE 802.3) or Token Ring (IEEE 802.5) Carrier Sense based technologies. Some support cable replacement schemes for Wang word processors and the IBM 3270 family of terminals. (A number of these more elaborate LAN implementations will be discussed further in this introduction.) Nonetheless, PBXs may not have gotten their fair shake as LANs for several reasons.

The first, and most obvious reason, is bureaucratic. Many institutions continue to maintain separate voice and data organizations and concurrence on inter-organizational technical approaches—regardless of the problem—are rare.

Second, most PBX LANs involve some kind of star typography. Data flow back and forth through a centralized processor complex. Obviously, if the central processor fails, the LAN crashes. But few on "the data side of the house" are aware that most large PBXs have redundant processors running in parallel and can be backed up with UPS and emergency generator support. Most PBX availability figures exceed 99.5 percent, which is usually 2 or 3 percentage points better than comparably large local area networks. Finally, with the most common outage, a brief but general power failure, the PBX and its switch-powered voice capabilities will continue to operate but the individual devices on the LAN will go down anyway unless the entire facility is provided with emergency power.

Third, the data rates offered by the current PBXs lack glamour; they are not in the multi-megabit range of Ethernet and IBM's IEEE 802.5 token ring LANs. This is a true but an exceedingly deceptive observation. To take a case in point, although Ethernet's digital broadcasts travel at 10 Mbps, only about 20 percent of the end-to-end transit time is on the cable, with 80 percent of the time spent buffering through relatively slow interfaces. Particularly in office environments, the costs of taps and transceivers are such that they are often shared by several devices, with each device

connected by RS-232C interfaces. Without being unduly critical regarding cable-based LANs, it is important to point out that as long as one's connection to a LAN is limited by RS-232C (limited to a maximum of 19.2 Kbps), that interface will determine the rate at which data appear on the workstation screen.

Fourth, there has been the concern about the cost of connection with PBXs. It it hard to generalize here, because most PBXs provide several methods to connect a workstation. The most expensive alternative involves an integrated voice/data set interconnecting with a specialized line card. This cost, in the neighborhood of $500, has declined rapidly with volume production and offshore manufacture. Nonetheless, it is comparable in price to technologies such as IBM's (Sytek's) PC Net and Token Ring, and does not require a dedicated workstation operating as a controller. Thus far, low network attachment costs have come in two directions. The first, already noted, involves sharing an expensive network attachment with a number of devices—topographically speaking, a baby star network off a bus network. The second is the bundled approach where LAN capabilities are built into the workstation, such as the Apple Macintosh and Appletalk, Sun workstations, and Ethernet.

Fifth, and increasingly a nonissue, is anxiety over deleterious effects that heavy data traffic may have on the PBX voice performance. As was noted previously, the newer switches have been designed for heavy data use and offer huge bandwidth in combination with a switching matrix that is nonblocking or capable of being engineered for nonblocking. Recent upgrades to earlier designs have been in the same direction, and as long as the error in forecasting data requirements is not too gross, degradation of voice processing owing to heavy data loads should not be a concern.

It may be that this introduction regarding the data LAN capabilities of PBXs has been too defensive. It is clear, that excepting some fringe LAN markets (the work group sharing a laser printer at the low end, and CAD/CAM and manufacturing control at the high end), the PBX offers most organizations a viable LAN alternative. It should be noted as well that it offers a number of salient advantages.

Chief among the advantages of implementing a PBX-based LAN is the multiple use, and effective configuration control, of the institution's twisted pair wiring resources. (Also, from the perspective of most cable LANs, the interbuilding bridges are part of the base system.) The centralized aspect of PBX LANs also allows one to pool expensive network resources such as protocol converters, gateways and modems. Finally, PBXs, and particularly data PBXs, offer superior administrative control and performance monitoring of the LAN. Classes of service can be assigned to LAN users. Available resources, particularly computer ports, can be contended for in a controllable manner. Usage patterns can be determined by feature "peg counts." Calls that exit a gateway generate station message detail records (SMDR). (In the case of many data PBXs, every call can generate a SMDR with called number, calling number, length of call, etc.) Moves and changes, though potentially circumscribed by the relative heterogeneity of the PBX station equipment mix, can often be effected in software.

Most of the foregoing discussion on PBX LAN capabilities has been oriented toward what is typically offered today: passing data by using a 64 Kbps time-division multiplexed slot. Note should be made, however, that a number of PBX manufacturers have contrived schemes to surpass the limitations of 64 Kbps switched circuits. In this area, five such PBX LAN technologies will be reviewed: hybrid switch, bandwidth agglomeration, attached packet processor, distributed circuit switch, and distributed multi-LAN.

A *hybrid switch* approach has been employed on GTE's Omni PBX. In the Omni, the line card serving a voice/data equipped station packetizes the voice traffic while a asynchronous (to 19.2 Kbps) or synchronous (to 64 Kbps) "packet manager" packetizes the data traffic. The voice packets are placed on a synchronous 64 Kbps voice bus while 13 byte data packets are dumped on a local, demand-assigned, 12 Kbps full-duplex bus and subsequently routed to a 6.176 Mbps high speed bus. The intent of the hybrid switch design is to accommodate, although not without additional cost and complexity, both the continuous nature of speech and the bursty nature of interactive data traffic.

The InteCom S/80 and S/80+ employ *bandwidth agglomeration* to support Ethernet. The implementation is quite complex. First, although the user believes he or she is transmitting randomly, the transmission of data packets is actually controlled by using the Request to Send/Clear to Send (RTS/CTS) protocol. Second, the packet is broadcast only if the path is unknown; in the usual case, the Ethernet address is translated into a "LandMark" address, and the packet is routed point to point. If a path is congested, LandMark may route some of the packets by a different route. Third, the actual transmission speed over the twisted pair and through the system is at 960 Kbps (fifteen 64 Kbps slots are agglomerated). With the combination of a "collisionless" Ethernet protocol and the point-to-point routing, the system's throughput generally exceeds all but the most lightly loaded cable Ethernets. Although expensive, in the maximum configura-

tion the S/80+ can support up to an awesome 12,288 Ethernet devices.

Both the Northern Telecom Meridian and AT&T Systems 75/85 employ variations of *attached processors*. In both cases, the attached processor is a high speed packet switch. Although the same twisted pairs used to carry voice are also used for data, the two are kept separate and are split off at the backplane of the PBX. The data traffic, which in both implementations is transmitted at over 2 Mbps, is then routed over a high speed link to the packet switch. AT&T uses a optical fiber link up to 2.2 kilometers; Northern uses one or more T-1 (1.544 Mbps) links. Both makers offer a large number of options for support of specific work stations and provide fiber optic bridging links to allow groups of workstations to be supported beyond the physical distance limits of twisted pairs. The chief advantages of the attached processor approach are that it permits the implementation of a high-speed LAN in conjunction with a voice switch and twisted pair wiring without affecting the performance of the voice switch. The chief disadvantage, other than extra cost, is that in by-passing the PBX, it also by-passes the PBX's usual features for security, administrative control, and accounting.

The Rolm CBX II 9000 and IBM 9751 CBX are both able to be configured as what can be characterized as physically *distributed circuit switches* to enable them to handle large amounts of data traffic. With the Rolm/IBM strategy, the data still use all or parts of a 64 Kbps voice channel but the switching resource can be physically situated, if need be, adjacent to the data processing host or hosts. This allows the distributed CBX to function as a port selector and other "data PBX roles" and minimize the impact of the data traffic on the other distributed switches. The caveat to this approach is that one needs to work with well-identified data quantities and flows, particularly when some of the data flow involve inter-node traffic. An environment featuring rapid growth and unpredictable traffic patterns could encounter blockage at the serving node, in the tie trunks that connect the nodes, at an exit node, or in accessing other networks. The advantage of such physically distributed circuit switches is that it allows for an economical placing of switching resources while partially segregating the effects of the data traffic.

The last approach discussed, dubbed *distributed multi-LAN,* was employed by several start-up companies, Ztel and CXC, whose overall architectures are discussed in chapter 5. In the case of these distributed approaches, nodes were to be interconnected by multiple LANs. Both planned a deterministically scheduled 50 Mbps LAN to carry voice traffic between nodes. In CXC's case, a broadband LAN allocated 33 Mbps to circuit switched voice and 16 Mbps (via a token ring) to signaling, while a physically separate, baseband 10 Mbps Ethernet handled data traffic. In the Ztel case, the development of the triple fiber optic ring LAN was overtaken by events—the company's demise—but the plans were to use circuit rings for voice and an IEEE 802.5 token ring for signaling and data traffic. A final observation regarding the five approaches outlined is that the various technologies employed are not mutually exclusive and may reappear in PBX LANs in the future. For instance, it would be conceivable that an attached processor could sit as one node on a physically distributed PBX system and communicate between nodes via a high speed LAN.

The papers that follow represent an attempt at several slants of the PBX-as-LAN. The first four are intended to present various views of proponents. The Thurber and Freeman paper, "An Introduction to Integrated Voice/Data PBX Systems," shows how PBXs may function in several LAN roles, and provides discussions of coding, switching, and blockage. Patrick, in "The Heat Is on for Phone Switches That Do a Lot of Fast Shuffling," pushes price/performance, arguing that the 19.2 Kbps asynchronous/64 Kbps synchronous data service offered by many PBXs provides sufficient throughput for most office applications and that twisted pair media is clearly less expensive than alternatives. The Harris, Sweeney, and Vonderohe paper, "New Niches for Switches," is a case study and describes the cutover experiences of a large (approximately 8,500 stations) Inte-Com PBX for voice and data services for the University of Chicago's 102 building campus. The Mier paper, "PBX

Trends and Technology Update: Following the Leaders," concentrates on the "big three" of the PBX market—AT&T, Northern Telecom, and Rolm—and, in a series of tables and schematic diagrams, shows the various ways they responded to user demands for data services.

The fifth paper, "Meridian DV-1: A Fast, Functional, and Flexible Data Voice System," by Murray and Carr, examines Northern Telecom's new Meridian DV-1 and shows (contrasting the Bir et. al. article on the SL-1 in Chapter 1) how a major maker radically augmented the basic architecture to meet the LAN challenge. The sixth and seventh articles present the case for an all-data PBX. Muller, in "Enhanced Data Switches May Outshine LAN, PBX Alternatives," presents an excellent discussion of the functionality and features offered by modern data PBXs, especially in the areas of control, security, and administration. Menta, in "Who Needs a LAN: Data PBXs Make Sense in Some Situations," from LAN Magazine, no less, admits that for average office environments a data PBX is hard to beat for price and features. The final paper, "The Evolution of Data Switching for PBXs," by Bhushan and Holger, basically argues the case for a hybrid switch combining circuit and packet switching. Citing the separation scheme used on GTE's OMNI PBX, the authors argue that the fundamental nature of voice and data traffic is so different that data traffic should be handled by virtual calls and packetization.

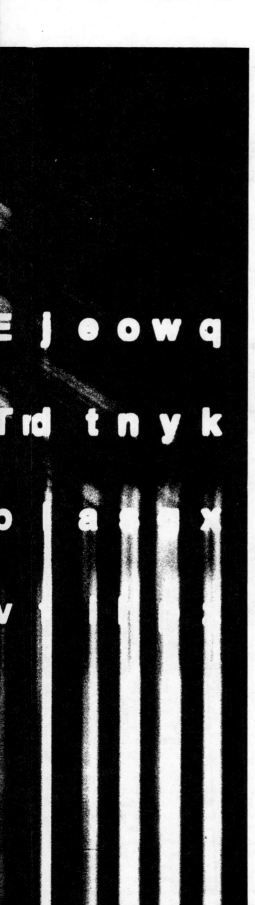

How the University of Chicago uses three digital PBXs for campus-wide communications.

Reprinted from <u>Datamation</u>, March 1983. Copyright © 1983 Cahners Publishing Company.

NEW NICHES FOR SWITCHES

by Fred H. Harris, Frederick L. Sweeney Jr., and Robert H. Vonderohe

The University of Chicago is now in the final phase of installing an integrated, PBX-based, digital telecommunications system. We now have approximately a year's experience with third generation private branch exchanges—PBXs that are fully digital, load independent, and have fully integrated voice and data. The system greatly enhances voice services and also provides the campus with a ubiquitous, high-speed data network. Moreover, compared to the continued use of a Centrex system with traditional voice services only, this pioneering installation will pay for itself and save the university an equal amount within a decade.

The University of Chicago is a private institution with an undergraduate college, four graduate divisions, six graduate professional schools, an extension division, and a major medical center with 12 hospitals. From its opening in 1892 the university has emphasized both research and teaching, and it is predominantly a graduate institution. Today there are approximately 1,000 faculty members, over 8,500 full-time employees, and approximately 8,000 students, of whom about 5,500 are in graduate and professional programs.

The university is located seven miles south of Chicago's Loop and housed in 102 buildings spread over approximately one square mile. There is a central steam plant on the southeast edge of the campus and a network of steam tunnels with easements and rights-of-way that provide fairly convenient access to almost all buildings.

In the winter of 1979, the Computation Center joined forces with the Office of Telecommunications, which is responsible for voice communications and related support services, to investigate the state of voice and data transmission facilities at the university and determine their future. The group anticipated several developments, including deregulation of traditional communication services, functional and economic gains from increased use of digital technology, and a proliferation of data and communicating data devices.

We were interested in cost savings and cost avoidance by means of the new technology becoming available and in achieving further management control. There were, moreover, growing difficulties with cabling and structural problems.

The new campus telecommunications system consists of three interconnected integrated business exchanges (IBXs) from Inte-Com Inc. (an Exxon affiliate), and each will service approximately one third of the university. University users in off-campus buildings continue to use Centrex service, and Illinois Bell remains the university's principal source of many special circuits, pay phones, and the various off-campus communications services. When completely installed, by the summer of 1983, the system will contain about 8,500 stations with about 15% initially equipped for simultaneous voice and data. The use of simultaneous voice and data is expected to increase to 30% within two years and 50% within five. Indeed, it is this level of data support that, to us, dictated that data services be as ubiquitous as voice.

A schematic diagram of the university's telecommunications system, including the microwave link for direct inward dialing (DID) service, appears in Fig. 1. The heart of each IBX, schematically illustrated in Fig. 2, is its central switch or master control unit (MCU). Inside the MCU, switching is performed by one of two totally redundant computers, master processor A or master processor B (MPA or MPB). These computers have databases containing information about each of the IBX's stations or phones—its numbers, the service features selected for it, the call group it's in, etc. Note that the master processors MPA and MPB do not share the workload; one is the reserve unit called into service only if the other should fail. Moreover, such backup is characteristic of every system component whose failure could imperil service for more than 16 ports.

Also part of the MCU are 16 switching networks (SNs). Each mediates communications between the master processor and an

interface multiplexor (IM) located at a remote site on campus. Each IM has 256 ports, each of which can be wired to:

• a universal connection block (UCB)—a two-slot wall jack for connecting a user's phone that may include computer terminal communications as well as voice;
• data access boards (DABs) for connecting to computer ports;
• an Illinois Bell trunk line for incoming (off-campus) calls;
• an Illinois Bell trunk for local outgoing (off-campus) calls;
• the university long distance network consisting of WATS, MCI, and other circuits for outgoing long distance calls.

Cabling between any phone and its respective IM consists of two twisted pairs of wires. These twisted pairs always are of the same type for all such connections throughout the system. The advantages of such modest cabling requirements become obvious when the 150-wire cable of a traditional call director phone is compared with the four-wire cable of the electronic telephone that functionally replaces and surpasses it.

Each interface multiplexor is connected to its respective switching network at the master control unit by coaxial or optical fiber cables. On either the coaxial or optical fiber cables, voice and data from the 256 ports in an IM are transmitted simultaneously at a combined speed of 44 MBps. Optical fiber is more economical for longer distances and has been used without problems from the outset.

BASIC TERMINAL EQUIPMENT

While the cabling between any phone and its respective IM is identical, there is a fundamental difference in how the two basic telephone instruments communicate. A piece of standard telephone equipment (STE) may be any industry-compatible push-button dual-tone multifrequency (DTMF) phone. This type of device communicates in standard analog voice fashion, through the twisted pairs to its interface card in the IM. The STE interface card digitizes the voice information and transmits on the 64Kbps portion of the 128Kbps allocated to that IM port. Of the remaining 64Kbps allocated for data and signaling, only the 8Kbps for signaling is used (switchhook flashing, etc.). Because STEs communicate back to the IM using the analog signals of standard telephony, users who have a dial-up modem or an acoustic coupler can use them just as they have in the past. Note that the flexibility to do so is important both for those with existing equipment who remain content with analog-level service and also for ease of transition to new digital transmission services.

The basic electronic phones available with the system, called integrated terminal equipment (ITE), have both the touch pad and 12 feature/function buttons. These buttons can be set via the system database to activate any feature or perform any function of the system. Other ITE models are available with additional feature/function buttons and with light-emitting diode displays to provide "to-from" calling information. Moreover, with an ITE there is a much greater range of data handling capabilities. Unlike the STE, the electronic phone uses the full 128Kbps allocated for the port to which it is attached. Since digitization takes place within the in-

strument, only digital signals are present on the two twisted pairs connecting it to the IM. This precludes the use of standard analog modems with an ITE, except for acoustic couplers. Instead of connecting a terminal to a modem, it is connected directly to an ITE containing a data option board (DOB).

The DOB, installed within the base of the ITE, provides a standard 25-pin RS232 EIA connector for the terminal connection. For intracampus calls to digital equipment the DOB replaces the need for modems since the transmissions are entirely digital. Further,

FIG. 1
UNIVERSITY OF CHICAGO TELECOMMUNICATIONS SYSTEM

FIG. 2
IBX SWITCH SCHEMATIC

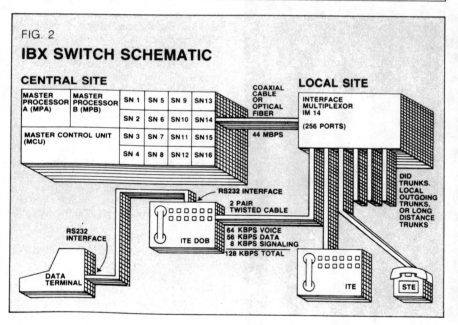

because of the preallocated bandwidth for both voice and data, an ITE, in which a DOB has been installed, continues to be available for voice calls while that DOB is being used on a data call.

Data option boards currently exist in two forms, with a third form announced for market. The DOB1 is an asynchronous device with various speeds up to and including 19,200 baud. The DOB2 is a synchronous communications device with the same operational speeds as the DOB1, while the DOB3 is intended to provide synchronous communication to 56,000 baud. In areas where a high concentration of data connections is required, such as with pools of computer ports, a data access board (DAB) can be used. The DAB, which comes in two varieties (DAB1 for async as above, and DAB2 for all synchronous speeds), is rack-mounted in a data cabinet similar to the IM card configurations supporting ITEs and STEs. By design, however, the DAB uses both the voice and data bandwidth of a port for data, permitting two independent computer connections to a single IBX port.

The Computation Center's two main computers, the DEC-20 and Amdahl, are connected to the IBX system by DAB1s as are the DEC-20 in the Graduate School of Business and the Microdata Sequel in the Registrar's Office. The IBX down-line loads the characteristics (e.g., speed) of the originating device (DOB1) to the answering device (DAB1) as the data call is initiated, thereby allowing maximum pooling of DAB1s associated with a given computer.

Although a DOB1 can send asynchronous data at rates as high as 19,200 baud, one is still limited by the communication speed of the device at the other end of the line. To accommodate the increased data handling capability of the IBX, the Computation Center now supports higher port speeds than in the past. Pools of 4,800 baud ports are now available, and pressure for 9,600 baud ports is increasing on the DEC-20 and on the Amdahl. In both cases, the ports are at fixed speeds

and not speed selectable, as are current DEC-20 and on the Amdahl. In both cases, the ports are at fixed speeds and not speed selectable, as are current DEC-20 300/1200 ports.

The center has installed and is now successfully using several synchronous DOB2s with "nailed" connections to replace leased Bell circuits that support high-speed RJE printers and terminal cluster multiplexors. Because these connections have proven to be "protocol transparent," we will be expanding synchronous data support to include controller communications for 3270s.

The Computation Center has installed a pool of modems to interface the digital transmission system with analog communication devices. For on- and off-campus calls from analog modems, the pool provides answer capability, thereby eliminating the need for separate analog modems. For calls from ITE-DOB1s to off-campus, the pool provides originate capability. Maximum transmission speed when using this modem pool is 1,200 baud. Terminals connecting to this pool via ITE/DOB1s must be set to match the characteristics of the answering device. Plans call for migrating toward a single modem pool capable of handling Bell 103 protocol at 300 baud, Bell 212A at 1,200 baud, and Vadic 3400 series at 1,200 baud for both originate and answer.

To implement modem pooling, the manufacturer has developed a modem interface card (MIC) that performs the analog equivalent of a data access board by allowing two analog paths into a single IBX port. Working in conjunction with a DAB, a MIC-DAB pair will support two analog modems. The modem connections consist of the standard modular cable to the MIC and the standard RS232 cable to the DAB. Data communication is totally digital on the DAB switch side and analog on the MIC switch side.

In addition, InteCom also offers Intenet packet controllers (IPCs) that can be added to the switch to provide additional data-related support. The keyboard option

IPC allows one to originate data calls from a terminal keyboard, the 3270 IPC provides 3270 emulation support for ASCII terminals, and the X.25 IPC is a gateway to public (or private) data networks supporting the X.25 protocol. We are evaluating the technical specifications for these to determine their applicability in our environment.

The majority of users will not be using the data capabilities of the IBX for the next few years. To them, the IBX is simply another telephone system, albeit one with extensive voice features. Those features include call forwarding, call conferencing, abbreviated dial, and numerous others. Users tell us that these features have increased office efficiency, especially when used to compensate for employee absences.

Among the users who have been using the IBX for data transmission for many months now, the consensus is that placing data calls through the IBX is as convenient as having a hardwired terminal. The data call can be placed with a few keystrokes on an ITE/DOB1. Also, when necessary, changing the transmission speed from the speed indicated in the database requires only two additional keystrokes.

There have been occasional periods of instability during which data transmission has dropped in midsession. Moreover, because there are so many points at which the problem could be occurring (the terminal, the phone, the IBX computer, the Gandalf PACX—private automated computer exchange—or the Amdahl or DEC), it has not always been possible to determine the exact source of the problem. The traditional multivendor coordination problems exist. When transmission is dropped by the IBX, however, now an infrequent occurrence, it is normally sufficient to place the call again.

SYSTEM USERS SCATTERED As anticipated in our original analysis, the use of the data capability is geographically scattered. A number of faculty and students at the Graduate School of Business are using the system to communicate with their DEC-20. Many individuals in the physical sciences and throughout the administrative departments are using the new network to communicate with the university's DEC-20 and the Amdahl, both of which are housed in the Computation Center. In addition, both the Graduate School of Business, the university library, and the college use the new network to obtain information from the student information system on the microdata sequel in the Office of the Registrar. Each week brings a new usage in an unexpected location, and the value of convenient ubiquitous access is substantiated.

While the hardware installation has been relatively problem free, such has not

"If it please the court, I would like to reenact the crime."

To accommodate the increased data handling capability of the IBX, the center now supports higher port speeds.

been the case with the systems software or with interconnect services from the local operating company. The software presented us with two different problems, one of which was anticipated: the bugs that come with any computer system. Being an early user of the IBX system and the only extensive data user to date, we apparently uncovered most of the latent bugs in the software. In addition, a number of enhancements were made to the system as a result of our experiences or at our request.

Corrections and enhancements have taken the two traditional forms, patches and new releases. Each of these forms has exhibited instabilities related to the manufacturer's inability to adequately test prior to implementation. These instabilities, though annoying, should not have been unexpected in a large complex distributed digital network. Because ours is a unique installation (both the largest and most heavily data oriented), no adequate environment existed in which to test extensively the corrections and enhancements. Indeed, even several days after installation, some problems did not occur until the appropriate circumstances appeared in coincidence. In this regard, our experience has been similar to that of installations using second or third generation digital communication switches of other manufacturers.

Moves and changes of equipment are not difficult. There are no wiring changes required for a situation where station wiring already exists, regardless of the type of change made. All moves and changes require database changes which in many cases are as simple as changing the port assignment by means of a keyboard entry. Where the analog standard telephone equipment is substituted for the digital integrated terminal equipment (or vice versa) a simple cross-connect is required at the main distribution frame located at the IM-site. This process is performed by a trained technician in five or 10 minutes.

The integrated digital data switch and transmission network is working well with respect to those data capabilities that have been installed. We are very pleased with asynchronous support, and can tell you from personal experience that once you've been to 4,800 baud for everyday, routine interactive terminal use, you will not want to go back to 300 (or even 1,200). And now, as stated earlier, pressure is mounting for 9,600 baud service on the Computation Center's computers. Indeed, the bottleneck has now shifted to the availability of higher speed computer ports.

We are equally satisfied with the effectiveness of synchronous data support. Using nailed connections with software-defined, fixed-end points, service has proven to be totally transparent to the protocol of the end-connected devices. In addition, moves and changes, in most cases, require little or

no lead time and minimal, if any, cost. Similarly, additions are treated as normal voice additions with the inclusion of database additions for the data line.

DRAWBACKS OF BEING A PIONEER Unfortunately, all enhanced data services are not yet available, though we were led to believe they would be by now; this is one of the drawbacks of being a pioneer. Paramount in this category are the services related to interswitch activity. (In this regard, being the first multi-switch installation has not been beneficial.) As a result of the manufacturer's emphasis on single switch development, InteCom's T-1 equivalent, called IXL (interswitch communication link), with its inherent interswitch transparency of data and some voice features, has not been delivered. Because of this we have had to devote a great deal of effort to providing interim interswitch data capabilities. In addition, some of the interswitch voice features, (e.g., LED display information), do not exist across switch boundaries.

The capital cost for the project with installation is approximately $1,100 per station equipped with standard telephone equipment. This compares quite favorably with the costs of alternative tariffed service from Bell operating companies. In the same vein, the incremental cost of an ITE compares quite favorably with the alternative tariffs for multibutton sets, call directors, etc. Indeed, when combined with the reduced operational costs for moves and changes, the savings in a decade in net present value terms—with payment of all capital and operating costs included—now exceeds the capital costs of the system and its installation. The incremental costs of a DOB1 with asynchronous range to 19,200 baud is comparable to 1,200 baud modems. Thus, there are no penalties or greater costs required to invoke use of the network for data.

While the university was not the first InteCom installation, it was the first multi-switch site, the first with extensive data requirements, and the first in a complex dispersed building environment. To our knowledge, there were no prior installations of a comparable nature or, for that matter, any fully integrated digital system. We were plowing fresh ground, for example, with the microwave linkage and with inter-master-control unit connections, and the soil has been rocky in places.

Nevertheless, suggestions for improvements and enhancements have been well received because of the nature of the system architecture and the relative youth of the vendor and its market. We have fewer fears today than at the outset about obsolescence because the supplier has now demonstrated an ability to extend the system and to

stay at or ahead of the leading edge.

The risks that we did not fully appreciate or enumerate at the outset are those associated with being a new user in a new industry—the interconnect industry. Our environment at the University of Chicago is strongly decentralized, and organizing to be an operating company has taken more attention and more time than we anticipated. Also, the interconnect industry is rapidly changing as deregulation movements take place. There are few skilled people with relevant experience in dealing with multiple vendors and new technology.

Finally, the projected savings (in addition to enhanced features and functions being obtained) made our selection process simple once we persuaded key people that our analysis was conservative and that the savings were not derived by sleight of hand. Since our first estimate of almost two years ago, the savings are even more substantial than originally anticipated, and the figure is growing with each new tariffed rate increase.

We now have a better appreciation of both the magnitude and complexity of installing an integrated campus telecommunications system. Nothing to date causes us to question the decision to do so. Had we to do it over, we would; but given our current, sometimes hard-learned knowledge, we would do it "smarter." ✳

Fred H. Harris is director of the Computation Center at the University of Chicago. He has degrees in physics from the University of North Carolina and Rice University and an MBA from the University of Chicago. He has over 20 years of experience in computing and data processing services.

Frederick L. Sweeney Jr. is director of operations for the University of Chicago. He has a BS in electrical engineering and a BA in economics from Tufts University, and an MBA from the University of Chicago. He has 20 years' experience in the use of digital computers for management information systems.

Robert H. Vonderohe is manager of communications services for the Computation Center at the University of Chicago. He has BS and MS degrees in electrical engineering from the University of Illinois. His background includes computer systems engineering and design, project management for hierarchical minicomputer network development.

An Introduction to Integrated Voice/Data PBX Systems

Kenneth J. Thurber and Harvey A. Freeman

The private branch exchange (PBX) or, as some people prefer to call it, the private automatic branch exchange (PABX), has been in place in major office buildings for years. For the most part, it has gone unnoticed: In the early days, most users did not even realize they were using something other than the "Bell System." Now that office automation has come about, however, everyone is sitting up and taking notice of these systems, which have put twisted-pair wiring in place throughout office complexes.

Traditionally, PBXs have had centralized control and were used to handle spurious connections usually lasting about 3 minutes. This short, "bursty" communication typically used more stations than trunk lines leased by Bell Operating Companies because it was unlikely that all would need to be connected simultaneously. This method was known as a blocking switch because, if the number of calls in progress and call attempts exceeded the number of trunk lines, the excess calls would be blocked, or not connected. In those exceptionally busy periods when the existing lines were filled, various queuing algorithms were used to accommodate the excess. Some stations that needed ready access to the lines (e.g., salespeople or executives) sometimes had their own lines tied directly to a trunk.

This conversion, coupled with the uneven quality of the lines, compromised the quality of the transmission and sometimes induced errors. The low quality resulted in slower transmission rates, and the errors created the need for retransmission.

Since PBXs were first developed, three types of systems have evolved. The systems that use normal telephone lines with modems as described above are called first-generation systems. Second-generation systems are of two types: (1) those in which an originally voice-only PBX overlays data on the lines and (2) those that were originally a computer communication network or a data PBX and which now incorporate voice transmission. The third-generation machines, which are truly integrated voice/data PBXs, are those that are designed to have both voice and data; the integration is not an afterthought.

In the future, through the Integrated Services Digital Network (ISDN) efforts developing nationwide—and even worldwide—telecommunications networks will be digital. Some private networks of this type already exist within large corporations that have incorporated tandem switches; however, the possibility of a public network stems almost entirely from the ISDN work. If this idea becomes reality, the data rates now possible on a public system will increase dramatically. Deployment of such a full-scale digital system is coming closer; all-digital connections already exist between some cities.

PBXs and Local Computer Networks

Unlike the PBX, local computer networks may have either centralized or distributed control and handle both short, bursty traffic (such as electronic mail) and long, steady connections (such as those required for interactive program development or for large file transfers). In building products for the automated office, some computer manufacturers have realized that voice communication is an effective and necessary way to conduct business. To control an office environment, these manufacturers realize that they need more than just shared resources between stand-alone computers: They need a complete communications system as well. The manufacturers who deal with integrated voice and data produce both baseband and broadband systems.

The ways in which a PBX may be used in conjunction with, or as, a local computer network are illustrated in Figures 1, 2, and 3. In the configuration shown in Figure 1, the PBX is used as the hub of office communications. Depending on the PBX chosen, this could be a viable solution, provided the attached mainframes did not have many data exchanges. If they did, a major bottleneck in the system would result because mainframe-to-mainframe communications are generally high-speed transfers, higher than those normally handled successfully by a PBX.

A second method of integrating the two communication schemes would be to use the PBX as a node on the network and have it act as a gateway to the modem pool and phone sets. Voice could be handled either off or on the network (see Figures 2a and 2b, respectively). Handling voice connections off the network makes the system easier to implement and is less demanding on the data network for real-time response. This method of configuration keeps the PBX virtually separate from all network operations, unless a long-haul link is needed. It also keeps the telephones operational when the network is down.

The third way in which the PBX has been envisioned for use in the office of the future is as a node on the network, acting as a gateway to dumb terminals, the modem pool, workstations, and telephone/intelligent station sets (see Figure 3). This scheme allows the PBX to handle only those data rates at which it is most efficient. In this configuration,

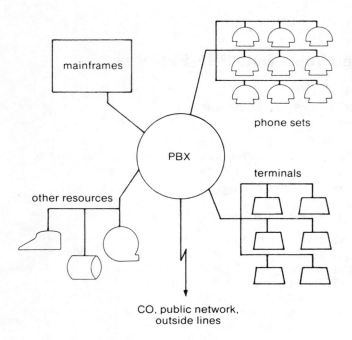

Figure 1: PBX as Hub

the network may or may not directly handle low-speed terminals. It does handle all voice and data traffic, as well as high-speed data transfers and computer-to-computer communication.

Some feasibility studies have even been conducted for voice capability on Ethernet. Ethernet transmits data bursts of packets, each containing the source and destination and a data field with multiple digitized voice samples or data. Access to the cable is by contention, using a carrier-sense multiple access with collision detection (CSMA/CD) technique. The access time to the network depends on the traffic load at that moment, and no allotment for any type of priority is possible. The results of these studies have shown that without placing restrictions on the traffic, instituting a priority scheme, and installing echo suppressors, voice transmission is not feasible on Ethernet alone. A partnership between Ethernet and a PBX is usually recommended in this case.

Digitizing Voice

To be able to carry voice signals over digital data lines, the analog voice signals must be converted to digital in a manner that allows the signal to be reconfigured at the receiver. Digitizing voice may be accomplished via waveform encoding or source coding.

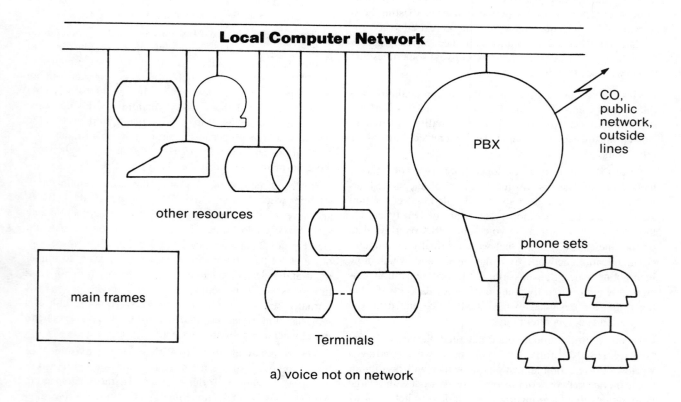

a) voice not on network

Figure 2a: PBX as Node on LAN—Voice Not on Network

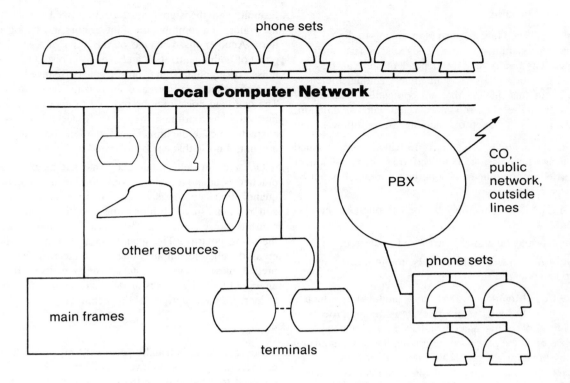

Figure 2b: PBX as Node on LAN—Voice on Network

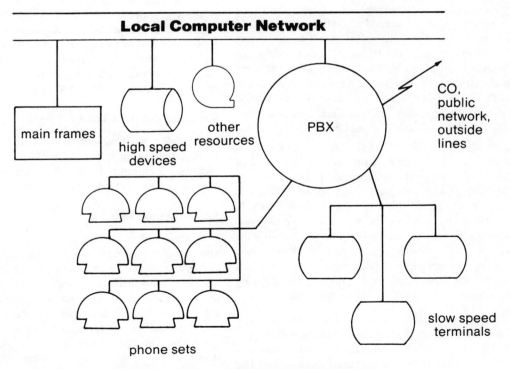

Figure 3: PBX and LAN Interconnected

Waveform Encoding

Waveform encoding performs an analog-to-digital conversion, representing the analog signals as accurately as possible. It is totally signal-independent and can be optimized to specific signals to obtain greater coding efficiency. Examples of this method that are compatible with digital systems are pulse code modulation and delta modulation. The advantages of waveform encoding are as follows:

- The signals may be regularly reshaped or regenerated during transmission, since information is no longer carried in continuous varying pulse amplitudes but by discrete symbols.

- All digital circuitry may be used throughout the system.

- Signals may be digitally processed as desired.

- Noise and interference may be minimized by appropriate coding of the signals.

Pulse Code Modulation: Pulse code modulation, which is most commonly found in binary systems, is the process of digitizing and coding analog signals into digital signals. Quantization is the process of digitizing the original analog signal by breaking the amplitudes into a prescribed number of discrete amplitude levels. The quantization level is determined by the lowest signal level to be encoded. The quantized and sampled signal pulse is encoded into an equivalent group or packet of binary pulses of fixed amplitude.

One type of pulse code modulation is linear 12-bit coding, in which the signal is sampled 12,00 times a second and each sample is coded into 12 pulses. Such linear coding makes conferencing easy; digital signals can be added together with little difficulty. Submultiplexing allows as many as 40 connections to be handled in the same system bandwidth as that of a single telephone conversation. In this way, higher data handling speeds can be obtained without compromising the voice communications system.

Another way coding is done is via a technique known as companding. In voice transmission, there is no specified peak value, and the signal level changes in a random manner. The voice signals may vary by as much as 40 dB in going from a whisper to a bellow. To obtain a relatively fixed signal over this dynamic range, signal compression is needed so that the soft speaker will not be penalized by the quantization level based on the loudest word of a loud speaker. To accomplish this, the signal-to-noise ratio is limited when the input signal power diminishes, allowing the quantization levels to vary as the sample value varies.

Delta Modulation: Delta modulation samples incoming analog data and outputs binary pulses whose polarity depends on the difference between the actual input signal and a quantized approximation of the preceding input signal. The output reflects changes in the input and approximates the differential (delta) of the analog signal. Unfortunately, from a fidelity standpoint, delta modulation acts in a nonlinear fashion, and the signal must be improved by adding an integrator, or a combination of integrators and analog amplifiers. Another disadvantage of this type of system is the effect of both quantization and overload noise. Quantization noise, or granular signal noise, can be reduced by either shortening the step interval or increasing the sampling rate. The overload noise may be reduced by increasing the step interval. The optimum step size is dependent upon the characteristics of the input signal, the sampling rate, and the amount of noise that can be tolerated in the system.

One type of delta modulation that has been adopted in practice is adaptive delta modulation. In this scheme, the quantization step size is varied dynamically. It begins small and remains small, limiting the quantization noise until overload noise is indicated by a series of output values with the same polarity. The quantization step size is then increased until the output values alternate in polarity for a predetermined number of steps. The step size is then once again reduced. The series of step size changes may be used to increase the adaptivity of the technique.

Source Coding

Source coding is usually performed by a vocoder (voice encoder/decoder; also known as a codec) on the line, located either at the station or on the port at the main switch. The techniques used with this method incorporate linear predictive coding (LPC) and other synthesis procedures. These procedures can be accomplished because of the consistency of the voice signals. The present input signal is used by algorithms to predict the next output to be synthesized. The signal is limited to a variance over a given time interval, but for voice signals, the encoding can be done much quicker than in waveform coding methods, and less bandwidth is required on the line (only 4,800 bps is necessary).

Switching Digitized Voice

A reliable switching system is needed to support digitized voice transmission in real-time. The techniques used for switching digitized voice may be space-division switching, time-division switching, or a combination of the two.

Space-Division Switching

Space-division switching, or space-division multiplexing, involves setting up a single path through one or more sets of crosspoint switches between the line and the trunk. The path, a physical connection, is used by the caller for the duration of the call. This path is released for use by other callers after the call is terminated. Because it is a circuit switch, or hard-wired connection, which could remain analog throughout, this connection need not be digital. The disadvantages of single-stage switching are as follows:

- The number of crosspoints required in a large system becomes prohibitive because only one connection may be made per crosspoint.

- The large number of inlets and outlets creates a capacitive load on the message path.
- One specific crosspoint is required for a particular connection which cannot be made if there is a failure in that crosspoint.
- Crosspoints are inefficiently used, since only one crosspoint in each row or column is in use when all lines are active.

A non-blocking switch incorporates space-division switching in the form of a multiple-stage switch. The advantages of a multiple-stage switch over a single-stage switch are that fewer crosspoints are needed, and those that are necessary are used more efficiently (i.e., crosspoints are shared so that more than one path is available for any potential connection, and alternate paths reduce blocking and provide protection against failures).

Time-Division Switching

Time-division switching is done by sampling the signal over time. For continuity of the signal, synchronous time-division multiplexing is usually incorporated in voice transmission, although asynchronous data transmissions or voice store-and-forward may be intermixed. In time-division multiplexing, a framing scheme in which the data can be encapsulated must be accepted. The frame, which is the smallest unit in which each station is serviced at least once, may be synchronous (fixed frame) or asynchronous (variable frame length). Each time slot within the frame is uniquely assigned to each data source, and a timing procedure is developed to sample each data source at the appropriate time interval. Control bits for framing and synchronization are inserted in the frame to enable the receiver to recognize the beginning of the frame, each slot within the frame, and each bit within the slot. Also, to reduce errors, provisions for small variations in the incoming bit rates to the multiplexer are handled by the system.

Two major classes of multiplexing are evident in long-haul communications: those that travel over voice-grade lines and those that travel at much higher data rates for data transmission service. The voice-grade channels are those designed to carry the range of frequencies that accommodate the human voice (up to 3.3 KHz). They may be dial-up or leased, regular or conditioned. The higher data rates are the type that evolved from the digital hierarchies created by AT&T for North America and Japan and by the CCITT for Europe; the two hierarchies are shown in Figure 4.

Hybrid Switching

Space- and time-division switching may be incorporated in two ways: The single- or multiple-stage space-division switch crosspoint outputs may be multiplexed on the same line, or else a multiplexed line may be switched through space-division switching. One of the two types of hybrid switching is necessary in larger digital switches.

Blocking Versus Non-Blocking Systems

Non-blocking is achieved in a multiple-stage switch in which each input may connect to an output. A blocking switch has more inputs and outputs than the number of simultaneous connections, which is actually a function of the type of call and its transmission rate. At first glance it appears that a non-blocking switch would be more desirable, but a non-blocking switch may not always allow a connection: A non-blocking switch will block a connection when all trunks are busy, when a DTMF receiver is not available, or when the processor (not the switch) is overloaded. In these cases, a blocking switch would be as equally effective (and sometimes more so because of queuing algorithms) as a non-blocking switch.

Efficiency

The efficiency of a PBX system is characterized by a number of factors. These include the maximum number of simultaneous calls that can be handled by the PBX, the line utilization of each connection to the PBX, the blocking factor if queuing occurs in the PBX, the ratio of trunks from the public switched network to/from the PBX, and the ratio of voice traffic to data traffic if both are to be carried by the same PBX.

The maximum number of calls that can be serviced by the PBX is simply the number of connections that can be made by the switching matrix of the system. This number should not be confused with the maximum number of connections as stated in the product literature: The maximum number of connections in a blocking system is always greater than the maximum number of simultaneous connections.

The line utilization of a PBX is usually measured in a unit of traffic load known as ccs, or 100 call-seconds per hour. The maximum traffic that can be handled on one line is one erlang, or 36 ccs (3,600 call-seconds/hour); at this level of traffic, the line will be continually occupied. For most business applications, less than 6 ccs is unacceptable.

The blocking factor of a blocking PBX is known as the P factor. Blocking switch matrices have more input and output ports than the maximum number of connections they can support. The P factor is the fraction of calls that will be blocked due to all switch paths being busy. A P factor of 0.02 indicates that 2 out of every 100 calls will be blocked and may be either queued or unserviced.

The ratio of trunks to the number of lines to/from the PBX must be adequate for the particular office it serves. If too few trunk lines are provided, the grade of service of the PBX may decrease. Even when the PBX is not overloaded or in a busy period, a person outside the company (or an employee) may not be able to access a trunk line from outside (or within) the organization. This usually causes poor user acceptance because the user feels the PBX is giving a low overall grade of service even though it is actually the incor-

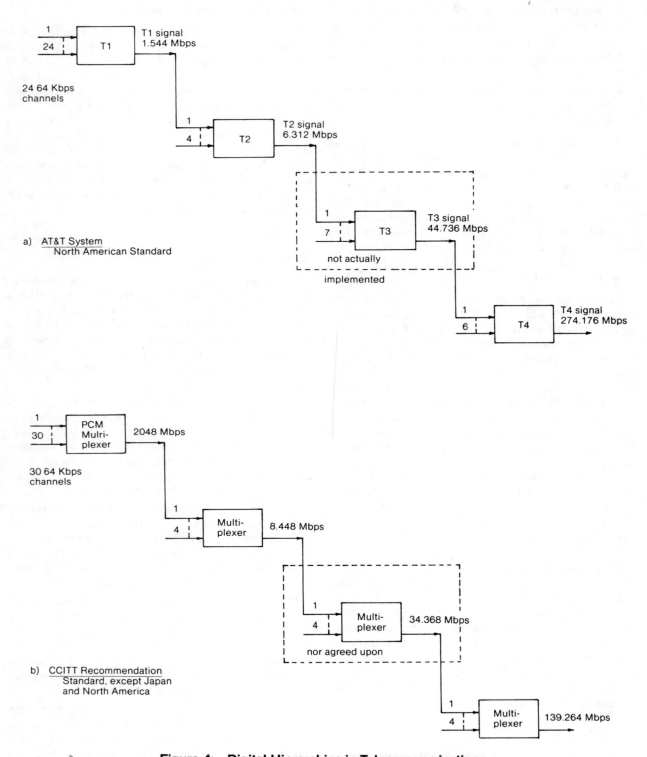

a) AT&T System
    North American Standard

b) CCITT Recommendation
    Standard, except Japan
    and North America

Figure 4:   Digital Hierarchies in Telecommunications

.ect setup and not the PBX itself that is causing the access problem.

The ratio of data traffic to voice traffic affects the PBX in that the two must be handled differently. The data calls have long holding times and low line bandwidth, as compared with the short holding times of voice calls and the high constant bandwidth needed for voice traffic. Also, voice calls must be handled immediately, so that the echo effect and propagation delay of the system remain transparent to the user. The multiplexing of the data traffic makes these delays irrelevant to the terminal user.

Traffic balancing may be necessary in some PBXs. As a system grows larger, it may be necessary to divide the users into subgroups. If many heavy telephone users are in one subgroup, they may not be able to get a connection through their subgroup's matrix, while light users in another subgroup may never use their subgroup matrix to its capacity. Traffic balancing mixes the light and heavy users so that each subgroup handles about the same amount of traffic which is within the capability of the system. Small systems typically do not need traffic balancing, because users are connected to the same matrix with equal access to all lines. In view of the increasing traffic loads produced by data circuits and facsimiles, it is clear that the ability to add ports without traffic balancing would eliminate a major planning problem.

The location of the voice integration also affects the efficiency of the system. If voice and data are integrated on the same line (sometimes two separate pairs are used, one for voice and one for data), the integration may occur at the phone/terminal, at the trunk lines, or at someplace in between, and is usually accomplished with a voice encoder/decoder.

If voice and data are integrated at the phone/terminal, a surplus capacity will result during off-peak hours and also during the idle periods inherent in voice transmission (about 50% unused capacity). The systems that integrate at the phone set or terminal require a separate codec (voice digitizer) at each station, which substantially increases the cost of the station.

If voice/data is integrated at the PBX, the spare capacity created by idle voice transmission can be used by other voice/data channels. The disadvantage of this method is that separate line interface ports are needed at the PBX for the voice and data lines. The voice signal is transmitted in analog form to the PBX, while frequency-separated subchannels on the twisted pair are used to transmit data.

Reliability

The reliability of the PBX is an important factor to consider when determining whether or not data should also be transmitted. Several systems offer redundancy, either as a standard component or as an option, possibly even to the board level of the system. Maintenance consoles with loopback testing are also common, so that problems may be detected before they become a nuisance. Some systems have self-checking and self-correcting software/hardware combinations, in which the software runs self-checks periodically, bypassing and reporting problems to the maintenance console. The sophistication of such an operation is dependent upon the system.

Component failures are normally lapses that can be handled in a less-than-critical manner; the major problems occur during a power failure, which is something no manufacturer can prevent. Some systems have backup tape or disks for volatile memory, which must be reloaded before the system is back in full operation; others use non-volatile read-only memory (ROM) and digital logic. Some systems offer backup power for varying amounts of time, which allows noninterrupted service to calls in progress.

Features

PBX features are usually implemented in the software. The more popular ones are, for the most part, universal among all manufacturers, while others are designed specifically for one manufacturer's product. Manufacturers may have a set package of features available on all phone sets, an optional set of features for the system and the user, or a complete write-your-own system-features package (from a list of all features that are available for the particular product).

System features are those available at the system level for maintenance, management, configuration, and control. Such features include ground start/loop-back testing, call detail recording, and a central control console which enables the telecommunications manager to configure and monitor the PBX system.

Console features may be available at a proprietary station set, at a standard telephone instrument, or at a data terminal through a data set. Such features include call waiting, call forwarding, camp on, do not disturb, and several others.

Integrated Voice/Data Products

Some of the systems available on the market can now meet all the requirements of the different organizations within a company. For example, the CXC Rose, Northern Telecom's Meridian SL line, and Intecom's IBX line can switch high-speed data transfers between computers and also handle voice effectively. All of these products have successfully integrated 1 to 10 Mbps local network technology into the basic switch. More of this can be expected in the future.

Michael W. Patrick, Ztel Inc., Wilmington, Mass.

The heat is on for phone switches that do a lot of fast shuffling

Modern PBXs and local networks present a range of alluring alternatives as voice/data solutions for automated office.

Since the first commercial exchange began operating, in New Haven, Conn., more than a century ago, the telephone has been a tool for business communications. In recent years the PBX has added a variety of functions—such as flexible calling, holding, and transferring—to office workers' desks. Lately, innovations in minicomputers, terminals, and microcomputers have brought technology to the point where it is economically feasible to provide desktop computer access to every office worker.

Now two new technologies are competing for a place in the foundation being set for improved communications services: voice/data PBXs and local networks. The new PBXs, by digitizing voice directly at the handset, readily support data rates at speeds up to 64 kbit/s (the standard data rate for digitized speech). The new local networks offer computer links able to carry data at speeds greater than 1 Mbit/s.

Not surprisingly, many information processing managers have trouble weighing the relative merits of PBXs and local networks. Requirements vary for voice and data communications in medium to large offices, and managers could use some criteria to compare the benefits and drawbacks of local networks and PBXs.

PBX voice features

Today the PBX is so ensconced in the office environment that almost no one disputes its ability to enhance voice communications. Modern PBXs provide upwards of 150 separately identified features, although many of them are not obvious to the typical user. One such PBX feature—responsible for major cost savings—is least-cost routing. This enhancement automatically selects the cheapest trunk for outgoing long-distance calls. Newer PBXs will permit bypassing the local telephone exchange entirely when connecting corporate facilities directly to the company's long-distance carrier of choice. The newer switches also permit tracking of each telephone call with the station message detailed recording function.

The more visible PBX features alleviate the major gripe with telephone calls—the frustration encountered when the called party is away from his or her phone, or is busy on another call. The ability to reroute calls automatically (call forwarding), have calls answered by another party (call pickup), and automatically redial an unanswered call (call-back, no answer) are all features that permit the caller some recourse other than hanging up when the called party does not answer.

Notification of incoming calls (call waiting), automatically redialing a called number when it becomes available (call-back when free), dialing a specified series of numbers in a group when no one answers on the original number (hunting), and a variety of "hold" options can put two parties in touch if the called party is busy on another call. Other popular features include call transfer and conferencing.

The most important—and most difficult—challenge in meeting an organization's data communications and processing needs is to determine the functional requirements of its users. Planning personnel often emphasize only the hardware required, concentrating on the most visible initial cost. They attempt to determine, say, how many minicomputers, disks, terminals, and microcomputers are required for a given organization. Unfortunately, most planners discover far too soon a variant of Parkinson's Law: The use of a resource expands to exceed its current capacity.

To circumvent such resource bottlenecks, planners should carefully analyze the application programs required by potential users. Thus, planners can anticipate the processing power, memory, disk storage, and communications capabilities that will be needed.

To determine the best way to meet users' needs, managers must first examine the job responsibilities of each user or user group. For example, organizations are typically divided into groups that perform special functions:

■ Engineering, responsible for the creation of new products.

■ Operations, in manufacturing firms, responsible for the assembly, testing, and shipment of products; and in service companies, responsible for the performance of the service offered.

■ Marketing, responsible for sales and customer support.

■ Control, responsible for financial accounting and analysis.

Also, to identify data communications requirements, it is useful to assemble all managers and secretaries into a separate functional group—managerial/secretarial, responsible for leadership and clerical support of other activities.

The application programs typically required for each function are shown in Table 1. Virtually all the applications listed deal with the manipulation of disk-based databases. Most other programs require random access for inquiry, entry, or update operations on a database. Such programs are considered transaction-based because the only information exchanged between a terminal and a program is a transaction request entered by hand plus the subsequent displayed data. Note that for all transaction-based programs the database involved requires access by several users. Therefore, some means of sharing that database is required.

Table 2 summarizes the results of a survey of business data communications network installations done by International Data Corp. of Framingham, Mass. The networks in this survey were usually dedicated to a particular purpose. They connected remote terminals or minicomputers to a host mainframe to run a particular set of applications.

Table 1: Computing requirements

FUNCTION	APPLICATIONS
ENGINEERING	CIRCUIT DRAWING, SIMULATION, TEXT EDITING, COMPILATION
OPERATIONS	MATERIALS RESOURCE PLANNING, PRODUCTION PLANNING, FACTORY MACHINE, CONTROL, SERVICE RESERVATIONS
MARKETING	ORDER ENTRY, SALES FORECASTING
CONTROL	EXPENSE CONTROL, ASSET MANAGEMENT, PORTFOLIO ANALYSIS, ACCOUNTS PAYABLE, ACCOUNTS RECEIVABLE, SPREADSHEET ANALYSIS, PERSONNEL RECORDS
MANAGERIAL/SECRETARIAL	WORD PROCESSING, PHONE MESSAGES, PROJECT PLANNING, EXPENSE CONTROL/FORECASTS, CAPITAL BUDGETING
ALL	INTEROFFICE MAIL, PHONE MESSAGES

Table 2: Data communications requirements

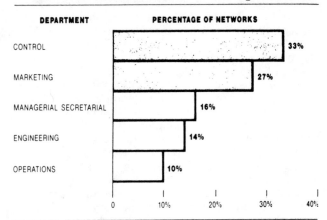

The survey results showed that data networks for the control functions of financial, accounting, and payroll applications constituted a third of the installed networks, though the personnel performing these functions formed less than 5 percent of the office workforce (according to the 1980 U. S. Census). Overall, the transaction-based networks for control, sales/marketing, and operations formed 70 percent of the data communications requirements. The key implications are that the bulk of traffic in an office's data communications network is for transaction-based database updates and not disk-based file transfers.

Common access

Local networks permit disk resources to be shared among several processor or memory devices with little consequent loss in throughput. For hard-disk drives (such as Winchester-based devices), the throughput between disk and memory is often limited by the standard defining the interface between the disk and controller. Most small computer hard-disk interfaces limit data transfer to 5 Mbit/s.

The transfer rate between a floppy disk and memory is limited by the disk itself and falls between 250 and 500 kbit/s. Hence, modern local networks, capable of providing burst throughput rates of 4 to 50 Mbit/s, introduce little delay in the transfer of disk storage to system memory. High-performance minicomputers (such as Digital Equipment Corp.'s recently announced Venus) that use a high-speed local network as the primary connection between CPU/memory and storage subsystems are now evolving. Data communications managers, then, should concentrate on the separate requirements for both terminal-to-program data flow and program-to-disk data flow.

The 64-kbit/s limit

For transaction-based programs, the throughput consists of terminal-to-program traffic of several information fields and, at most, one full terminal screen at a time. Because human entry and review is involved, a maximum of about two transactions per second should be supported (typical scrolling speed when perusing a file). Conversely, the transaction-based program should have a turnaround of less than 0.5 seconds. Assuming a 24-line, 80-character-per-line display sent every half-second, the terminal requires up to 30.72 kbit/s. Hence, the 64-kbit/s limitation of PBX-based terminal connections easily surpasses the needs of transaction-based applications.

Exceptions to this requirement for shared access to one database are text-editing programs. In this case, users develop their own documents and then distribute them in either electronic or paper form. Microcomputers are ideal for this application because the text-processing CPU time and disk activity required for development of the document need not affect other users. If microcomputers are not used, a text-editing program on a shared minicomputer or mainframe suffices. The network must handle the communications between the terminal and the text-editing program. In that case, the load of text-editing programs is similar to the load required by transaction-based programs.

Most organizations, however, do not provide terminals at every potential reader's desk, so most electronically stored documents are eventually printed. Because of the high-density one-way traffic, document printing then places the greatest load on an internal communications network. Yet the fastest line printers available, which provide speeds of approximately 1,200 lines per minute, require only 19.2 kbit/s.

The new laser printers, which rely on stored fonts, can print 100 picture elements, or pixels, per inch but are nevertheless limited to transmitting the standard 80 characters per line. Still, a 66-line page requires 42.24 kbit/s (66 lines multiplied by 80 characters multiplied by 8 bits per character), so a PBX link at 64 kbit/s limits a laser printer to less than two pages per second. Standard blank compression techniques, however, can push that limit to approximately four pages per second.

Even so, high-speed printers are quite expensive and are usually justified only for the high-output requirements of a mainframe computer room. For economic reasons, the more common laser printers distributed throughout a work area will be limited to throughput of 1 to 10 seconds per page for several more years. Hence, the 64-kbit/s maximum throughput of a PBX data link will suffice for distributed printer applications.

Most memoranda are limited to less than two pages. Program source-code modules (following generally accepted programming principles) are limited to only a few pages as well. Based on an estimate of 66 lines of 80 characters per page with no blank suppression, a data rate of 64 kbit/s transfers over 1.5 pages per second. Therefore, most edits can be transferred in a few seconds. Even a 1-Mbyte file is transferred in only two minutes.

The most frequently cited requirement for high-speed local networks is the connection of intelligent engineering workstations. In a hardware engineering department (whether electrical, mechanical, or chemical), the primary documents produced are graphic drawings and their accompanying text. Engineering workstations with displays providing 1,024-by-1,024-pixel resolution are becoming economical for individual use in large design engineering departments. A unit with such high-resolution display on a black-and-white screen requires approximately 1 Mbit/s of display memory; a color display requires about four times that amount.

Graphics

The display memory size is important because the entire memory must be updated when the screen is redrawn. Studies have shown that user productivity significantly drops and dissatisfaction results when screen redraw times exceed one second. Therefore, graphics terminals require at least 1-Mbit/s channels between the programs that generate the display data and the display memory. But an intelligent workstation includes a program to generate graphics data; and the workstation display memory resides in the high-bandwidth system bus of a program in the terminal itself. It is not necessary, therefore, that an intelligent graphics-based terminal require a megabit-per-second connection to a host computer running a graphics application program.

On graphics terminal applications, usually there is a language of higher-level primitives to define certain graphics operations (such as a single command to draw a line). Furthermore, many graphics terminals define custom fonts, or character representations, within the terminal and then select an 8-bit character code to designate that font.

Based on a benchmark of 1,024-by-1,024 pixels per 8.5-by-11-inch page, a video screen showing a third of a page could display 24 lines of 80 characters each, where each character defined 12 pixels across and 16 pixels down. Thus only 19,200 characters would have to be transmitted to fill the screen. That is the number required by today's terminals.

In such applications, the most annoying factor is actually downloading the font from the host to the terminal. In the example, each character requires 192 (12 times 16) bits; a font of 256 characters therefore requires that 49.152 kbit/s be sent. At the de facto standard of 9.6 kbit/s using asynchronous protocols, the font download requires more than 6 seconds. With the newest PBXs, however, graphics terminals can transmit synchronously to their hosts at up to 64 kbit/s for a font download time of only three-quarters of a second. Even for graphics terminals, then, PBXs' 64-kbit/s transfer rates are acceptable.

Cost considerations

The costs of an office data communications network can be divided into three areas—equipment costs, wiring costs, and adapter costs. The equipment costs include the terminal or microcomputer cost as well as that of the equipment required at the host to connect the terminal. Wiring costs include the cost of any physical media and attendant installation costs. Adapter costs include the expense of any adapter required between desk equipment and the data communications network.

The communications equipment typically found on an office desk includes a telephone and one or more electronic keyboards. Keyboard costs vary widely, depending on the type of equipment they are attached to:

Coaxial-cable-connected
display head: $2,400−$3,000
Word processor: $2,000−$6,000
ASCII terminal: $500−$2,000
Microcomputer: $1,600−$9,000

The industry-standard IBM 3270 display uses coaxial connections between dumb display heads (models 3278, 3279, 3178, 3179) or printers (model 3287) and an intelligent terminal controller (models 3274 and 3276). The terminal controller connects either directly to the mainframe through a channel attachment or remotely with a synchronous communications link. Other mainframe manufacturers offer similar coaxially connected display systems. The price tag of $2,400 to $3,000 includes not only the cost of the terminal but also a prorated fraction of the controller's cost.

Word processors have traditionally been standalone items, but several products (Wang Laboratories', for example) offer a coaxial connection to a common control element for communications with other displays or printers. The cost per station can vary widely, depending on the number of per common control elements and the type of printer selected.

ASCII terminals are the primary interface to minicomputers and microcomputers. They transmit data asynchronously using either RS-232-C or current-loop protocols. The so-called dumb terminals with few display and scrolling options fall at the lower end of the cost range. Terminals offering multiple fonts and sophisticated features fall at the high end. Because minicomputer application programs often take advantage of all terminal characteristics, many terminal manufacturers offer modes that emulate the full-feature terminals of the minicomputer manufacturers (DEC's VT100, for example).

Microcomputer prices primarily reflect the amount of disk storage they provide. Units that accommodate just one floppy disk fall at one end of the price range. Those with 10-Mbyte Winchesters tend toward the high end. Because disk storage and printer peripherals represent the bulk of a microcomputer's cost, a cost-effective data communications network permits several microcomputers to share a disk or printer. The most popular configuration attaches the microcomputers to a common minicomputer or mainframe, which in turn controls the common disk and printer resources.

Wiring costs
Today, computer terminals are typically hardwired to a computer. When an office worker moves, the host connection (and the terminal itself) has to move with him or her. Whenever a new application is required, existing terminal equipment becomes obsolete. As more office workers require electronic keyboards, the need for flexible means of attaching terminals to other equipment grows. Otherwise the cost of physically rewiring a terminal connection over several years can exceed the cost of the terminal itself. This situation contrasts with telephone wiring, which is routed to each workplace when a building is constructed or before office furniture is put in place.

Data communications wiring installed after initial building construction has been completed is usually set in raised ceilings and other environmental plenums. Because of this, smoke-resistant insulation such as Teflon is required for the media, adding significantly (as much as three times that of preconstruction wiring) to wiring costs.

Most terminal equipment is wired in a star fashion, with the terminal directly wired to a host computer or controller. Telephones are wired in the same manner; all wires are brought to a central wiring closet. Within an office, the average length of the wire between a workplace and a host computer is about 150 feet. Using industry-standard conventions for the placement of wiring closets, an IBM study demonstrated an average workplace-to-closet wiring distance of 85 feet. The latter figure is relevant because telephone wiring must be bundled into cables containing 25 to 300 pairs for connections between closets. The labor charges for laying such a bundled cable must be prorated across each connection.

Table 3 lists five of the most popular data wiring options and compares the cost of an average connection. PBX-based front-end wiring consists of quad wire from the wiring closet to each workplace. The connection from the wiring closet (the intermediate distribution frame) to the PBX (main distribution frame) would normally be by 50-pair cables. Because the labor cost of laying the 50-pair cable is shared by the 25 stations thus connected, the labor costs for PBX wiring are low.

As shown, telephone wiring is by far the least expensive of wiring options. And it becomes even more attractive in light of the fact that all other front-end connection schemes (except the IBM Cabling System) must also add the cost of connecting telephone wiring to their initial costs. If four-wire telephone connections are already set up, a company can take advantage of considerable savings by using existing wire rather than installing new media.

For connection to minicomputers, such as the popular DEC VAX, RS-232-C-based wiring is used. In general, only four connections are required for an asynchronous terminal connection: send data, receive data, signal ground, and protective ground. Data rates up to 19.2 kbit/s across these wires are possible, but 9.6 kbit/s is typical. Wiring for a minicomputer connection costs more than $100 in parts and labor.

Coaxial cable is used for most mainframe terminals, including IBM 3270 displays. For local connections, the 3274 or 3276 controller is attached to the CPU in the computer room, requiring coaxial cable to run from the computer room all the way to the workplace. This type of wiring scheme costs approximately $117 per connection (Table 3). For remote connections, the 3274 or 3276 controller may be placed in the intermediate distribution frame, saving about $14 in cable costs and $35 in labor.

Unlike other schemes, Ethernet connections are not

wired in a star fashion. Instead, the main Ethernet cable, a specially developed coaxial cable, is routed throughout a building. For each station to be attached to the network, a transceiver unit is tapped into the cable. The station is then attached to the transceiver unit by a separate transceiver cable (with four twisted pairs). This analysis assumes a main cable is routed once through each area of 10 feet by 10 feet (10 feet of main cable per user).

Based on recent average list prices, a fixed cost of $90 for the transceiver cable (including connectors and the labor required to make the cable) is used. If a building is prewired with main cable and transceiver cable (but no transceiver tap) for each desk, the cost of wiring to each desk will be about $186 per workstation. In this case, when a user moves, a new tap must be placed on the main cable. If, for convenience, a tap is provided for each user, another $270 must be spent, bringing wiring costs to $456 per workplace.

IBM recently announced the IBM Cabling System. With this approach, a combination voice/data wire is routed between each workplace and a wiring closet at the time of building construction. The cable used

includes four twisted-pair wires for each telephone connection and two twisted-pair wires within a metal shield for data. At the wiring closet, the telephone wires and data wires are separated and brought out to separate patch panels. The shielded twisted pairs, or data-grade media, can replace virtually any coaxial connection within a building.

A data-grade media cable is used to connect the patch panel of the wiring closet to a terminal controller or host in the computer room. With this scheme, both voice and data cabling networks can be established at a cost of $186 per workplace. Although the IBM Cabling System can be used to provide universal connection for coaxial-based terminals, IBM has also announced that they intend to provide a token-ring-based local network using the data-grade media within two to three years.

Adapter costs

The adapter costs depend on the type of communications network used. A point-to-point connection between terminals and computers incurs no additional adapter cost because the circuitry to transmit and receive the data is built into the terminal and host equipment. If a local network is used, a separate adapter to the network is required at both the terminal and host sides.

The average cost of a single connection to a local network is currently more than $900. To lower the per-connection price, many local network adapters (terminal servers) allow eight to 24 terminals to connect to the local network, providing transparent RS-232-C connection across it. Such terminal servers can provide a per-terminal cost in the range of $400 to $500 but require an unwieldy and inflexible scheme combining both RS-232-C and local network wiring.

A PBX data adapter, which adds data capability to a telephone connection, however, costs only $200 to $500 and can be attached to any workplace. The difference between local network adapter and PBX data adapter costs—a factor of two—is due primarily to the following:
■ The local network connection requires additional memory to buffer the high-speed data between the terminal's system memory and the local network.
■ The local network adapter requires additional processing power to implement mechanisms for distributed error detection, reporting, and recovery.
■ The local network adapter requires additional logic to implement the access control mechanisms to the local network.

Only the latter is amenable to cost reduction through the use of local network chips. Also, because of the plethora of local networks and the widespread acceptance of RS-232-C, lower-speed UARTs/USARTs will always enjoy a significantly higher volume of shipment. For these reasons, it is expected that low-speed (less than 64-kbit/s) terminal connections will remain significantly cheaper than local network connections, and that the economic justification for a separate front-end network involving point-to-point terminal wiring connections will remain sound.

Table 3: Average data-wiring costs

FRONT-END NETWORK	WIRING	WIRE COST[1]	CONNECTOR COST[2]	LABOR COST[3]	TOTAL COST
PBX	QUAD WIRE	$3.62	$3.00	$45.74	$52.36
RS-232-C	4-WIRE, 22-GAUGE UNSHIELDED	18.76	4.00	80.00	102.76
COAXIAL	RG58	31.36	6.00	80.00	117.36
ETHERNET	STANDARD ETHERNET AND TRANSCEIVER CABLE	100.00	6.00	80.00	186.00
IBM CABLING SYSTEM	TYPE 2 TO WIRING CLOSET; TYPE 1 TO CPU	81.25	26.00	80.00	187.25

NOTES: 1. BASED ON A DISTANCE OF 85 FEET FROM WORKPLACE TO INTERMEDIATE DISTRIBUTION FRAME AND 65 FEET FROM THE INTERMEDIATE DISTRIBUTION FRAME TO THE MAIN DISTRIBUTION FRAME, WITH NONPLENUM WIRING COSTS AS FOLLOWS:

	COST/1,000 FEET
PBX QUAD WIRE	$36
RS-232-C 4 CONDUCTOR	125
RG58 COAXIAL	209
IBM VOICE/DATA	650
DATA ONLY	400
ETHERNET MAIN CABLE	985
TRANSCEIVER CABLE	90 (50-FOOT DROP)

THESE COSTS ARE BASED ON THE MANUFACTURER'S RETAIL PRICE FOR JUNE 1984. THE ACTUAL PRICES PAID UNDER A VOLUME CONTRACT WILL PROBABLY BE LOWER.

2. WIRING CONNECTORS, FACEPLATES, WALL MOUNTS, AND THE LIKE; DOES NOT INCLUDE TRANSCEIVER FOR ETHERNET STATION ($270).

3. TWO HOURS FOR CABLE CONNECTOR ASSEMBLY, LAYING, AND DISTRIBUTION FRAME WIRING AT $40 AN HOUR CONTRACT LABOR.

Five in one. Host computers and PBXs may be connected in several ways, including: via common data banks, by direct T1 links, and through terminal servers.

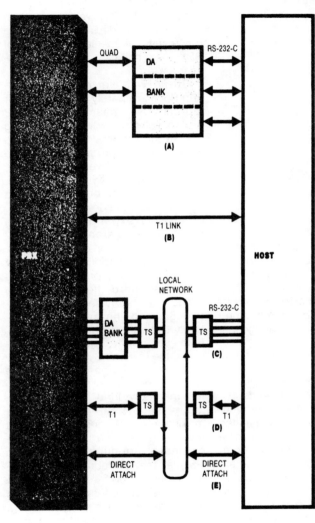

DA = DATA ADAPTER   TS = TERMINAL SERVER

While local networks are not cost-effective as front-end networks, they are necessary for the interconnecting computers with disk resources, major shared peripherals, and other computer peripherals. The backbone local network connects the mainframes, minicomputers, and any smaller department-specific local networks. It is required for the high-speed file transfer between hosts. The backbone local network is also necessary to accomplish the transparent connection of applications running in one host accessing files in disks attached to another.

A fiber-based local network, such as the IEEE 802.5 ring supported by IBM, is optimum for this purpose. Other possibilities include the 100-Mbit/s fiber distributed data interface local network under development by the ANSI X3T9.5 committee. Fiber-optic connection is desirable for its high speed and immunity to radiated noise. Further, because it is immune from electromagnetic interference, it is an excellent choice for connection between buildings.

One problem managers encounter is how to make the connection between the front-end network provided by digital PBXs and an organization's computers and local networks.

How to connect
The accompanying figure shows five separate mechanisms for interconnecting computers or local networks with PBXs. Today, the primary mechanism is a direct connection between a PBX and a host through a bank of data adapters that transform the PBX's internal 64-kbit/s stream to an asynchronous or synchronous stream at a standard data rate of 300 bit/s to 19.2 kbit/s (Fig. 1A). Each lower-speed terminal link is then separately attached to a serial port of the host.

To reduce the cost of this solution, two standards have been proposed to multiplex up to 24 terminal links on a T1-carrier link between a PBX and host (Fig. 1B). Northern Telecom and DEC have proposed the Computer-to-PBX interface, and AT&T and Hewlett-Packard Co. have spearheaded the development of the Digital Multiplexed Interface (DMI). One mode of DMI conforms to the provisions of common channel signaling specified in the emerging ISDN standards.

When a local network is involved, the PBX can provide connections for its terminals in one of three ways. A proprietary terminal server connection, such as those provided by Ungermann/Bass or Interlan, can be used to provide several transparent data streams through the local network. The PBX then attaches to those streams with each individual terminal link (Fig. 1C). Combining the approaches of Figures 1B and 1C, a T1 link could connect the PBX and host to the proprietary terminal servers (Fig. 1D). The most cost-effective approach, of course, is for the PBXs to directly attach to the local network (Fig. 1E). This scheme, however, requires a standard for providing virtual terminal service on a local network, a standard that does not yet exist.

Next to come
Because no standard exists for combining both voice and data on a high-speed local network, separate networks are required for voice and data connections. Future local network standardization work should address this problem by including both types of terminal traffic, thus allowing a single medium to be used for both networks. In addition, when designations of the geographical right-of-way do not exist, or the distances between buildings are greater than those handled by a local network, a fourth-generation PBX is able to provide full transparency of voice and data across a wide-area public network. ∎

Michael W. Patrick, senior research associate at Ztel, received his B. S. and M. S. in computer science from Massachusetts Institute of Technology. Formerly, Patrick worked for Texas Instruments as a systems architect for the company's microprocessor division.

Special report

PBX trends and technology update: Following the leaders

Edwin E. Mier, Data Communications

You've pored laboriously over the responses to your company's request for proposals. The list of semifinalists is down to perhaps a half dozen. If you're like the majority of communications managers about to make that final cut, statistics show your list includes proposals from AT&T Information Systems (ATTIS), Northern Telecom Inc., and Rolm Corp.

These vendors alone now account for some three-fourths of the entire installed base of PBXs. For new installations of digital PBXs handling more than a couple of hundred lines, their collective slice of the pie is even more. And it is a healthy three-way tie for first place. Indeed, in the fiercely competitive race for the multibillion-dollar PBX market, the customer truly still has a choice.

The mentality that has pervaded the computer industry for years—epitomized in the oft-heard statement "You'll never get fired for buy-

ing IBM''—simply does not carry over to the office communications and information network sector. In fact, it seems that IBM is having a problem getting its subsidiary, PBX manufacturer Rolm, to fit comfortably into the Big Blue scheme.

To be sure, support for IBM devices, protocols, and interfaces is an important concern in selecting a PBX. But so is the networking of all the other vendors' computer gear, as well as the provision of basic and reliable voice communications. In addition, communications managers have the all-important bottom line to consider.

Still, a company's voice and data communications network can be its lifeblood, and top management is becoming increasingly hesitant to sacrifice this for the sake of a quarterly profit margin. Even the U. S. government has deviated from its sacrosanct lowest-bidder policy recently in the award of contracts for new PBX and communications switching equipment. Nobody wants to buy cheap just to learn that they have purchased technological obsolescence.

Managers need to be able to look beyond the marketing hype of the individual PBX vendors—and their bare-bones bottom-line costs—and draw valid points of comparison. The three PBX leaders have very different offerings—especially regarding the data capabilities they currently offer and in their long- and

short-term technological directions. Just as important, however, is what current users of these PBXs have to say about their equipment, about hardware reliability, vendor support, and so on. Included in this report are the results of the latest survey of PBX users released by Datapro Research Corp.

But AT&T, Northern Telecom, and Rolm are hardly resting on their laurels. Over the next 12 to 18 months, all three will be announcing or delivering major enhancements to their respective product lines. These enhancements serve either to make smaller, and thereby less expensive, the individual components required to add lines, data devices, and such, or else to expand transmission bandwidths, distances, features, and interconnection capabilities.

■ **Systems 75 and 85.** It is clear that the original development of AT&T's three vanguard digital switching products—the System 75, the System 85, and the Information Systems Network (ISN)—was undertaken separately. Though ATTIS unveiled the low-end System 75 barely 18 months after the System 85, the common equipment packaging differs markedly. In fact, if a user outgrows the System 75 (currently a maximum of 800 lines, which itself represents nearly a doubling of the originally announced maximum size), all the common control equipment must be discarded to move up to a System 85. Users can retain and reuse the station gear and wiring, but all central equipment must go.

Obviously a high-priority development project, ATTIS has endeavored to ensure that the two PBXs work together in large networks. And the battery of new enhancements announced by ATTIS a few months ago will help achieve this. New software will allow the 75 to become a node in an Electronic Tandem Network (ETN) configuration, and new T1/DS-1 trunk modules will facilitate the high-speed interconnection of the two.

In addition, a new set of modules will allow IBM 3270 users to replace their coaxial cable with twisted pair (Fig. 1). The modules will also allow 3270 operators to switch connections through the System 75/85 among multiple 3270 cluster controllers. To accomplish this, ATTIS will offer a family of modules that convert the high-speed coaxial signaling rate of the 3270 displays (2.358 Mbit/s) to a 64-kbit/s channel carried in ATTIS's standard digital communications protocol (DCP) format over a single twisted pair and through the switch.

It is noteworthy that ATTIS has succeeded in reducing the size and amount of the circuitry required for its digital station equipment to handle the DCP multiplexing and transmission of data, while at the same time extending the single-wire-pair transmission distance of DCP. A new digital telephone set, the 7404, which integrally includes one of the new data-handling and multiplexing modules, sells for $700 and operates over a single pair of 26-gauge wire at up to 4,000 feet. Earlier digital sets, by comparison, required an add-on data module that alone cost $800, more than the complete 7404, and that limited DCP transmission to no more than about 2,200 feet. Still, ATTIS's digital station equipment is expensive, much more so than either Rolm's or Northern Telecom's. An ATTIS insider confides that the PBX configurations that AT&T ships typically consist of 70 percent or more plain-old analog station telephones, which tend to absorb and distribute the high cost of its digital station equipment.

In addition to high digital station prices, it is also much more expensive to add data devices to a System 75 or 85 than with either Rolm or Northern Telecom (Table 1). Note that ATTIS does not now offer the variety of data add-on modules that its competitors do, which means that a System 75/85 user must either run data through a digital telephone set (the least expensive one, the 7404, is used in Table 1), or else carry voice-grade asynchronous data, via modem, to an analog line card at the switch.

For higher speed and synchronous data switching, and in all cases above 19.2 kbit/s, users must buy one of ATTIS's modular processor data modules (MPDMs), which makes the incremental cost of switching such a device through the System 75/85 at least double that of either Rolm or Northern Telecom.

Note, however, that an advantage for AT&T is in the price of its T1/DS-1 interface, which will be available later this year. The same compact unit (a single plug-in module) is used either for T1-formatted transmission or for local multiplexed PBX-to-host links working with AT&T's Digital Multiplexed Interface (DMI). The difference in operation is under software control, which is included in the next PBX software version.

Informed sources report that ATTIS is indeed selling and shipping its data-oriented ISN in combined packages with its digital PBXs. And the new features soon to be available with the ISN make its array of data interfaces and capabilities truly awesome. In fact, one industry analyst privately confides that the battery of new features that ATTIS unveiled for ISN a few months ago—especially for synchronous data support and 3270 terminal switching—has caused considerable consternation within IBM and is even responsible for delaying IBM's formal unveiling of its token ring local network.

Based on the recent study completed by Datapro, users tend to be more satisfied with System 75 than with either System 85, Rolm's CBX-II, or Northern Telecom's SL-1. This apparently reflects superb maintenance of the equipment by ATTIS (rated a perfect 4.0 by the Datapro survey respondents) and the ease with which it is installed and customized (3.67, versus 3.06 for Northern's SL-1, 3.0 for System 85, and 2.85 for Rolm's CBX-II). ATTIS's training program and materials for System 75 users are lacking, however, according to the Datapro users surveyed (Table 2).

■ **SL-1 and Meridian.** Northern Telecom was the last of the Big Three PBX vendors to embrace digital telephone sets. It announced this earlier this year as part of its flashy unveiling of Meridian, a multiprocessor, high-speed packet-transport add-on for the SL-1 PBX. In fact, some industry watchers believe that Northern might have in mind for Meridian to eventually replace the SL-1, especially since certain basic functions, such as voice digitization and protocol conversion, are replicated to some degree in the Meridian unit.

Still, Meridian is being marketed to the large SL-1 installed base as a fancy add-on unit, with which customers can gain impressive capabilities, including: integrated digital voice and data; file storage, manipulation, and editing; graphics; and two-pair-wire local networking at 2.56 Mbit/s, which it calls LanLink (Fig. 2). But all data and voice connections out of Meridian must go through the circuit-switched SL-1 and, therefore, must be reduced to the lowest common denominator of 64 kbit/s. One or more local T1 channels are used to connect the Meridian with the SL-1.

Northern has finally conceded that certain economies of scale could be realized with digital telephone sets that multiplex voice and data to a single port on the SL-1. To be available later this year, the digital sets will cost from $325 to $700, plus $300 extra for a plug-in module required to connect a data device. (The data module that provides the same capability with its analog feature sets currently costs $500.) In addition, Northern will effectively quadruple the port capacity of its line cards with the new digital station equipment, from a card supporting either four data ports or two voice and two data ports, to an eight-port card, with each capable of supporting a multiplexed voice and data bit stream. This also will reduce the per-port cost from $300 currently for a voice or data port, to $350 for a voice and data port.

As one of the first digital PBX vendors, along with Rolm, Northern has developed a mature product line of data adapters and, more recently, interfaces for IBM PC, 3270, and other synchronous devices. A Northern spokesman claims that it has shipped "over 1,000 ports" of its 3270 terminal-to-cluster-controller coax elimination and switching product, which AT&T is effectively copying with its new 3270 modules. Northern also claims to have shipped over 10,000 of its $375 asynchronous interface modules (AIMs), which shows a clear need that AT&T does not yet address. Also, Northern claims shipments of 1,000 of its multicard T1/DS-1 interfaces, but only a few dozen of its Computer-to-PBX Interfaces (CPIs).

Special report

Table 1: Cost of adding data*

CBX	AT&T SYSTEM 75		NORTHERN TELECOM SL-1		ROLM XVZ-II	
WIRING PLAN	4-PAIR WIRE (TO 26 GAUGE)		3-PAIR WIRE (TO 26 GAUGE)		3-PAIR WIRE (24 GAUGE OR LARGER PREFERRED)	
INCREMENTAL COST OF ADDING ON:						
AN IBM PC	DIGITAL PHONE** WITH DATA MODULE	$700	PC ADAPTER CARD	$200	PC ADAPTER CARD	$407
	PORT ON-LINE CARD (1/8 OF CARD)	200	PORT ON-LINE CARD (1/4 OF CARD)	300	PORT ON-LINE CARD (1/16 OF CARD)	97
	TOTAL	$900	TOTAL	$500	TOTAL	$504
A STANDALONE ASYNCHRONOUS TERMINAL	DIGITAL PHONE** WITH DATA MODULE	$700	AIM	$375	DTI	$324
	PORT ON-LINE CARD	200	PORTION OF LINE CARD	300	PORTION OF LINE CARD	97
	TOTAL	$900	TOTAL	$675	TOTAL	$421
A HIGH-SPEED (56-KBIT/S) SYNCHRONOUS DATA DEVICE	MPDM	$1,385	ASIM	$500	ROLMPHONE 100 WITH DATA MODULE	$330
	PORTION OF LINE CARD (1/8)	200	PORTION OF LINE CARD (1/4)	300	PORTION OF ROLMLINK LINE CARD (1/16)	142
	TOTAL	$1,585	TOTAL	$800	TOTAL	$472
A DATA MODULE TO DIGITAL PHONE SET	ADD-ON DATA MODULE FOR EARLY DIGITAL SETS (7405, 7403)	$800	ADD-ON DATA MODULE TO CURRENT ANALOG FEATURE PHONE	$425	TO ANY ROLMPHONE, ASYNCHRONOUS TO 19.2 KBIT/S; SYNCH- RONOUS TO 64 KBIT/S	$230
	INTERNAL DATA MODULE FOR NEWER DIGITAL SETS (7407)	320	PLUG-IN DATA MODULE FOR DIGITAL PHONE (AVAILABLE 4TH QTR, 1985)	295		
A 1.544-MBIT/S T1 (DS-1) TRUNK INTERFACE	SINGLE-MODULE T1 PLUG-IN TRUNK UNIT (T1/DS-1)	$2,520	2-CARD T1 PROCESSOR SET	$3,000	10-CARD T1 PROCESSOR CARD SET FOR MIXED VOICE AND DATA	$13,000
			6-LINE CARDS (4 VOICE/DATA PORTS PER CARD)	7,200		
			TOTAL	$10,200		

*COSTS ARE CURRENT PUBLISHED OR VENDOR-SUPPLIED PRICES AND COULD VARY UP OR DOWN BY AS MUCH AS 20 PERCENT, DEPENDING ON WHETHER COMPONENTS ARE FACTORY-INSTALLED WITH ORIGINAL PBX CONFIGURATION OR ADDED ON LATER (TYPICALLY AT LIST PRICE).

**IN THESE CASES, AT&T DOES NOT NOW OFFER MODULAR UNITS TO SUPPORT THESE DEVICES. THE PRICES SHOW THE LEAST EXPENSIVE EQUIPMENT REQUIRED TO ACCOMPLISH CONNECTION—THE 7404 DIGITAL SET, WHICH INCLUDES AN INTEGRATED DATA MODULE. USING MODEMS TO ANALOG LINE CARDS WOULD BE ANOTHER, AND IN SOME CASES LESS EXPENSIVE, ALTERNATIVE.

AIM = ASYNCHRONOUS INTERFACE MODULE
ASIM = ASYNCHRONOUS/SYNCHRONOUS INTERFACE MODULE
DTI = DATA TERMINAL INTERFACE
MPDM = MODULAR PROCESSOR DATA MODULE

Special report

This fuels speculation that CPI may be losing ground to AT&T's DMI.

Based on Datapro's user survey, the Northern Telecom SL-1 is held in generally higher regard by its users than are those with Rolm's CBX-II— across-the-board in every category. The SL-1 edges out AT&T's System 75 and System 85 in user ease of operation but trails slightly behind the System 75 for ease of installation or customization, hardware reliability, and overall user satisfaction. The SL-1 fares better against the System 85, however, being rated by users as better for hardware reliability, ease of installation, and ease of user operation. Still, System 75 and System 85 users are overall more satisfied than are SL-1 users, and users rate AT&T's maintenance and troubleshooting considerably higher than Northern Telecom's.

■ **CBX-II.** Since the introduction two years ago of the Cypress and Juniper, which afforded its users impressive, though expensive, IBM PC integration with the CBX, new features from Rolm in the areas of local networking and digital trunking have been noticeably absent.

There are two possible explanations for this. As far as local networking goes, IBM is believed to be holding new announcements from Rolm in abeyance while it figures out how to tie the CBX into its token ring, and vice versa.

Rolm is said to be preoccupied with getting its high-speed inter-PBX backbone network, dubbed the Rolmbus 295, to work.

The CBX is nevertheless a potent player in the digital PBX arena, and one to be reckoned with. Rolm has a full range of data interfaces available to its CBX customers, but there are some noticeable holes. Unlike Northern Telecom, and most recently AT&T, Rolm has no comparable 3270-to-cluster-controller interface product. The consensus among industry watchers is that IBM's token ring is designed to do this, and offering it via the CBX would rob IBM's long-overdue token ring of a prime application.

Then there is IBM's ill-fated Cabling System, into which Rolm's three-pair unshielded twisted-pair wiring plan simply does not fit. Whereas Rolm's wiring requirements can generally be met with existing in-house telephone wiring, even for its digital telephone sets and the Cedar, Cypress, and Juniper workstations, IBM's cannot.

Nor has Rolm yet followed its competition with an add-on coprocessor for its CBX. AT&T's add-ons involved assorted application processors and later its ISN, while Northern Telecom's essentially supported Meridian. In the two latter cases, the packet-oriented add-on facilitates a wide spectrum of new data-oriented features and local networking. Again, IBM may intend to fill this gap with its token ring.

Rolm has reportedly been shipping more of its digital telephone sets in CBX configurations than AT&T has (though the leader is reportedly Intecom). Rolm's digital sets are much cheaper than AT&T's, and the integrated data module for asynchronous data to 19.2 kbit/s or synchronous data to 64 kbit/s is extremely affordable at $230 (Table 1). In addition, dense packing of voice/data ports on line cards (16 per card) permits Rolm's CBX customers to add data devices at the lowest per-connection cost (Fig. 3).

The problems with the fiber-optic-based Rolmbus 295, announced prematurely two years ago, reportedly lie in supporting software and in the architecture of the bidirectional bus extension. Rolm is now shipping its Rolmbus 74, which is a typically coaxial-based 73.728-Mbit/s extension of the main CBX backplane bus that is designed to make separate CBX processors work together as if colocated in the same cabinet.

It is unclear how the Rolmbus 295 will fit in with IBM's high-speed backbone-type token ring plans. IBM's high-speed token bus, also fiber-optic-based, reportedly operates at 16 Mbit/s, which is a pittance compared with the Rolmbus 295. Still, it is possible that IBM may perceive the two as conflicting for the same market.

One thing seems sure: The CBX is not likely to use token ring technology for the transmission of real-time voice traffic. Others who have tried it have found that token rings cannot accommodate data and real-time voice, in addition to the token control and overhead associated with managing the ring.

Based on Datapro's user survey (Table 2), Rolm trails its two leading competitors in overall user satisfaction and hardware reliability. Users rated Rolm's maintenance service and troubleshooting capability slightly below Northern Telecom's and considerably below AT&T's. Still, Rolm's new identity as the PBX division of IBM will warrant close scrutiny in the months ahead, and a number of new integration announcements were rumored recently to be forthcoming.

■ **User survey.** The PBX products of AT&T, Northern Telecom, and Rolm are by no means the only ones that warrant consideration by potential buyers. Depending on line size and other features sought, especially for data handling, there may be a dozen or so other top contenders.

As mentioned earlier, Intecom has developed a solid reputation as delivering the greatest percentage of data lines in its PBX configurations installed and has wrapped its local network enhancements around compatibility with the IEEE 802.3 Ethernet standard.

Also, one of the fastest-growing influences in the U. S. digital PBX market is NEC, with its NEAX 2400 PBX. But while NEC has captured many contracts with per-line prices lower than many U. S. competitors, some NEC users have complained of the unavailability of data interfaces and features that have long been promised. It is noteworthy that, according to the Datapro user survey, NEAX 2400 users are overall less satisfied than are Centrex users. (The same is true, incidentally, for users with PBX equipment from GTE Communications Systems, Harris Digital Telephones, Oki Telecom, and United Technologies Lexar.)

The overall satisfaction rating in the Datapro User Survey of PBX/Centrex Systems (Table 2) is 3.09—based on weighted averages from 1.0 (poor) to 4.0 (excellent). The ratings are based on responses from 465 PBX users and 90 Centrex customers.

The complete report is available from Datapro Research Corp., 1805 Underwood Blvd., Delran, N. J. 08705.

Special report

Table 2: User ratings of PBX CENTREX systems

MANUFACTURER AND MODEL	NUMBER OF RESPONSES	WEIGHTED AVERAGES AND RESPONSE COUNTS OVERALL SATISFACTION					TRAINING PROGRAMS AND MATERIALS					SYSTEM AND ATTENDANT DOCUMENTATION					EASE OF OPERATION, ATTENDANT*				
		WA	E	G	F	P	WA	E	G	F	P	WA	E	G	F	P	WA	E	G	F	P
AT&T																					
DIMENSION 400	33	3.00	4	25	3	1	2.63	2	17	9	2	2.81	5	14	13	1	3.16	7	22	3	0
DIMENSION 600	5	2.97	0	4	0	0	2.63	2	2	1	0	2.70	0	3	2	0	3.13	2	3	0	0
DIMENSION 2000	83	3.00	18	52	9	1	3.20	16	26	32	7	2.60	13	47	14	7	3.40	26	46	6	4
SYSTEM 75	3	3.50	1	1	0	0	2.33	1	0	1	1	3.00	1	1	1	0	3.33	1	2	0	0
SYSTEM 85	10	3.44	4	5	0	0	2.90	3	4	2	1	3.00	2	6	2	0	3.20	5	3	1	1
PRELUDE	4	3.00	0	3	0	0	2.50	0	2	2	0	2.25	0	1	3	0	3.25	1	3	0	0
OTHER	1	2.00	0	0	1	0	0	0	0	0	0	1.00	0	0	0	1	2.00	0	0	1	0
SUBTOTAL	139	3.08	27	90	13	2	2.66	24	51	47	11	2.77	21	72	35	9	3.16	42	79	11	5
DIGITAL TRANSMISSION ALL MODELS	4	3.00	1	2	1	0	2.25	0	2	1	1	2.75	0	3	1	0	3.00	1	2	1	0
ERICSSON ALL MODELS	4	3.00	0	3	0	0	2.50	0	2	2	0	2.75	0	3	1	0	3.50	2	2	0	0
GTI COMMUNICATION SYSTEMS CORP. G1D SERIES	19	2.83	2	12	3	1	2.63	2	11	3	3	2.84	4	11	1	3	2.84	6	8	1	4
HARRIS DIGITAL TELEPHONES ALL MODELS	5	2.80	1	2	2	0	2.25	0	2	1	1	2.60	1	1	3	0	2.60	0	4	0	1
INTECOM IBX SERIES	7	3.17	1	5	0	0	3.00	1	4	1	0	2.67	0	4	2	0	3.33	2	4	0	0
IPC COMMUNICATIONS ALL MODELS	5	3.00	2	2	0	1	3.20	2	2	1	0	3.20	2	2	1	0	3.40	3	1	1	0
ITT TELECOM ALL MODELS	5	3.00	1	2	1	0	3.00	1	2	1	0	3.20	2	2	1	0	3.20	2	2	1	0
MITEL																					
SX 100/-200	11	3.64	7	4	0	0	3.40	6	3	0	1	3.18	6	1	4	0	3.90	9	1	0	0
SX 2000	4	3.75	3	1	0	0	3.50	2	2	0	0	3.50	2	2	0	0	3.75	3	1	0	0
SUBTOTAL	15	3.67	10	5	0	0	3.43	8	5	0	1	3.27	8	3	4	0	3.86	12	2	0	0
NEC TELEPHONES																					
NEAX 2400	11	2.91	3	5	2	1	2.36	2	1	7	1	2.64	2	3	6	0	3.18	4	6	1	0
OTHER MODELS	8	2.71	1	4	1	1	2.43	1	3	1	2	2.43	0	4	2	1	2.86	0	6	1	0
SUBTOTAL	19	2.83	4	9	3	2	2.39	3	4	8	3	2.56	2	7	8	1	3.11	4	12	2	0
NORTHERN TELECOM SL SERIES	107	3.23	33	63	9	0	2.81	21	42	30	5	2.96	28	50	26	3	3.40	53	43	7	2
OKI TELECOM ALL MODELS	4	2.00	0	1	2	1	1.67	0	1	0	2	1.75	0	1	1	2	2.26	0	2	1	1
ROLM CBX/CBX II	106	3.13	33	54	12	4	2.81	17	50	30	7	2.96	27	36	31	12	3.39	54	40	7	3
SIEMENS COMMUNICATION SYSTEMS ALL MODELS	7	3.00	2	4	0	1	3.00	1	3	1	0	3.14	1	6	0	0	3.43	3	4	0	0
SOLID STATE SYSTEMS ALL MODELS	3	3.33	1	2	0	0	3.00	0	2	0	0	3.50	1	1	0	0	3.50	1	1	0	0
UNITED TECHNOLOGIES LEXAR ALL MODELS	7	2.71	1	4	1	1	2.57	0	5	1	1	2.14	0	4	0	3	2.86	3	1	2	1
OTHER PBX VENDORS ALL MODELS	9	2.78	3	3	1	2	2.78	2	4	2	1	2.89	2	5	1	1	2.89	3	4	0	2
TOTAL PBX VENDORS ALL MODELS	465	3.09	122	264	48	16	2.73	82	193	130	36	2.82	99	213	116	34	3.25	191	212	33	21
CENTREX ALL TYPES	90	2.92	11	57	12	3	2.43	6	38	25	14	2.34	3	37	34	13	3.09	22	51	10	2

*RATED ONLY IF PROVIDED BY EQUIPMENT VENDOR.

WEIGHTED AVERAGE (WA) IS BASED ON ASSIGNING A WEIGHT OF 4 TO EACH USER RATING OF EXCELLENT (E), 3 TO GOOD (G), 2 TO FAIR (F), AND 1 TO POOR (P).

WEIGHTED AVERAGES AND RESPONSE COUNTS

MANUFACTURER AND MODEL	EASE OF OPERATION, STATION USERS					EASE OF INSTALLATION, CUTOVER					HARDWARE RELIABILITY					MAINTENANCE SERVICE*					TROUBLE-SHOOTING*				
	WA	E	G	F	P	WA	E	G	F	P	WA	E	G	F	P	WA	E	G	F	P	WA	E	G	F	P
AT&T																									
DIMENSION 400	2.94	3	25	3	2	2.81	7	15	7	3	3.16	14	17	1	1	3.31	12	14	5	1	3.07	9	15	7	1
DIMENSION 600	2.88	0	4	1	0	2.81	2	3	0	0	3.33	3	1	0	0	3.16	3	1	0	0	3.00	2	1	1	0
DIMENSION 2000	2.80	15	48	18	1	3.40	21	32	16	10	3.75	33	37	2	9	3.75	34	39	5	2	3.25	24	42	10	4
SYSTEM 75	3.33	2	0	1	0	3.67	2	1	0	0	3.50	1	1	0	0	4.00	2	0	0	0	3.50	1	1	0	0
SYSTEM 85	2.90	3	4	2	1	3.00	4	3	2	1	3.22	4	4	1	0	3.67	6	3	0	0	3.33	5	2	2	0
PRELUDE	3.25	1	3	0	0	3.00	0	3	0	0	3.00	0	4	0	0	3.50	2	2	0	0	3.25	1	3	0	0
OTHER	2.00	0	0	1	0	0	0	0	0	0	4.00	1	0	0	0	2.00	0	0	1	0	2.00	0	0	1	0
SUBTOTAL	2.93	24	84	26	4	2.87	36	57	25	14	3.23	56	64	4	10	3.32	59	59	11	3	3.08	42	64	21	5
DIGITAL TRANSMISSION ALL MODELS	3.00	1	2	1	0	3.00	0	4	0	0	2.75	1	1	2	0	3.00	1	2	1	0	2.75	1	2	0	1
ERICSSON ALL MODELS	3.00	0	4	0	0	3.00	0	4	0	0	2.67	0	2	1	0	3.00	0	2	0	0	3.00	0	2	0	0
GTI COMMUNICATION SYSTEMS G1D SERIES	2.53	2	8	7	2	2.58	4	7	4	4	2.78	5	7	3	3	2.75	3	9	1	3	2.72	2	11	3	2
HARRIS DIGITAL TELEPHONES ALL MODELS	2.40	0	3	1	1	2.40	1	1	2	1	2.20	0	2	2	1	2.75	0	3	1	0	2.67	0	2	1	0
INTECOM IBX SERIES	3.60	3	3	0	0	2.83	0	5	1	0	3.60	3	3	0	0	3.00	1	4	1	0	2.67	0	4	2	0
IPC COMMUNICATIONS ALL MODELS	3.80	4	1	0	0	3.20	3	1	0	1	3.60	3	2	0	0	3.40	3	1	1	0	3.00	2	1	2	0
ITT TELECOM ALL MODELS	3.00	2	1	2	0	3.20	2	2	1	0	3.25	1	3	0	0	2.67	1	1	0	1	2.67	1	1	0	1
MITEL																									
SX 100/-200	3.73	8	3	0	0	3.55	7	3	1	0	3.82	9	2	0	0	3.86	6	1	0	0	3.50	3	3	0	0
SX 2000	3.75	3	1	0	0	3.25	1	3	0	0	4.00	4	0	0	0	3.75	3	1	0	0	3.25	1	3	0	0
SUBTOTAL	3.73	11	4	0	0	3.47	8	6	1	0	3.87	13	2	0	0	3.82	9	2	0	0	3.40	4	6	0	0
NEC TELEPHONES																									
NEAX 2400	3.18	4	6	1	0	2.36	4	1	1	5	3.20	5	3	1	1	2.67	2	4	1	2	2.66	2	3	2	2
OTHER MODELS	3.14	2	4	1	0	2.57	0	4	3	0	3.29	3	3	1	0	2.29	2	2	1	0	2.20	0	2	2	1
SUBTOTAL	3.22	6	10	2	0	2.44	4	5	4	5	3.24	8	6	2	1	2.86	4	6	2	2	2.43	2	5	4	3
NORTHERN TELECOM SL SERIES	3.34	46	53	6	2	3.06	33	55	12	7	3.42	60	32	13	1	3.08	27	34	15	3	2.98	23	36	19	3
OKI TELECOM ALL MODELS	2.50	0	3	0	1	2.50	1	1	1	1	2.50	0	3	0	1	1.00	0	0	0	2	1.50	0	0	1	1
ROLM CBX/CBX II	3.05	29	59	12	6	2.85	23	54	19	10	3.21	44	43	10	6	2.92	36	26	26	9	2.85	29	35	28	8
SIEMENS COMMUNICATION SYSTEMS ALL MODELS	3.29	3	3	1	0	3.33	3	2	1	0	3.14	3	3	0	1	4.00	2	0	0	0	3.50	1	1	0	0
SOLID STATE SYSTEMS ALL MODELS	3.50	1	1	0	0	2.50	0	1	1	0	3.67	2	1	0	0	3.67	2	1	0	0	3.33	1	2	0	0
UNITED TECHNOLOGIES LEXAR ALL MODELS	2.57	1	3	2	1	2.57	2	2	1	2	2.86	4	0	1	2	3.00	0	1	0	0	3.00	0	1	0	0
OTHER PBX VENDORS ALL MODELS	2.78	3	3	1	2	2.63	2	3	1	2	2.89	3	3	2	1	2.83	1	4	2	1	3.00	1	3	2	1
TOTAL PBX VENDORS ALL MODELS	3.07	136	246	60	21	2.89	122	210	74	48	3.24	206	177	40	29	3.10	150	155	61	25	2.94	109	178	83	25
CENTREX ALL TYPES	3.01	21	49	14	3	3.01	25	40	16	4	3.64	55	23	3	0	3.18	30	35	11	3	2.85	22	33	20	3

Special report

Summary of data options—AT&T

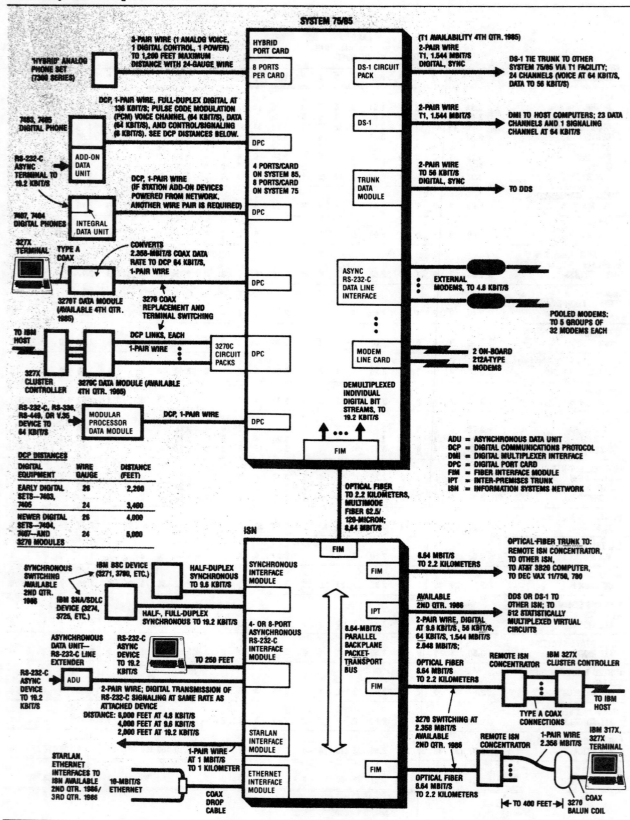

SYSTEM 75/85

3-PAIR WIRE (1 ANALOG VOICE, 1 DIGITAL CONTROL, 1 POWER) TO 1,200 FEET MAXIMUM DISTANCE WITH 24-GAUGE WIRE

'HYBRID' ANALOG PHONE SET (7300 SERIES)

HYBRID PORT CARD — 8 PORTS PER CARD

(T1 AVAILABILITY 4TH QTR. 1985)
2-PAIR WIRE T1, 1.544 MBIT/S DIGITAL, SYNC

DS-1 CIRCUIT PACK

DS-1 TIE TRUNK TO OTHER SYSTEM 75/85 VIA T1 FACILITY; 24 CHANNELS (VOICE AT 64 KBIT/S, DATA TO 56 KBIT/S)

DCP, 1-PAIR WIRE, FULL-DUPLEX DIGITAL AT 136 KBIT/S; PULSE CODE MODULATION (PCM) VOICE CHANNEL (64 KBIT/S), DATA (64 KBIT/S), AND CONTROL/SIGNALING (8 KBIT/S). SEE DCP DISTANCES BELOW.

7403, 7405 DIGITAL PHONE

ADD-ON DATA UNIT

DPC

2-PAIR WIRE T1, 1.544 MBIT/S

DS-1

DMI TO HOST COMPUTERS; 23 DATA CHANNELS AND 1 SIGNALING CHANNEL AT 64 KBIT/S

RS-232-C ASYNC TERMINAL TO 19.2 KBIT/S

4 PORTS/CARD ON SYSTEM 85, 8 PORTS/CARD ON SYSTEM 75

DCP, 1-PAIR WIRE (IF STATION ADD-ON DEVICES POWERED FROM NETWORK, ANOTHER WIRE PAIR IS REQUIRED)

7407, 7404 DIGITAL PHONES

INTEGRAL DATA UNIT

DPC

2-PAIR WIRE TO 56 KBIT/S DIGITAL, SYNC

TRUNK DATA MODULE

TO DDS

327X TERMINAL

TYPE A COAX

CONVERTS 2.358-MBIT/S COAX DATA RATE TO DCP 64 KBIT/S, 1-PAIR WIRE

DPC

ASYNC RS-232-C DATA LINE INTERFACE

EXTERNAL MODEMS, TO 4.8 KBIT/S

3270T DATA MODULE (AVAILABLE 4TH QTR. 1985)

3270 COAX REPLACEMENT AND TERMINAL SWITCHING

POOLED MODEMS: TO 5 GROUPS OF 32 MODEMS EACH

TO IBM HOST

DCP LINKS, EACH 1-PAIR WIRE

3270C CIRCUIT PACKS

DPC

MODEM LINE CARD

2 ON-BOARD 212A-TYPE MODEMS

327X CLUSTER CONTROLLER

3270C DATA MODULE (AVAILABLE 4TH QTR. 1985)

DEMULTIPLEXED INDIVIDUAL DIGITAL BIT STREAMS, TO 19.2 KBIT/S

RS-232-C, RS-336, RS-449, OR V.35 DEVICE TO 64 KBIT/S

MODULAR PROCESSOR DATA MODULE

DCP, 1-PAIR WIRE

DPC

FIM

ADU = ASYNCHRONOUS DATA UNIT
DCP = DIGITAL COMMUNICATIONS PROTOCOL
DMI = DIGITAL MULTIPLEXER INTERFACE
DPC = DIGITAL PORT CARD
FIM = FIBER INTERFACE MODULE
IPT = INTER-PREMISES TRUNK
ISN = INFORMATION SYSTEMS NETWORK

DCP DISTANCES

DIGITAL EQUIPMENT	WIRE GAUGE	DISTANCE (FEET)
EARLY DIGITAL SETS—7403, 7405	26	2,200
	24	3,400
NEWER DIGITAL SETS—7404, 7407—AND 3270 MODULES	26	4,000
	24	5,000

OPTICAL FIBER TO 2.2 KILOMETERS, MULTIMODE FIBER 62.5/120-MICRON; 8.64 MBIT/S

ISN

FIM

SYNCHRONOUS SWITCHING AVAILABLE 2ND QTR. 1986

IBM BSC DEVICE (3271, 3780, ETC.)

HALF-DUPLEX SYNCHRONOUS TO 9.6 KBIT/S

SYNCHRONOUS INTERFACE MODULE

8.64 MBIT/S TO 2.2 KILOMETERS

FIM

OPTICAL-FIBER TRUNK TO: REMOTE ISN CONCENTRATOR, TO OTHER ISN, TO AT&T 3B20 COMPUTER, TO DEC VAX 11/750, 780

IBM SNA/SDLC DEVICE (3274, 3725, ETC.)

HALF-, FULL-DUPLEX SYNCHRONOUS TO 19.2 KBIT/S

IPT

AVAILABLE 2ND QTR. 1986

DDS OR DS-1 TO OTHER ISN; TO 512 STATISTICALLY MULTIPLEXED VIRTUAL CIRCUITS

ASYNCHRONOUS DATA UNIT— RS-232-C LINE EXTENDER

RS-232-C ASYNC DEVICE TO 19.2 KBIT/S

TO 250 FEET

4- OR 8-PORT ASYNCHRONOUS RS-232-C INTERFACE MODULE

8.64-MBIT/S PARALLEL BACKPLANE PACKET-TRANSPORT BUS

2-PAIR WIRE, DIGITAL AT 9.6 KBIT/S, 56 KBIT/S, 64 KBIT/S, 1.544 MBIT/S, 2.048 MBIT/S;

RS-232-C ASYNC DEVICE TO 19.2 KBIT/S

ADU

2-PAIR WIRE; DIGITAL TRANSMISSION OF RS-232-C SIGNALING AT SAME RATE AS ATTACHED DEVICE
DISTANCE: 6,000 FEET AT 4.8 KBIT/S, 4,000 FEET AT 9.6 KBIT/S, 2,000 FEET AT 19.2 KBIT/S

OPTICAL FIBER 8.64 MBIT/S TO 2.2 KILOMETERS

FIM

REMOTE ISN CONCENTRATOR

IBM 327X CLUSTER CONTROLLER

TYPE A COAX CONNECTIONS

TO IBM HOST

3270 SWITCHING AT 2.358 MBIT/S AVAILABLE 2ND QTR. 1986

REMOTE ISN CONCENTRATOR

1-PAIR WIRE 2.358 MBIT/S

IBM 317X, 327X TERMINAL

STARLAN, ETHERNET INTERFACES TO ISN AVAILABLE 2ND QTR. 1986/ 3RD QTR. 1986

1-PAIR WIRE AT 1 MBIT/S TO 1 KILOMETER

STARLAN INTERFACE MODULE

10-MBIT/S ETHERNET

COAX DROP CABLE

ETHERNET INTERFACE MODULE

OPTICAL FIBER 8.64 MBIT/S TO 2.2 KILOMETERS

FIM

TO 400 FEET

3270 BALUN COIL

COAX

Special report

Summary of data options—Northern Telecom

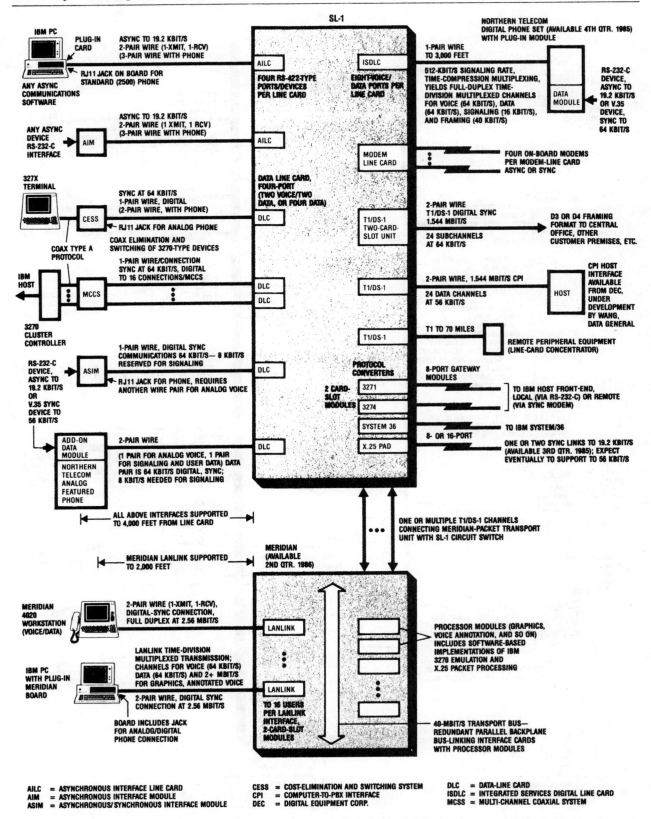

AILC	= ASYNCHRONOUS INTERFACE LINE CARD	CESS	= COST-ELIMINATION AND SWITCHING SYSTEM
AIM	= ASYNCHRONOUS INTERFACE MODULE	CPI	= COMPUTER-TO-PBX INTERFACE
ASIM	= ASYNCHRONOUS/SYNCHRONOUS INTERFACE MODULE	DEC	= DIGITAL EQUIPMENT CORP.

DLC	= DATA-LINE CARD
ISDLC	= INTEGRATED SERVICES DIGITAL LINE CARD
MCSS	= MULTI-CHANNEL COAXIAL SYSTEM

Summary of data options—Rolm

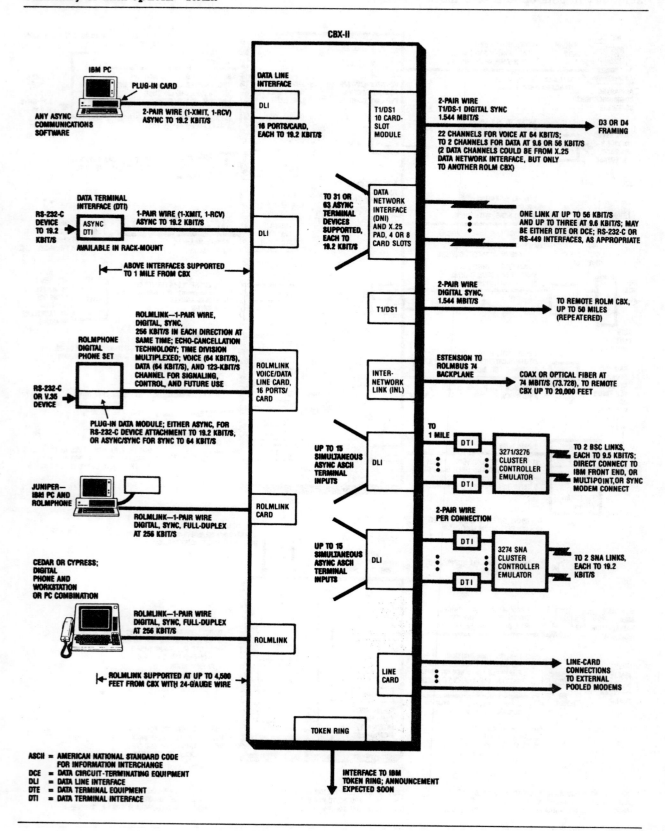

CBX-II

IBM PC
PLUG-IN CARD
ANY ASYNC COMMUNICATIONS SOFTWARE

DATA LINE INTERFACE
DLI
16 PORTS/CARD, EACH TO 19.2 KBIT/S

2-PAIR WIRE (1-XMIT, 1-RCV) ASYNC TO 19.2 KBIT/S

T1/DS1 10 CARD-SLOT MODULE

2-PAIR WIRE T1/DS-1 DIGITAL SYNC 1.544 MBIT/S
D3 OR D4 FRAMING

22 CHANNELS FOR VOICE AT 64 KBIT/S; TO 2 CHANNELS FOR DATA AT 9.6 OR 56 KBIT/S (2 DATA CHANNELS COULD BE FROM X.25 DATA NETWORK INTERFACE, BUT ONLY TO ANOTHER ROLM CBX)

DATA TERMINAL INTERFACE (DTI)
RS-232-C DEVICE TO 19.2 KBIT/S
ASYNC DTI
AVAILABLE IN RACK-MOUNT
1-PAIR WIRE (1-XMIT, 1-RCV) ASYNC TO 19.2 KBIT/S
DLI

TO 31 OR 63 ASYNC TERMINAL DEVICES SUPPORTED, EACH TO 19.2 KBIT/S

DATA NETWORK INTERFACE (DNI) AND X.25 PAD, 4 OR 8 CARD SLOTS

ONE LINK AT UP TO 56 KBIT/S AND UP TO THREE AT 9.6 KBIT/S; MAY BE EITHER DTE OR DCE; RS-232-C OR RS-449 INTERFACES, AS APPROPRIATE

ABOVE INTERFACES SUPPORTED TO 1 MILE FROM CBX

ROLMLINK—1-PAIR WIRE, DIGITAL, SYNC, 256 KBIT/S IN EACH DIRECTION AT SAME TIME; ECHO-CANCELLATION TECHNOLOGY; TIME DIVISION MULTIPLEXED; VOICE (64 KBIT/S), DATA (64 KBIT/S), AND 123-KBIT/S CHANNEL FOR SIGNALING, CONTROL, AND FUTURE USE

T1/DS1
2-PAIR WIRE DIGITAL SYNC, 1.544 MBIT/S
TO REMOTE ROLM CBX, UP TO 50 MILES (REPEATERED)

ROLMPHONE DIGITAL PHONE SET
RS-232-C OR V.35 DEVICE

ROLMLINK VOICE/DATA LINE CARD, 16 PORTS/CARD

INTER-NETWORK LINK (INL)
ESTENSION TO ROLMBUS 74 BACKPLANE
COAX OR OPTICAL FIBER AT 74 MBIT/S (73.728), TO REMOTE CBX UP TO 20,000 FEET

PLUG-IN DATA MODULE; EITHER ASYNC, FOR RS-232-C DEVICE ATTACHMENT TO 19.2 KBIT/S, OR ASYNC/SYNC FOR SYNC TO 64 KBIT/S

JUNIPER—IBM PC AND ROLMPHONE
ROLMLINK—1-PAIR WIRE DIGITAL, SYNC, FULL-DUPLEX AT 256 KBIT/S
ROLMLINK CARD

DLI
UP TO 15 SIMULTANEOUS ASYNC ASCH TERMINAL INPUTS
TO 1 MILE
DTI
DTI
3271/3276 CLUSTER CONTROLLER EMULATOR
TO 2 BSC LINKS, EACH TO 9.5 KBIT/S; DIRECT CONNECT TO IBM FRONT END, OR MULTIPOINT, OR SYNC MODEM CONNECT

CEDAR OR CYPRESS; DIGITAL PHONE AND WORKSTATION OR PC COMBINATION
DLI
UP TO 15 SIMULTANEOUS ASYNC ASCH TERMINAL INPUTS
2-PAIR WIRE PER CONNECTION
DTI
DTI
3274 SNA CLUSTER CONTROLLER EMULATOR
TO 2 SNA LINKS, EACH TO 19.2 KBIT/S

ROLMLINK—1-PAIR WIRE DIGITAL, SYNC, FULL-DUPLEX AT 256 KBIT/S
ROLMLINK

ROLMLINK SUPPORTED AT UP TO 4,500 FEET FROM CBX WITH 24-GAUGE WIRE

LINE CARD
LINE-CARD CONNECTIONS TO EXTERNAL POOLED MODEMS

TOKEN RING
INTERFACE TO IBM TOKEN RING; ANNOUNCEMENT EXPECTED SOON

ASCII = AMERICAN NATIONAL STANDARD CODE FOR INFORMATION INTERCHANGE
DCE = DATA CIRCUIT-TERMINATING EQUIPMENT
DLI = DATA LINE INTERFACE
DTE = DATA TERMINAL EQUIPMENT
DTI = DATA TERMINAL INTERFACE

Meridian DV-1: a Fast, Functional and Flexible Data Voice System

Cecil Murray
William Carr

Northern Telecom's Meridian was developed by Bell Northern Research and provides an integrated voice and data system that offers connectivity and concurrency

O ffice environments today are often complicated networks of standalone word processors, personal computers, telephones, and electronic and paper filing systems. Sitting in the midst of this chaos are knowledge workers—individuals who must make decisions based on the analysis of data and information they obtain from a variety of people and machines, and who must cope with frequent interruptions (Fig. 1).

The Northern Telecom Meridian DV-1 was developed by Bell Northern Research to satisfy the complex communications and computing needs of these professional multi-function workers [1,2]. A modularized system, the DV-1 streamlines the operation of the modern office by integrating voice and data functions into one system, thus serving as an efficient departmental integrated local area network. As such, it handles, in addition to telephony, information storage and retrieval, personal computing, word processing, high resolution graphics, and electronic mail. Because such a system did not already exist in the marketplace, the research team was treading new ground.

For this reason, the system's designers gave themselves two basic principles in developing a system that would support the needs of multi-function workers in their complex and interrupt-intensive environments: namely, connectivity and concurrency.

Connectivity

The DV-1 system:

1) Allows the user to access multiple applications executing in multiple system environment from a single multi-media workstation.
2) Provides this connectivity by being integrated into the corporate voice and data networks.

The DV-1 is positioned as a work group system, designed to communicate up and down the corporate computing hierarchy and interconnect with standard communication facilities. It can communicate with other DV-1 systems, PBX's and public voice or data networks, as well as with corporate host computers, such as IBM mainframes, and with desktop workstations, such as the Meridian M4020 IVD workstation, IBM and IBM-compatible personal computers, and ASCII terminals.

Concurrency

In addition to providing enhanced connectivity, this system was designed to provide concurrent access to multiple applications in several execution environments.

Unfortunately, most work group minicomputers and personal computers introduce significant delays when users change from one task to another—or, if these systems do support task switching, they do not continue execution of the background task. The DV-1, however, allows knowledge workers to move quickly from activity to activity without incurring delays and without forcing them to cope with operating system command sequences.

For example, workers can undertake a database search under Unix, a popular operating system, while

Reprinted from IEEE Communications Magazine, December 1986, pages 36-42. Copyright © 1986 pages by The Institute of Electrical and Electronics Engineers, Inc. All rights reserved.

243

EHO282-4/89/0000/0243$01.00 © 1986 IEEE

directing their attention to an on-line IBM application through a DV-1 based 3274 emulation. Simultaneously, they can turn their attention to a telephone call and access their personal calendar without disturbing other applications. Finally, they are able to return to the results of the database search.

Expressed in more general terms, concurrency allows knowledge workers to juggle several on-going applications without start, stop, and restart delays and allows them to continue a time-consuming task while their attention is directed simultaneously to other tasks.

Because of its connectivity and concurrency capabilities, the system has the presentation power and bandwidth to provide a responsive, multimedia, user-interactive system. One example is the basic "Share" function: "Share" allows for *ad hoc* conferencing of both voice and data. For example, two workers may be in a phone conversation on the same or different DV-1 systems when they decide to jointly update a spreadsheet. Furthermore, since "Share" is a system function, it applies to all DV-1 applications such as SNA 3274/3778 emulation.

The "Share" feature also illustrates a key value of an integrated voice and data system. Since the system maintains a single directory service for both the voice and data connections, the shared data connection is established based on the information known about the existing voice connection.

To meet these two criteria—connectivity and concurrency, the Meridian DV-1 was designed as a central core of computing and communications equipment connected to terminals and telephone sets using a twisted-pair wiring scheme. Northern Telecom's technology allows the twisted pair to carry either ordinary

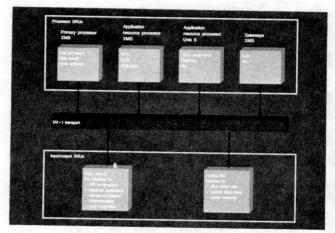

Fig. 2.   DV-1 System Block Diagram.

analog telephone traffic, or digital voice and data traffic at rates of up to 2.56 Mb/s.

The central core has been designed in a modular fashion, so that the entire system can be configured in various sizes as well as tailored to meet particular customer needs.

Overview of System Hardware Architecture

The central core of the DV-1 consists of a set of cooperating, yet independent modules called shared resource units (SRU's). These modules are housed in an extendable group of low cabinets. Each cabinet can house up to 8 SRU's, in two rows of four.

The SRU's are the basic building blocks of the system. They slide into the cabinets and connect to a common bus. This bus provides separate interfaces that allow voice and data signals to be exchanged between the SRU's. It also distributes a bulk 32 V power supply (Fig. 2).

Bus for Voice and Data

SRU's are connected by the Meridian DV-1 bus that runs along the cabinet backplanes. The bus provides transport for both voice and data, though separate techniques are used for each (Fig. 3).

The voice bus is a time-division multiplexing bus, eight bits wide, which is standard for digitized voice. A one-way voice connection requires one sample to be carried every 125 microseconds (μs), again a standard value. The bus can carry up to 320 such samples in a 125 μs frame. This sequence is repeated for the next frame for a total of 2,560,000 samples per second. Any continuous voice connection occupies the same relative position—or timeslot—in each frame, so there are 320 timeslots available. This scheme is efficient for voice, because control intervention is necessary only when setting up or taking down a connection. Simple low level hardware can continue to use the allocated timeslot as long as the connection is held.

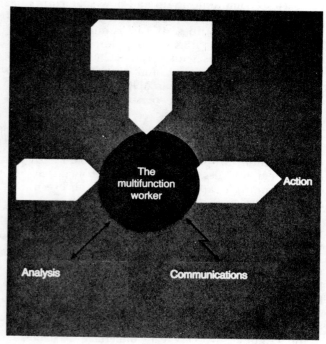

Fig. 1.

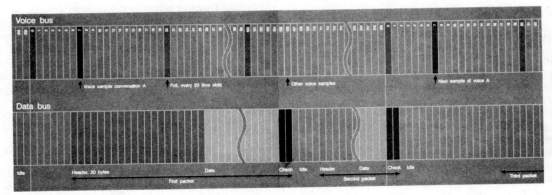

Fig. 3. The DV-1 Bus.

Every voice SRU is designed so that it can use any timeslot on command. Thus, a voice connection can go directly from a source SRU to a destination SRU, simply by commanding them both to use the same timeslot. This capability was a fundamental design decision for this system. If timeslot allocations had been fixed in advance—for example, if a particular telephone set always connected to timeslot 42—then the voice SRU need not have been designed to vary the timeslots. However, any speech connection made from this set would involve copying voice samples from timeslot 42 to the timeslot used by the telephone set at the far end. This being the case, additional hardware, called a timeswitch, would be needed to handle it. Furthermore, such a design would use two timeslots instead of one for each one-way voice connection, thus reducing the maximum number of conversations that could be carried on the bus.

Data connections are not like voice connections. Instead of a steady flow of samples at regular 125 μs intervals, data flow between SRUs usually occurs in short bursts. Each burst is sent at a high rate, but there are long idle periods between bursts. To accommodate this situation, the data part of the bus is arranged as a packet bus. Physically, this part of the bus is very similar to the voice bus; specifically, it is eight bits wide, and it transports up to 2,560,000 eight-bit samples per second.

When an SRU wants to transmit, it waits for an opportunity to use the bus. When permission is granted, the SRU sends a packet of data at the full bus rate of 2.56 Mb/s. The first few bytes of such a packet specify the address of the destination SRU. In fact, the packet is available at all the SRU positions connected to the bus, but only the destination SRU recognizes it and collects it.

However, some fair way of allocating bus usage must be found, to ensure that only one SRU at a time will try to transmit. Therefore, a bus controller has been incorporated into one of the SRUs. The bus controller asks each SRU in turn whether it wishes to transmit a packet and, if so, allocates the bus as required. This polling process is designed to use some of the timeslots on the voice bus, so that while one data packet is being transmitted, the polling continues over the voice bus to preselect the next SRU to transmit a packet. Polling takes place at regular intervals, using only a small number of the 320 timeslots on the voice bus. Only a few timeslots are needed for this purpose because, for example, if 20 timeslots are used, it is possible to poll 160,000 times per second, well in excess of the rate needed to allocate the bus efficiently.

Shared Resource Units (SRU's)

Consistent with the objective of module independence, each SRU is designed as a closed box, communicating only through the defined interface. Each box has a complete metal skin, shielding it from radiated electrical interference and, at the same time, helping to ensure that the unit meets electromagnetic interference regulations. Power converters within the SRU generate any voltages needed inside, using a common bulk power supply that is carried on the bus. In most cases, an SRU can be plugged into the cabinet, or unplugged, without requiring a power shutdown and without disturbing the operation of other SRUs.

There are four types of SRU: power, processing, storage, and communication.

Power SRU's

The power supply SRU provides bulk DC power for the system. It accepts 115 VAC and distributes unregulated 32 VDC power over the bus. System current is provided by a number of SRUs that are connected in parallel. Alternative designs of this power supply module can be used where the incoming supply is at a different voltage or frequency; for example, in Europe or in telephone central office, a 48 VDC supply is standard.

An internal bulk power supply, is used in the SRUs as well. There are two good reasons for doing so. First, with minimal redesign, DV-1 systems can run on various primary power supplies. Second, and even more important, it is unsafe to have high voltages on the cabinet backplanes, and 32 volts is accepted internationally as a safe voltage.

Processing SRU's: The Primary Processor

The primary processor SRU is the heart of the system. It contains a Motorola 68010 processor and up to four megabytes of memory. These execute the base system software described in the next section of this article. The SRU also contains a disk controller that interfaces to a 5 1/4-inch Winchester hard disk mounted in a separate SRU. In addition, the primary processor SRU contains circuit for controlling the system bus.

A system at the low end of the size range can be configured without other processing SRUs, but most configurations will have one or more application processors described below.

Processing SRU's: Application Resource Processor

The application processor SRU, as the name suggests, hosts various applications. Most DV-1 installations have one or more of these. The primary processor and some of the 68010 processors support the XMS operating system [3]. XMS (extended multiprocessor system) is a custom-built, state-of-the-art computing technology providing a common base for Northern Telecom's telecommunications and office systems.

The DV-1 application processors also run real-time system software for such activities as call processing, which manages telephone sets, drives terminals and trunks, and provides screens on Northern Telecom's Meridian M4000 integrated terminals. Other 68010 processors run the popular Unix system, which supports third-party software, such as word processing and spreadsheet packages.

The architecture allows other SRUs to use different hardware environments. For example, BNR has developed laboratory models of an Intel applications processor SRU, which is built around the Intel multibus card set and consists of an Intel 80286 processor with one or two Mbytes of random access memory. It accepts standard Intel boards for interfacing with existing peripheral devices and networks. The Intel processor could host such operating systems as Concurrent DOS by Digital Research Inc.

Storage SRU's

There are several types of storage SRUs. All provide storage on non-removable magnetic disks, for personal and system data or programs. The disks are controlled by file server software.

In small DV-1 systems, the file server runs on the primary processor. This processor has a direct connection to a 5 1/4-inch Winchester disk storage SRU. The initial release contains 80 mbytes of disk storage. In larger system, the file server will run on a dedicated resource processor, with its own dedicated connection to DV-1 cabinets containing larger 8 in. Winchester disks providing several 100 Mbytes of storage.

Some form of removable magnetic medium is needed so that data can be loaded to the system disks, or copies for archival purposes. This need is satisfied by quarter-inch cartridge tapes. One of these cartridges holds up to 40 Mbytes of data. For small systems, a cartridge tape drive is incorporated into a 5 1/4-inch disk storage SRU. For larger systems, a cartridge tape drive will come as a separate module.

External Communications—LANLink SRU's

Connections between Meridian DV-1 and the outside world can be made in various ways. All have a common plan—an SRU plugs into the DV-1 bus, and connects also to some external wiring or fiber optic cable. Relatively few types of external connections are needed at present, though new types will be developed as market needs arise, or as new technology permits.

Perhaps the LANLink is the most important SRU that connects to the external world. The LANLink SRU logically extends the DV-1 bus over 12 high-speed lines to remote units, enabling the units to communicate over a network similar to a local area network (LAN). Each line has a total bandwidth of 2.56 Mb/s—enough for two voice channels and more than 2 Mbit/s of packetized data. This SRU multiplexes and buffers the voice and data connections. In addition, it provides high-speed connections to desktop devices, such as Northern Telecom's Meridian M4020 integrated terminal and IBM's personal computer (PC) at a distance of up to 600 m (approximately 2000 ft.) from the DV-1.

To achieve this 2.56 Mb/s capability, novel circuit techniques were needed in the areas of clock recovery, signal detection, and line equalization. A new chip, the X31, was designed [4]. The X31 contains two main paths—transmit and receive. The transmit path combines clock, data, and frame information into a single signal. The receive path extracts the clock, data, and frame information from a line signal (Fig. 4). By connecting appropriate logic signals to two external pins, the X31 can be configured to operate in system or remote mode, as required. In system mode, the chip operates at the central switch location and uses the system clock, data, and frame sources to encode the signal for transmission onto the line. In remote mode, it operates at a distant location; recovers data, frame, and a local clock from the line signal, and uses the recovered clock and frame to transmit remote data back to the central switch.

The 2.56 Mb/s yields a raw rate of 320 bits in every 125 μs. Thirty two of these bits are reserved to provide overall control and two eight-bit voice channels. The remaining 288 bits provide 32 bytes of packet data, each with an additional one-bit flag for indicating whether the data byte is valid. The maximum data rate is thus 2.048 kilobits per second or 256 kilobytes per second. However, like most packet data stream, the link is idle most of the time.

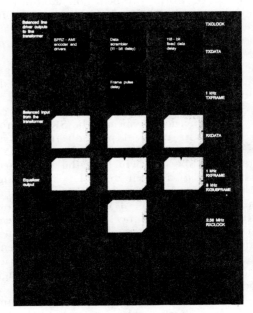

Fig. 4. The X31 Chip.

LAN Interface Units (LIU's)

Devices that operate at lower speeds, such as printers and tape drives, can interface via a LAN interface unit (LIU). The LIU is a buffered, programmable adapter that has been designed to operate with the LANLink over twisted pairs. It can be loaded with various configurations of software and, in some cases, can be fitted with optional hardware features to serve a variety of purposes. The basic LIU supports one parallel printer and two asynchronous SDLC or bisynchronous data communications connections. In many systems, only one LIU is needed to connect to the departmental printer and to interface to a remote IBM host via SNA. The LIU is also designed to accommodate other types of peripherals. Two new types of LIUs will be introduced in 1986, the magnetic tape LIU and the asynchronous multiplexer LIU.

The magnetic tape LIU connects the DV-1 to a nine-track, half-inch magnetic tape unit by means of an industry standard interface. Rather than connect a normally bulky peripheral directly to the bus, a high-speed twisted-pair wire distribution proves a convenient method for making the connection. This method allows the tape unit to be located near the system administrator, who may or may not be close to the DV-1 system.

The asynchronous multiplexer LIU allows a cluster of eight ASCII terminals to be connected remotely to the DV-1 system. Each terminal appears to be operating at 19 kb/s although a single 2.56 Mb/s line is used to connect back to the DV-1 system. The same asynchronous LIU can also connect with up to 10 asynchronous modems to provide a modem pool for other remote computers or dial-in terminals.

Telephone Access SRU's

Another important type of SRU provides access to the world of telephon both in its analog form and in forms allowing digital access. The analog line module (ALM) connects to conventional 500 (dial) or 2500 (pushbutton) telephone sets. The ALM also connects to trunk lines, to a private branch exchange, or to the public network. A companion module—the network services module (NSM)—provides tones, such as dial tone and digit tones. Call processing software, running on one of the processors in the DV-1, drives these telephony modules by sending commands to them over the data part of the DV-1 bus. The T1 module connects to digital T1 lines, which can carry up to 24 two-way voice connections in digital form.

Other SRU's

There are still more SRU's to provide links between DV-1 systems, or between DV-1 and other Northern Telecom systems. A simple Campus Net will use the 2.56 Mb/s twisted-pair technology to connect data between DV-1 systems. A forthcoming fiber optics interface, FXNET, will provide both voice and data interconnection at rates up to 80 Mb/s.

The DV-1 hardware elements thus allow the system builder considerable flexibility. Preserving this flexibility while creating a convenient user interface presented a major challenge for the system software designers.

System Software

The typical DV-1 system user rarely needs to be aware of the system software. This software operates in the primary processor, as needed, to manage the system on the user's behalf. The major software components are summarized in the following paragraphs.

System initialization is a start-up procedure orchestrated by the primary processor SRU. When power is first applied to the system, the primary processor begins to execute a program stored in permanent memory (ROM). This program loads from disk a bootstrap program that initializes the processor and the bus control circuits. The bootstrap next reads in a configuration table from disk, then proceeds to read programs and download software for the other SRUs, attached terminals, and other equipment.

The screen activity manager (SAM) controls the presentation on each user's terminal. It presents the system application menus, introduces the terminal to the applications selected, and manages the context changes from one concurrent application to another. SAM can be viewed as the system receptionist—always available to help users find the applications they require. When a user at a terminal selects a Unix application from the main menu, SAM gets the appropriate network addresses from the name server (see below) and initiates the connection dialog between the Unix application processor and the requesting terminal. Once the address is known, the terminal passes data packets directly to a user port on the Unix system.

Each terminal can "know" several addresses and can, thereby, conduct several conversations concurrently.

The DV-1 system accommodates system file servers—initially as part of the first processor SRU and later as separate SRU's. The file server controls the use of the disk system. It allows application programs to regard the disk as a collection of files that may be written or read, that may be extended or modified, without the application needing to be concerned about the allocation of disk space or its recovery and reuse when files are removed. The file server is a key element of the system's software; it allows files written by one application to be made available to another application. For example, Unix files can be accessed and transferred by XMS-based applications, such as the 2780 RJE emulator. Even more importantly, the file server ensures that one application, or one user, cannot damage—or even access—files belonging to another user. It can support one or more disks and, on smaller systems, can be run on the primary processor.

To maintain actual system addresses for the various logical units within the system a name server is used. For example, when the screen activity manager needs the address of the Unix SRU, it obtains the address from the name server. (For more details, see "Modularity" below.)

The printer access manager (PAM) provides printer services for the application processors and associated terminals. It also provides multiple printer queues for multiple printers and printer types. Each printer, whether it is located on an LIU or connected to a Meridian M4020 terminal, may be configured either as a system or dedicated resource and may be supported by one or several print queues.

The system administration services (SAS) are a collection of utilities necessary for the installation, maintenance, and day-to-day support of the DV-1 Data Voice System. They can be run from any terminal as long as the user logs on as the system administrator—though most users will never need to know anything about these utilities.

The preceding paragraphs highlight only some of the more significant system software modules; the full list is too long to cover in an article of this nature. The intent here is simply to give a feeling for the extent and function of the DV-1 system software—allowing access to multiple services in various execution environments, while at the same time shielding the user from the complexities of the access process.

Application Software: Making the System Work

The DV-1 system supports a number of diverse application areas, including call processing, data communications, and third-party software. The software structure is correspondingly diverse, and, in many ways, it matches the hardware concepts. A common base acts as a framework for modules and module communication. Various types of modules can be built on it, as long as their interfaces are consistent.

The framework is formed by the link level software, which controls operation of the data bus. This software must use a uniform protocol, or modules cannot communicate with one another. The protocol for the link level software is known as the virtual transport protocol (VTP). A VTP message corresponds closely to the physical packets sent on the bus.

The message has three parts: header, message content, and a check code at the end. The header defines the message source and destination. Four bytes specify a hardware source and destination. In each case, a single byte is needed to specify the SRU number or, more exactly, its physical position in the system's cabinets. For remote units connected by the LANLink, a second byte specifies which of the 12 twisted-pair lines connects to a given unit. The remaining two bytes are not used within the DV-1 system, but are available when multiple DV-1s or other Northern Telecom products are connected by a LAN. The use of a four-byte address allows units on many systems to have universal addresses, in the same way that ten-digit telephone numbers are universal in North America; for instance, your telephone number does not change as a function of the calling location. Within a single system, only two bytes are needed. To continue the analogy, these shorter addresses correspond to seven-digit, local telephone numbers.

Most system processors support multiple software tasks, running independently. Thus, both the source and destination SRU must be specified in the task. In each case, four bytes are used, and this has proved to be ample address space.

Because the protocol differs for the various application environments which can be run on DV-1, the message header requires another four bytes which contain a code that specifies which protocol is being used. In the normal VTP protocol, they also contain a message serial number and a message length.

Following the header is the data part of the message, which can be up to 1024 bytes long. Next comes a two-byte check code. Normally, each message is checked and acknowledged, so that if some electrical interference or some other fault corrupts a message, the check will fail, and the message will be retransmitted. The serial code ensures that such retransmission cannot result in the same message being acted upon twice.

Software Environments

The software environment for an application is built directly on top of the link level software. DV-1 supports several types of these environments, both for proprietary operating systems and for commercially available systems such as Unix. The main proprietary system is the XMS operating system. Programmed in BNR Pascal—an extension of standard Pascal—it offers support for real-time multitasking. The software environment used in BNR is called the standard kernel. An abbreviated version—known as the micro kernel—is used in some hardware-intensive applications, where the full kernel capability is not needed, making it possible to program the Meridian M4020 terminal, and many of the peripheral controllers, in BNR Pascal. Traditional assembler code is used only in dedicated, time-sensitive cases, such as the low level scanners and digit collectors of analog telephony units.

Modularity

Critical to the success of any system is its ability to be configured with varying collections of software to meet the individual needs of different organizations. Equally important is the requirement that the organization itself be able to add extra units without expert assistance, and without having to recompile the complete system. The concept of modularity, therefore, is central to the design of this system.

By allowing diverse software modules to communicate via messages, the link level helps the modularity concept considerably. Software can be written to send messages to defined interfaces for other software—so long as the universal address of the destination is available.

Naturally, it is not possible to code the universal address of one software module into another, independently designed one. To do this, every module's whereabouts in the system would have to be known in advance, thereby defeating any notion of easy system configuration. Instead, the DV-1 software uses a scheme that needs advance knowledge of only a few fixed addresses—for some initialization and maintenance routines, and a name server or resource manager.

When the power to a hardware module is first turned on, code in the module sends a message to the system initialization software, asking to be loaded with software. This message is delivered to a fixed address, already built into the ROM. Once the operational software is loaded into the module, it registers itself by sending messages to a few more fixed addresses. Some of these messages are concerned with system maintenance, thus ensuring that all units are periodically checked, and that they report good health. After the operational software has been loaded, all subsequent addresses are obtained from the name server or resource manager.

The name server plays an important part in this process. Each operational software unit can send messages to the name server. Each message contains a name, in the form of a string of characters, and a universal address. The name is the term agreed upon for the software module, or for some service it provides. The address is the destination to which messages must be sent in order to obtain the named service. Subsequently, other software can ask the name server to find the address to use when a particular service is needed. This scheme enhances the modularity of the system. The address of one piece of software need not be imbedded in any other place; only the names need be defined, since they are fixed and do not vary between one configuration and another. This arrangement even allows optional units to link themselves into a menu system.

The name server technique is, of course, not new. However, there are systems that use nothing but pre-allocated addresses, thereby limiting their flexibility.

A resource manager is merely a more exotic version of the name server. By itself, the name server is not fault-tolerant: if one of the named devices fails, or is removed, users find this out the hard way, in due course. To protect against such event, a resource manager is linked into the maintenance system. Now, if a unit fails, or is removed, the resource manager is informed, and it, in turn, informs all those units that had previously enquired about the service. One way that the resource manager gets this information is through regular audits to determine which units are operational. Protection such as this provides the basis for a more automatic system, appropriate for an office environment, where the system is expected to look after itself.

Conclusion

The system we have described, the Meridian DV-1 Data Voice System, follows a modular design. Extra functionality, or additional computing power, can be added simply by plugging in more units. The software structure is also modular. It can be updated with minimal disturbance to existing systems, and extra features—not foreseen when the system was first installed—can easily be added to the DV-1 when desired. Thus, this is the DV-1, a modular expandable system designed to meet the computing and communications needs at the work group level for the business of today and tomorrow.

References

[1] N. Asam, "Meridian DV-1: a sophisticated data voice system," *Telesis*, vol. 12, no. 2, pp. 34–39, 1985.
[2] N. Asam and B. Williams, "The Meridian DV-1: system architecture," *Telesis*, vol. 12, no. 3, pp. 13–19, 1985.
[3] N. Gammage, et al., "XMS system concepts and architecture," *Telesis*, vol. 11, no. 3, pp. 20–25, 1984.
[4] A. Anderson, et al., "X31: delivering high-speed data to the desk top," *Telesis*, vol. 12, no. 2, pp. 40–46, 1985.

W.A. (Bill) Carr is Manager, Office Application Projects, Office Products Group, Bell-Northern Research. He was appointed to this position in February 1983 with the responsibility of planning, implementing and managing user-driven "Office Learning" trials with BNR. Mr. Carr has been managing an internal Meridian DV-1 Trial since 1985. This program provides systematic input to designers on product requirements and enhancements from a user perspective.

Prior to this appointment, he was Manager in the New Services Development Department where he helped to identify and quantify the business and product opportunities for Value Added Services.

Before joining BNR in 1979, Mr. Carr had experience in marketing and engineering management positions at Hewlett Packard and GEC-Marconi. He is a graduate in Applied Physics from Strathclyde University in Glasgow, Scotland.

Cecil Murray received his B.S.E. in Electrical and Computer Engineering from the University of Michigan in 1972. Prior to joining BNR in 1985, Mr. Murray was President of T and B Computing, Inc., in Ann Arbor. Between 1972 and 1979 he was a senior research engineer at the University of Michigan. He is currently a director in the Ann Arbor Laboratory and is responsible for the development of the recently introduced DV-1 and DNG-1 integrated voice and data products. ■

Who Needs A LAN?

DATA PBXs MAKE SENSE IN SOME SITUATIONS.

by Suketu Mehta

A long, long time ago, before LAN ever walked the face of this earth, there existed a network called the PBX. Many people think LANs have replaced data PBXs. Not so. The data PBX lives! Here are some very good reasons to consider investing in this ancient, but still spry technology.

Take this advice from *LAN* Magazine: you don't always need a LAN. Yes, we're probably killing some of our circulation with this statement. And yes, I will probably be impaled on the door of Harry's office in the morning for all the other writers to see, but the truth must out.

There is an alternative which is just fine if you have specific needs. It's been around for 16 years, so you know it's reliable; the hardware and software are in place, and plenty of instructional materials exist on the subject. Most of all, it's cheap, cheap, cheap: about a quarter of what it costs to buy a LAN.

It is known by many names: port selector, data switch, intelligent switch. But in the land of the gearheads, it is most often called the **Data PBX** (for Private Branch eXchange). A PBX is a system which, in response to dynamic demand, establishes communications paths between the devices terminated on its input/output ports by receiving, processing and transmitting electrical signals.

Devices connected to a PBX can include telephones, data terminals, integrated voice/data workstations, computers and peripherals, gateways to public and private voice or data networks, and other PBXs. PBXs are installed on the premises of the customers they serve (hence "private"). We of the LAN world are concerned most closely with one type of PBX, the digital data PBX.

In its simplest form, a data PBX is like a patch board for computers and terminals. Not everyone is connected at the same time. When you want to be connected to something through a patch board, you go to the patchboard and plug yourself in.

With a data PBX, you don't have to get up from your desk and move the RS-232 connectors. You just type in an acronym for whatever device you wish to access, and the PBX switches you to that device. So it reduces the clutter on your desk as you don't need six different terminals to access six different computers. Everything can be reached through that one little dumb terminal sitting quietly on your desk.

	DATA PBX	LAN
SPEED	Up to 19.2 Kbps async per channel, 64 Kbps sync.	Usually from 1 Mbps to 10 Mbps, total throughput. Token ring LAN is 4 Mbps.
COST	From $100 to $250 per connection.	From $500 to $800 per connection.
CONTENTION	Access ports of device until all ports full. Queue users for busy device.	Slows as more people get on. Multi-user software for more than one user.
SELECTION	Use easy acronym to access device. Usually must log off one device to log on to another.	Access connected devices as virtual drives. Run software on compatible devices.
FILE TRANSFER	May be too slow for lots of data transfer.	Good for file transfer; get highest use of bandwidth when you send large chunks of data.
SECURITY	Passwords for users and devices. Security levels for users and devices. Can have dial-back.	Restrict access to certain devices. Passwords. Encryption.
DISTANCE	Devices up to 10 kilometers apart.	Varies by speed and bandwidth. Ethernet up to about 1 Km. Fiber, realistically, up to about 10 Km.
CABLING	Usually twisted pair. As few as three wires.	Usually coax or fiber.

LAN vs. data PBX—Here's a quick comparison of the salient features of two networking alternatives.

But a PBX also does more. It can convert for communications parameters like baud rate and parity. It can have a protocol converter built in so that asynchronous ASCII terminals can access synchronous mainframes. Most data PBXs, though, are used to connect dumb terminals to async mainframes.

Here's the biggest difference between a data PBX and a LAN. If you have mostly PCs and engineering workstations, you need a cable-based LAN, with megabits-per-second speed for transferring files. If you're going to have lots of CPU-to-CPU communications, you need a LAN.

But if you have mostly dumb terminals running at 9600 bits per second tied in to a minicomputer, a PBX with a 19.2K bits-per-second transmission rate is much more cost-efficient. If you have lots of interactive users frequently using small bits of data, and you need a consistent, albeit low, speed at a low cost, a PBX will do fine.

Transmission

The biggest difference in data transmission technique between a LAN and a PBX is that most LANs use packet switching and most PBXs use circuit switching. In terminal-to-mainframe LANs, the terminal server adds routing information and wraps data from the terminal in "packets."

Whether the terminal sends a single comma or a full page out of "War and Peace," the same packet size is sent over the network. This can and does result in packets that are up to 98% empty being sailed out onto the LAN, clogging up precious bandwidth. Also, some terminals need to pack much more controlling information into their packets than others, so that many more packets have to be sent because less data can fit into each one.

The contention method on Ethernet LANs creates another problem. Ethernet uses CSMA/CD (Carrier Sense Multiple Access with Collision Detection). Under this method, only one device on the network is allowed to transmit at a time. If any two begin transmitting at the same time, their packets collide, and both must begin again after a predetermined time which is different for each device.

So as the number of devices transmitting on the LAN increase, the number of collisions and retransmission increase correspondingly, leading to decreased throughput (data successfully transmitted) during rush hour on the LAN.

All this means that there is a huge gap between the ideal and the actual speed of a LAN. For example, Ethernet is specified at running at 10M bits per second, but that's just the data rate on the cable. According to the original blue book spec put out by Xerox Intel, and DEC, an Ethernet's real throughput will be less than 4M bits per second, under optimum conditions.

Data PBXs use time division multiplexing (TDM) to create a permanent circuit for each device. TDM slices up contending transmissions, letting one go for a period of time, putting it on hold and sending another for a while before switching back. Of course, it is all done in fractions of seconds, but the principles are the same.

Data PBXs have a non-blocking architecture, on the terminal side. That is, every terminal will succeed in getting to the PBX, but from there it is hit or miss if they make it to another end device. There's no contention for bandwidth and no drop in performance on a data PBX as devices are added. Once connected, you get all the bandwidth. If you aren't connected, you get no bandwidth.

If more terminals want access to a device such as a printer or a VAX than the device has ports for, the data PBX tells them to wait. Waiting terminals are assigned a number in a queue, and are kept updated periodically as their position in the queue advances. The fact that all users of the PBX have the maximum bandwidth available to them at all times (while they are attached) means that the speed of traffic on a PBX is as predictable as that of a token-ring LAN. The total throughput on a data PBX, if you add all the channels, may be similar to the throughput on a LAN.

Wiring

Data PBXs use coax or twisted-pair cable. Lately, twisted pair has been the medium of choice as it permits telephone-like installation, often using modular RJ-11 or similar jacks. The typical wiring plan is this: there would be two sets of twisted-pair cable, one from your telephone and one from your terminal, which would go into a double wall jack and merge into one sheath. From there, the wire runs to a punch block on each floor, and into another RJ-11 jack, through a convertor and out at the other end through a DB-25 connector. Then it's the old telco connector technology of 50-pin connectors and 25-pair cable which takes your data from the punch block to the PBX.

Actually, you may already own much of this twisted-pair wiring and not even know it. According to Doug Noble of Equinox, telephone companies frequently leave lots of extra telephone wire lying around in the wiring closet, or in the walls or ceilings. This is because they anticipate that you may need to expand your phone connections.

There's nothing to stop you from using this wire for your data PBX. In fact, you can use the same telephone installer that so lovingly put your phone system in place to wire your data PBX, and save the expense of a LAN installer.

THE BENEFITS OF INTEGRATION: VOICE/DATA PBXs

We've sold you on the idea of data PBXs. Now we'll confuse the issue with **voice/data PBXs.**

If your company's large enough to need a data PBX, it is certainly large enough to need a voice PBX. An integrated voice/data PBX makes sense for a number of reasons. They both run over the same twisted-pair wiring, saving you separate wiring costs. System administration can be performed from one location. The speed of a voice/data PBX is higher — it's non-blocking at 64 Kbps, not 19.2 Kbps as in the case of a data PBX. Plus, in an integrated PBX, a data device can be as much as 4,000 feet away from the switch, a bigger distance than most data PBXs are capable of.

The benefits of centralization are many and glorious. You need only one connection at a desk, need to learn only one set of procedures to transmit voice and data, save space by having only one PBX instead of two, take up only one port on the PBX and make moves and changes easier than if you had two, use the same trunk for voice and data and maybe even justify the cost of a T-1 link, etc. You get the point. One is better than two.

The cost of an integrated PBX, though, is huge: $800 and up per line. Remember though, that this includes lots of features you'd have to pay extra for in a data PBX, such as least cost routing and call accounting.

The voice/data PBX gathers and processes call data. And considering the cost of the extra cable you'd have to run if you bought separate voice and data PBXs, in addition to duplicating the equipment, installation, training and maintenance costs, you're much better off buying the voice/data PBX. Many vendors who make data PBXs also make integrated ones.

If you need to actually buy more wire, fear not. It's reliable and cheap at less than 10 cents per foot.

Security

PBXs are generally more secure than LANs because they use a star configuration. All communications in such a configuration go through the hub of the system (the data PBX), which makes it easier to detect an intruder than on a LAN, where a foreign device can insert itself simply by tapping into the LAN cable. And, because on the LAN all devices see all traffic, that intruder can get hold of information without having to first log on to the network as in a PBX.

In a PBX, there are a series of steps the user must take before he can access the device of his choice. They involve logging on and furnishing a password. Various security levels can be assigned to different users. Some can only access, say, the company's VAX while others are empowered to work on everything up to the mainframe.

Since PBXs use predominantly dumb terminals, those terminals are too stupid to steal. You cannot just insert a floppy disk in a PBX terminal and walk away with the company's skeletons.

Some PBXs feature callback security — after you log on, the PBX tells you to hang up. Then it calls the number which has been assigned to your device and establishes a connection. So you cannot phone in from a remote location and pretend to be Terminal B in the Pentagon.

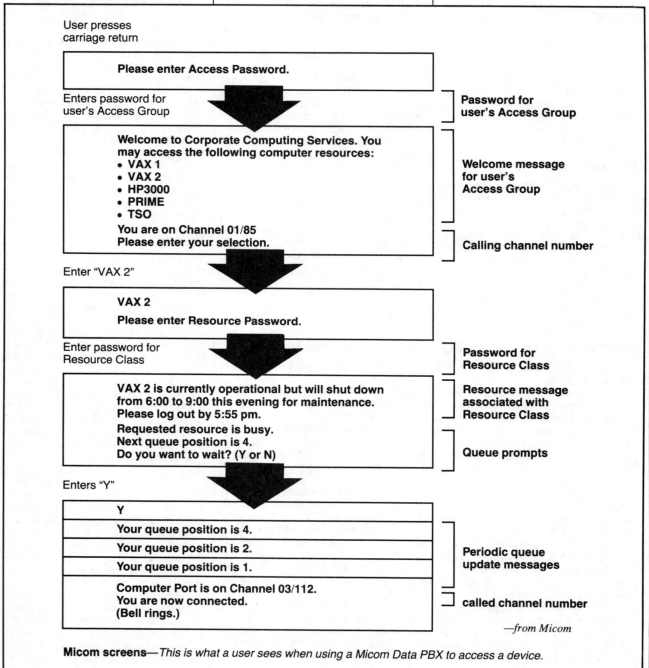

Micom screens—*This is what a user sees when using a Micom Data PBX to access a device.*

PBXs also feature call accounting. A log is made of calls placed, calls connected, as well as unauthorized attempts to access a device. You get valuable information for managing your network. it's also a lot easier to get your information in one place than to poll 10 different mainframes. Compared to a LAN, network management, maintenance and diagnostics on data PBXs are light-years ahead because data PBXs take advantage of many of the advances in network management made by the voice network industry.

Buying a PBX

What should you look for when buying a PBX? Cost per line and number of ports served are major considerations. If the cost per line exceeds $250, the PBX probably has some special features, like Infotron INX's distributed switching capacity, that you may or may not need.

Infotron's PBX is the biggest in the market, weighing in at a huge 4,000 ports. In the other corner we have Equinox's MDX, at 16 ports.

Ease of configuration is another factor. A PBX shouldn't cost half as much to install and operate and train its users as a LAN. Although you'll probably need a professional installer to lay the cable for you, you shouldn't have to hire specialists to help you configure a basic PBX. Do it yourself. Equinox's PBXs have 200 pages of help text built in, so you don't even need a user manual.

Connectivity is also something to look for. Micom's Instanet PBXs feature integrated protocol converters for IBM SNA protocols and integrated statistical multiplexers. They are able to connect to an X.25 network or T-1 multiplexer, for more remote communications. ❏

LAN AND DATA PBX

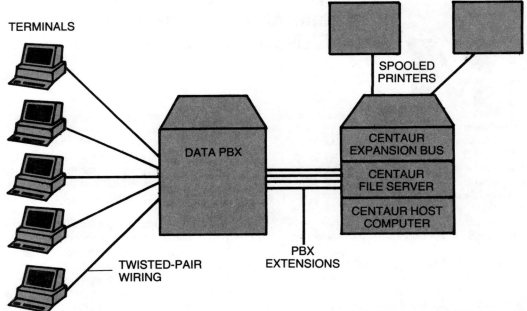

A company called Data Voice Solutions Corp. (DVSC) in Newport Beach, CA, has a product that might be called a compromise between a LAN and a data PBX. Technically it is called a closely coupled LAN. Actually, it is called Centaur.

Centaur incorporates three main components: an IBM-compatible Centaur Host Computer, a Centaur File Server containing Centaur Processor/Memory (PM) circuit cards, and an optional Centaur Expansion Bus containing additional PM cards. Each of these components is contained in a separate PC-type chassis which can be stacked one upon the other. Together the three pieces attach to a data PBX using the extensions it provides.

Each of the PM cards is a PC-on-a-card and attaches to the PBX via its RS-232 connection. These are available as hosts to the terminals on the network. The file server is just that. It stores programs and runs the control software which provides file and record locking, system administration, print spooling and security control. NetWare and other LAN operating systems can be used also.

Terminals access the PC cards and then communicate with the other resources using them. In a sense, the system is a multi-user PC attached to a PBX. Except that the terminals are not dedicated to the PC cards. Rather, they are allocated to them via the data PBX on a port contention basis. This means that when someone is not using a PC card, it is available to someone else, saving computing resources.

The shared resources are not only the PC cards themselves, but also the hard disk drives and printers which are attached to Centaur. These provide file service and print spooling.

In another sense, Centaur is like a minicomputer because all of this is being done in one place. Instead of one processor, however, there are several which communicate as they would on a LAN.

Centaur functions as an add-on applications processor for the data PBX. It is compatible not only with most data PBXs, but can also be used with integrated data/voice PBXs to allow both data and voice communications over standard phone lines. Centaur can be connected to PBXs from Gandalf, ROLM/IBM, Northern Telecom, AT&T, NEC, Mitel and others.

Nathan J. Muller, Telecom Planning & Analysis, Huntsville, Ala.

Enhanced data switches may outshine LAN, PBX alternatives

Adding redundancy voids the old single-point-of-failure argument; new-generation data switches give PBXs, LANs a run for the money.

Many communications managers are rediscovering the basic data switch—and finding a few new surprises in the process. In addition to offering greatly improved port-contention and access-control facilities, as well as fail-safe backup and security options, data switches now on the market can deliver true nonblocking, switched data communications at speeds up to 19.2 kbit/s. In addition, advances in technology now allow totally modular design. What's more, typical configurations today break the $100-per-port barrier with as few as 30 lines.

Data switches, which are also known as port-selection or port-contention devices, have been in use for many years in synchronous networks. There, the data switches allow a large number of users to share a smaller number of computer ports—on one or more hosts—or associated peripherals.

While many vendors are predicting flat data-switch sales over the next few years—some, such as M/A-Com Linkabit, are even letting products like their IPX 3000 gracefully die—a handful of manufacturers have made strides in turning the basic data switch into a powerful, multifunction network-management tool. Such sophistication could even threaten the long-predicted onslaught of integrated voice-data PBXs and LANs.

Sequel Data Communications (Raleigh, N. C.), Equinox (Miami, Fla.), Gandalf (Wheaton, Ill.), Comdesign Inc. (Santa Barbara, Calif.), and Micom (Simi Valley, Calif.) are among the manufacturers that have introduced products with enough redundancy to virtually eliminate the concern that data switches constitute a single point of network failure.

Not only has control logic migrated from a hard-wired matrix to a single processor card, but special redundant logic can allow some data switches to automatically activate an optional secondary processor card if the first one fails. Configuration instructions are automatically cop-

ied into the standby card upon cutover, eliminating the need for manual reentry.

In addition, some manufacturers are offering redundant power supplies that plug into the switch and start up automatically when the primary power supply unit fails. To prevent unnecessary downtime, these plug-in power modules may even be replaced while the data switch is in operation (Fig. 1).

A redundant backplane or "split backup lane" protects the switch from damage that may occur from the failure of other components connected to the bus. In the event of a failure on the bus, the switch automatically switches over to the backplane to maintain uninterrupted operation.

These levels of redundancy are especially important for data switches because they are usually configured in the center of star networks. Without such protection, the data switch could well be the only single point of failure that could bring down an entire network. Together, redundant logic, redundant power supply, and redundant backplane assure communications managers of a more or less fail-safe configuration for uninterrupted service.

Expanding a one-chassis data-switch configuration beyond its maximum line capacity usually requires a networking card that plugs into each of several chassis that are linked together. In this manner, the total data-switch network capacity can grow to beyond 1,500 lines. What's more, remotely distributed chassis may be linked together via statistical multiplexer cards and connected over leased lines. Such an arrangement can typically support a total port capacity in excess of 3,000 lines.

The emergence of universal backplanes (distinct from the redundant backplanes mentioned earlier), which allow any vacant slot to be filled with any option board, greatly enhances the operating characteristics and configuration flexibility of existing data switches. One such board offers a statistical multiplexing capability that compresses up to

36 channels over a modem line or Dataphone digital service facility for long-distance networking between data switches. Other options, including a security callback capability, X.25 interface, and multiswitch networking, may be added through separate boards that can also be inserted into any vacant slot.

The degree of contention for a particular computer port or other resource is generally determined by the communications manager in keeping with organizational needs. For example, the switch configurator may decide to group ports with similar characteristics and capabilities under a class name to permit more efficient sharing (Fig. 2). Even smaller switches permit upwards of 20 password-protected classes to be assigned, which both restricts access and enhances network security.

When a user enters the appropriate connect command and personal password, the switch attempts to complete the requested connection. If the requested port is busy, the user is put into queue and notified of changes in queue status via a screen message. Certain users may be assigned a higher queue priority than others. When a high-

*1. **Redundancy.** Three levels of redundancy—the power supply (A), logic (B), and backplane (C)—eliminate concerns about single point of network failure.*

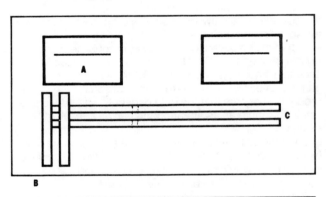

priority user attempts to access a busy port, he or she bumps other waiting users and automatically assumes first place in the queue. In this case, other contenders for that port are then notified of their new status.

Some data switches are able to continuously perform background diagnostics so that faulty channels may be immediately, and automatically, disabled. If a requested port is out of service, a switch-generated message typically notifies the user that the port has been made to appear busy. And when the faulty channel board is replaced, the data switch will automatically re-enable it. The communications manager still retains the option of using a terminal keyboard to disable any port from any terminal for any reason.

A popular new feature—single sign-on—is now becoming available. It allows users to reach anywhere on a multinode switch network without performing cumbersome manual reroutes if the primary route is busy. Often also referred to as dynamic rerouting, single sign-on works much like the direct-dial telephone network: When a call is dialed, it is automatically routed over the most direct route;

if that route is blocked, the call is then rerouted.

Contemporary data switches work much the same way to transport data over multinode networks, and the process is similarly transparent to the user. Not only do some switches support alternate configurations per chassis, but they can also automatically switch from one configuration to another under control of a time-of-day clock or a microcomputer program. A late-night crew of programmers, for example, can operate under one configuration that gives them maximum access to the main computer from 6 P.M. to 8 A.M. Then, when the normal workday begins at 8 A.M., the time-of-day clock changes the data switch back to the primary configuration for general access—and without interrupting existing connections (Fig. 3). The alternate-configuration capability enhances overall network performance because scarce resources may be reallocated during the 24-hour day for optimum usage based on the varying needs of different classes of users.

Another new feature—session toggling—also enhances operator productivity by permitting two connections per user: one primary and the other secondary. The operator can toggle back and forth between connections to perform multiple tasks simultaneously. A batch-file transfer can be in progress over the secondary link, for example, while a real-time database search is being performed over the primary link.

If one of the destinations is busy during an attempt to connect via the toggling feature, the user will usually be placed in queue and then may still use the primary connection. An audible signal will notify the user that the connection has been made to the alternate port.

Some data-switch makers also offer a third-party connect feature, which allows the network manager to connect two switch ports together from a terminal connected to a third port. The ability to establish a connection between two called channels via a terminal keyboard, rather than manually jumping the connection at a patch panel in the computer room, can greatly simplify a host of network-management duties.

Flow control

With a data-rate-conversion capability that is now common, the network manager can set the data rate for each computer port at its highest operational speed. The data switch then performs the rate conversion for any device that attempts communications with the computer at either a faster or slower data rate. For reliable data-rate conversion, the connecting devices should also support independent flow control. Otherwise, there is a risk of losing data.

When Request To Send or Xon/Xoff are used for flow control, the switch buffer will be prevented from overflowing during data-rate conversion. These flow-control techniques are also used to reformat character structures, enabling devices made by different manufacturers to communicate with each other through the data switch.

Some switch manufacturers offer limited Xon/Xoff flow control only through specific ports. A few offer data-rate conversion through all ports. One switch maker, Sequel, has taken data-rate conversion a step further by supporting Tandem Computers' T-Pause flow control.

Instead of being access-restricted to one workstation, new-generation data switches allow the network manager to log on to the switch control intelligence from any terminal operating at any data rate. This enables the manager to conveniently reconfigure ports while making rounds or responding to service calls.

A broadcast feature allows the network manager to transmit messages to individual users at preset times, with delivery controlled by the time-of-day clock. The same message can be sent to the same user every hour, or a different message can be sent to different users at the same time. In addition, a special link feature lets the network manager establish permanent connections between any particular terminal and a port. Sometimes called "nail-up," this feature allows the permanent connection of devices such as printers, which have no keyboards, so they can be logged on or off the network from a remote location or otherwise reconfigured. The fixed logical address assigned via the nail-up capability facilitates this.

With a "force disconnect" feature, the network manager can unconditionally disconnect any port at any time for any reason. Open files can even be closed automatically with the proper disconnect sequences, including any required control characters. This graceful close-down capability is also found in a time-out feature, which improves efficiency by automatically disconnecting idle ports after a pre-defined period.

The network manager can assist inexperienced users by monitoring any port's activity and watching what individual operators are doing, character by character. To learn the status of all switch users—who is connected where and who is waiting for a port—the manager can call up a constantly updated network-status display.

On most new data switches, configuration details need to be entered only once; primary and secondary configurations are stored in battery-backed random access memory or else in nonvolatile, electronically alterable read only memory. Configuration data may also be developed and stored in a microcomputer for later downloading to the data switch through an RS-232-C interface. When equipped with an activity-logging capability, the data switch can provide a complete record of port connects and disconnects—a feature that can delight corporate security officers. And this feature can also log activity via security call-back ports, providing a precise audit trail of connections, as well as the users who made them.

Antihacker

The security of computer data continues to be a legitimate concern of corporate management. And some data-switch manufacturers have made great progress in addressing security issues. Certain new security features cannot be disabled by sophisticated users for the sake of operational expedience.

And while data-switch makers have made certain security functions easy to use so that log-on procedures are not overly complicated, this does not mean that penetration by intruders has been made easier. Aside from the security features already mentioned—passwords, class names, alternate configurations, and connection logging—other new options specifically address dial-in access.

When a user calls in through a modem, the switch may disconnect the line after it receives the proper password and then redial the associated telephone number to estab-

lish the connection. Some manufacturers have equipped this feature to support as many as 1,000 passwords and telephone numbers.

Several different security scenarios, designed to guard against unauthorized dial access and confound the most tenacious hackers, can be employed:

■ Bare-bones security, where connection to the network is permitted only when a dial-up user inputs a valid password. If the password is invalid, the switch will pause for 15 or more seconds before allowing the user to try again. A maximum of three tries is usually allowed before the switch hangs up.

■ A more sophisticated approach allows a user to call up over an autodial modem and input a valid password. The switch will then hang up and call the user back over that modem. If an invalid password is entered, the data switch simply hangs up and never calls back.

■ A dial-up user can input a valid password, after which the switch hangs up and calls the user back on a different line or line group. As before, if the password is invalid, the switch does not call back.

Local area networks

When CXC and Ztel announced plans to introduce integrated voice-data PBXs a few years ago, industry pundits and media literati praised the coming of such products as the quintessential networking solution. But the promise of the so-called integrated approach to networking failed to live up to expectations. Data calls often could not pass reliably through the PBX: Users often could not readily ascertain if a line had hung up or if a file was lost or garbled. Even today, the performance of some PBX products suggests that the data capability was added to a voice PBX more as an afterthought than as a result of design.

Enter the LAN. As recently as a year ago industry gurus were touting the LAN as the definitive connectivity solution. Indeed, data switches were slighted as outmoded technology that would have to be quickly replaced .

But many of the promises of LAN technology failed to materialize, in part because of the lack of standards and disagreement over high-level protocols. LANs are not non-blocking, and wait states, particularly on token-passing schemes, can seriously affect throughput. LANs promise a simple networking solution, but users often find themselves adding black boxes throughout their networks to enable LAN interconnection and access.

The cost of LAN implementation, still hundreds of dollars per connection, far outweighs the benefits LANs have to offer. In simplest terms, a LAN is a conduit, or family of conduits, linking various communications devices. The conduit usually consists of a backbone coaxial cable, possibly with a fiber link between buildings. Some local networks, like AT&T's Starlan, now use existing copper wire or twisted pairs. Protocol conversion is accomplished at the network interface units, enabling incompatible asynchronous and synchronous devices to communicate with each other over this same conduit.

LANs promised substantial cost savings in installation, maintenance, and expansion. Since a single cable invariably is used for a variety of applications, no additional cabling would be necessary to accommodate either new

2. Class. *The data switch may be partitioned through class names so that only users within a class may access resources allocated to them. This makes network manage-ment and security easier. Class divisions may also have ben-eficial effects on network economics. Ports may be grouped by application, department, or technical specifications.*

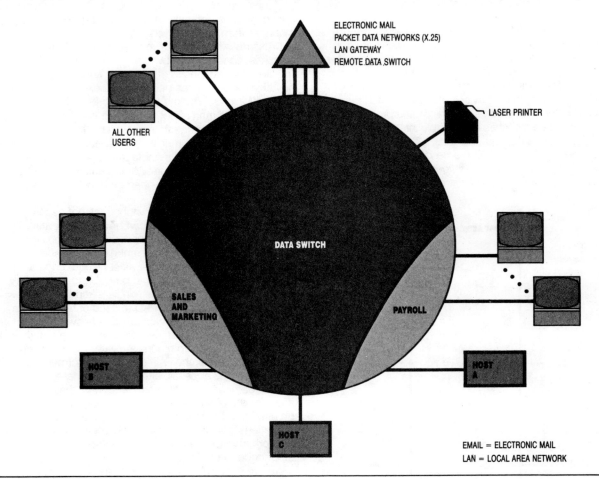

applications or any changes in a user's requirements.

Other benefits of LANs include:

■ Distributed control, which can prevent a single point of failure from stranding large numbers of users.

■ Low-cost network expansion, since additional terminals can often be easily plugged into the LAN.

■ Low-cost reconnection for the installed base of terminals through local access points, instead of wiring each terminal to a central switch.

■ Consistent throughput, despite the variety of applications or number of active terminals on the network.

■ Higher transmission speeds and lower error rates than with data switches, though distances may be limited to within a building or local building complex.

■ Support of value-added services, such as electronic mail and packet-switched services, through bridges or gateways.

■ Decentralized terminal distribution to the most convenient locations, rather than forced distribution within a centralized area.

■ Protocol conversion, which would permit devices from different manufacturers to communicate with each other on the same network.

■ Enhanced network management capabilities, since any authorized terminal on the LAN may be used to collect usage and performance data.

Because so much publicity had been bestowed upon LANs, many users have come to understand that such products are now the only viable solution to local area networking problems. But data switches perform many of the functions of LANs, resulting in many of the same efficiencies that LANs claim. Cost, however, is usually much less with the data switch.

What's more, networking over long distances is per-formed better by data switches than by LANs—and at a lower cost. For example, linking separated LANs is much more complicated and expensive than linking disparate data switches. LANs require the use of expensive gate-ways; data switches do not. And routing data between LANs is not transparent to the user, as is the case with data switches.

Even though LANs are touted for their ability to link different types of computer equipment into a single, cohesive network, LANs are still plagued by the lack of standards to make such schemes fully workable. Because data switches invariably use the elementary RS-232-C standard for data interchange, compatibility with virtually all vendors' products is assured. With the protocol conversion, flow control, and data-rate conversion capabilities of full-function data switches, questions of compatibility rarely arise.

Data switches offer many other advantages over LANs. In a homogeneous network—where all devices are from the same vendor—port selection among a small number of terminals or microcomputers can be more efficiently handled by a data switch than a LAN. On a mixed-vendor network of incompatible devices, the data switch may be an even better choice because it facilitates protocol conversion. Data switches offer stable interfaces and time-

---

3. Port assignments. *The same data switch can implement two entirely different configurations through either a time-of-day clock or a stored microcomputer program.*

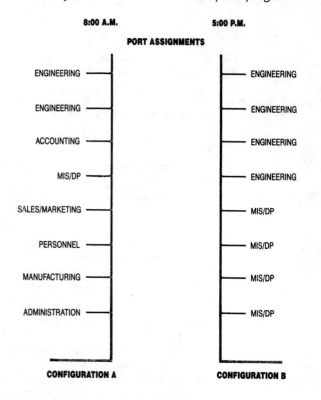

MIS/DP = MANAGEMENT INFORMATION SYSTEMS/DATA PROCESSING

---

tested maintenance and control procedures.

Although network configuration considerations (number of users to be supported, types of applications, and so forth) should be hammered out before the purchase and installation of a LAN, configuring a local network for optimum performance still involves trial and error. Configuring a LAN oftentimes requires extensive vendor support or the services of outside consultants because it involves specific knowledge of the operating systems involved (to determine such things as cache buffer sizes, time-slicing parameters, and the amount of memory to allocate for caching files).

With data switches, the task of configuration is relatively

simple: The communications manager usually just steps through a series of menus and supplies the required information to properly configure the network. Some data switches also allow the use of abbreviated commands for configuring the network, which saves the experienced manager from several levels of prompting.

LANs do offer one relative advantage over data switches: speed, or bandwidth. LANs such as Ethernet transmit data at speeds of 10 Mbit/s and more; the most innovative data switches deliver nonblocking speeds of only about 19.2 kbit/s. But the functions offered by data switches more than offset this speed limitation.

In fact, only the largest users with the heaviest traffic requirements—those running high-speed graphics, computer-aided design and manufacturing, or transmitting voluminous files along with bursty data traffic among 30 to 40 or more microcomputers—need be concerned about bandwidth.

Actually, LANs and data switches are not necessarily mutually exclusive, and they can be complementary. For example, recent improvements in packet assembler/disassembler (PAD) products make data switches especially useful and cost-effective as servers to bridges or gateways between LANs, or to broadband, proprietary, and public data networks. Terminals linked to a data switch may be multiplexed to the PAD to send or receive data from the packet data network. And data switches that are optionally equipped with the X.25 packetizing software do not require PADs to communicate with packet data networks. In this case, the PAD function, performed inside the switch, is made accessible through designated ports.

As a company's LAN expands to meet the increasing need for data communications, modular data switches can be used to add management oversight and to control operations, saving both time and money. Data switches' network management functions operate on top of any other underlying network management schemes, and are therefore especially useful in gateway situations.

When the need for a local area network is fully justified, data switches still can play a subordinate but important role as servers. Thus, the capital outlay for the LAN is minimized and the original investment in data switches is totally protected.

The purchase of even a small data switch will not be wasted if an organization outgrows it. If the expansion cannot be adequately served by adding incrementally to a small, modular data switch, and immediate movement to a high-end data switch of much greater capacity is called for, then the smaller unit can still be put to use as a terminal server to the larger switch.

What's more, today's data switches offer numerous self-healing schemes that come from multiple levels of redundancy. And finally, data switches use a variety of common cabling schemes, including existing copper wire. They may be unpacked, tested, and configured for operation in a single afternoon. ∎

Nathan J. Muller is an independent telecommunications consultant in Huntsville, Ala. He specializes in network planning as well as in hardware and software evaluation. His 15 years in the computer and telecommunications industries have included positions in engineering, operations, and field service.

The Evolution of Data Switching for PBX's

BRIJ BHUSHAN AND HOLGER OPDERBECK

Abstract —This paper compares and contrasts the different traffic requirements of both voice and data communications. The role of evolving PBX technology in solving data communication needs is outlined, beginning with the analog PBX's of the 1970's and the digital-switching technology of the 1980's. The evolution of packet switching, and its superiority over traditional circuit switching for solving data communications needs, is analyzed.

The paper concludes by describing a PBX implementation that takes advantage of circuit- and packet-switching technologies and thus offers a truly integrated multipurpose communications switch.

I. Introduction

THE PBX has long held immense potential as the most economical and workable solution to meet the rapidly growing voice and data communications needs of business. This potential has not gone unrecognized, but until now it has gone largely unrealized. For years, manufacturers and users of PBX's have seen that their central voice switch would also be the ideal candidate for accommodating their data-switching requirements. The points in favor of the PBX as a data switch are numerous. First, the PBX provides an existing network of installed lines that runs from every office to the central switch, creating a de facto local area network. Second, the PBX can be connected directly to long-haul transmission networks via T1 interfaces using media such as conventional cable pairs, coaxial cables, fiber optics, and satellite and terrestrial microwave links. These interfaces provide high-speed access to public and private communication networks. Third, PBX's have always provided the most sophisticated administration and maintenance capabilities which are required to optimally manage complex telecommunication systems.

The problem, therefore, was one of technology, of fulfilling the potential of the PBX in the area of data communications. Over the last five to eight years, the attempts to produce a completely integrated voice/data PBX have brought it through three basic phases of development. With each phase, the data capabilities of the PBX improved, and the true integration of voice and data moved a step closer to reality.

II. Requirements of Voice/Data Traffic

The characteristics of voice and the demand it imposes on the switch have been fairly well known for a long time

and are well documented. A summary of these is shown in Table I (for details see [1]).

Data traffic typically has three attributes that differ from voice attributes. Unlike voice, data communication generally does not occur between any two devices, e.g., data devices always tend to reach applications at a prefixed number of hosts. Thus, there is a tremendous amount of concentration that is required. This attribute has led to a number of manufacturers making concentrators, multiplexors, etc. With the advent of personal computers, this trend will change when these personal computers will generate traffic by connecting to each other and occasionally connecting to a host computer.

The second attribute of data is its bursty nature. There can be inactivity on the line for quite some time and then suddenly a burst of data at the rated speed will come through followed by a period of silence. In contrast to this phenomenon, the voice traffic generally is continuous once a connection is made.

The third is the time sensitivity of the information. Unlike voice, where one copes with busy modes (station busy or congestion busy) and attempts to either call back or queue the call, the data call demands instant connection since the information retrieval from the application/database is time-critical for the end user in some applications. This last attribute translates into a requirement of "nonblocking" switch for data applications.

Data traffic is extremely diverse and is generally application dependent. The common characteristics are outlined in Table II and can be classified into four broad categories,

TABLE I
VOICE-TRAFFIC CHARACTERIZATION AND PBX REQUIREMENTS

Holding Time	120–180 s
Traffic/Port	6–12 CCS/Station; 18–28 CCS/Trunk
Busy Hour Call Attempt	4–5/Port
Dial Tone Delay	< 3 s 98.5 Percent of the Time
Busy Hour	20 Percent of Daily Average Traffic

PBX REQUIREMENTS FOR VOICE TRAFFIC

In Switch Blocking Characteristics	P.01
	Nonblocking P0.0
	Essentially Nonblocking (At a Certain Traffic)
Intra-PBX Traffic	10–40 Percent
Inter-PBX Traffic	30–50 Percent
Off-net Traffic (For network of PBX's)	20–40 Percent

Manuscript received July 9, 1984; revised March 25, 1985.

B. Bhushan is with GTE Telenet Communications Corporation, Reston, VA 22096.

H. Opderbeck is with Opderbeck Communications Consultants, Vienna, VA 22180.

Reprinted from IEEE Journal on Selected Areas in Communications, July 1985, pages 569-573. Copyright © 1985 by The Institute of Electrical and Electronics Engineers, Inc. All rights reserved.

TABLE II
DATA-TRAFFIC CHARACTERIZATION AND PBS REQUIREMENTS

Call Holding Time	8 s–15 h
Bit-Error Rate 10**−5 to 10**−8	End-to-End
(Or Error-Free Seconds)	
Speed and Type of Transmission	
• Asynchronous and Synchronous	
• Half-Duplex versus Full-Duplex	
• Codes ASCII/EDBDIC/	
BAUDOT, etc.	
Throughput of switch	
• Bits	
• Packets	

e.g., call-holding time, speed and type of transmission, bit-error rate (or error-free seconds) and the throughput of the switch.

Call-holding times are very much dependent upon the application. The typical values may range from 8–15 s for an interactive transaction-type application to 8–15 h a day for remote job entry applications. These applications could be run on asynchronous or synchronous terminals in full- or half-duplex mode ranging in speed from 300 bits/s to 56 kbits/s and demand bit error rates of 10**−5–10**−8.

III. ANALOG APPROACH TO DATA SWITCHING IN A PBX

The majority of PBX's that were designed and built in the late 1970's were analog PBX's, and were primarily designed for switching voice. These PBX's were built for call duration of only a few minutes which is typical for voice calls in the office environment. Although data could be switched through these early PBX's, the data stream had to be converted to a format that made it look to the PBX like a voice call, e.g., typically modem data. This conversion to voice format at the input, and back to data format at the output, was accomplished by modems. Thus, PBX handled data in the same way as the public-switched network regardless of its destination, e.g., intra-PBX or inter-PBX.

The call establishment and tear-down process was accomplished via telephone sets. These PBX's thus were unable to provide error correction or concentration of data into high-speed lines or ports. This implied that all data had to be switched on a direct line-to-line low-speed access basis (up to 9.6 kbits/s) as shown in Fig. 1. These constraints not only limit the value of the PBX for data communication capability; this cost was further compounded by the cost of lost voice capacity. In short, running data communications on the voice-only PBX of the late 1970's was an inefficient and expensive proposition.

IV. DIGITAL APPROACH TO DATA SWITCHING IN A PBX

The 1980's saw the beginning of integrated voice/data communications. When a digital local loop is used, the expensive modems are replaced by digital devices like the

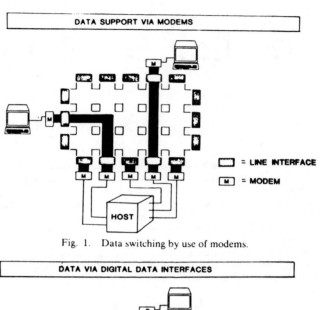

Fig. 1. Data switching by use of modems.

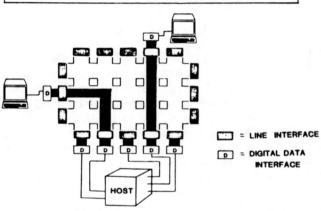

Fig. 2. Data switching utilizing DDI.

digital-device data interface (DDI) shown in Fig. 2. DDI's are capable of connecting to data terminal equipment in the same manner as the modems. The received data bits are transmitted digitally to the PBX where the bit stream is switched the same way as voice. This digital interface, however, is capable of signaling to the PBX via a message-oriented format. Thus, for the first time, the switch must be aware of the fact that it is not dealing with a standard analog device, but a more sophisticated device with different requirements.

These digital interfaces can be stand-alone devices or integrated into the new generation of station equipment. In some cases, this integration is only a physical integration, i.e., housing, power supply, etc., being shared, and there is no integration on the access lines. In this arrangement, two or more wire pairs running to the PBX support voice and data on distinctly separate pairs terminating at different line interfaces. Thus, the PBX treats them as two distinct connections, resulting in added cost of lost voice capacity as in the case outlined earlier.

A more cost-effective solution is realized if data and voice devices share the line card at the PBX as well as the local-access wires. This local-access integration also results in the phone and DDI sharing electronics, thus making this

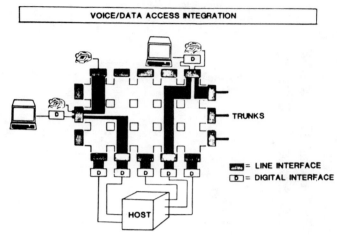

Fig. 3. Local access-integrated PBX.

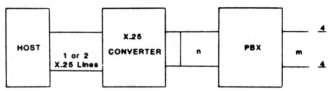

Fig. 4. Use of X.25 converter box with PBX.

an attractive implementation as shown in Fig. 3. This implementation yields an increase in port capacity since the new line cards can now handle voice and data ports simultaneously. The data, however, are still switched in the circuit mode and the PBX switching bandwidth is not efficiently utilized (a 9.6 kbit/s connection through the 64 kbit/s PCM stream yields 15 percent efficiency). A more balanced approach to voice/data switching requires the introduction of true packet-switching capabilities integrated in a PBX. These capabilities are now emerging in the third phase of voice/data integration.

The integration of packet switching in a PBX brings with it the concept of a virtual connection. A virtual connection is a logical association of two endpoints. The two endpoints can exchange information any time they desire to do so after a virtual connection. Therefore, the long intervals of idle time, which are so typical for data communication, do not consume any switching capacity. Moreover, the bandwidth of the local-access line can be allocated to multiple connections as required, instead of being dedicated to serve just one voice and one data call.

Virtual connections have several attractive features. They do not care what application they are being used for. They represent a universal transport mechanism which is independent of format, speed, or content. In principle, they can be used for voice, data, text, facsimile, and new applications with still unknown service characteristics. They consume transmission and switching resources only when required. This dynamic allocation of resources reduces the total resource requirements substantially.

The dynamic resource allocation introduces potential delays due to the competition for resources among multiple connections. In the data world, these delay considerations are well understood and efficient system engineering guidelines exist to minimize their impact.

The integration in the PBX can be effected by separating the voice PCM samples from data stream at the line-card level and switching it via conventional switching means—a noncontention PCM bus, whereas the data stream can be switched in a contention mode via a data bus.

This dual-bus architecture, with one bus for the switch-

ing of voice in a dedicated mode and one bus for switching of data in a contention mode, provides the user with the best of both worlds. It guarantees voice quality due to its priority handling of PCM samples and it brings the flexibility of virtual connections to every station connected to the PBX and the problem of lost voice capacity is thus resolved.

V. Data Communications Using Packet Switching

With the advent of packet switching and subsequent standardization of the applicable X series of recommendations (X.3, X.25, X.28, and X.29) a formal standardized method was made available offering, among other things, a multiplexing method, as well as universal applicability. Despite its initial slow acceptance, it is now provided by all DP equipment vendors. The standardization of X.25 also brought with it entrepreneurs that will afford retrofittability to the installed equipment by offering hardware that can concentrate traffic in one X.25 pipe as well as allow asynchronous terminals to talk to the host machines as shown in Fig. 4.

The equipment configuration could still be engineered to provide the level of service that the customer desired. This approach meant that the number of host ports went down at the cost of the X.25 converter boxes.

ROLM and INTECOM have attempted to integrate the X.25 functions in their PBX's in different ways. Both implementations are capable of supporting X.25, but the number of ports in each implementation is limited and this addition has an effect on the voice-carrying capacity of the PBX [2], [3].

The GTE OMNI SII and SIII series of PBX's have implemented the X.25 packet-switching approach in an evolutionary and innovative way. The main goals of the architecture are to
● use the standard OMNI-PBX cabinets, power, and backplane;
● provide X.25 based capabilities;
● minimize impact on the existing OMNI-PBX hardware and software;
● minimize the effect on the OMNI-PBX's voice-handling capability;
● present the PABX as a "unified" system to the user;
● provide an architecture that is extensible, for future product evolution.

The architecture chosen is shown in Fig. 5. To provide the communication capabilities needed to support data switching without affecting voice capacity, a packet trans-

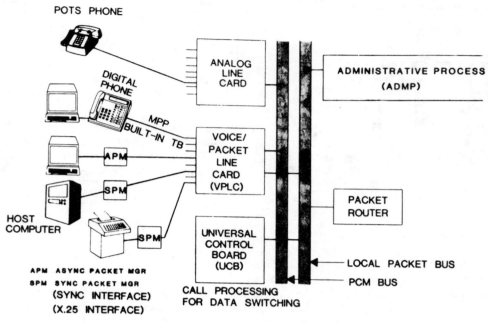

Fig. 5. Dual-bus architecture of OMNI PBX's.

port system (PTS) has been designed. The PTS consists of a packet router (PR) and the appropriate number of local packet buses (LPB). This bus structure has been added to the backplane in the OMNI PBX. Communication within the PTS is in the form of self-routing entities known as minipackets. Each uniquely addressable device, communicating via the PTS, is assigned a packet line address (PLA). The PLA provides the internal "address" of the component to which it is assigned.

The assignment of a specific PLA is performed automatically by the data subsystem, and is based on the hardware configuration of the switch. It has no relationship to the "subscriber number" which the user sees. Devices communicating via the PTS use a GTE proprietary protocol, known as minipacket protocol (MPP) to ensure reliable error-free communication.

Attached to the LPB are universal controller boards (UCB's) and an administrative and maintenance processor (ADMP). These cards provide the processing power needed to supply the various features required of the data option, such as data-call setup/takedown. The modularity of this design allows additional cards to be added to a system as the need arises and provides an evolution path for future capabilities.

In addition to the UCB's and ADMP's, line cards, known as voice-packet line cards (VPLC's) attach to the LPB. These provide the ability for remote processing devices (RP's) to communicate via the PTS. "Remote Processor" is the name given to a family of devices which are located remotely from the central OMNI PBX cabinet, and which interface directly with the subscriber equipment.

Asynchronous devices (terminals, hosts, and modems) connect to the data option via remote processors, known as Asynchronous Packet Managers (APM's). APM's are

colocated with their asynchronous devices. Except for call-control-related events (call setup, disconnect, etc.), APM's contain all of the processing and transmission capabilities necessary to support transmission of asynchronous data. In addition, the APM provides an $X.3$ packet assembler/disassembler (PAD) function. This converts asynchronous data into $X.25$ packets and allows a user to enter signaling information (commands) via the keyboard on the attached asynchronous device. Commands are supplied using an $X.28$ type command language. APM's are addressed using the $X.121$ numbering scheme.

Each APM supports, at most, one single active data call (known as a virtual circuit), and will interface to full-duplex devices running at speeds up to 19.2 kbits.

Each APM is connected, via a single twisted-wire pair, to a voice-packet line card (VPLC) located within the OMNI switch. APM's communicate via the packet transport system and are assigned a unique $X.121$ address. APM's are locally (wall) powered or powered over a separate wire pair from an auxiliary power supply. Each APM is connected to a host, terminal, or modem with an RS-232C connector and may be configured as a DTE or DCE.

PBM is a remote processor known as a Synchronous Packet Manager (SPM). SPM's are colocated with their $X.25$ hosts, and connect to them via an RS-232C or $V.35$ connector. The SPM may be configured as a DTE or DCE. Each SPM supports $X.25$ level 3 (packet level) and $X.25$ level 2 (link level, LAPB). SPM's, like APM's, are locally (wall) powered or powered over a separate wire pair from an auxiliary power supply. They interface via a single twisted-wire pair to a VPLC card located within the OMNI PBX.

Data call-processing functions are completely handled by a universal controller board which is configured with data call-processing (DCP) software. The OMNI PBX voice switch does not participate in data call procedures.

Most administrative functions for the data option are handled by an administrative and maintenance processor (ADMP), although some functions are controlled by the voice portion of the PBX. ADMP is also responsible for integrating the voice and data functions from an administrative user's point of view.

Every OMNI PBX running the data option requires a disk as its resident mass-storage device. This mass-storage device is shared by the voice and data processing portions of the switch. The disk will provide storage for code and tables used within the entire OMNI PBX.

This approach has led to typical traffic capabilities of supporting up to 256 devices all running at 9.6 kbits/s full-duplex. This traffic capacity can be further augmented by adding more buses and more processor power in a modular fashion.

VI. Conclusions

During the last few years, the PBX has gone through an evolution of data switching which turned it into a sophisticated data-switching device. With the emergence of new standards like ISDN, the integration of voice and data will be merged into the integration of other communications services. The result of this integration will be a truly integrated multipurpose communication switch.

References

[1] *Functional Product Class Criteria — PBX*, AT&T Pub. 48002.
[2] ROLM product brochures.
[3] INTECOM IBX product brochures.

Brij Bhushan was born in India. He received the B.S. and M.S. degrees in electrical engineering from Auburn University, Auburn, AL.

He did advanced graduate work in electrical engineering at Concordia University, Montreal, P.Q., Canada, and taught various undergraduate courses in electrical engineering prior to joining the communications industry. From 1973 to 1980 he was with the Computer Communications Group of Bell Canada, where he was responsible for the Datapac Services planning. He has also served as consultant to Satellite Business Systems for satellite-based voice/data communication services. Currently, he is responsible for business planning activities for GTE Telenet, Reston, VA. His prior responsibilities at GTE included product planning for customer-premise equipment and product evolutions for evolving ISDN standards.

Mr. Bhushan is a member of the Association of Professional Engineers, Ontario, Canada, the Association for Computing Machinery, and Eta Kappa Nu.

Holger Opderbeck received the M.S. degree in physics from the University of Munich, West Germany, and the M.S. and Ph.D. degrees in computer science, from the University of California at Los Angeles.

He began his career in communications as head of the Arpanet Measurement Project at the UCLA Network Measurement Center. In 1975 he joined GTE Telenet, a common carrier company providing the first public packet-switched data communications service in the United States. There he became Director of Software Systems, responsible for the design and development of all Telenet software products. In 1979 he was promoted to Vice President of Network Products, with full responsibility for all of GTE Telenet's product development. GTE Telenet's new family of multimicroprocessor-based communications products, the first of their kind in the industry, and the public electronic mail system, Telemail, were developed under his leadership and direction. In 1981 he transferred to GTE Business Communication Systems as Vice President of Advanced Communications Development. There he was responsible for the development of all GTE business terminals, office automation products, PBX data-switching products, and the next generation of integrated voice/data switches. Currently, he is President of Opderbeck Communications Consultants (OCC), a company specializing in the design, selection, and evaluation of integrated voice/data systems.

Chapter 5: Distributed Architectures

In the early 1980s there appeared a group of private branch exchanges (PBXs) that challenged the conventional wisdom in PBX hardware design, topographical distribution, environmental requirements, the relative roles of voice and data, and switch software and its development. They came from start up companies such as CXC and Ztel, from a strong niche player like Mitel, and from an old-line Swedish manufacturer, Ericsson.

From the perspective of 1988 their problems appear more salient than their promise. Ztel went Chapter 11, CXC retreated into a less ambitious niche market, Mitel, a Canadian firm, was first courted and then abandoned by IBM and subsequently sold to a British firm, and Ericsson, finding the U.S. market too costly to penetrate, changed its plans and returned to the European market. Nonetheless, from the perspective of the early 1980s, the opportunity had looked promising. Deregulation created seven new Bell Operating Companies (BOCs) whose deregulated divisions were forming marketing alliances with firms with attractive products to compete head-to-head against what were perceived as the lackluster offerings of AT&T Technologies. Also, deregulation had repriced Centrex in such a way as to shift the cost advantage to customer premise-based systems. Finally, the new products were genuinely new, and they promised important advantages to the customer over what was then on the market.

Why devote a section in such an overview to PBXs that conspicuously failed in the marketplace of the 1980s? The argument for doing so is that although the particular products were business failures—the companies were inadequately capitalized, they lacked strategic alliances for marketing and support, they rushed incomplete products to market—the designs portend the future directions in PBX design. In one aspect, they are worthy of attention in the same way that an auto buff regards the VW Beetle, the Cord, the Chrysler Airflow, and the 1954 Corvette. In another aspect, the approach taken by Rolm in its CBX II 9000 (and carried over into the IBM 9750), whereby an older single node, star-oriented design was creatively upgraded to multi-systems networking represents a direct intellectual debt to CXC, Ztel, and Ericsson. In the following paragraphs, some of the distinguishing features of these designs will be discussed.

First, these designs redefined the standards for compactness and modularity. The basic unit was a node, targeted for the low end of the PBX market, that would support about 60 lines. The node intensively employed the latest in microprocessor technology with multiple main processors, usually of the Motorola 680X0 or Intel 808X family, configured for fault tolerance. The node was air cooled, ran on 115 volt AC power with self-contained UPS, and required no extra air conditioning. Ericsson, with its bookshelf design, went to the point of eliminating fans through ceramic mounted components. Extremely compact, they were designed for small spaces, such as conventional wiring closets. If more capacity were needed, Ztel's nodes could be stacked. In sum, these manufacturers, through advanced design, created modular systems that could become very large by building on packaging techniques that had been largely limited to the small systems end of the PBX market.

Second, for larger environments, they incorporated one or more high speed local area networks (LANs) as part of their distributed architecture. Often employing fiber optic links, the multiple LANs were to enable a distributed, digital architecture serving heavy voice and data requirements while providing link redundancy. CXC and Ztel opted for high speed rings, Ericsson for a logical ring imposed over a physical star. Distribution of nodes promised new levels of survivability via trunking several nodes to the central office (or multiple central offices) and designating one node the network master with an explicit succession sequence. In implementing these distributed schemes, fiber optic cable has become the media of choice. Not only does it offer very high transmission speeds, but, frequently of more importance, its small diameter allows the use of existing conduits.

Third, because the distributed system was bus-oriented, it was viewed that extra functions, such as voice or electronic mail processors, gateways or bridges, station message detail records (SMDR) processors, and the like, would be added through third-party vendors by means of either well-known interfaces (particularly RS-232C) or attached processors. This choice has had a number of important consequences. Allowing for easy inter-connection emerges as an important technique for keeping PBXs relevant to the needs of today's office. PBXs are usually treated as physical plant equipment with expected lives of 5 or 7 years. Many of the electronic devices in today's offices—personal computers (PCs), facsimile machines, printers—should properly be considered consumer electronics, with product cycles of approximately 2 years and expected lives of barely double that. Possessing a mechanism to accommodate these product life disparities is of considerable utility to both PBX makers and users. Too, the general ability to incorporate other manufacturers' equipment such as voice mail systems, attached processors for network planning, protocol converters, network bridges under a PBX services umbrella allows a single source for maintenance, a significant advantage vis-a-vis most cable LANs.

Fourth, software was to be written in a high-level language so to be more maintainable, better able to accommodate extensions, and to allow it to be easily portable to the latest members of the microprocessor family. In practice, this often meant that the software was written in the C programming language on a DEC VAX, cross assembled, and the resulting object code downloaded to the PBX's microprocessors. In retrospect, it was in software that the new designs most overreached themselves. They very much underestimated the time to code and test complex, real-time call processing software. None of the designs ever worked satisfactorily in a multinode, commercial environment. Forced to push out product to meet marketing and financial schedules, a series of premature beta tests revealed buggy software and resulted in unfavorable publicity in the trade press.

To view the difficulties of a CXC or Ztel with applying conventional higher-level software languages and modern programming practices as an indictment of the software technology misses the point. People who are prone to push at time/technolgy barriers can overreach themselves just as easily with assembly or proprietary languages. And large resources, such as available to defense contractors developing large weapons systems, provides no protection from failure. Nor are software bugs limited to new machines. Central office switches, which in former times often provided the yardstick by which reliable software was compared, have recently suffered a series of well-publized outages traced to recently suffered a series of well-publized outages traced to adding new features to the software. Clearly, if higher-level languages and the good software engineerir.g practices can produce code that meets the performance demands of real-time systems, and at the same time prove easier to maintain and extend and be capable of migration across microprocessor generations, then the economic advantages would be enormous.

The papers, included here, by necessity a mixture of papers and excerpts from system descriptions, reflect the relative newness of PBXs designed from the ground up to be physically distributed. Coover and Kane, in "Notes From Mid-Revolution: Searching for the Perfect PBX," optimistically contend that, sooner or later, small, energy efficient, distributed PBXs will change the look—and cost structure—of the PBX market. McKay in "New Wave Coming in Data/Voice Switching," reviews GTE-sponsored research in burst and packet switching, PBX-on-a-chip, LAN-linked distributed switches, and other new technologies. Karavatos describes Ztel Corporation's Private Network Exchange or "PNX," in "The Next Generation in Business Communications." It consists of a triple ring (voice, data, spare) IEEE 802.5 LAN with two kinds of node processors (ring, applications) and offers both circuit-switched voice and packet-switched data. The Frank article is a *Business Communications Review* product announcement describing the very advanced Ztel and CXC PBXs. These so-called "fourth-generation" systems attempted to marry the complimentary functions of LANs and digital PBXs for the "office of the future." Their approaches toward electronic messaging were particularly innovative. The final selection is a selection from a "System Overview" of the Ericsson MD 110. Note in particular the unique packaging of its Line Interface Modules (LIMs) and how the LIM software comes in both "regional" and "central" modules.

Edwin R. Coover and Michael J. Kane, Mitre Corp., McLean, Va.

Notes from mid-revolution: Searching for the perfect PBX

The Washington, D. C., operations arm of an engineering firm embarks on a reconnaissance mission for its new in-house PBX.

Late in 1983, Mitre Corp. began planning. Mitre, an independent, not-for-profit organization that works in a variety of areas — including computers, communications, public health, and environmental protection — counts among its major clients the Department of Defense and the Federal Aviation Administration. Although over 18 months distant at the time, the convergence of two events — the expiration of the lease on its AT&T Dimension PBX (Private Branch Exchange) and the scheduled completion of a new, 300,000-square-foot building — demanded action.

Mitre's 2,300-person staff in Washington, D. C., would be consolidated from nine present buildings into four closely grouped buildings requiring voice, data, and video services. On the data side, the organization's Sytek LocalNet, a broadband local network, would have to be extended and reconfigured. Many of the 900 terminals, hundreds of microcomputers, two IBM mainframes, 10 VAXs, and other specialized processors would be affected by the relocation. IBM PCs are the most common microcomputer used by Mitre, followed closely by Apple Macintoshes. The corporation has something of everything: symbolic Lisp machines, Digitial Equipment Corp. (DEC) Rainbows, Apple IIs, Sun workstations, and so on. On the voice side, Mitre's aging Dimension switch and a satellite voice / data Rolm CBX II needed to be replaced, along with all station equipment. Mitre's new switch needed data capabilities, but a broadband coaxial cable would have to form the primary communications network for ASCII terminal-to-host transmissions. Mitre wanted the flexibility to use broadband, local networks, or PBXs.

Mitre's approach was systematic. A Mitre-wide advisory group was formed with members from every division who were then surveyed and interviewed. The data about their requirements was consolidated, reviewed, and revised. Vendor product and capabilities information was culled from sources such as Datapro Research on PBX original equipment manufacturers (OEMs) who sold switches in the 2,000-4,000 line range. The data was digested, manufacturer representatives interviewed, and a Lotus 1-2-3 database created. Management was briefed, an RFP (request for proposal) written, and a vendor meeting held in which 14 different organizations were present, with four different bidders representing NEC-America equipment. At a subsequent meeting, building blueprints were distributed and a physical inspection of extant wiring took place. Mitre is currently awaiting proposals.

Although the replacement and procurement process is proceeding routinely enough, the quest itself turned out to be an awakening. Mitre's investigation revealed that in the last half-dozen years substantive changes have occurred on the digital PBX scene. New PBXs have radical designs that, if commercially successful, will, by their superior cost / performance margins, redefine PBX economics.

Microcomputers and data terminals can now plug into the back of streamlined telephone sets. (Perhaps, more ambitiously, hybrid microcomputer / telephone devices will become common.) Furthermore, in the future it will be less likely that PBXs will appear to the uninitiated as a DEC PDP-11 minicomputer look-alike stood on its head, with cables flowing out the top, and ensconced in a large, wire-crammed, heavily air-conditioned room in the basement. Instead, it is probable that a discreet peek into a telephone closet will reveal a compact device — often no larger than an under-the-counter refrigerator — that, in concert with similar switching nodes, will handle thousands of voice or voice / data devices.

What has happened, and why now? How may one

broadly characterize these new devices?

■ *Microprocessor-based.* Where conventional PBXs were once built around 16-bit and, later, 32-bit minicomputers serving as principal processors, high-performance microprocessors have usurped these functions. As with other similarly replaced equipment, the substitution has generated a benign ripple effect: greater performance, less power, less heat, drastically less space, and, most compelling, less cost. The 32-bit Motorola 68010s used in the Ericsson MD110, CXC Rose/Western Union Vega, and Mitel SX-2000 cost approximately $128 each.

■ *Parallelism and hierarchy.* The cost/performance advantages of high-performance microcomputers have allowed PBX designers to further increase parallelism in their switch architectures. Up until recently, conventional PBXs usually featured 8-bit microcomputers servicing the station equipment (analogous to cluster controllers in orthodox data processing shops). A fairly large minicomputer ran the more complex applications, such as least-cost routing and the central directory. Usually, a second standby minicomputer was offered as part of the package in arrangements of 400 lines or greater. The extra minicomputer served as a hot standby or alternated in regular use with the primary minicomputer.

The new generation of microcomputers allow creative extensions to this parallel, hierarchical architecture. The low cost of the microprocessor allows cold or hot standby processors at both the line cards and the application processors, thus augmenting fault tolerance. The proliferation of bus-connected 16- and 32-bit microprocessors results in significant capacity for parallel processing. Even more than that, however, they form high-speed proprietary local networks.

In the future, PBXs are likely to develop that have multiple processor complexes executing in lockstep and vote in a manner similar to that of Stratus Computer's fault-tolerant computers. This means that the processors are synchronized to a particular clock and execute instructions in parallel, comparing the results. Further, the bus designs allow the attachment of application processors for specialized (and usually extra cost) optional functions, such as voice and text mail, local network control, traffic analysis, protocol conversion, and the like.

■ *Local network architecture.* Fundamental to the newest designs (Ericsson MD110, CXC Rose/Western Union Vega) is that they are, or are capable of becoming, processing nodes on a geographically extensive local network. Among the older switches, Rolm's has implemented this capability in software, although the space and air-conditioning requirements of PBXs call into question its practicality.

In distributed cases, the PBX architecture resembles a packet network. Standing alone, the node is usually capable of handling about 200 ports. If the growth is in a contiguous area, the node grows by adding colocated processors. If the growth is not contiguous, a local network is implemented, connecting a network of physically distributed, homogeneous processing nodes. This can occur in either a physical ring or a physical star. In either configuration, the node forms the hub of the wheel in a hub-and-spoke connection with the station equipment.

By contrast, in the traditional data processing context, local networks have been most frequently used to replace terminal wiring, as in the typical installation of physically distributed Ethernet or IBM PC Network. Less frequent are local networks with multiple processors. Where network control and homogeneous multinode operating systems are involved, the processing nodes are more likely to be contiguous—as in Tandem's 6700 Fiber Optic Extension and DEC's Massbus—due both to the requirements for computer-room operating environments and to the signal/media limits. Much less frequent is the combination of network control and heterogeneous operating systems, as in Network Systems Corp.'s Hyperchannel and Hyperbus or host-to-host Ethernets.

Finally, among the distributed PBXs, the preferred local network medium to link the nodes is fiber-optic cable. In addition to fiber's environmentally resistant qualities, it is able to support the very high speeds (up to 50 Mbit/s) employed in internode communications.

■ *Local/global software.* Where the local network physically implements the distributed architecture, local/global software conceals the local network from the user. By "local/global," it is meant that typically all routine call-handling software is present in every node; specialized functions, such as calculating the least-cost supplier for long-distance calls, may not be present in every node. Take the case of a multinode network on a campus or a multibuilding corporation. If the user dials within the building, the call will be processed by the local node; if the user dials outside the building but within the campus, the call is processed both by the local node and, through the local network, by the node in the distant building; if the user dials long-distance, the independently trunked local node normally runs the routing program and processes the call; if the user dials long-distance and the local node's central office trunks are busy, the call can be routed to the next node through outside trunks for processing according to a pre-established succession and depending on the class of service assigned.

A secondary, but important, aspect of local/global software and nodal trunking is resistance to failure. When the errant backhoe severs one set of trunks, calls can be alternately routed to surviving trunks. An equipment-room fire cannot damage the critical main distribution board because there is none. Should the node maintaining the network's directory fail, its predesignated successor assumes the task. Although ultimately dependent on building power and batteries, the distributed switching nodes can deliver a more robust PBX by eliminating single points of failure.

■ *Digital throughout.* The appearance of the digital Northern Telecom SL-1 and Rolm CBX in 1975 pointed the PBX market in a direction from which it has not retreated. Nonetheless, these and subsequent switches have had to accommodate, primarily for reasons of cost, analog station devices. The inevitability of the analog/digital conversion, performed by codecs at the

1. Power requirements. *The actual figures in kilowatts for the 3,200-line switches in this diagram are SL-1, 7.5; CBX, 16.45; IBX, 45; MD 110, 8.5; NEAX 12.12; Vega,* 22.7; 20-20, 10. *Battery needs and types of power, not shown in this diagram, should also figure into switch cost-benefit analysis.*

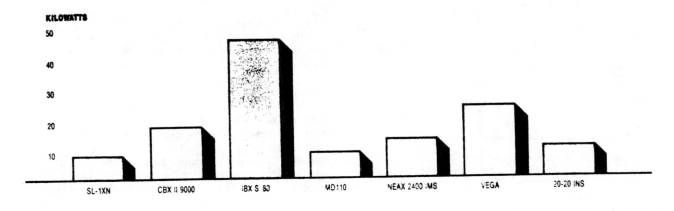

line cards, forced their designs to accommodate this mixed mode. With the falling cost of codecs, and of electronic station equipment in general, the newer switches are designed to routinely service digitized voice and data—usually, but not always, multiplexed—from the stations. While all manufacturers continue to support analog devices (commonly the ubiquitous 2500-type instrument), increasingly these devices are perceived as a technological burden that not only require special and sometimes costly accommodation at the node but also impose subtle penalties in operational flexibility. For instance, a Mitre study concluded that retaining 2500-type instruments can be costly if the overall mix of station instruments necessitates switch-side rewiring upon moves and changes. Further, each PBX OEM has handled the analog versus digital line-card design and cost, and the line-card impact on switch size and cost, in individualistic and unpredictable ways.

■ *Packet switching.* Similarly evolving is the technique of packet switching, which in the context of a digital switch performing time-division multiplexing (TDM) can be seen as further multiplexing the time-division multiplexing. The earliest accommodation of circuit switches to packet-switch technology appeared with gateways. Northern Telecom, Rolm, InteCom, and others ran circuit-switched, TDM data into a protocol converter to let terminals at user stations access X.25 public data networks. More recently, the Western Union Vega, Rolm CBX, and AT&T System 85 have used packetization to combine what is usually a digitized voice and ASCII data stream from station to switch. Next, apparently already in test with GTE and Western Union, is a hybrid packet/circuit switch where the digital voice is packetized and switched in a datagram fashion (DATA COMMUNICATIONS, "Packetized voice could move out of the laboratory soon," February, p. 45).

The data is handled similarly, internally employing—within the PBX network—a stripped-down high-level data link control packet that retains a virtual-circuit operations model. Although increasingly complex con-

ceptually, it may not be appreciably more costly given the falling price of microprocessor-based PADs (packet assembler-disassemblers). The returns are real: higher and higher data rates, fewer data errors, and most importantly, dynamic and highly efficient bandwidth employment.

■ *High-level languages, intensive software.* For performance reasons, the earlier minicomputer-based PBX software was usually written, all or in part, in assembly language. The use of assembly language kept software development, maintenance, and enhancement costs high. The coincidence of the high-performance microprocessors, the availability of powerful high-level languages such as Pascal and C, and the availability of integrated software development packages have effected dramatic changes.

Although technological conservatism is still present—AT&T largely rehosted its Dimension feature package from its analog pulse amplitude modulation (PAM) switch to the digital TDM hardware on the System 85—many of the new switches makers (CXC/Western Union, Mitel) choose to develop their software in these new languages because they come equipped with a variety of programming tools. These tool-intensive programming languages employ a full array of routine libraries, run-time debuggers, test data generators, cross-assemblers, and software configuration support aids. In the typical case this involves a programmer working on a DEC VAX running Unix or VMS coding in C and then downloading the object code to the Motorola 68010s. At least in theory, these new application software suites should be more sturdy, require less maintenance, and be easier to enhance than the older models. The real lesson, however, is probably different—that there is no substitute for thorough testing.

■ *Remote service.* At present, most PBX makers employ service technicians, either their own or trained by their company, at the customer site. These technicians are augmented by time-shared remote diagnostic services. Almost needless to note, these technicians

3. _Centralized space requirements._ _Where office space is more expensive than $17 per square foot, differences in bulk may be an important factor._

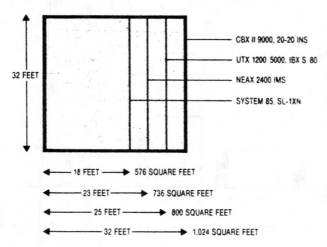

32 FEET

CBX II 9000, 20-20 INS
UTX 1200 5000, IBX S 80
NEAX 2400 IMS
SYSTEM 85, SL-1XN

18 FEET → 576 SQUARE FEET
23 FEET → 736 SQUARE FEET
25 FEET → 800 SQUARE FEET
32 FEET → 1,024 SQUARE FEET

COST OF OFFICE FLOOR SPACE:
WORST: 1,024 SQUARE FEET AT $17 = $17,408 PER YEAR
BEST: 576 SQUARE FEET AT $17 = $9,792 PER YEAR
DIFFERENCE: $7,616 PER YEAR

USUALLY HEAVY AIR-CONDITIONING; SQUARE FOOTAGE INCLUDES SPACE FOR BATTERIES

have automatically activated emergency generators, getting rid of the need for extra generators dedicated to serving a PBX, some also demand venting for batteries, ignoring the existence of the new, nonventing batteries.

In Figures 3 and 4, the actual square footage consumed by the multiple nodes can, in some cases, exceed that consumed by a single central switch. The reason is that several of the distributed designs (CXC Western Union) require front and rear access. Interestingly, in virtually all the distributed switches the door swings require more space than the switch itself, raising the possibility that future, perhaps military, versions may come with sliding doors. Third, the lower

power, low air-conditioning, and small space requirements of the newest distributed switches are aimed at coexisting with office workers. Sites where building management routinely powers down their air-conditioning on weekends may find calculating the difference in savings complex: keeping the air-conditioning on to sustain a cost-saving PBX or switching it off and running the risk of damaging the switch is not a simple trade-off.

User interface

The least revolutionary aspect of the new switches is often the user interface. Part of the reason is cost. The large number of units associated with station equipment leads to a large line item and, often, a go-slow attitude. Some will undoubtedly continue using 2500-type station equipment for these reasons, though it is hoped that few will wonder why switchhook-initiated features are seldom used. Perhaps no failure has been so complete as the effort to piggyback advanced features on the 2500 telephone. Well-meaning lectures, slide shows, and faceplates aside, people who, from their earliest days, learned that pressing the switchhook resulted in a dial tone are hard to convince that it is part of the proper procedure to press the button to transfer the Boss's "very important call."

At the high-end, the voice station cum microcomputer has several entries with GTE's icon-directed Omni-Action, a friendlier approach than Rolm's Cedar (a midget IBM PC) and Northern Telecom's Displayphone, a shrunken smart terminal.

The major failure, however, has been in imagination. Station features continue to be added in an add-a-feature, add-a-button manner, with more and more buttons—often of small size or doing dual duty and with tiny print—appearing in a reduced, desk-amenable "footprint." Unfortunately, the employment of all these features is not intuitive nor is it likely that potential users will willingly digest a user manual to access them. To employ another parallel from data processing, the add-a-function, add-a-button approach most resembles the IBM microcomputer and its large and complex keyboard. Contrast it with the software-driven, icon-and-mouse user interface em-

4. _Distributed space requirements._ _Some switches are so compact that their greatest need is to have adequate walk-around space to reach cabling in the back._

The Ericsson MD 110 solves this problem by pulling its cabling through the top. Compact PBXs might eliminate this need through the use of sliding doors.

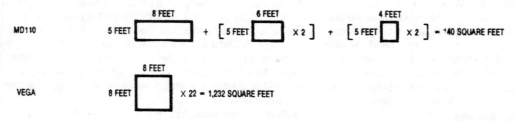

MD110   8 FEET
5 FEET [   ] + [5 FEET [   ] X 2] + [5 FEET [ ] X 2] = 140 SQUARE FEET
          6 FEET   4 FEET

VEGA   8 FEET
8 FEET [   ] X 22 = 1,232 SQUARE FEET

(OFFICE ENVIRONMENT, INCLUDES SPACE FOR BATTERIES)

APPROACH:
● DISTRIBUTED PRIVATE BRANCH EXCHANGES PLACED IN TELEPHONE CLOSETS
● SAVINGS AT $17 PER SQUARE FOOT RANGE FROM $9,792 TO $17,408 YEARLY

ployed in the Apple Macintosh; other alleged merits and defects aside, the Apple product clearly has the superior user interface.

In their rush toward baby buttons and midget screens, the station equipment designers have overlooked the operant natural language: the spoken word. Although it required a great deal of intelligence at the switch (a human being), the earliest telephone networks were of the signal-and-voice-recognition variety. Later, five digits had to be remembered with two additional digits artfully camouflaged by easy-to-remember—and sometimes famous—exchange names (BUtterfield, MUrray Hill). At the moment, a 14-digit international direct-dial call seems unfriendly enough. Although incorporating an immediately unlikely economic development scenario, can one imagine the numbers unleashed should the First World personal communications trend in the United States, Western Europe, and Japan—a phone at home, at work, and in the car—extend to India and China?

Despite the fact that they have not as yet been embraced by a major vendor, the best prospects for an easier to use, more successful user interface seem to lie in the area of incorporating some kind of voice response/voice recognition capability. Already some major PBXs use voice response "helps" when using station equipment features. Current voice recognition gear is more limited. Although they permit one to say "Call...," they can only be programmed to recognize one voice, that of the instrument's habitual user. This excessively personalizes the device; in experimental military applications using voice recognition, some machines will recognize voices that imitate the original users, or at least make a statisticallly accurate guess about who within the switch's voice-recognition directory the speaker's voice most closely resembles. This is a far cry, though, from a commerically available general recognition capability based on an entire company's staff directory.

A chaotic environment

Some aspects of the new, postdivestiture PBX world have already been alluded to—Western Union, a large common carrier, adding data features to CXC's Rose and marketing it as the Vega. The present PBX vendor marketplace can be characterized as anarchistic at best. Large OEMs such as AT&T, Northern Telecom, and Rolm directly market and service the customer through regional organizations. Ex-Bell operating companies like Bell Atlantic distribute other PBXs through OEM arrangements (InteCom and NEAX 2400 in Bell Atlantic's case) while their regulated side pushes the PBX-like capabilities of the latest upgrade (Centrex III) of their central-office location switches. And everywhere loom relatively small local distributors, typically with low overheads, that market switches obtained directly from the OEMs at large discounts. Not infrequently, the local distributor's combination of low overhead and aggressive pricing provides the buyer with the same switch at lower cost than from the OEM. Nor is it clear who is likely to provide the best service support: a small, aggressive local company, an ex-Bell regional company with multiple products and services, or a large, nationally or internationally based original equipment manufacturer

Unfortunately, the most technologically sophisticated PBX, particularly one launched by a startup company, may not be the one that wins bids. Since the financial status of the company, the ability to write a good proposal, and an established service record may often determine who wins a PBX sale, the Darwinian law of the jungle may be mitigated with PBX sales. Those who survive may not be the technologically fittest. More and more prevalent throughout the computer industry, this twist on the old truism is especially likely to apply to the PBX market. Just about every feature is available from every vendor, or can be bought off the shelf and added to any switch that appears lacking to a prospective customer. Comparison of features and functions becomes moot. PBX purchases become business decisions and not technology decisions, unless use of new technology drastically lowers prices: for example, the NEC switch is based on U. S.-designed Intel 8086 microprocessors manufactured in Asia and assembled by nonunion labor in Texas.

From the perspective of the first half of 1985, it is not clear where the revolution in PBXs is leading. Arguably, the U. S. PBX market is of insufficient size for so many switch vendors to make money. Nonetheless, a number of these switches (NEC, Ericsson, Mitel, Harris) are aimed at other national or world markets; success in the large U. S. market would be a boon but is not absolutely essential. Although the newest designs (Ericsson, NEC, CXC, Mitel) offer advanced features and should be cheaper to manufacture than conventional architectures, it is debatable to what extent they will carve out market share based on superior price/performance. Despite the fact that its present dominance remains largely based on the aging Dimension switch, the U. S. PBX marketplace remains an AT&T preserve, and new vendors must survive the IBM-like pull of AT&T. Nor is it self-evident who will best sell and service the switches. All three approaches—OEM, ex-BOCs, and distributors—offer both advantages and disadvantages to the buyer.

As for the revolution, the path of PBX technology seems most promising; what remains to be seen is whether the disorderly PBX market of today will postpone or accelerate it. As of this writing, Mitre expects proposals representing three different approaches: a Centrex III spin-off, minicomputer-based switches from vendors such as Rolm and AT&T, and microcomputer-based products such as CXC and NEC. ∎

Edwin R. Coover, who holds B. A. and M. A. degrees from the University of Virginia and a Ph.D. from the University of Minnesota, has been with Mitre Corp. since 1982 and is currently on staff in the Systems and Information Department. Michael J. Kane, who has been with Mitre since 1983, is a member of the technical staff of the Systems and Information Department. Kane earned a B. A. from La Salle College and an M. S. from American University.

NEW WAVE COMING IN DATA/VOICE SWITCHING

Frank G. McKay

TELEPHONE
ENGINEER & MANAGEMENT
October 15, 1984

Researchers are blending various developing technologies to produce, in the very near future, new switch capabilities.

New wave coming in data/voice switching

By Frank G. McKay

The 1980s are exciting times for researchers and engineers working in the telephone switching area.

Historically, telephone calls have been routed and connected by switching systems located in a central building in each community. And usually two different networks are needed—one for voice and one for data transmission.

But advancing technology and changing user demands are stimulating technologists to consider new ways of connecting telephone calls.

Much of the research being conducted by GTE today has changed our traditional assumptions about how calls can be switched. New developments in semiconductor technology and fiber optics will change the future structure of the telecommunications network. The result will be faster, smaller and more efficient systems for switching and transmitting calls.

Switching is one of the central technologies for GTE, because the company provides local telephone service in portions of 31 states, two Canadian provinces and in the Dominican Republic. It also offers long distance telephone service through GTE Sprint and

Frank G. McKay is vice president-product management, for GTE Communication Systems, marketing division.

worldwide data communications through GTE Telenet. In addition, GTE Communication Systems markets throughout the world a broad line of switching systems and other types of telecommunications equipment.

In 1983, GTE ranked second in the United States telecommunications industry in company-sponsored research and development expenditures, spending more than $262 million. To achieve and maintain leadership in telecommunications innovation, switching research plays an important role in GTE's overall research and development program.

Researchers at GTE Laboratories in Waltham, Mass., are exploring a variety of new frontiers in switching. Some are investigating optical switching, possibly the next step after electronic technology. Others are researching the switching properties of organic and inorganic chemicals, including polymers, and conducting mathematical analyses of network theory that may lead to rules for optimizing the design of switching systems.

Three switching research projects at GTE Laboratories offer especially exciting opportunities for the future:

• A 40-line private automatic branch exchange (PABX) on a single silicon chip no larger than a baby's fingernail.

• A "burst" switching system that can handle voice and data calls 20 times faster than conventional switching.

• An office information network that transmits voice, data and video messages with eight times the capacity of present networks.

In addition to GTE Laboratories, other GTE companies are conducting research specifically related to their products and services.

GTE Communication Systems, headquartered in Phoenix, Ariz., primarily conducts product-related research and development in the switching, transmission and business systems areas, in addition to marketing a full range of telecommunications equipment worldwide.

Current research in the switching area is focused on innovations and enhancements to the GTD-5 Electronic Automatic Exchange (EAX), a family of digital telephone switching systems, and to GTE's Omni family of PABXs.

Sending in bursts

Not long ago, data transmission was only a small part of the total telecommunications traffic. With increasing use of everything from home computers to credit card verification systems, data transmission needs are growing. Switching systems may require increased capacity to cope with these growing needs.

Two different types of networks often are used for voice and data communication because voice and data calls are frequently switched in different ways.

For voice calls, a communications channel is dedicated to the conversation for the duration of the call. This is called circuit switching.

On the other hand, data signals often are transmitted by packet switching. With this technique, data are divided into sections or "packets" of information that are fixed in length. Each packet is given a "header," with information of the data's origin and destination, and an "end," which terminates the packet. The packet then is sent to its destination. In packet switching, one channel is not dedicated to a call for its duration. Instead, data communications channels are shared by many users, because packets from many different sources can travel together like cars on a freeway.

GTE Laboratories is experimenting with a "burst" switching system which sends both voice and data messages in packet-like bursts. Unlike packet switching, however, burst switching packets have no limits to their length. And in burst switching, unlike circuit switching, a communications channel is dedicated to each burst of information only as long as information is being sent. As soon as there is voice or data silence, the channel becomes available for another message. In a three-minute conversation, the burst switching system might send 500 voice bursts.

The average voice message is two-thirds silence. By freeing communications channels during periods of voice silence alone, a burst system could send three times the number of voice and data messages possible today. In fact, the burst switching system can handle as many simultaneous calls as there are channels in the switching system.

A burst switching system consists of a number of link switches, installed in customers' neighborhoods, each handling 16 telephone lines. In turn, link switches connect to each other and to a hub switch, which connects all the channels in a given area. The link switch makes three types of switching decisions: it originates information and sends it through a channel; takes information from an incoming channel and switches it through to an outgoing channel, or terminates an incoming channel of information. Through this decision-making ability, the link switch determines a call's path through the communications network every time a burst is sent.

Not all voice and data bursts can be sent as soon as they are formed. Voice bursts have priority over data, since any delay in voice transmission interrupts and confuses the message. Thus, some data bursts wait briefly in a queue until a channel becomes available for transmission. Data bursts have low priority, since data can be somewhat delayed and stored in the switch's memory without harming its delivery.

Burst switching currently is available only in an experimental model at GTE Laboratories. The hub switch in the model is currently switching 1344 channels. This could provide service to more than 14,000 customers. Once additional development work is done, the burst switching concept should have a significant impact on the future telecommunications network. Parts of the burst switching technology applied to current GTE products could be available within the next five years.

A private automatic branch exchange (PABX) can provide service to facilities with a substantial number of telephone users and is usually about the size of a bookcase. In a research program at GTE Labs, technologists are developing a 40-line PABX on a single silicon chip. They expect their development work will progress to the point where they will be able to make a laboratory version of the chip by early 1985. In the future, the PABX on a chip might be used in a variety of different ways. For example, it could be placed in a telephone to provide switching functions and features. This "thinking" telephone would make decisions every time you place a call.

The PABX on a chip can be produced through the very large scale integration (VLSI) technique which allows hundreds of thousands of electronic components to be placed on a small semiconductor chip. Processor and memory capabilities that once required many electronic components on a large printed circuit board can now be shrunk to fit on a single chip.

Even with new VLSI techniques, several other hurdles will have to be overcome to build the PABX on a chip. First, in designing the chip, researchers have to work around space limitations. A complex structure of components and interconnections must fit into a small amount of space.

And with such a small amount of space, designers also face the problem

of actually fabricating a chip that is very complicated.

Lastly, the chip has to be "testable." Unlike a printed circuit board, the chip is so small that it is impossible to use a probe to check connections between components as is possible with solder connections on a printed circuit board.

GTE Labs is in the process of addressing these challenges, and will be building even smaller versions of the PABX on a chip until its goal is attained.

Researchers have created two innovative designs to handle other problems caused by the size of the semiconductor chip. To provide power to the 40-line chip, 40-line drivers are required. If traditional line drivers were used, they would send in enough power to burn up the chip. To prevent the problem, researchers developed an open-circuit line driver that provides a thousand times less power.

Space required for memory was another problem. To make switching decisions, the chip needs information about the destination of each message. This information must be stored in the chip's memory. Researchers developed a memory structure that will provide all the needed information and take up less space than other memory systems.

Once the GTE Labs researchers find answers to remaining design questions, they will experiment with uses for the PABX on a chip. The trend in telecom-

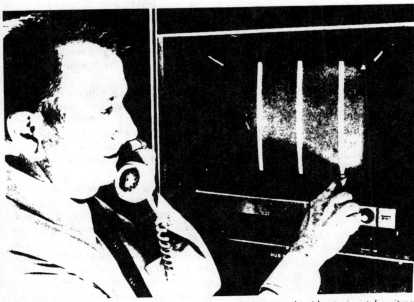

At left, Maurice Tierney, technical associate at GTE Laboratories, works with experimental equipment used to study principles of burst switching. In photo above, Tierney demonstrates a principle of burst switching on a display unit of the research model equipment.

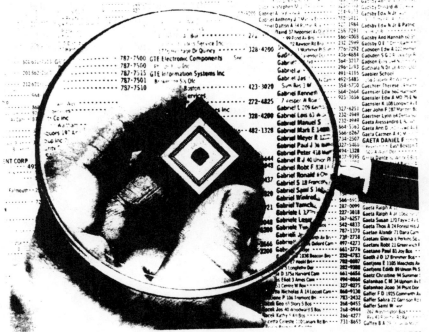

This chip, housed in a leadless package, some day will provide voice and data services to 40 PABX subscribers, based on research on single chip systems at GTE Laboratories.

munications networks is toward distributed control. That is, instead of putting a system's entire intelligence into a computer at one central location, capabilities are added to the rest of the system.

By using the PABX on a chip with switching or fiber optic transmission systems, it should be possible to increase communication speed. Distributed control also will offer greater survivability. Thus, if problems occur at the central location, the rest of the system won't be substantially affected.

The office information network being designed at GTE Labs would be a way to transmit voice, data and eventually video signals within an office, school, hospital or other large building.

Within this network, fiber optic links interconnect major office areas, called nodes, which in turn are connected to telephone data terminals and other facilities by conventional wire. Fiber optics require less space, are easy to connect and are not susceptible to electromagnetic interference, a common problem with telephone cables in factories. In the future, infrared or microwave technology could replace wires in connecting terminals and telephones to the fiber optic network.

Microwaves or infrared light waves could make possible the "wireless office" of the future. Each terminal and phone would be equipped to receive the waves and a source on the ceiling would send the signal throughout the room. By eliminating the wires to the terminals, it would be possible to substantially reduce the cost of putting in a PABX to transmit both voice and data.

Researchers have developed a new way to connect the nodes, or major office areas, in the network. Nodes are linked together by two optical fibers in the shape of a ring. Because of this dual fiber ring, messages are automatically rerouted over the remaining fiber if one fiber breaks. Even a bad node does not result in total system failure. By using a principle called optical continuity, made possible by a unique GTE design, signals automatically bypass faulty nodes.

It should be possible to insert the GTE 40-line PABX on a chip at each node to provide distributed voice and data switching. For example, 10 chips might serve 400 phones and data terminals in an office. The PABX on a chip might also take over some of the local switching functions of a centralized PABX, allowing GTE's office information network to serve a wide variety of communication needs.

GTE Communication Systems engineers in Phoenix, Ariz., and Reston, Va., are researching packet switching as an added capability for the GTD-5 EAX and for the OMNI series of private automatic branch exchanges (PABXs). Packet switching will allow these systems to simultaneously switch data and voice signals, using circuit-switching techniques to switch voice traffic and packet-switching techniques to switch data traffic. Packet switching technology uses switching facilities more efficiently, thereby providing cost savings and better use of switching capabilities.

In addition, packet switching enables the PABX to act as a go-between for a variety of different terminals and systems that may not be directly compatible with each other.

The PABX with packet switching capabilities will be an integrated business communication system to handle all types of voice, data and text communication generated by a wide variety of terminals.

The GTD-5 EAX with packet switching capabilities, currently under development, is one step toward an integrated services digital network—a public telephone network capable of satisfying the need for data information in the future.

The Communication Systems Division of GTE Government Systems develops and manufactures command, control and communication systems for government, military and commercial use worldwide. This includes equipment required to gather and process information for military commanders and relay command orders through the military organization in the field.

Switching systems for the military have to be mobile enough to be carried by trucks, airplanes, ships or people. They must offer privacy and security against wiretapping. And they must be reliable under less than ideal circumstances.

To achieve these goals, GTE Communication Systems Division is researching a number of technologies to make its message switching system, the AN/TYC-39, smaller and smarter. Artificial intelligence is one area being investigated. By writing software programs to "teach" computers to infer, research and think, the computer can prevent human errors.

If a computer can be taught to be very intelligent, it can defend itself by correcting the mistakes that people are bound to make. The computer can also help its users by employing its extensive memory to diagnose problems and make decisions. The ultimate goal is to make the computer think like a human being, not just do what it is told to do.

To expand and advance today's communications technologies, research is of primary importance to the telecommunications industry. These research projects and many others at GTE are providing new technological developments, processes and techniques that will bring about major improvements in the ways we communicate tomorrow. ☎

Reprinted courtesy of <u>Telecommunications Magazine</u>, August 1983. James Budway, Publisher.

The Next Generation in Business Communications

WILLIAM P. KARAVATOS
Ztel, Inc.
Andover, Massachusetts

Information has become an invaluable resource to any firm trying to survive profitably in today's highly competitive business world. Nevertheless, cost-conscious managers who need to integrate voice and data remain hesitant to make what amounts to a short-term investment in today's PBX. Rapidly changing technologies make products obsolete even as they are being installed.

The ever-present question is: Which type of system would be the best investment for now *and* for the future? Even when the best possible *combination* of voice and data equipment is purchased, the individual system may prevent the user from upgrading his system in the future. Existing voice/data PBX products of the centralized-architecture variety simply do not provide the necessary flexibility even for today's needs — and certainly not for tomorrow's.

The need to retrieve, share, and enter data is following a growth pattern similar to that of telephone systems, in terms of networking, features, and use requirements. Advances in telecommunications and computerized networking provide the means to build an integrated voice and data information-delivery network that combines the features and functions of a voice-switched computerized branch exchange (CBX) with the strengths and economies of a packet-switched data local area network (LAN). This new type of network will meet all voice, data, and image information requirements, grow as the company expands, and lower overhead expenses.

Ideally, this communication system would have an architecture which, when in place, would be essentially technology-independent. Thus, users would require only equipment additions to obtain any type of new, more beneficial capability. Instead of reinventing basic architectures every seven or eight years, communication-equipment manufacturers could concentrate on solving real worker problems.

This integrated network concept has been labeled by Ztel as a PNX, standing for Private Network Exchange.

THE PNX CONCEPT

By integrating the best features of the PBX/CBX with the best of the data LAN at the lowest, internal level of a system, Ztel's PNX provides more efficient communications — better voice operations than the traditional PBX/CBX and better data handling than a separate LAN or data switch. In addition, by applying innovative system design principles, the cost of providing this kind of service is similar to the traditional second and third generation PBX/CBX — in many instances, it is less.

The technology selected for moving information on the "backbone" network, or between distribution nodes, is the baseband token-ring LAN architecture as defined by the IEEE-802.5 Standards Committee. Ztel selected this architecture after evaluating both CSMA/CD (Ethernet) and token-bus standards; both were rejected because of performance inadequacies and/or reliability problems.

Figure 1 is a conceptual overview of the PNX architecture as it may be applied to a campus of three buildings or a single, large structure. Each LAN ring can be five miles in length, using standard coaxial cable. For longer distances or severe environments, low-grade fiber cable can be directly accommodated.

Software Architecture

The PNX's software architecture follows the International Standards Organization's Open System Interconnection (OSI) layered model for distributed networking. This means that real-time voice-switching applications are implemented with the method recommended and recognized for switching LAN data communications.

Other significant benefits associated with the software architecture include an easy interface with other local networks and the ease of software maintenance and update due to the extreme modularity and layering of the software structure. Software development was implemented in the high-level C language; this encourages value-added re-marketers to customize the basic Ztel system to meet other requirements.

The PBX in the PNX

Ztel uses PBX functionality as the cornerstone for the PNX, because reliable voice communication is the most critical element in business operations. The full set of standard PBX features have been implemented. Many additional packages — such as an online directory, call-detail recording, least-cost routing, messaging, system management, and attendant operations — are also included.

The PNX architecture is particularly appropriate for such real-time processes as interactive voice. Since token-ring LAN's have the unique characteristic of being highly deterministic, the worst-case delay for any access is fixed. Thus, performance under any load condition is known. Networks can be installed with complete assurance of the quality of service that each user will obtain.

The LAN in the PNX

The availability of true local area networking, which enables users to attach a desired vendor's product to the PNX, is a primary difference between this concept and other voice/data PBX's. The availability of chip sets for the IEEE-802.5 token-ring LAN in the near future, and IBM's probable announcement of the use of this standard with its forthcoming LAN, will mean wide availability of devices that use this networking method.

The representative ring LAN in **Figure 1** can be used to transport data from any non-Ztel user devices that are attached directly to any ring. Simple wall connectors, defined by the 802 standard, are used.

The desirability of the LAN data rates — which are greater than the 19.2-kbps maximum limitation of RS-232C connections — is apparent when one considers such basic applications as graphics and file services to workstations and personal computers. To transmit a typical 100-kbyte file from a file server to a personal computer for processing requires at least 3.3 minutes if the popular 4800-bps transmission speed is used. The PNX can deliver the same information in about one tenth of a second.

Other vendors' 802 ring LAN's are integrated into the PNX at the lowest internal level, the media level, merely by directly connecting the two networks. No costly gateways (with a resulting loss of functionality) are required. Thus, devices in each of the networks not only can communicate within their respective networks, but also have direct access to the features resident in the PNX's system or its applications software. This constitutes total "feature transparency" across the network. Devices that are attached directly to a PNX ring also have this same capability.

The PNX has numerous other attachment methods for the full range of data devices currently available, and for the types of environments encountered in businesses. Ztel's digital telephones have an optional port for attaching conventional devices at rates of up to 56 kbps. Remote data servers attach devices via telephone wire to the PNX network. Remote multiplexers concentrate dense configurations of data devices onto a telephone-wire connection to the PNX. The devices thus connected have different physical characteristics than those connected directly to a ring LAN. The PNX converts their transmission characteristics as required.

INSTALLATION FACTORS

The PNX approach has been

designed to provide a high "comfort" factor to buyers in today's marketplace. Those users in the forefront of business automation embrace LAN concepts and are quite comfortable with them. However, a reasonable percentage of users need the reinforcement of total marketplace acceptance of LAN's before embarking in that direction.

For those users, the PNX can be installed to appear physically as today's second and third generation voice/data PBX/CBX. The LAN rings and function processors (FP's) can be incorporated into central equipment cabinets located in the

telecommunication room. All building wiring is then accomplished by the familiar quad telephone wire via main distribution frames and intermediate distribution frames, using existing or new wire. The "insurance" for the user is that the LAN and FP's can be redistributed at any future time, simply by moving subsystems and adding some cable.

FUNCTION PROCESSORS

The PNX system consists of a series of token-passing rings, function processors that are attached to one or more rings, and a variety of telsets, workstations, and terminals. A function processor is the generic designation for a node, which can operate either as a distribution point for user stations and transmission facilities or as a provider of user-level applications.

The desired level of "non-blocking" is achieved by adding rings into the network. Since the network is self-adapting to the quantity of rings present, malfunctioning rings are automatically ignored by FP's and traffic is directed through oth-

ers. To further increase network reliability, FP's can self-configure to function as complete, stand-alone, voice/data PBX's if all interconnections between nodes become unavailable.

The added benefit of this type of architecture is that it will not disrupt services during growth. A PNX is cost-effective from 50 lines up to the largest installation, due to the extreme modularity of the architecture. No large-scale equipment replacement is ever required to expand beyond certain line-sized barriers, as is often the case with some second and third generation systems.

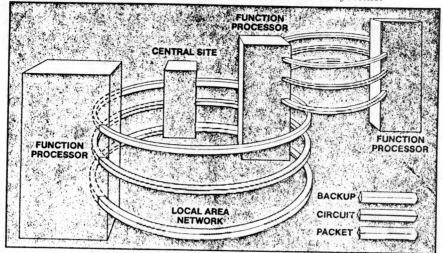

Fig. 1 PBX/LAN topology.

Two types of function processor control the Ztel PNX — application processors (AP) and ring processors (RP) (**Figure 2**). Ring processors provide interfaces to the network and local switching for both telecommunications and data communications equipment. The application processor is the vehicle for providing value-added services. Users can also add their own type of function processor directly to the network if any unique operations are required. The processors can be located anywhere in the network, and virtually any quantity can be accommodated.

In contrast, the star architecture of the previous generation PBX (see **Figure 3** for a comparison of architectures) requires that any new features be added to a single central processor. Software integration becomes difficult, and the central processor eventually loses the ability to perform its primary tasks.

The ring processor is a switching node which supports general-purpose ports, which may accommodate lines, trunks, or service circuits. Each RP contains its own redun-

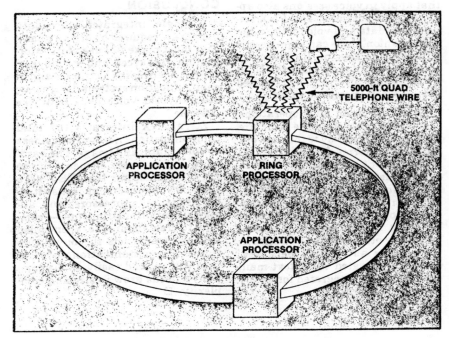

Fig. 2 Function processors for the PNX.

dant processing elements for supporting the switching functions. The RP can also function as a stand-alone voice/data PBX.

Attributes, or class-of-service, of each port are controlled by a relational data-base management system, which provides automatic configuration, source/destination port-type conversion, and resource control. Each physical port of an RP attaches to one 5000-foot standard quad telephone wire, which simultaneously carries voice and data.

The AP functions as a shared resource to numerous RP's and their attachments. Any number of AP's may be included in the system. They may be supplied by Ztel or by another party to solve a problem unique to an industry. Since the software has been written in the C language and since the OSI model is rigorously followed, OEM buyers can program specific applications easily. Examples of Ztel-supplied functions in AP's are least-cost routing, mail, call-detail recording, online directory, network management, X.25 packet assembly/disassembly (PAD), etc.

RELIABILITY

Ztel provides three levels of reliability. The "basic" level contains sufficient redundancy to attain minimum required system availability. At the "advanced" level a malfunction anywhere in the entire system can disable only 16 ports; everything else continues to operate at specified

levels. At the "extended" level, used only for critical portions of the network, any single malfunction anywhere in the system will not cause a service interruption to the ports for which this level of protection was installed.

The PNX utilizes a unique form of load-sharing to maintain system operation in case of any failure; there is protection even if multiple items fail. For example, if an AP fails, there are others that are operational and are sharing the processing load, typically not at a high utilization level; these immediately accept the added service requests for the failed unit. With this concept, multiple AP's can fail and all other available resources will be used to maintain operation.

The customer, therefore, has the option of installing whatever protection level is judged appropriate for his business.

To make sure that the system is available at all times, an extensive set of diagnostic aids is supplied for detection of hardware and software faults. In case of equipment malfunction, repairs are made online. Maintenance support is included in the PNX to allow fault isolation and correction locally or remotely from either an on-network or an off-network site. Audible and visual indicators alert users to the existence of major and minor system faults.

TELEPHONE STATIONS

Since 80 percent of all communication in the office environment is voice communication, Ztel has paid specific attention to the design of the telsets available with the PNX. A family of stations serves the full requirements found in businesses.

The telephones retain the familiar key-set capabilities to which users have become accustomed. Additionally, single-button access to calling features — call forwarding, call back, speed dial, last number redial, etc. — are standard. As a further aid to administration of the system, the user can implement or change any calling feature at his telset, including line assignments.

A unique feature offered with the manager-level telset is the ability to eliminate totally the "pink slip" messages that flood the typical office. The set contains a 40-character display and 28 feature keys. By assigning some keys to such popular messages as "call me back" or "returned your call," a caller can cause the selected message to be sent with a single key depression. Also, a message center, or anyone with a full

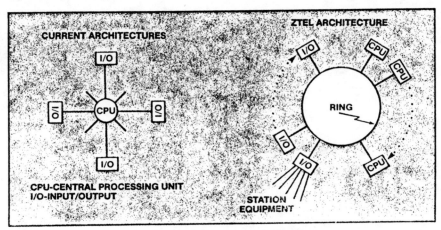

Fig. 3 Comparison of system architectures.

279

keyboard/CRT connected to the network, can construct custom messages that are scrolled on the receiver's 40-character display. The display is also used for such aids as announcing the calling party by name, call duration, etc.

A second version of the digital telset contains 12 user-assignable feature keys and no display. Either of the two feature-key instruments can contain as an option an RS-232C or RS-449 data interface for transmission at speeds of 19.2 kbps or 56 kbps asynchronously. A third version of the digital telset provides four user-assignable feature keys. The transformation of the data into packets begins in the telsets.

DATA OPERATIONS

Most station calling processes that apply to voice sessions also apply to data sessions. The PNX implements data and voice calls in the same fashion. Calls can be processed either by the keys on the telset, by the keyboard of an attached CRT, or by a text stream from an attached system's communication port.

The basic set of data operations provides the user with the structure required to add safely to the network all the data forms found in typical businesses. Examples are X.25 PAD, data call-detail-recording processing, line concentration, conferencing, level 1 protocol conversion, messaging, auto-baud and code detection, least-cost routing, external access, and many others. With the basics in place, other automation packages, such as integrated voice/text mail, can be directly integrated for use by all attached users.

CONCLUSION

A single, integrated voice/data network is required by any company that must make full use of its information resources and employees' time. Advances in electronics and in communication systems design now make hybrid voice/data networks possible.

Ztel's PNX integrates the total communications needs of a business — voice, data, and image. The hardware and software elements of the PNX — including the IEEE-802.5 LAN, OSI structure, packet switching, function processors, and all-digital station instruments — combine to create a unique, open-ended network which lets the buyer configure a communications system that will expand easily as his company's needs grow in size and complexity. □

Reprinted with permission from Business Communications Review, January-February 1984, pages 9-19. Copyright © by Business Communications Review Enterprises, Inc. 950 York Road, Hinsdale, IL 60521.

NEW PRODUCT

Ztel, Inc. and CXC Corp. Announce Voice/Data Switches

by Ronald A. Frank

For several years users have been hearing that the carrier system of tomorrow is the Integrated Services Digital Network. This promised digital pipe reportedly will support a variety of business communications needs with plenty of bandwidth and applications flexibility.

If an on-site equivalent exists for the ISDN, it may well be the new PBXs now under development by two small firms at either end of the country: Ztel Inc. of Andover, MA, and CXC Corp. of Irvine, CA. Both of these venture capital, technology-driven companies are convinced that they can bring distributed, integrated voice/data switches to the user before the established PBX suppliers.

Their systems will be all-digital from end-to-end within the customer's premises. They will incorporate local area networks into the switching capability, and they will handle voice and data traffic over facilities without the blocking and other architectural limitations of older PBXs.

Both CXC Corp. and Ztel Inc. started their product development in 1981 and both are slated to have systems for first delivery late this year. Despite press introductions and plans for prototype system demonstrations, neither company has installed an operational system at a customer site.

Even though the companies have a limited track record, the features claimed for the switches sound impressive. Both systems are being custom designed with the latest chips and proprietary integrated circuitry. Ztel has installed an extensive CAD/CAM system, which is credited with savings as much as two years over more conventional circuit design methods.

The CXC Rose and Ztel Private Network Exchange (PNX) come with a variety of digital telsets. CXC calls theirs the Personal Teleterminal (see Figure 1). There are two models which include a variety of control keys, indicator lights, etc. The larger model handles eight separate lines and has an 80-character LCD display for messages. Ztel has three telsets with call forwarding, speed dial, and similar features. The "management level" telset has a 40-character LCD display and also handles messages (see Figure 2). All telsets include RS-232 interfaces and can switch back and forth between data and voice calls, with both types of traffic handled simultaneously when required.

Both switches allow a user to start with standard analog phones, but obviously the digital features can be fully implemented only after a customer switches to the proprietary telsets that come with each system. These proprietary telsets operate over standard two pair cabling like the

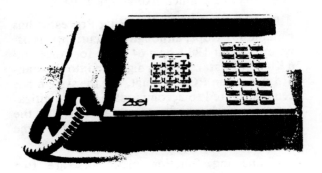

Figure 1. The CXC Personal Teleterminal.

Figure 2. The Ztel management level telset.

existing generation of intelligent phones such as the Displayphone from Northern Telecom. A substantial percentage of digital telsets will increase the cost of the system, but Ztel spokesmen claim that station for station the cost of their switch will be lower than earlier generation PBXs. Further they say that the new PBX system will have a longer useful life, ranging from ten to twenty years, thus making the installation more cost effective.

The Ztel PNX uses a baseband token ring local area network architecture that is compatible with the IEEE 802 standard. Information is sent around the ring in sequential order and devices transmit and receive information as tokens circulate through the network. Each Ztel ring operates at 10 Mbps and multiple rings can be interconnected.

A typical PNX configuration is controlled by function processors that are used as distribution points for user stations and transmission facilities, and as processors for applications. Two types of function processors are used. The Ring Processor (RP) interfaces up to 512 information devices, which can include Ztel digital telsets, standard analog telephones, personal computers, data terminals, mainframe computers, facsimile units, "industrial processes," and compressed video. Something like 256 Ring Processors can be connected to the system. The Application Processor (AP) supports system management, voice/data call processing, on-line directory, electronic mail, least cost routing, call detail recording, protocol conversion, and X.25 packet assembly/disassembly (PAD).

A system requires at least one Ring Processor and one Applications Procssor. In small systems, these may be housed in the same cabinet, and connected via rings also located within the cabinet. In this configuration, the PNX looks very much like a typical PBX.

Each termination on the Ring Processor has available 192 Kbps, although Ztel currently is using only about 128 Kbps: 64 Kbps for voice and 56 Kbps for data. Voice is digitized into the standard American PCM format, but the system has been configured to accommodate lower digitization rates if this becomes cost-effective in the future. Connections betwen devices on the same Ring Processor are non-blocking.

Connecting function processors are two types of coaxial rings: a circuit switched ring and

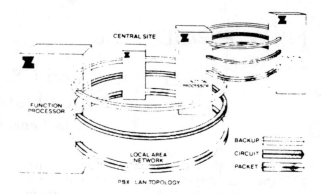

Figure 3. Ztel PNX function processors and ring architecture.

a packet switched ring as shown in Figure 3. The rings provide universal communication paths between the various terminals, host computers, Applications Processors and other devices on the system.

The circuit switched ring is used for transmitting voice and higher speed data in a transparent mode up to 56 Kbps between devices connected to different Ring Processors. This ring uses a packet switched protocol with fixed length packets that is very similar to a time division multiplexing approach. The ring can accommodate 110 "time slots" or simultaneous conversations, and multiple rings can be configured if more capacity is needed. For the system to be totally non-blocking, half as many time slots as the total number of voice and circuit switched data terminals on all Ring Processors would be needed.

The second ring is an 802 token passing ring for transporting packetized data and system control information. Data can be from either RS-232C or RS-449 devices or IEEE 802.5 LAN Units. Data is packetized at the digital telsets, which serve as network entry points, thus saving considerable bandwidth, according to Ztel. The company does not foresee needing more than 10Mbps and, thus, multiple packet rings to handle traffic requirements on the system.

The PBX is configured with a third "backup" ring that can be used as either a circuit switched or packet switched ring if one of the primary rings fails.

The CXC Rose has a more complex networking structure. It consists of two LAN systems that employ different technologies. A 33 Mbps circuit-switched 802 token passing ring serves as the main network. This broadband ring actually has a total bandwidth of 50 Mbps but 16 Mbps is reserved for "future token ring implementation." The Rose system also has a 10 Mbps Ethernet baseband system that is used for signaling, for control interfaces between nodes, and for packet switched data. From the user's standpoint, the CXC switch is primarily a broadband system that can handle 512 simultaneous voice conversations or over 4,000 full-duplex data sessions at 8 Kbps or less. Using a proprietary RF modem and standard CATV cable technology, the bandwidth is dynamically allocated for either voice and data in 8 Kbps increments. Figure 4 is a schematic diagram of the system.

Individual Personal Teleterminal telsets are supported in a processing node which can handle up to 192 non-blocking ports. Each full-duplex port can handle up to 192 Kbps transmission for simultaneous digitized voice and data over standard two pair cable. Associated with each port is a VLSI switch that "connects" the port to the 33 Mbps broadband ring for communicating between nodes. The signaling information to operate the switches is passed over the baseband

ring, so that the full capacity of the broadband ring is available for information. There is no overhead loss. In addition, because the information is switched off the ring at the destination node, broadband ring capacity for any conversation is required only between the originating and terminating nodes.

A broadband ring can support up to 64 single nodes, or 32 fully redundant nodes in a cluster. Thus, a full-blown system can accommodate up to 12,288 ports. Clusters can be bridged together to support more than 50,000 subscribers in a single integrated voice and data environment.

A node contains four 68000 microprocessor based CPUs, and individual nodes can be configured as gateways to interface to other networks. These can include analog telephone connections, X.25 packet nets, digital communications links, and Telex TWX facilities. A microprocessor controlled Network Interface Module (NIM) dynamically allocates bandwidth in 8 Kbps increments on the broadband ring and it can vary the ratio of intra-node and inter-node voice conversations and data sessions. CXC claims that this technology allows the Rose

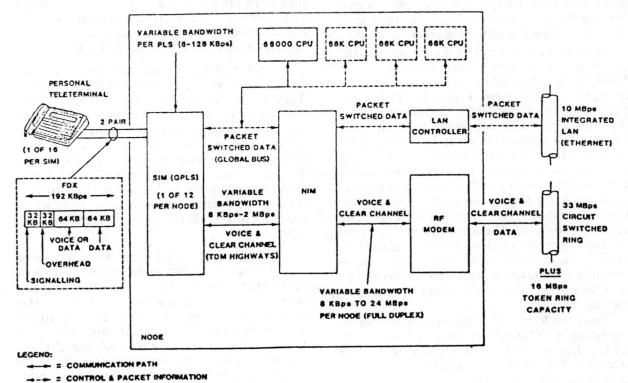

Figure 4. CXC Rose node schematic diagram.

ZTEL/CXC Comparison Chart

	Ztel PNX	CXC Rose
Bandwidth	10Mbps ring	33Mbps ring, 10Mbps Ethernet
Throughput at telset	56Kbps data, 64Kbps voice/data	64Kbps data, 64Kbps voice/data
Voice Encoding	PCM	PCM
Switching	circuit, packet	circuit, packet
LAN medium	coax, fiber	coax
Interfaces	RS-232, 449, X.25 PAD	RS-232, X.25 PAD*
IBM LAN compatible	yes	yes
Sales channels	end user, OEM	OEM
Data speeds (max.) at telset	19.2Kbps asynch, 56Kbps synch	19.2Kbps asynch, 128Kbps synch

*Telset has built-in PAD

system to have fully distributed communications while retaining the advantages of a centralized PBX.

Although CXC has embraced both Ethernet and ring implementations, the ultimate implementation may concentrate on the ring topology using the available spare broadband capacity. Whether Ethernet remains may depend to a large extent on what the IBM LAN product looks like. Both CXC and Ztel emphasize that full interface compatibility with the IBM local network will be key to their market strategy.

Both of these switches have architectures and bandwidth to support a range of applications. Some of these are still in the future and depend on technology advances. Facsimile will be compatible via digital fax devices and teleconferencing must wait for commercially available video compression products that can be interfaced effectively. Thus, the first installations will stress conventional telephone and data applications. CXC and Ztel each feel that messaging is a key application for the business user. Ztel concentrates on text messages but CXC has both voice and text capability. The Rose system is designed to allow text annotations to be appended to voice mesages stored in the system. The success of this type of advanced function will depend on user education and a willingness to try new message formats.

Both the Rose and PNX stress user friendly operation. CXC combines the use of icons on its LCD display with prompts to assist the operator. Ztel makes a point of single button access to complex operations.

Each system embraces the software structure of the seven-layer Open Systems Interconnection model favored by CCITT and other international standards-making agencies. The Rose can be programmed using Pascal while the PNX uses C language. Each apparently uses a Digital Equipment Corp. VAX processor for software development. Nevertheless, these systems appear to have complex software requirements and applications development will probably be limited to OEMs such as telcos or interconnect companies. CXC expects new software to be industry specific packages, such as the hotel features now available on many PBXs.

In place of traditional PBX attendant consoles, these systems employ CRT terminals. They include the ability to answer calls at a central location when called parties are not available. In such situations, these systems emphasize the generation of electronic messages, rather than the standard written "pink slips," as Ztel describes them. Additionally the digital switches provide each telset user with an array of message generation options, including telset-to-telset messages, that bypass the attendant console.

Both CXC and Ztel claim extended useful life through modular architecture, high level redundancy, and flexibility to adapt to changing user needs. Only a few installations of either system will be completed this year. Significant user feedback will have to wait until 1984. It will be interesting to see whether CXC and Ztel can get their systems up and running without the substantial problems experienced by other manufacturers who have announced new PBXs in recent years.

ERICSSON MD110
INTELLIGENT NETWORK

The dependence of modern business on information systems has caused a reevaluation of all information processing elements, particularly communications systems. One of the greatest areas of interest is the immediate potential for productivity gains and hard dollar savings inherent in an efficient telecommunications system. Successful organizations are realizing that effective information handling requires planning and implementation with a total communications system approach which

- considers both voice and data transmission,
- anticipates growth and change,
- integrates existing equipment, such as PCs, and
- facilitates control of communications costs.

The MD110 Intelligent Network is a digital communications system designed to function in business environments and to serve the needs of organizations today and in the future. The MD110 offers many benefits, including:

- streamlined voice communications and enhanced data communications over a single twisted pair of wires,
- flexibility through modular hardware and fully distributed functions and control,
- design which increases reliability and survivability,
- cost containment through control of station features, traffic routing, and usage of existing equipment and lines,
- dramatically reduced power and environmental costs leading to significantly reduced life cycle costs, and
- expansion of capabilities through standard interfaces (ISDN compatibility).

The MD110 offers the intelligent solution to today's communications requirements.

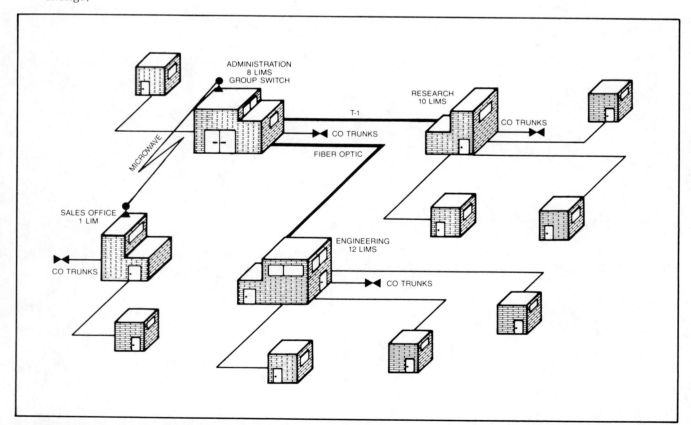

The MD110 Intelligent Network features fully distributed architecture.

ERICSSON MD110 INTELLIGENT NETWORK

The MD110 Intelligent Network is a totally modular, digital communications system capable of transmitting voice and data simultaneously over a single pair of wires. Simplified system design, higher efficiency, and significantly reduced operating costs are all benefits of the MD110. The MD110 is not tied to a specific system configuration. Instead, its flexibility allows each customer to build a system to meet their unique communication requirements with the versatility to change and grow in both size and functionality. The inherent flexibility of the MD110 allows efficient use of existing wiring, terminal equipment, and transmission facilities. The small footprint of the MD110 and its dramatically low power requirements result in an overall lower life cycle cost of the system. The MD110 Intelligent Network provides a cost effective, reliable communications network with processing power located where it is needed.

Modularity

The MD110 uses modular hardware and software in a building block approach that provides flexibility of system configuration. In its simplest form the MD110 consists of a single entity called a Line Interface Module (LIM). As the system grows in response to traffic demand, more LIMs are added and interconnected by a Group Switch. The interconnection is accomplished easily using links carrying a standard Pulse Code Modulated (PCM) format. The PCM link media can include copper wire, coaxial cable, fiber optic, or microwave (CCITT or North American T-1 format). The LIM and Group Switch can be located wherever required, building a single trans-parent communications system from 200 to 10,000 lines distributed over a geographic area as wide as 1200 miles.

Software Sophistication

The MD110 software is modularized into functionally-related Program Units for optimized memory usage and simplified database changes. Each LIM is equipped with the software required for fully independent call processing within that LIM. Additional Program Units are installed in each LIM to build multiple connections between LIMs and to access operating/service software on an as-needed basis. Each Program Unit has its own database so that software faults can be isolated to individual program modules. Changes in software modules can be implemented without impacting the entire operating system. The MD110 modular software allows installation of new features by simply entering appropriate commands into the MD110 database.

Physical Simplicity

Operating overhead normally associated with a communications system is significantly reduced with the MD110. Each LIM requires less than two square feet of floor space, a fraction of the size of other systems on the market. The LIM is cooled by natural convection and requires no special air conditioning. Electrical operating costs are further reduced by the low power requirement of only 2 watts per port. The reduced floor space and operating requirements of the MD110 result in a lower initial capital expenditure and dramatically reduced life cycle costs of the system.

Networking

The MD110 is ideally positioned to address sophisticated networking demands of organizations through networking capabilities which incorporate technological advances and industry standards.

The MD110 offers future networking applications through extensive ISDN capabilities. The first ISDN release will be available in 1989 including features such as Uniform Numbering Plan, basic call and routing over the Primary Rate Interface, private network access, and digital subscriber line.

An additional MD110 network offering that is ETN compatibility will protect investments in existing communication networks. ETN features included in this release are Traveling Class Marks, Facilities Restriction Levels, Network Class of Service, Authorization Codes and Queueing on distant trunk groups.

Through active participation in CCITT and other telecommunication standards committees, Ericsson will continue to deliver product enhancements ensuring that the MD110 Intelligent Network remains on the leading edge of technology.

SYSTEM ARCHITECTURE

Line Interface Module (LIM)

The LIM is a microprocessor-controlled module equipped with all elements (processors, memory circuits, service circuits, line circuits, and trunk circuits) necessary for call processing. Switching is performed by an internally non-blocking, time division switch using 512 time slots with a throughput of 32.8 Mbps. A LIM can function as an autonomous communications system or as an integral part of a larger MD110 Intelligent Network.

The LIM consists of the following five subsystems:

- The Line Signaling Subsystem contains interface circuits (connecting the LIM with voice/data devices) and the service circuits for call processing functions.

- The Switching Subsystem permits two-way communication between the stations, trunks, and other equipment.

- The Processor Subsystem directs all functions of the LIM processor.

- The Input/Output Subsystem provides the man-machine interface for system administration and maintenance.

- The Service/Maintenance Subsystem monitors system hardware and software, detects faults, generates alarms, and aids in fault clearing.

Each LIM supports up to 172 voice and 172 data ports. The universal backplane wiring of the LIM allows any available card slot in the cabinet to be equipped with any type of line or trunk circuit card. Analog and/or digital lines and trunks can be connected to any LIM in the system. Each LIM can have up to four PCM links (120 channels) connecting it to the Group Switch.

LIM operations are controlled by the LIM processor, which is directed by the software Program Units loaded into memory during system initialization. The independent processing of each LIM in the MD110 yields more efficient processing and increased call handling capability. Multiple calls can be set up simultaneously and, if required, the LIM processor can access resources (i.e., trunks, service circuits) under the control of another LIM processor. At maximum capacity, a LIM handles up to 2,160 busy hour call attempts.

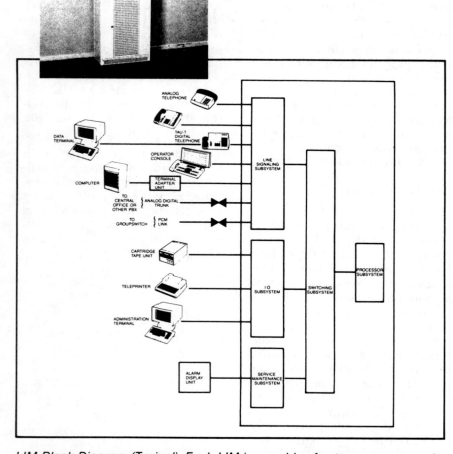

LIM Block Diagram (Typical). Each LIM is capable of autonomous operation.

Group Switch

When a system consists of three or more LIMs, a Group Switch is required to connect the LIMs. The Group Switch is a nonblocking, time division switching matrix which interconnects the time slots on the PCM links that it serves. Thus, the Group Switch makes it possible for a device served by one LIM to communicate with a device served by another LIM. The Group Switch is a passive device which is controlled by the LIMs. A fully-equipped Group Switch consists of eight cabinets (Group Switch Modules) and can accommodate up to 248 PCM links.

PCM Links

The PCM links provide up to 32 serial channels that carry PCM encoded voice or data in a CCITT format at the rate of 2.048 Mbps. A full 64 Kbps of encoded voice or data signals are carried on each of the thirty channels. The remaining two channels are reserved for synchronization and signaling. The PCM links enable the LIMs, which are essentially standalone nodes, to function as an integral system with full feature transparency. Inter-LIM traffic is dependent upon the number of PCM links provided between the LIM and the Group Switch.

Reliability

MD110 reliability is attained through distributed processing and modular redundancy. Since full processing power is distributed to all LIMs, a hardware or software fault impacts service only within the LIM where the malfunction occurs rather than affecting the entire system. In addition, each LIM is capable of operating

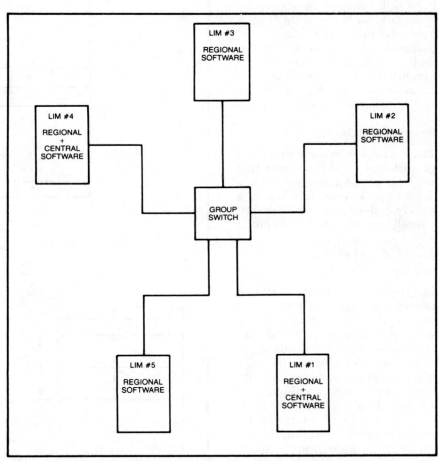

MD110 reliability is enhanced through redundant software.

autonomously in the event that it is cut off from the system. MD110 software is divided into central and regional operating segments. Each LIM contains all the necessary regional software to ensure complete control of call processing within the LIM (intra-LIM communications). Inter-LIM communications are controlled by central

software, which is accessed as required by a LIM originating call processing to another LIM. Central software is duplicated in multiple LIMs for added reliability.

In addition to redundant software modules, the MD110 system builds modular redundancy into all hardware units including the LIM,

PCM links, and Group Switch. The LIM is equipped with multiple PCM links each terminating in a different Group Switch Module. Loss of a Group Switch Module or PCM link interface device results in an automatic rerouting of calls over alternate paths. Distributed processing and modular redundancy effectively ensures survivability of the system and reduces outage impact to the smallest number of users.

Environmentals

The MD110 hardware is contained in uniformly sized cabinets that can be arranged in single or double (back-to-back) rows to adapt to a variety of floor plan requirements. Overall cabinet dimensions are:

> 23.6" wide
> 11.8" deep
> 82.7" high

Electrical components with low heat generating characteristics and low power requirements are selected to ensure cool and efficient operation of the MD110. A fully configured LIM requires only 360 watts of power. Stringent component selection combined with advanced cabinet design eliminates the need for cooling fans for heat dissipation, resulting in virtually silent operation. In short, the MD110 is packaged in attractive cabinetry for installation in the typical office environment. Compact size and low operating costs are distinct advantages which lower the long term costs of running the system.

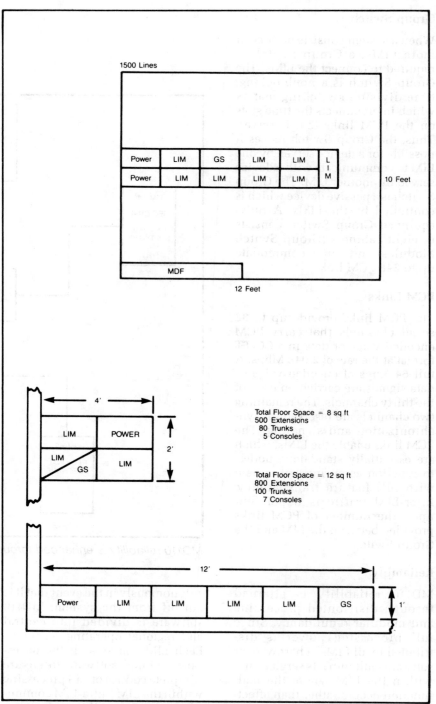

The MD110's small footprint allows customized equipment layouts.

INTELLIGENT SOLUTIONS FOR VOICE COMMUNICATION

While the MD110 Intelligent Network offers many features, voice communication remains a primary function of the system. The MD110 offers a spectrum of user features designed for more efficient communications and increased productivity. Statistically, 15% of all telephone calls are not completed on the first attempt because the extension is busy. Features such as Automatic Call Back, Executive Intrusion, or Call Waiting offer a solution to the frustrating busy signal. An additional 20% of all calls are not answered on the first try because the person is away from the desk. Activating Call Diversion, Follow-Me, Paging, or Automatic Call Back-No Answer provides alternatives to unanswered telephone calls. The MD110 user features are easily activated by a single key on a digital telephone instrument or through the keypad of an analog telephone set. Calls are initiated and completed efficiently, saving time and improving employee productivity.

Efficient utilization of MD110 features and facilities is guaranteed through full feature transparency and shared resources. The MD110 ensures that transmission and feature activation are transparent to all users regardless of their geographic location. With a uniform station numbering plan and a uniform dialing plan, the user can dial the same feature access codes from any extension on the system. System facilities such as voice messaging and dial dictation are shared and universally available throughout the MD110 using peripheral equipment interfaced to a single LIM. Operators can be centralized or decentralized throughout a system.

Operator Console

The MD110 operator console consists of a Visual Display Unit (VDU), a detachable keyboard, and a handset or headset. The keyboard is equipped with non-locking keys for selection of the various call handling capabilities available to the operator. The VDU has three 40-character Liquid Crystal Displays (LCDs) which provide call progress information, incoming call queue status, alarm data, and console status information. A single pair of wires connects the console to a standard digital line circuit in the LIM, providing both power and two-way digital transmission.

Telephone Instruments

A complete line of telephones is available for use with the MD110, ranging from standard analog telephone sets to state-of-the-art digital telephone instruments. All Ericsson telephone sets feature fully modular instrument design from the handset to the wall receptacle. Ericsson analog and digital telephone sets transmit over a single pair of wires using line supplied voltage.

E100

The Ericsson 100 family of 2500 type (DTMF) telephones offers a choice of colors, with optional tap button and message waiting lamp. Both desk model and wall mounted versions are available.

E200

The Ericsson 200 is a basic digital telephone which supports up to three voice lines. The E200 is equipped with five programmable feature keys, Light Emitting Diode (LED) indicators, a transfer key, and on-hook dialing capability. Additional features, such as Last Number Redial and Abbreviated Dialing, are accessed by dialing the system access code.

E350

The Ericsson 350 is a digital telephone instrument equipped with twelve buttons located at the right side of the telephone. Each button is non-locking with an associated LED. Eight buttons are available for assignment to features or line appearances. The remaining four, as well as the four adjacent to the handset, have permanently assigned functions. A microphone, loudspeaker, and volume control provide speakerphone operation. An LED located near the volume control illuminates when the E350 is in handsfree mode. A 16-character LCD provides call details on both voice and data calls.

The Ericsson 350 can be upgraded by adding a Terminal Adapter Unit for Telephones (TAU-T) for data communications.

E450

The Ericsson 450 is a digital telephone instrument equipped with 36 non-locking buttons with associated LEDs arranged in three columns of twelve buttons. Thirty-two buttons are available for assignment to features or line appearances. A microphone, loudspeaker, and volume control provide speakerphone operation. An LED located near the volume control illuminates when the E450 is in handsfree mode. A 16-character LCD provides both voice and data call details.

Adding a TAU-T to the E450 allows simultaneous voice and data communications over a single pair of wires.

The MD110 was specifically developed to accommodate the rapidly expanding requirements for switched data as well as voice communications. Internally the MD110 makes no distinction between data and voice transmissions. Voice and data communications are performed independently or simultaneously using only a single twisted pair of wires. Data devices and digital telephone sets utilize the same digital line card. As a result, data equipment relocations are made as quickly, easily, and inexpensively as telephone instrument rearrangements. This integrated approach offers more efficient use of existing switching equipment and provides a smooth transition to full office automation.

The bit transparent architecture of the MD110 supports both asynchronous and synchronous data transmission independent of protocol. Data users can directly access various MD110 features such as host port contention, domain switching, and destination queueing. In addition, data call origination options including telephone-keypad dialing, smart modem command, menu selection, single button access, and hotline connection are available to the data user.

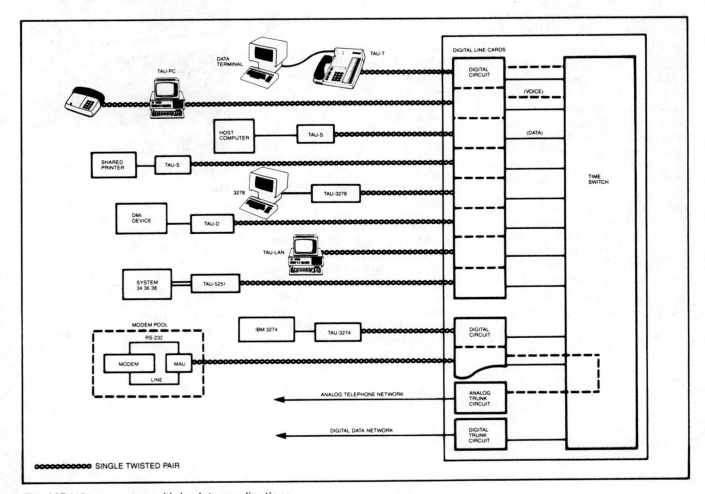

The MD110 supports multiple data applications.

Terminal Adapter Units

Using a Terminal Adapter Unit (TAU), data communications capability can be added or relocated without affecting the integrity or operation of the MD110 Intelligent Network. TAUs are available in numerous customer configurations to support asynchronous operations up to 19.2 Kbps and synchronous operations up to 64 Kbps. Digital-to-analog or analog-to-digital conversion is not necessary for internal data switching. Modems are eliminated for on-net communications since the connection is entirely digital in format.

Each TAU has the flexibility to enable the appropriate signals on the RS-232-C (or V.35) interface on data terminal equipment with various control requirements. In addition, the TAU provides the proper characteristics such as speed, number of start/stop bits, interface type, and modes of transmission. All TAUs provide asynchronous and synchronous operation, either full or half duplex modes, with visual indicators to monitor the progress and status of calls. A local test button is provided for user testing and problem isolation.

TAU-T

The Terminal Adapter Unit for Telephones (TAU-T) is designed to easily attach to the rear of the Ericsson 350 or 450 digital telephone. An electrical connection is established between the units, allowing the data interface to communicate with the MD110 through the telephone's 2-wire extension line. The TAU-T provides asynchronous and synchronous data capabilities with speeds ranging up to 9.6 Kbps through an RS-232-C interface.

This TAU-T/telephone integration features the ability to program, save, and automatically recall 8 data communication parameters. Initiation, acceptance, and disconnection of data calls are performed through various methods, including hotline/ auto-answer, data carrier detection, and standardized voice telephony procedures.

The TAU-T displays call progress information, data traffic status, and protocol selection through the digital telephone LCD. In addition, a test button and LED indicators are provided for power, data terminal ready, receive data, transmit data, and carrier detect.

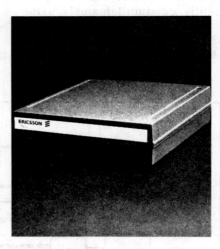

TAU-S

The Terminal Adapter Unit for Standalone operations (TAU-S) is designed for applications where voice communication is not required. Such situations include the MD110 connection to shared printers, computer ports, and isolated terminals. The TAU-S supports transmission speeds up to 19.2 Kbps for asynchronous applications and 48 Kbps for synchronous applications through an RS-232-C interface.

Four program buttons with associated LEDs on the TAU-S front panel allow user selection of predefined functions and call destinations. Power, test, receive data, transmit data, and data terminal ready indications are provided by LEDs. In addition, a two-digit display shows call progress information.

The TAU-S is interconnected to the MD110 through a single twisted pair of wires. Fault location and loopback testing are initiated from the front panel test button. User options for the TAU-S are programmed from the MD110 administration terminal.

TAU-D

The Terminal Adapter Unit for Digital Multiplexed Interface (DMI) operations (TAU-D) is a high-speed data interface offering several switch selectable protocols. Through interconnection with other TAU-Ds or multiplexed over the MD110 DS-1 Digital Trunk Card, the TAU-D supports the DMI standard.

Packaged like the TAU-S, the TAU-D supports both asynchronous/synchronous operations to 48 Kbps through an RS-232-C interface and synchronous operations to 64 Kbps through a V.35 interface. A rear panel switch allows the user to select from the following protocols:

TAU-S Protocol supporting asynchronous operation to 19.2 Kbps and synchronous operation to 48 Kbps through the RS-232-C interface.

ECMA Protocol supporting clear-channel synchronous speeds of 48/56/64 Kbps through the V.35 interface.

DMI Protocol supporting:
Mode 0 – 64 Kbps synchronous V.35 interface
Mode 1 – 56 Kbps asynchronous V.35 interface
Mode 2 – 19.2 Kbps asynchronous with inband handshaking, RS-232-C interface.

ANSI Menu Interface with autobaud detection supporting keyboard selection of the asynchronous modes.

The TAU-D allows direct terminal keyboard access to MD110 data call processing features such as Auto Dial, Speed Call, Connect-When-Free, and Autobaud.

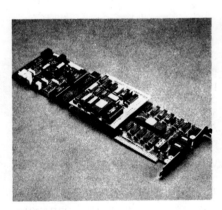

TAU-PC

The Terminal Adapter Unit for Personal Computers (TAU-PC) allows IBM PC/XT/ATs and their compatibles to interface directly with the MD110 and communicate with other data equipment using the telephone's 2-wire extension line. The TAU-PC installs in the PC and operates as a serial communications port, supporting standard communications software packages.

The TAU-PC can be used alone or with any MD110 digital telephone. When interconnected with a telephone set, the user has access to all the capabilities available with the TAU-T interface. Modular phone jacks are provided on the rear of the TAU-PC for the MD110 and telephone extension lines.

The TAU-PC supports both asynchronous and synchronous data communications at speeds up to 9.6 Kbps. Advanced features such as autobaud detection, "AT" smart modem call-control for voice and data calls, and screen-displayed call progress messages are available through the TAU-PC.

TAU-LAN

The Terminal Adapter Unit for Local Area Networking (TAU-LAN) is a 64 Kbps token-ring network interface card for IBM PC/XT/ATs and compatibles. This option offers a fully integrated data network utilizing standard MD110 hardware and the existing 2-wire voice cabling, complying with MS/DOS and IEEE 802.5 standards. Networked applications and most single-user applications function without modification.

Coupled with networking software from 3 COM, the TAU-LAN gives the user the freedom to create from 1 to 64 logical LANs within one MD110 system with up to 20 workstations in each LAN. Since this is a logical virtual-network, the LAN can be redesigned at any time without altering the hardware. It is as simple as moving a telephone extension. Additionally, networks within the MD110 can be bridged using PCs configured as servers.

The MD110 monitors the data network in the same way that it monitors the voice network. Any breaks in the token-ring (a powered-off PC for instance) are instantly bypassed to guarantee uptime for your data communications. When the bypassed unit becomes functional, it is automatically reinserted in the network.

TAU-3278/3274/3287

The Terminal Adapter for 3270 Terminals (TAU-3278) is a 3270 type A coax to asynchronous protocol converter for selected IBM 3270 terminals. This allows 3270 users access to switched async host facilities through the MD110 Intelligent Network.

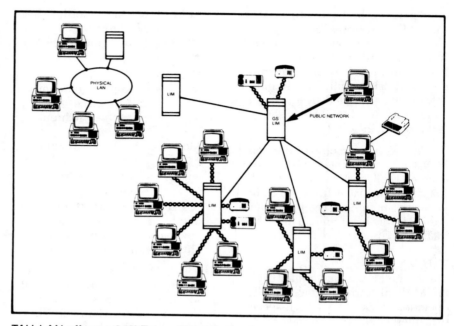

TAU-LAN offers a 64K Token-Ring Network.

The Terminal Adapter Unit for 3274 Controllers (TAU-3274) is an asynchronous to 3270 type A coax protocol converter for IBM 3274 cluster controllers. The TAU-3274 allows users of general purpose ASCII terminals or personal computers access to IBM hosts.

When interconnected through the MD110, the TAU-3274 and TAU-3278 provide coax elimination over single twisted-pair wiring for IBM host connections. The MD110 automatically recognizes that a TAU-3278 and TAU-3274 are interconnected and an extended VT-100 communications protocol is initiated. This protocol supports 3270 features such as extended attributes, graphics, etc. By using MD110 data switching capabilities, the number of 3274 ports dedicated to casual users can be reduced.

The Terminal Adapter Unit for 3287 Printer Emulation (TAU-3287) allows the connection of asynchronous serial printers to IBM 3274 cluster controllers through the MD110. This interface provides 3287 printer emulation in either hotline or dedicated connection modes, as well as switched access for asynchronous printing.

TAU-5251

The Terminal Adapter for 5251 Communications (TAU-5251) allows a general purpose ASCII terminal or the personal computer user to access synchronous IBM Systems 34, 36, and 38 computers. The TAU-5251 offers host port contention, domain switching, single twisted-pair distribution, and MD110 data call processing capabilities.

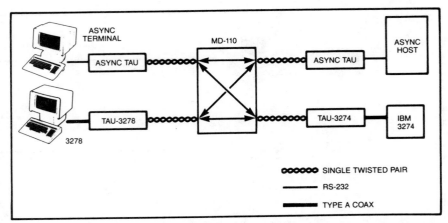

3270 Coax Elimination and Protocol Conversion using Terminal Adapter Units.

Modem Pooling

The design of the MD110 facilitates transparent data transmission connections between on-net and off-net data equipment. When communicating with remote devices over analog facilities, a modem becomes an integral part of the communications link. With the MD110, modems may be grouped in modem pools or configured as dedicated devices.

Modem pools allow modems to be shared by numerous data users, eliminating the need to dedicate a modem and trunk to each terminal. Fewer modems are required and trunk facilities are more efficiently utilized resulting in substantial cost savings.

Modems may be centralized in a single LIM or distributed among multiple LIMs. Class-of-service determines which modem or modem pool is accessed by the data user, regardless of the equipment's physical location. Specific modem pools can be selected either automatically or manually. Free modems in the group are selected on a load-sharing basis

and the MD110 provides queueing when a modem or trunk is unavailable. Modems can also be selected individually for testing purposes.

To provide further savings on communications, Least Cost Routing may be used to place data calls. Additionally, established voice calls can be switched into data connections without having to redial.

Modem Access Unit

Each modem is connected to the MD110 via a Modem Access Unit (MAU). Both the analog (trunk) and the digital (RS-232-C) sides of the modem are connected to the MAU providing system selection of the appropriate configuration for each data call. Once a connection is established, the MD110 together with the TAU and the MAU are totally transparent to the data equipment and modems.

A single twisted pair of wires connects the MAU/modem to the MD110, allowing for distribution from the LIM. For leased-line networks, the external trunk can be permanently connected to the

MAU/modem. This facility can now be shared by all data users.

The MD110 allows data transmission in both directions through a modem pool based on the characteristics of each modem. The MAU conforms to the modem characteristics, adapting to modems from different manufacturers. Additionally, the MAU is capable of accepting control signals from the TAU and using these signals to control the modems and their handshaking.

The MD110 supports both asynchronous and synchronous modems varying in speeds up to 4800 bps for switched line applications and 19.2 Kbps for leased line applications. For synchronous applications, the MAU accepts the transmit/receive clocking from the modem.

The MAU/modem combination is designed for operation on an unattended basis. The MD110 has the capability to scan, locate, and isolate faulty modems. A pushbutton and LEDs on the MAU front panel provide for testing and status displays.

SYSTEM MANAGEMENT AND CONTROL

The MD110 Intelligent Network simplifies system control and management functions through dedicated devices connected to the LIM. The administration terminal provides a simple man-machine interface for system parameter alterations, software upgrades, maintenance tasks, and configuration changes. The MD110 supports additional Input/Output (I/O) devices such as printers and Cartridge Tape Units (CTU) for system reports and database back-up. I/O software and hardware is installed in two LIMs for reliability. Administrative and maintenance tasks can be performed remotely via a dial-up modem through an I/O port. Connection of all I/O devices is made via a standard RS-232-C interface at selectable signaling speeds of 300 to 9600 baud with a maximum of six devices active simultaneously.

Security

The MD110 ensures security through a two-level scheme using both password protection and Authority Classes. Eight Authority Classes define the levels of system access allowed. In addition, a unique password must be entered and validated before the MD110 will acknowledge further commands. Failure of the system to acknowledge unauthorized commands ensures software integrity and protects the system parameters from being inadvertently altered, for example, by the communications clerk entering station relocation data.

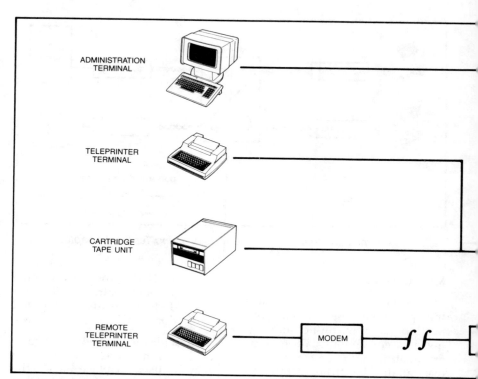

ADMINISTRATION TERMINAL

TELEPRINTER TERMINAL

CARTRIDGE TAPE UNIT

REMOTE TELEPRINTER TERMINAL

MODEM

Input/output devices support local or remote maintenance and administration.

System Administration

MD110 system administration is accomplished by entering a series of command messages from an I/O terminal. Progress messages are printed or displayed by the system indicating the status of command processing, i.e., EXECUTED or INVALID ENTRY. Administrative functions such as station relocations (adds, moves, changes), collection of traffic data, and system software reloads are executed through command messages by the Administration Terminal operator.

The MD110 provides both traffic measurements and Station Message Detail Recording (SMDR). MD110 traffic measurement reports detail the traffic load carried by various elements of the system and serve as a useful management tool for evaluating the efficiency of system operation. Traffic measurement commands specify parameters including:

- start/stop time
- recording frequency
- system measurement points
- method of output (printer or CTU).

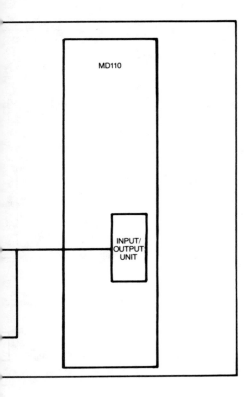

```
┌─────────────────────────────┐
│ ┌──────────┐                │
│ │  MD110   │                │
│ │          │                │
│ │          │                │
│ │          │                │
│ │  ┌────┐  │                │
│ │  │INPUT/│ │                │
│─┼──┤OUTPUT│ │                │
│ │  │UNIT │  │                │
│─┼──┤     │  │                │
│ │  └────┘  │                │
│ └──────────┘                │
└─────────────────────────────┘
```

System traffic measurement points include:

- outgoing/incoming routes
- operator positions
- operator groups
- extension groups
- hunting groups
- common bell groups
- call pickup groups
- data groups
- modem groups
- PCM links
- tone senders/receivers
- conference equipment.

SMDR provides information on station activity such as calling number, called number, and duration of call.

SMDR data is formatted according to Bell Standard FP15 and output via an RS-232-C connection for further processing by the support system. Sophisticated support systems for the MD110 address the requirements of cost-conscious companies. SMDR continuously monitors call detail and traffic data to record activities down to the individual station level. SMDR provides the vital information needed for cost allocation to individuals or departments within an organization and provides the communications manager with the tools to ensure cost-effective use of the MD110.

Maintenance

The MD110 Intelligent Network's primary maintenance utilities are designed to provide early fault detection and system action to minimize the disruptive effects of a fault. Simultaneously, the system initiates alarms and messages to alert maintenance personnel of a malfunction. The Service Supervision feature provides continuous monitoring of system operations for immediate fault detection. Overall system performance is monitored for proper switching matrix connections, system synchronization, normal power supply output, and line/trunk circuit operation.

If a major malfunction is detected, the system automatically reinitiates processing by restarting call processing or reloading software. If restart/reload attempts fail, the affected hardware is assumed to be faulty and is blocked from further use until the fault condition is corrected. The system continues call processing utilizing other available resources.

There are five classes of alarms assigned to various fault conditions ranging from Class 0 Information Only-No Action Required to Class 4 Critical-Immediate Action Required. The alarm condition and its severity are displayed on the central alarm panel and on the operator console. An alarm log, printed on command, records the following information:

- time of alarm
- fault type
- program in effect when fault occurred
- LIM location of fault
- suspected failed unit
- confinement procedures executed by the system.

Using the alarm log data, maintenance personnel can quickly locate the faulty subassembly, replace it with a spare, and return the system to normal operation.

Chapter 6: Positioning for ISDN

Integrated services digital networks (ISDN) will impel substantial changes for most private branch exchange (PBX) manufacturers. On the central office side, 23B+D "primary" services, that is, 23 64 Kbps bearer channels and one 64 Kbps control channel, will not be that disruptive because modern PBXs handle a similar T1 (1.544 Mbps) channelization today. On the user side, however, the ISDN "basic" services, 2B+D, two 64 Kbps channels and one 16 Kbps channel, differ from virtually all proprietary schemes employed by PBX vendors. Further, the 2B+D basic services open up vast possibilities for new combined services. For instance, one PBX manufacturer may choose to stay with the 64 Kbps voice standard and offer 56 Kbps data service, reserving the entire D channel for signaling. Another may move to the 32 Kbps voice standard, run circuit switched data at 19.2 Kbps on the remainder of the B channel, run 64 Kbps compressed video on the other B channel, and use 9.6Kbps of the D channel for packet data. The combinations of services "mix and match" within the basic 2B+D are, at least in theory, unlimited.

How would these services actually appear to the user? Although problematic, it is not too hard to imagine that they could appear as additional "windows" in a work station similar to that of the Sun/Apollo/Macintosh II variety. Because of "B" channel restrictions, the video exchange would probably have to occur in alternating half-duplex mode. The visual effect would likely be that of rapidly changing freeze frames. The work station could incorporate an auto pan, auto focus camera to capture the outgoing images with a dedicated processor handling the compression. The received image would be decompressed from 64 Kbps, and would appear as a user-sized, overlayed window for the duration of the 64 Kbps telephone call. The telephone could be built into the work station, in the same manner as today's "hands-free" sets, or, in environments with high background noise, a conventional handset could be used. Electronic mail, using a portion of the 16 Kbps "D" channel, could be delivered in real time via an announcement strip similar to that used on commercial television, or, less obtrusively, a screen icon resembling a rural mail box could be posted. Given the growing complexity of work station produced documents (different fonts, letterheads, tables, pictures, diagrams), the easiest form of transfer would likely be to have the work station digitize them and send them by CCITT Category IV facsimile via the packet facility. Again, a received diagram could appear as a strip notification ("facsimile received at date/time") or as a mailbox indicator.

The reaction of many to this sort of "omni-workstation" for the office will be extreme skepticism because of its apparent high cost. Today, if one views it as an $8,000 work station cum $2,000 camera cum $500 voice set cum $3,000 facsimile machine, it would appear so. But many of these items contain logically redundant components, and the 2B+D ISDN standarization for station sets would be a major force, one that is conspicuous in its absence today. With a packet switch mail standard (X.25, X.400), a packet facsimile standard (Category IV), and the 64 Kbps PCM voice standard, the only major deficiency is the lack of a digital 64 Kbps video standard. It is not unlikely that a digital 64 Kbps video standard is attainable by the time ISDN becomes—similar to cellular phones—a ubiquitous urban sevice. Once that occurs, it is likely to follow the well-known consumer electronics/mass distribution curves traced by radio, television, video cassette recorders, personal computers, compact disks, and cellular phones. As such, an omni-workstation is likely to appear initially as a high-end product ($5,000-10,000), standardization is crucial. Just as standarization problems have plagued the video disk market, and limited the acceptance of cassette recorders, instant cameras, home video recorders, and microwave ovens, the architecture of omni-work stations must be both open and standard.

Initially at least, cost-conscious users will opt for an evolutionary transition from separate devices already on hand. In which case, ISDN breakout boxes, which enable users to both connect various extant devices (phones, personal computers (PCs), terminals, facsimile machines) to the 2B+D multiplexed twisted pairs and incorporate some kind of services selector, will play an important role in the transition to more integrated end-user devices.

Continuing the declension from the possible to the probable, probably the greatest uncertainty surrounds cost. All the way through the transmission process, ISDN, mostly because it is new and different, will add cost. Already noted is how, just within the PBX sphere, it will result in changes to extant switch trunk cards, signaling software, and user interfaces. In some environments, wire plants may have to be upgraded so to reliably deliver the 192 Kbps (144 Kbps + 48 Kbps of overhead) data rate for basic services. In the end, users must be willing to pay for the expanded services offered. Although it is likely that in many high-volume, high profit areas such as financial services, combined voice/data/video would be perceived as adding substantial value to telecommunications services, its dissemination over broader areas of the business community, much less the home and mobile markets, is less assured.

Where the preceding paragraphs have been largely optimistic vis-a-vis the spread of ISDN-enabled PBX services, a balanced treatment should at least survey the negative aspects. Increased costs and an uncertain dissemination have already been noted. Others have noted that the choice of a 64 Kbps "B" channel, although convenient, is probably an anachronism. Many long-haul voice networks already use 32 Kbps adoptive differential pulse-code modulation (ADPCM) coding and convert it to 64 Kbps PCM at the long distance vendor's point of presence (POP). Also, although there are digital video standards currently at 384 Kbps, 765 Kbps, and 1.54 Mbps, one for 64 Kbps, although obtainable, may prove both expensive and of unacceptable quality.

Probably the constituency most left out in the cold by the ISDN standards (both basic and primary) is that of computer graphics and imaging. Even with substantial compression, it is hard to see how even the quality graphics of today (1024 picture elements or pixels × 1024 × 16 or 24 bits of color × n refreshes per second) can be transmitted. Even the recommended resolution for X-rays (2048 pixels × 2048 pixels × 8 bits for gray scale) would be difficult to accommodate.

It would appear that most of these applications must await a subsequent high-bandwidth ISDN standard that builds on fiber optic technology. This development is clouded by the unlikelihood of significant price decreases in bulk bandwidth in the near future. Despite the proliferation of available bandwidth, particularly from satellites and several vendors' fiber optic networks, the outcome of deregulation has been to create an environment where the dominant vendor (AT&T) targets long haul prices and several very small competitors, who are so highly leveraged that they cannot afford to sustain major price cuts. The result is a communications environment that is unlikely to see major reductions in the costs of long haul bandwidth in the near future.

The selected papers cover a number of user "views" of ISDN. The first, by Anderson, "Transition to the AT&T-IS Integrated Private Network Architecture, provides something of an historical perspective, describing the evolution of analog and digital ETN networks—and the role that PBXs play in them—to ISDN. More marketing oriented, Tang, of IBM/ROLM, identifies three fundamental principles embodied in ISDN (digitization with high bandwidth, worldwide standards, and integrated functionality/services) and speculates on their likely effects on information-intensive business enterprises in his "ISDN: New Vistas in Information Processing."

Robin's paper, "Customer Installations for the ISDN," focuses on ISDN service delivery and the changes—particularly with PBXs—necessary to bring them to business environments. The Decina and Scace paper, "CCITT Recommendations on the ISDN: A Review," overviews both the CCITT's role in the standards process and the CCITT's I-Series of Recommendations on ISDN. (The I-Series of Recommendations outline the most important standard features of ISDN regarding service capability, network architectures, and user-network interfaces.)

Turner's paper, "Design of an Integrated Services *Packet Network*," is an exciting critique of the current ISDN plan. He argues that ISDN is unnecessarily dependent on a circuit switching mentality and will result in a less-than-satisfactory hybrid circuit/packet switched solution. He believes that for adaptability to changing traffic, a truly integrated internal architecture, and transmission efficiency, a high performance *packet* network would have been the better direction.

The last two selections look to the future. Browne's paper, "Network of the Future," presents a view of the ISDN future where synergistic relations between flexible and distributed network structures, industry standards at the interfaces, and standard user interfaces and signaling procedures will generate new information services for the "Information Age." The Vickers and Vilmansen paper, "The Evolution of Telecommunications Technology," begins with a wide-ranging assessment of achieveable software and hardware technologies and ends with a visionary, post-ISDN telecommunications future dominated by highly intelligent and adaptable "soft technologies."

ISDN—New Vistas in Information Processing

W. Victor Tang

This article presents a set of perspectives on why ISDN is important for the advancement of voice and data communications. It begins by analyzing the fundamental principles of ISDN: digitization with high bandwidth, worldwide standards, and integrated functionality/services. It discusses the implications of these principles on the information worker and on the communicating desk-top devices. Next, the article identifies the key information processing requirements of the enterprise where users work and addresses the ISDN implications. It then examines the role of the PBX in that environment. The article concludes that ISDN provides a sound basis to support emerging customer requirements for potentially more cost-effective information processing and communications

Fundamental Principles of ISDN

There are three fundamental principles embodied in ISDN. They are:

- digitization with high bandwidth
- worldwide standards
- integrated functionality/services

We will analyze these principles and discuss why they can provide a robust basis to support emerging customer requirements for more cost-effective information processing and communications. We will show that they are, therefore, important elements that have the potential to enable industry expansion to the benefit of users and vendors alike.

Without loss of generality, we adopt a familiar and simplified layer model for communications systems, as shown in Fig. 1. Recall that the bottom layers focus on the functions of transport and connectivity. The middle layers concentrate on the functions of networking, while the top layers focus on the content and agenda of end-to-end communications. Critical to communications systems are reliable transport, connectivity, and a robust system structure to provide the capabilities customers/users want. Standards and architecture enable these elements to come together in a cohesive and consistent manner. It is for these reasons that IBM developed SNA and for these same reasons that IBM supports standards. ISDN is important because it will enable the use of a common transport and attachment to a variety of networks and services.

Focusing on the lower layers of our communications model, we see increased bandwidth and digitization. Network capacity is growing rapidly, fueled by alternative technologies and an increasingly competitive environment. The rapid growth of fiber links, high bandwidth T1 channels, satellite and microwave communications are all evidence of these phenomena. Simultaneously, increased digitization is taking place in the transport systems. Products in customer premises, local loop, long distance, and central office are all increasingly digital capable. This increased digital capacity is facilitating users to share information across networks, where the information is needed and the processing required.

In industrialized countries, studies show that voice utilization grows at a modest, but steady rate. Because network capacity is growing so rapidly, there will be increasing capacity available for non-voice usage, such as data and image. Consequently, the transport infrastructure will facilitate users' ability to communicate large volumes of voice and data traffic for their information processing requirements.

These developments are taking place in a very dynamic environment. In addition to the accelerating pace of technology, there is increased privatization, deregulation, and competitive participation. The effects are structural changes in the economics of communications; tariffs reflect these dynamics. The tariffs for leased and switched lines and the separation of transport and access costs are typical examples.

Digitization with high bandwidth provides a common denominator for information of all forms—voice, data,

Reprinted with permission from the Proceedings of the International Conference, ISDN, vol. 1, Europe, London, 1986. Online Publications, Pinner, Middlesex, UK.

303

COMMUNICATIONS ARCHITECTURE

	SNA DOCUMENT
CONTENT	• ARCHITECTURE & SERVICES
NETWORK	• SNA • OSI
TRANSPORT	• SDLC, X.25, LAN, ISDN

STANDARDS . . . CRITICAL

Fig. 1

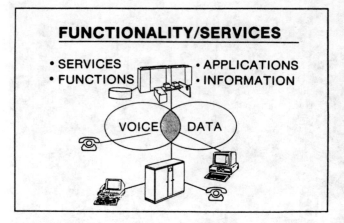

Fig. 3.

image, video—to take advantage of increased network capacity. (See Fig. 2.) Therefore, using digital carrier networks, users can share information. This same digitization allows carriers and voice products to use the effectiveness of stored programs and computers to provide functionality and services. (see Fig. 3.) For example, the customer can use equipment, such as a PBX, or a front-end communications processor, to develop the most functional and cost-effective means to manage his communications requirements. Consequently, the ability to control and manage both cost and functionality is emerging as an important customer requirement. (See Fig. 4.)

Digitization is facilitating the convergence of voice and data. (Fig. 5). Historically, the major thrust of voice has been the provision of connectivity, but now the focus is on functionality. Increased functionality is being provided by carriers; for example, 800 calling, improved centrex, and so forth. There is also improved functionality in customer equipment, such as a PBX, providing least cost routing, phone mail, call screening, and so on. On the other hand, data has emphasized functionality and now customers want improved connectivity. Computer networking, local area networking, terminal and work station controllers are all examples of

connectivity in the data environment. ISDN's fundamental principles: digitization with high bandwidth, integrated function and services, and worldwide standards act as facilitators to all participants in the industry. In an increasingly competitive environment, it enables all participants to address the requirements of func-

CUSTOMER CONTROL & MANAGEMENT

• COST
• PERFORMANCE
• FUNCTIONALITY & SERVICE

Fig. 4.

INFORMATION

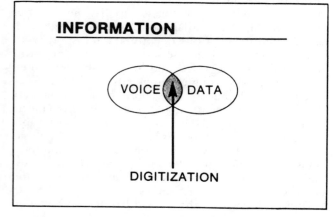

Fig. 2.

CONVERGENCE

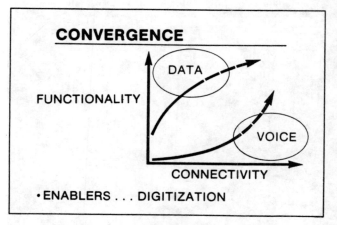

Fig. 5.

tionality and connectivity in the most cost-effective manner to benefit users and customers.

These are the reasons why IBM/ROLM are studying ISDN technology, participating in national and international standards bodies, and trying to understand the customers' requirements for future ISDN equipment and services so that ISDN can be supported, when customer requirements/benefits become clear and as the standards stabilize.

Users and ISDN Implications

According to the U.S. Bureau of Labor Statistics, over 50 percent of the U.S. labor force handles information for a livelihood. The users are, therefore, the "information workers." In this section, we will analyze what they do, what information they use, the desktop devices they use, and the ISDN implications.

The tasks users perform are changing from the relatively structured and repetitive to the more stochastic and specialized. (Fig. 6). For example, one of the first commercial on-line applications was a reservation system for commercial airlines. In this type of application, the user performed essentially standardized and deterministic transactions against a centralized database. While these applications continue to be important, emerging applications expand the range of capabilities, and many have become considerably more sophisticated. For example, CAD/CAM graphics or financial analysis using spreadsheets are less deterministic, require more information, and are targeted for the specialist to use.

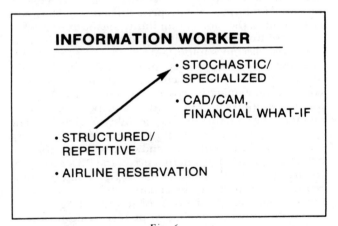

Fig. 6.

The volume, the forms, the sources, and destinations of information are changing (see Fig. 7.) Initially, users communicated, one line at a time, with one computer. Then they started communicating with full screens of data, in color, doing more complex tasks. More recently, engineers are using graphics representation of physical objects to design and simulate their operation. Circuit design and simulation, heat transfer, and stress analysis are typical examples. Not only is information more voluminous and more complex, but the origins and destinations of information are more varied and far-flung. The user wants access to multiple computers concurrently and, through "windowing," wants concur-

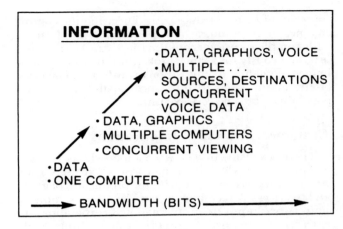

Fig. 7.

rent viewing, for example, comparing physical deformation of an object under stress, graphical analysis of different financial scenarios, and so on.

Voice communications will continue to be important; therefore, considering the total information needs, concurrency of voice and data is also a requirement. ISDN is designed to meet this need for concurrency by supporting multiple digital channels through a standard attachment with twisted pair wiring.

Products are evolving to meet these user needs. They range from simple data terminals, to advanced work stations, to advanced voice-data based work stations, to communicating desktops. Their capabilities include the handling of many forms of information, connecting to different applications and different hosts either directly or through a network.

The user has benefitted, and will continue to benefit, from the synergistic interaction of applications, communicating desktop devices, information of all forms, and connectivity among users and information, as shown in Fig. 8. The equal access of a control channel to the customer out of band, can further expand this synergy. For example, by using a PBX or a front-end multiplexor, the customer can have the capabilities to dynamically manage expensive communications re-

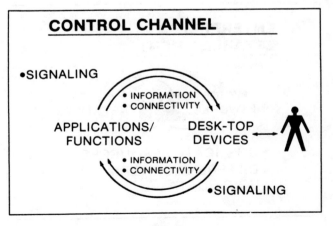

Fig. 8.

sources and thereby reduce costs. Or, using the control channel information, a multiple-window application with complete customer information, can be triggered instantaneously for a stock broker. Another example is immediate address identification in medical emergencies or fire alarms. Add to this voice capability, and the possibilities become very exciting.

Customers and ISDN Implications

There is a distinction between a user who is a person and a customer which is an enterprise where users work. An enterprise is concerned with the production of goods and services. An enterprise can be, for example, a university campus or an international corporation. An enterprise is frequently comprised of multiple establishments, that can be as small as a department or a work group, or as large as a corporate headquarters. An enterprise can also be a single establishment such as a hotel or a lawyer's office.

Traditionally, customer's typical enterprise data and voice transport networks have been physically separate and managed separately. (See Fig. 9.) Frequently, the networks have redundant resources. Voice networks have been principally justified on the basis of cost effectiveness to meet service-level criteria. Data networks are typically driven by application functionality, information requirements, and reliable transport to meet information processing and performance requirements.

Customers have to examine voice/data traffic needs to meet differing business needs, as well as the voice/data technical and economic criteria. They also have to be constantly alert for opportunities to reduce cost and to improve functionality and service. To do this, the customer must consider alternative technologies, the service and tariff implications, and must then choose from many competitive suppliers and vendors. Simultaneously, the customer must remain flexible to change. In this environment of considerable flux, the PBX can play a key role, as shown in Fig. 10.

For application functionality, the PBX provides connectivity with hosts, departmental or establishment computers to access databases and applications. Within

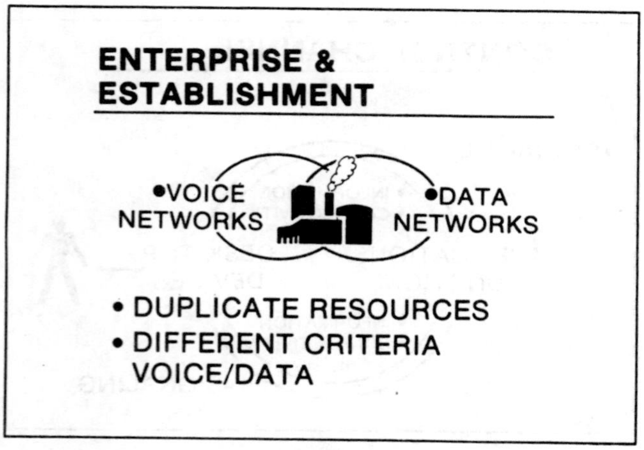

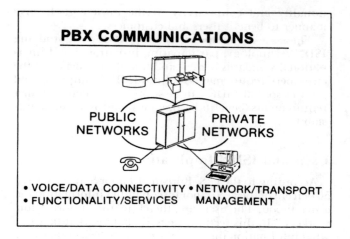

Fig. 10.

the establishment, using PBX voice/data connectivity the customer can provide users the capability of information sharing and communications via communicating desktops. Among establishments at the enterprise level, the customer can access a variety of public and private network capabilities, such as switched lines, leased lines, high bandwidth transport, and so forth, through the PBX.

In this manner, the customer is able to choose from the most cost-effective vendors to meet user demands and provide bandwidth, grade of service, and reliability. With the availability of increased choice comes the potential for improved enterprise efficiency and business effectiveness; therefore, tailorability and optimization of these resources and capabilities, to meet the needs of the enterprise, becomes important.

These needs can be met through PBX intelligence, available directly to the customer to tailor, optimize, control, and manage for cost and functionality. As tariffs and technologies for information transport continue to change, the intelligence is flexible to change. The customer has the benefit of managing and using whatever mix of capabilities and services are economical and appropriate. The intelligence in the PBX can adapt to change through software use. ISDN addresses these requirements, it provides standard network attachments and customer access of a control and signaling channel that can be used by intelligent premise equipment.

PBX intelligence provides visible and cost-effective information transport management functions. (See Fig. 11.) Among the first enterprise/establishment tools were Call Detail Recording and Call Management capabilities. They provide detailed information on the use of the communications equipment and transport facilities. The PBX then introduced a greater degree of flexibility and adaptiveness via Network Optimization and Route Optimization algorithms. In the current environment, where customers have many competitive alternatives, Equal Access software and Multiple Interexchange capabilities provide improved access to transport and networks. Many high bandwidth offerings can provide attractive economies-of-scale to the customer. A proven and effective technique to benefit from these economies

ENTERPRISE & ESTABLISHMENT

• VOICE NETWORKS   • DATA NETWORKS

• DUPLICATE RESOURCES
• DIFFERENT CRITERIA VOICE/DATA

Fig. 9.

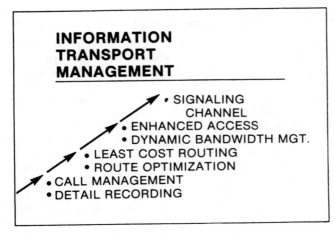

INFORMATION TRANSPORT MANAGEMENT

- SIGNALING CHANNEL
- ENHANCED ACCESS
- DYNAMIC BANDWIDTH MGT.
- LEAST COST ROUTING
- ROUTE OPTIMIZATION
- CALL MANAGEMENT
- DETAIL RECORDING

Fig. 11.

is dynamic bandwidth allocation. ISDN common channel signaling can be used to dynamically assign channels for voice/data over an integrated facility. This signaling capability offers the potential for the customer to optimize utilization of transport facilities for voice/data.

A similar evolution can be observed in functionality and services for voice/data, as shown in Fig. 12. Starting with voice and data connectivity on the PBX, increasing functionality is provided to the communicating desktops, they include such functions as phone mail, messaging between PBX's, and computer office applications. They also include connectivity such as gateways, bridges, and translation functions such as protocol conversion. ISDN common channel signaling and increased deployment of computer software will enable increasing levels of dynamic intelligence to be available to the customer. Dynamic information transport management is a precursor of the potential for comprehensive network management functions.

Two key enterprise communications expenditures are customer equipment and network services. Intelligent customer premise equipment, such as a PBX, can be an effective customer lever to reduce overall costs. The customer can manage and use whatever mix of capabili-

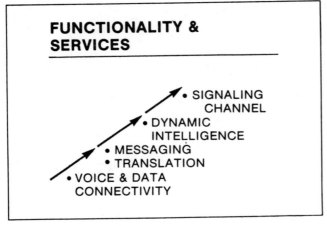

FUNCTIONALITY & SERVICES

- SIGNALING CHANNEL
- DYNAMIC INTELLIGENCE
- MESSAGING
- TRANSLATION
- VOICE & DATA CONNECTIVITY

Fig. 12.

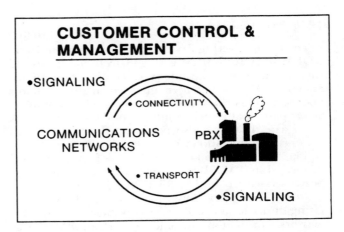

CUSTOMER CONTROL & MANAGEMENT

- SIGNALING

COMMUNICATIONS NETWORKS

- CONNECTIVITY
- TRANSPORT

PBX

- SIGNALING

Fig. 13.

ties and services that are economical and appropriate. The customer can adapt the intelligence to the changing needs of the enterprise. Access of the ISDN common channel signaling facilitates intelligent customer equipment to call upon the network capabilities to support network management. (See Fig. 13.)

ISDN Principles

Examination of the fundamental principles embodied in ISDN, worldwide standards, integrated functionality/services, and a high bandwidth digital-transport system, shows that they can provide a robust basis to support users'/customers' requirements for information processing and communications systems. Customers and users have increasing needs for information of all forms, in greater volume, and from multiple dispersed sources. Increased digitization of information, particularly in the transport system, can facilitate customer's ability to share information of all forms wherever it is needed. Moreover, to be effective and productive, more customers are using information processing for increasingly complex tasks. Therefore, increased functionality and services can improve the customers' and users' effectiveness. The provision of these capabilities in a competitive environment will ensure continuing innovation and price/performance advantages to customers and users.

Remaining ISDN Challenges

Although a bright and exciting promise, the ISDN agenda is incomplete. Increased customer involvement is required for an accurate understanding of the range of functions and services, products and systems that customers need. The providers of ISDN transport need to describe when, where, and how ISDN is going to be provided. Most importantly, the provision has to be with a standard, consistent implementation. The users want cost-effective solutions for their information needs. To assess the cost effectiveness of ISDN, they need to know what the tariff assumptions are. Namely, what are costs going to be like? Addressing these unfinished areas will help customers achieve the effectiveness they want and industry participants the opportunities they seek.

IBM Involvement

Today, engineering development teams within IBM/ROLM are actively investigating ISDN technologies and their implications on future products and applications. IBM/ROLM also has been involved in the standards bodies, such as the CCITT and the ANSI committee, active participation will continue in the future.

To date, IBM has also participated in two tests in Europe. In cooperation with the Belgian Telecommunication Administration (RTT), a team of IBM engineers connected an IBM ISDN terminal equipment to an ISDN bearer service provided by RTT; Bell Telephone Manufacturing Company supplied the central exchange. Using the basic access, terminals operated with an RTT network termination equipment (NT1), an IBM PC was used as the data terminal. In the next phase of this test, IBM will use an IBM graphics workstation. In the UK, in cooperation with British Telecom, IBM connected an IBM PC to an ISDN interface provided by British Telecom using the basic access. An actual IBM host VM application was used in this test. IBM and BT developed their test products, based on the CCITT standards, and the ISDN connectivity was made without major difficulties.

Conclusions and Summary

The promise of ISDN is that it will:

- simplify attachment to networks
- smooth the way for growth in communications applications
- evolve into a worldwide network and communications standard
- provide consistency and flexibility
- improve customer's productivity

- enable innovation and price performance improvements

We believe that these are the customer's requirements and that it will take all of us in this industry to deliver these benefits to our customers. Furthermore, we think that a competitive environment will ensure that these benefits will be continually available to customers.

The economist who said, ". . . telecommunications . . . has become key to the biggest industrial change of the next few decades: the developed world's shift to an information economy," stated it well. ISDN has the potential to become a pillar of that information economy, and IBM/ROLM will be among the key players in the ISDN arena as providers of both equipment and services.

Acknowledgment

The author would like to acknowledge the insightful leadership of Mr. Tom Furey Jr. and discussions provided by Mr. Gary Beckstrom from IBM/ROLM in the preparation of this article.

Victor Tang is Manager of Strategy and Business Development in the IBM Rochester Development Laboratory of the System Products Division. Prior to this assignment, he managed the Strategy Development Group in the Information Systems and Communications Group focusing on communications systems. He has been on the Corporate Planning Staff concentrating on workstations and software. He was also a member of the new venture team that formed IBM's graphics business unit. He has held a number of management positions in the areas of communications systems, communications software, and high-function work stations. He has a B.S.E.E., an M.S. in Mathematics from Purdue University, and an M.B.A. from Columbia University. ∎

Transition to the AT&T-IS Integrated Private Network Architecture

GEORGE M. ANDERSON, SENIOR MEMBER, IEEE

Abstract—Private corporate networks are evolving toward all-digital networks that combine voice and data transmission on a single network. The functional and economic advantages of those networks over their analog antecendents motivate this evolution. This paper describes AT&T Information Systems integrated private network architecture and its functional and economic advantages.

INTRODUCTION

PRIVATE networks are an important element of modern corporate communications. They have evolved over more than 50 years from primitive beginnings based on tandem tie trunk technology. For most of this period their justification was economic—they offered less expensive service than the public network. Their performance did not match that of the public network, and dialing required user familiarity with the network topology. Customers accepted these limitations in order to achieve lower communication costs.

The economic advantages of private networks have changed as a result of recent regulatory changes. Costs of short-range voice-grade private lines have increased significantly and are often not competitive with public network services. Some people even forecast the complete demise of private networks. This judgement overlooks the economic impact of digital technology and the explosive growth of data transmission. Digital technology has the potential to revitalize private networks and lead to an era of vastly expanded private corporate communications. Twin technological forces driving this evolution are digital switching and digital transmission. Taken together, they will provide voice and data networks that can accommodate much of the corporate communication traffic.

This paper will discuss the transition to the AT&T-IS integrated voice/data private network architecture. The primary emphasis is on circuit-switched PBX systems that are interconnected with wide-bandwidth digital facilities. Both voice and data may be switched at the nodes and carried by the backbone digital transport. Packet-switched private networks, of course, share the use of these facilities as well. From a larger perspective the network concepts described here are integrated into AT&T-IS's Information System Architecture (ISA). A discussion of ISA is beyond the scope of this paper.

The target architecture projects the use of digital circuit switches interconnected with digital transmission facilities. An important element is the provision of an orderly and smooth transition from the current analog world to the digital environment. The present capabilities of the private Electronic Tandem Network (ETN) and the Distributed Communication System (DCS) are retained and evolved. As these latter systems represent major current capabilities resident in the target architectural plan, they are described as a precursor to that plan.

AT&T Information Systems switching systems used in private networks include the DIMENSION PBX, a time-division pulse-amplitude modulation system, and DIMENSION System 75 and 85, time-division digital switches.[1] The DIMENSION PBX is widely deployed in corporate networks in the U.S. System 85 [1],[2], introduced in 1982, is the first of a line of AT&T-IS digital-switching vehicles. System 75, the second member of the digital product family, was introduced in April 1984. System 75 is described in a companion paper by A. Feiner in this issue.

ELECTRONIC TANDEM NETWORK (ETN)

ETN, introduced in 1978 [3], provides a private network architecture with sophisticated routing, management, and control capabilities. It can be configured using a wide range of existing PBX's including electromechanical, nondigital electronic, and digital PBX's. It uses both private lines and public facilities for interlocation communications.

The major capabilities provided by ETN include
a uniform numbering plan,
automatic alternate call routing,
authorization screening,
subnetwork trunking, and
system management.

The uniform numbering plan provides a unique address for each station on the network. Calls to that number use a dialing sequence independent of the point of origin of the call. Routing, a network function, is screened from the user. Automatic alternate routing provides optimum (least cost) use of network facilities and alternate routes for congested conditions. The other features are described later.

Manuscript received June 15, 1984; revised February 14, 1985.
The author is with AT&T Information Systems, Lincroft, NJ 07738.

[1] DIMENSION is a registered trademark of AT&T.

ETN employs intelligent switches[2] at network tandem nodes interconnected in a four-level hierarchical architecture. An illustrative configuration is shown in Fig. 1. High-usage groups are used between level 3 tandem nodes and other network nodes where the direct traffic-flow warrants. Final groups are used to interconnect level 3 and level 4 nodes and level 4 nodes with each other. Access lines connect subtending PBX's to network nodes.

Alternate-routed traffic overflows from direct high-usage groups to other high-usage groups or final groups. A simple routing rule is employed to avoid traffic circulation. Traffic cannot be routed between fourth-level nodes via intermediate third-level nodes. This rule is similar to the public network Direct Distance Dialing (DDD) routing rules.

Pooling of traffic at the tandem switches achieves traffic efficiencies in the utilization of the intertandem tie trunks and off-net facilities such as Wide Area Telephone Service (WATS), Foreign Exchange Trunks (FX), and DDD trunks. Alternate routing improves private network performance by providing backup facilities (i.e., second-, third-, and fourth-choice routes) when the first-choice route is busy.

ETN utilizes a four- to seven-digit number. The first two or three digits, *RN* or *RNX*, are the location code followed by a two- to four-digit station code.[3] Tandem, main, and tributary PBX's are identified with location codes. Satellite PBX's carry the location code of the associated main PBX.[4] Subnetwork trunking provides the sophisticated routing capabilities of the tandem switches to subtending main PBX's.

Alternate routing is provided by tables in the tandem PBX. Fig. 2 illustrates their content. The route for the destination *RNX*, 741 in the example, is found in route list #6. This list, which may contain up to four trunk groups, indicates first- (181) and second- (182) choice intermachine trunk codes, a third-choice route, a WATS group, and finally the DDD network. WATS and DDD codes provide the destination NPA code and the appropriate office code. Up to 180 route lists are available.

Facility restriction levels (FRL's) for authorization screening are associated with each route. Up to eight levels are provided. The caller's FRL is compared to the route FRL and a call is routed over a facility only if the user FRL is equal to, or higher than, the facility FRL. The caller's FRL is transmitted between tandem switches as a digit appended to the address digits. This digit, known as a traveling-class mark, is used by succeeding tandem switches to apply the call treatment appropriate to the customer's class of service. Authorization screening provides the network administrator with an ability to control network costs by denying expensive network facilities to users on a selective basis.

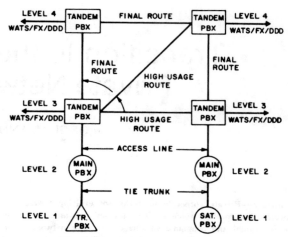

Fig. 1.   Simplified ETN configuration.

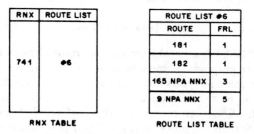

Fig. 2.   On-net ETN routing tables.

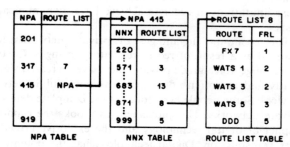

Fig. 3.   Off-net ETN routing tables.

Off-net least cost routing is provided in a similar manner using routing tables. The user dials the AAR access code followed by the 10-digit DDD number. The system screens the first three digits and marks the call on- or off-net. If the call is off-net, call processing accesses a second set of tables as illustrated in Fig. 3. If an FX trunk is provided to the NPA, the first six digits of the address must be translated. The pointer in the NPA table directs call processing to a second table containing the *NNX*'s for the NPA. *NNX* codes within an NPA are combined into a maximum of four route lists. In Fig. 3, a call with dialed number 415-871-*XXXX* is routed using route list 8. The route list table, number 8 in this case, indicates the serving trunk groups and the order of access. The capability permits using an FX group to route calls to a class five office (*NNX* codes) within the free calling area of the office terminating the FX group, and routing calls outside this area over other facilities. Up to 64 route lists for off-net

---

[2]DIMENSION or DIMENSION System 85 PBX's. System 75 functions as an intelligent main PBX.

[3]*R* is a digit 2–9, *N* is a digit 1–9, *X* is a digit 0–9.

[4]Tributary PBX's have attendants and a unique *RN* or *RNX* code. Satellite PBX's have no attendants and use the *RN* or *RNX* code of the associated main PBX.

calling can be provided with each list containing up to 10 trunk groups.

If no FX group is available to the destination NPA, translation of the *NNX* code is not required and call processing is directed toward the route list from the NPA code. In Fig. 3, for example, calls to NPA 317 are directed to route list number 7 (not shown).

The routing treatments described can be varied based on the time of day. Up to three patterns are available for both on-net and off-net calling.

Both on-hook and off-hook queueing are available at the tandem switches. This capability can improve the utilization of expensive trunk facilities and insure that service is provided on an order-of-arrival basis.

System management capabilities are provided to administer the network and provide detailed billing for cost allocation. Terminal and facility changes can be administered centrally using an associated applications processor. Customer account codes and call detail are recorded on magnetic tape and utilized in later processing for billing and cost allocation.

Distributed Communication System (DCS)

Business customers today are seeking ever-improved means of achieving higher levels of productivity and efficiency of their office organizations. This trend is leading to more sophisticated communication systems offering improved functionality [4].

Many enhanced communication capabilities are currently provided on stand-alone PBX systems. There is a need to further integrate geographically distributed systems so that these enhanced capabilities can be made available to a network of PBX's. Such a network ideally would appear to the user as a single system. There are basically two approaches to enhanced network integration that employ either centralized or distributed control.

The centralized method uses processor(s) at a central location to handle call processing for all switching entities on the network. This method requires high-speed communication on the signaling network, usually at multimegabit rates, between locations. Fiber links are sometimes employed. This method is very effective in campus environments or for short distances, particularly where a right of way is available. The high cost of ultrawide-band tariffed facilities makes this method less attractive beyond a few miles. Its major advantage is that it provides a single system with full functionality. Vulnerability, however, to a failure of the interprocessor signaling network, which causes a total failure at one or more network locations, is the principal limitation of this method.

The distributed-control alternative provides switching and call processing at each location. Enhanced network functionality is provided by connecting call processors on the individual switches by means of a data network. The advantages of this method are greatly reduced bandwidth requirements on the signaling network between locations, and much reduced vulnerability to a data link failure.

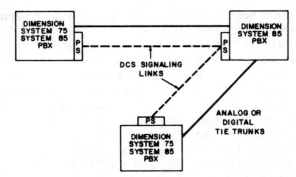

Fig. 4.   Illustrative distributed communication system network.

Additionally, since interlocation-control signaling costs are greatly reduced relative to centralized control, the distributed system can serve a large geographical area. The disadvantage of the method is that it usually provides less functional integration than does centralized control.

The DCS architecture utilizes the distributed method of control to achieve enhanced system functionality. Intelligent PBX's, providing local switching and network access, are interconnected by transmission and signaling private line facilities. These may be voice-grade trunks provided over analog or digital facilities, or alternate voice/data digital trunks. DIMENSION, System 75, and System 85 PBX's support DCS network configurations. Fig. 4 shows an illustrative configuration.

Signaling for enhanced functionality is effected by a packet-switched network using the X.25 protocol. Permanent virtual circuits interconnect the switch processors. Signaling rates range from 1.2 to 19.2 kbits/s with the rate being chosen to meet performance objectives. The transmission network is used for address signaling and call supervision. This design leaves intact basic services if there is a signaling link failure.

Tandem switching of message packets is provided by the DIMENSION PBX, and System 85. System 75 functions as an end point on a DCS network configuration. The DIMENSION signaling interface provides up to four signaling network ports. This capability, coupled with the original requirement of no more than one tandem point per path, leads to a maximum of twelve network nodes. System 85 provides up to eight signaling network ports. In addition, experience with the performance of the signaling network has lead to an increase in the number of allowed tandem points per path to two. These two changes have greatly increased the number of network nodes.

Since signaling messages may be tandem switched, there is no need for full interconnectivity. The signaling network may also utilize a different configuration than the associated transmission network.

DCS provides a logical enhancement to ETN. For a selected set of important features, the distributed network of switches functions as a single system. Four- or five-digit dialing, depending on system size, is provided for station-to-station calling on the DCS network. A planned enhancement is the capability for station users to retain their station number for moves between locations. Attendant features permit centralizing attendants for the economies such operation provides.

Features that operate essentially the same in DCS as they do on a single-switch system are said to be transparent features. The list of these features can clearly be extended to approach full transparency. This expansion will be facilitated by the growing availability of digital connectivity between nodes that will provide high-speed signaling, i.e., 64 kbits/s, on the data link.

DCS configurations can be integrated into ETN configurations. In these networks the DCS functions as a node on the ETN.

INTEGRATED SERVICES DIGITAL NETWORK (ISDN)

As digital telephony becomes pervasive in communication networks, the profusion of specialized networks will be replaced by integrated networks meeting the requirements of a wide variety of services. The integrated services digital network (ISDN) utilizes a small set of standard interfaces to access general-purpose circuit and packet-switched networks [5]. Both public and private ISDN's will evolve, although private versions are likely to appear first because of the lower embedded investment.

Private circuit-switched digital networks are available today that can provide for many of a customer's voice and data requirements. These networks employ DS-1 transmission facilities[5] and digital circuit switches. The AT&T-IS digital switches, Systems 75 and 85, utilize 8 kbit/s sampling with 8 bits/sample μ-law encoding, resulting in 64 kbits/s/channel. This encoding, which is compatible with the North American transmission standard, permits direct connection to DS-1 facilities without the need for gateways.

Two types of digital trunk terminations are provided on Systems 75/85 (see Fig. 5). The first of these provides a means of interconnecting Systems 75/85 and a nondigital switch, e.g., a DIMENSION PBX, utilizing a DS-1 facility that is terminated in a D-4 channel bank. This arrangement provides 24 voice-grade tie trunks between PBX's that are used for voice or voiceband data services. So-called "robbed-bit signaling" is used to pass supervisory information. The robbed bits are taken from every sixth frame, a rate that is sufficiently low so as not to interfere with speech or voiceband data.

A second completely digital interface is provided for DS-1 connectivity between two System 75/85's. This capability provides 23 64 kbit/s alternate voice data (AVD) channels with one channel reserved for multiplex signaling data.

Systems 75 and 85 employ the digital communications protocol (DCP) for multiplexing digital voice/data, both at 64 kbits/s, and signaling at 8 kbits/s between the switch and the station on two pairs of building wiring [6]. Data formatted using DCP can be switched digitally by Systems

---

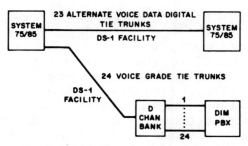

Fig. 5.   DS-1 facility and equipment options.

75/85 and transported to a remote host computer or terminal over DS-1 AVD facilities. A high-speed interface to host computers, the digital multiplexed interface (DMI) [7] provides 23 64 kbit/s data channels in the DCP format, plus a common-channel signaling circuit. DMI is very similar to the DS-1 AVD interface.

The AVD and DMI interfaces are consistent with the planned CCITT ISDN standard for primary digital network access. The current implementation of the common-signaling channel does not now employ the protocol expected in the standard. AT&T-IS expects to introduce the standard protocol when it has been finalized.

At the present time the network imposes a ones (1's) density requirement on T-1 facilities. No more than 15 zeros (0's) can be transmitted sequentially or repeater clocks lose synchronization. Systems 75 and 85 meet this requirement on a per-channel basis, i.e., at least one 1 must be present in every eight bits. If the data stream fails this requirement, the interface enforces it by replacing the least significant bit in an all 0 octet. D-4 channel banks provide a similar capability. The DCP protocol is designed to insure compliance with this requirement and thereby avoid data errors due to interface action.

Digital transmission is more cost-effective than Multipoint Private Line (MPL) service if the facility is adequately loaded. For short distances, using current common-carrier tariffs in the U. S., this translates into a requirement for 10–12 trunks [8]. If private line digital data requirements are combined with voice tie trunk requirements, and the combined traffic is carried over the DS-1 facility, DS-1 proves in even more rapidly.

Further transmission economies are attainable using digital facilities. Voice has traditionally been transmitted using 64 kbit/s channels. Recent progress in digital processing has made available robust low bit-rate transmission at 32 kbits/s using Adaptive Differential Pulse Code Modulation [9]. The commercial embodiment, known as a channel expansion multiplexer (CEM), has three options. The first of these provides up to forty-four 32 kbit/s channels over one DS-1 facility, as illustrated in Fig. 6. A 64 kbit/s signaling channel is provided for each of the AVD DS-1 inputs. These signaling channels are uncompressed by the CEM. The capability exists to flexibly mix compressed and uncompressed channels. Digital data circuits can be passed through the CEM uncompressed. Of course, they require the equivalent of two compressed channels on the interoffice facility.

---

[5] The DS-1 standard provides for 24 64 kbit/s DS-0 channels carried over T1 carrier at 1.544 Mbits/s.

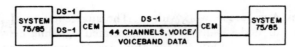

Fig. 6.   Illustrative channel expansion multiplexer configuration.

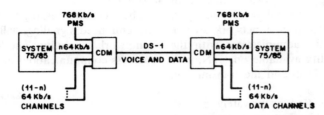

Fig. 7.   Illustrative channel division multiplexer configuration.

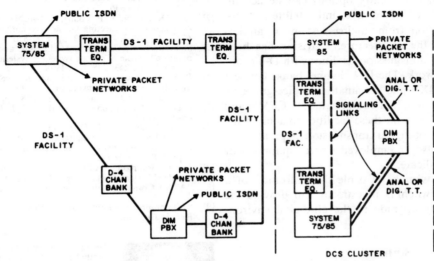

Fig. 8.   Private integrated voice/data electronic tandem network
architecture.

If there are no digital data requirements, or if one of the PBX's is an analog system, the CEM can accept the D4 format with robbed-bit signaling. Two options are available that provide either 44 or 48 compressed channels.

The compression algorithm accommodates voice and voiceband data up to 4.8 kbits/s. It permits any number of 64/32/64 coding/decoding cycles without degradation of voice or voiceband data if there are no analog links. Introduction of analog links reduces the number of 32 kbit/s links to three for data or five for voice.

The integration of private line data with the network is readily accomplished through the use of a serial device in the DS-1 facility known as a Channel Division Multiplexer (CDM). Fig. 7 shows an illustrative configuration. In the example, 12 channels are used for picturephone meeting service (PMS) video, n for switched voice and/or data, and 11-n for nonswitched data. The tradeoff between switched and nonswitched channels is completely flexible.

A similar integration of private packet networks with the circuit-switched network can also be effected. In this case, the circuit switches, perhaps with nailed-up connections, and the transparent facilities provide the desired connectivity between nodes. The packet-switch interface may be either at the CDM or directly on the PBX.

Voice and data channel capacity that have been preempted for private line data or PMS using CDM can be effectively restored using a CEM. These equipments are synergistic and provide customers with greatly improved transmission economies and flexibility in the use of transmission bandwidth.

NETWORK ARCHITECTURE

The AT&T-IS private network architecture integrates, as shown in Fig. 8, DCS regional clusters into the ETN. In its ultimate deployment, network switching is performed with digital switches interconnected with DS-1 facilities. One or more nodes of the DCS cluster function(s) as a tandem switch on the ETN. Transmission-terminal equipment, i.e., CEM's and CDM's, are employed as required for improved transmission economics and network flexibility.

The architecture allows stand-alone ETN or DCS configurations as well as integrated configurations, as illustrated in Fig. 8.

In the target architecture, the DCS nodes are interconnected by digital facilities, and signaling between DCIU's is carried on a common signaling channel.

Connectivity to both public and private networks is

provided at network tandem nodes. A mature network with full digital connectivity between all nodes will provide end-to-end digital connectivity at 64 kbits/s for terminal–host and host–host data communications everywhere on the network. Improved system management capability will be provided from a central location using an application processor associated with a network switch. This capability will provide the administrator with the system management functionality available on ETN, plus new capabilities to administer DCS and the transmission terminal equipment.

Digital facilities may be tariffed or privately owned. Since the DS-1 interface is a North American standard, many digital-transmission facility options can be accommodated, including fiber, microwave, and satellite systems.

An orderly evolution from analog systems is an important property of the architecture. Provision has been made to utilize DS-1 transmission facilities with D-4 channel banks in the network. The D-4 channel banks connect to DIMENSION PBX's and terminate 24 voice-grade tie trunks. Transmission terminal equipment, i.e., CEM's and CDM's, can be employed on these facilities to achieve the economic advantages of digital transmission. Separate analog tie trunks may also be employed where equipment or facility options so indicate.

These arrangements make possible the utilization of a customer's existing switching investment and provide the means for an orderly upgrade to the all-digital environment of the future.

Summary

Digital technology is providing the impetus for integration of business networks. Customers' separate voice and data networks will be integrated to provide more capability at lower cost. Digital transmission equipment will provide customers with flexible means to utilize their networks, sharing them over a variety of services.

The augmentation of the transmission network with an associated signaling network will greatly enhance the feature capabilities of the network. The users of such networks will be provided with features previously available only on a single system.

These capabilities, digital transmission, digital switching, and enhanced signaling, will provide a significant increase in network capabilities and lead to improved efficiencies in business communications at lower cost.

References

[1] H. O. Burton and T. G. Lewis, "DIMENSION AIS/System 85—System architecture and design," in *Proc. ICC '83*, vol. 2, June 1983, pp. 831–836.

[2] D. A. Keller and F. P. Young, "DIMENSION AIS/System 85—The next generation meeting business communications needs," in *Proc. ICC '83*, vol. 2, June 1983, pp. 826–830.

[3] S. E. Bush, R. Carlsen, I. M. Lifchus, and M. J. McPheters, "Expanding the role of private switching systems," *Bell Lab. Record*, vol. 57, no. 9, pp. 243–248, Oct. 1979.

[4] R. S. Divakaruni, G. E. Saltus, and R. B. Savage, "New dimensions in enhanced voice networking," in *Proc. 6th Int. Conf. Comput. Commun.*, London, England, Sept. 1982, pp. 362–369.

[5] I. Dorros, "Telephone nets go digital," *IEEE Spectrum*, vol. 20, pp. 48–53, April 1983.

[6] G. M. Anderson, J. F. Day, and L. A. Spindel, "A communications protocol for integrated digital voice and data services in the business office," in *Proc. 6th Int. Conf. Comput. Commun.*, London, England, Sept. 1982, pp. 367–371.

[7] A. R. Severson, "AT&T's proposed PBX-to-computer interface standard," *Data Commun.*, pp. 157–162, April 1984.

[8] "AT&T Communication's (proposed) private line tariff," FCC Tariff 3.

[9] "Bit compression multiplexing," AT&T Commun. Preliminary Tech. Ref. 54070, Jan. 1984.

George M. Anderson (S'42–M'49–SM'52) is currently head of the Networking Systems Engineering Department at AT&T Information Systems, Lincroft, NJ. His experience with the prior Bell System spans 20 years and includes experience with Bellcomm on the Skylab program. Prior to joining the Bell System, he served as Assistant Professor at Carnegie-Mellon University, Pittsburgh, PA, and for ten years was Director of Engineering at the Thomas A. Edison Laboratory, West Orange, NJ.

Network of the Future

THOMAS E. BROWNE, MEMBER, IEEE

Invited Paper

Telecommunications networks of the future will exploit two new network architecture concepts that are currently being implemented, or soon will be. These are the Intelligent Network and ISDN, the Integrated Services Digital Network, which together will support a full range of voice, data, and image services that Information Age telecommunications users will demand. These new network architectures, operating synergistically with intelligence in terminal systems, will constitute a framework in which users and service providers will link together standardized functional components to create customized services. These components, along with interfaces and signaling protocols at the interfaces and within the network will result from continuing national and international standardization efforts.

In the planning of these new architectures, a few major goals are of paramount importance:

- *the achievement of a flexible network structure in which functionality is distributed among the network components in a way which supports the timely and economic introduction of new services in response to user needs;*
- *the establishment of industry standards at the interfaces between network elements such that service suppliers can choose among a set of available systems products in building their networks and avoid dependence on a small set of suppliers;*
- *the development of standard user interfaces supporting signaling procedures which can provide the user with increased control of, and access to, services to satisfy his needs;*

Achievement of these goals will result in the realization of an Open Network Architecture.

The ISDN and Intelligent Network architecture concepts are described in this paper.

INTRODUCTION

The characteristics of telecommunications networks largely reflect the technology employed by users connected to the network, and which send and receive information through it. In today's telecommunications networks, the user technology is dominated by the telephone station set. Accordingly, the network interfaces, and the internal network structure and operation reflect the dominance of the circuit-switched service used for telephony.

The dramatic growth of data communications in the last decade has relied on the use of modems at each data

Manuscript received February 28, 1986; revised May 1, 1986.
The author is with Bell Communications Research, Inc., Livingston, NJ 07039, USA.

terminal to convert digital data into an analog form suitable for transmission on the switched telephone network. To be sure, the present telephone network utilizes vast amounts of digital technology in the provision of telephone and data services, but its application has been motivated principally by the favorable economic characteristics of the technology, and its digital nature has not been fully exploited in a functional sense.

However, it is important to recognize that the characteristics of telecommunications networks, and the capabilities they provide for users to manage the generation, processing, distribution, storage, and retrieval of information are one of the limiting parameters of the information technology infrastructure supporting society. For the "Information Age" user, the limitations of the present networks could have a stifling effect, and these limitations must be removed. The availability of sophisticated digital technology for both business and residential users to apply in their information intensive work makes it possible for the user to change the media used in his communication from voice to data, to images, and to combinations of these very rapidly, and he will expect the communications network to carry it all.

New Network Architecture Concepts

Two new network architecture concepts that are now emerging will shape the future of telecommunications networks. These are the introduction of database-derived controls of the switching functions in the network, known as the "Intelligent Network," and the Integrated Services Digital Network, ISDN. These architectures exploit the digital switching and transmission technology that has been growing since digital carrier systems were introduced in the 1960s. The full potential of these architectures to satisfy the emerging needs of an information-driven economy will be realized with the application of fiber-optic transmission technology in all parts of the network, especially in the customer access.

These network architectures are responsive to both user's needs and to the availability of new user technology, as well as to the user's demand for control of network facilities and services. They also recognize the initiatives of the network providers in applying new technology to improve service, and manage their costs, and the emergence of a new structure for the telecommunications industry in the U.S. and in several other countries around the world.

In this paper, the architecture principles embodied in the present telecommunications networks are described, together with these new architectures which will constitute the network of the future. The development of national and international standards for user interfaces and the interfaces between the major network elements is also discussed.

PRESENT NETWORK ARCHITECTURE

The major elements of the present (largely digital) network are depicted in Fig. 1. Telephones and data communications terminals are connected to the circuit-switched network. Also, data terminals may be connected to packet-switched networks and through leased circuits (non-switched) to access databases supporting information storage and distribution services. In essence, the present telecommunications network consists of several separate networks, each having specific capabilities for providing certain services. In general, each of these networks has its own topology and terminal identification scheme, resulting in the terminals connected to each having network-specific addresses, access procedures, and signaling protocol.

Each of these service-specific networks have the internal structure shown in Fig. 2. Users in a geographical area are connected to an "end-office" which provides for interconnection between user terminals connected to that end-office, and concentrates the traffic destined for users served from other end-offices. End-offices may be connected together directly, or through a tandem office, depending on the amount of traffic between them. The network is de-signed to achieve minimum total cost by trading off the cost of switching and transmission investment, resulting in the sizing and placement of switches in relation to the location of customers and the intensity of demand for service. Another important aspect of this network topology is that the service characteristics of each network are those that are wired, or programmed into the switches in the network.

In the United States, changes in public policy during the last several years, and the divestiture of the local exchange network operation by AT&T in 1984 have given rise to another feature in the present networks that has special significance. These events have resulted in many of the service-specific networks discussed above also having the structure depicted in Fig. 3. In this figure, the user-to-user connection is seen to be comprised of five service elements, each furnished by a different supplier.

The terminal equipment at each users premises is connected to the network of a local service provider, which may be a local exchange carrier (e.g., a Bell Operating Company) or another service provider such as a CATV network. These local networks are, in turn, interconnected through the networks of long-distance service providers. This five-part connection makes it mandatory that well-defined standards exist for the interfaces between networks, and at the user interfaces to assure that the service received by the user complies with his requirements and applicable standards for service quality. The differing business plans of industry participants, the varying technology plans among equipment manufacturers, and the still-developing regulatory framework for the industry make the establishment of these standards both a high priority and a technical challenge which will be achieved only with the full cooperation of all the participating organizations, and

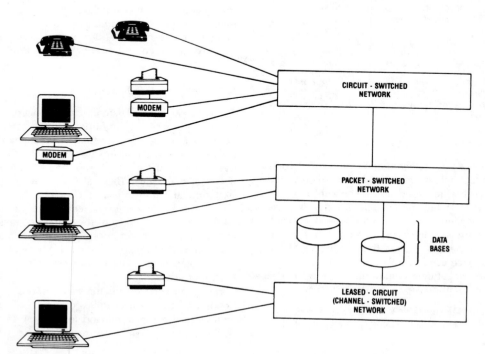

Fig. 1. Present network architecture.

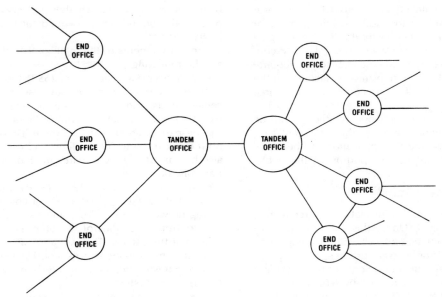

Fig. 2. Service network structure.

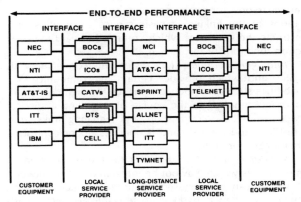

Fig. 3. Telecommunications industry interfaces.

with the highest dedication of the professionals that are involved in it.

This policy environment, and these technical and economic characteristics of present telecommunications networks, result in users generally connecting to several networks to obtain the services they need to satisfy their telecommunications needs. In most cases, the user's access line is a two-wired twisted pair, which for telephone service is time-shared between signaling and end-to-end signal transmission. Data communications devices using the telephone network emulate telephone sets (using voice-band modems) to communicate across the network. Also, computers emulate the human telephone user when directing the network to establish connections. User access lines to packet-switched and leased-circuit networks are generally analog telephone lines, although in recent years digital technology has found application in this area where higher data rates are desired by the users.

The characteristics of present networks discussed above can be summarized as follows:

- The user has a specific network interface for each kind of service: for example, a "plain-old-telephone" (POTS)

interface, a digital private line (e.g., DDS) interface, a packet service interface which may use a digital or voice-grade private line for access, and perhaps a wide-band interface for video service (today a private-line service also).

- The service available at each of these interfaces is exactly what the interface is specified to support, and the user cannot change the service without the network provider also changing the facilities connecting the user to the network, a labor-intensive (and possibly also capital-intensive) operation.

- The switching and signaling capabilities available at an interface are those that are equipped in the end-office that the access line is connected to, and the provision of new services to the customers served by a given end-office requires the upgrading of that end-office by the network provider before the services can be offered.

- The bandwidth of customer access facilities is generally limited to a few kilobits per second, or where T-carrier facilities are used to 1544 kbits/s, and will not conveniently carry video signals on a single access line without the use of costly signal processing devices in the terminal.

These characteristics of the present network architecture support a rich mix of services to the user population, and have supported the growth of the telecommunications industry for the past century. However, as users become accustomed to the sophistication of emerging information technologies, they will expect (demand) more flexibility and control of network resources than the current service-specific interfaces allow. Also, network providers will look for ways to gain independence from equipment manufacturers in the implementation of new services and capabilities to respond to their customers in a timely way. Finally, as users perceive the value of communicating in images as well as voice and digital data, the bandwidth limitations of the facilities connecting them to the network will be seen as an impediment to the growth of business or family activities. This is evident already in the proliferation of local area networks using wider bandwidth facilities than commonly available in public networks for the sharing of digital data among information systems within a relatively confined geographical area. This user need places a requirement on the bandwidth of the transmission facilities in the public networks to provide for interconnecting local area networks.

These limitations of today's network architecture will be removed with the implementation of the Intelligent Network architecture, ISDN, and the application of fiber-optic technology in the customer access portion of the public networks.

INTELLIGENT NETWORK ARCHITECTURE

The elements of the "Intelligent Network" architecture plan are depicted in Fig. 4. Under this concept, the allocation of network functions and call control among network elements departs from present practice, resulting in a new distribution of these functions. Some of this is evident in the "800 Service" presently offered by AT & T. It is also evident in the some of the "800 Service" plans of the local telephone companies. In this service, the customer-dialed number is not the network address of the terminating line. Rather, the dialed number is used as an index into a database which contains the terminating line network address, and perhaps also the identity of a long-distance service provider which the "800 Service" customer has selected to carry the long-distance portion of the calls. Also, the "800" customer will be able to specify other parameters of the service to suit his business plans. These parameters would allow, for example, the customer to specify a different geographic location for the call to terminate at depending on the location of the calling customer, time of day, or day of the week.

In the present network architecture, the concentration of functionality in the software programs and hardware of the switching systems has, at times, been an impediment to the rapid introduction of new services by network providers. Also, it has complicated planning services where there has been uncertainty about whether the service falls within or outside the boundaries prescribed for the carrier by regulatory authorities.

The key elements of the Intelligent Network architecture are the following:

• A *Service Switching Point* (*SSP*) function which controls the interconnection of transmission facilities, and receives, sends, and processes signals originating at either a user terminal or a Service Control Point. The SSP function can be part of an end-office, where it provides service for both trunk transmission facilities and user terminals. It can also be part of a tandem switch interconnecting only trunk transmission facilities.

• A *Service Control Point* (*SCP*) function interconnected with switching systems by a common-channel signaling network. Service logic and data at the SCP and the SSP control the actions of the other network elements in providing the service. The SCP also can collect traffic data and other measurements to support network management and administration, and it interfaces with the Service Management System.

• A *Service Manager* (*SM*) provides administration, coordination, and control of databases associated with the various services supported by the network. It provides the interface to operating personnel and users to specify services in terms of the linking of basic service capabilities, and constructs the databases which control the real-time connection functions of the service.

• *Intelligent Peripheral* (*IP*) system interconnected with an SSP, and which provides certain telecommunications capabilities such as announcements, DTMF digit collection, and others required in the course of providing a specific service. An IP has a transport and signaling links with end-offices and the SSP, and it operates under control of either an SCP or an SSP.

• *Vendor Feature Nodes* (*VFN*) for the distribution of some services which may not be integrated in the network, but which can be accessed by users connected to the network. This new network element is interconnected with the switch by means of a standard interface (e.g., the ISDN

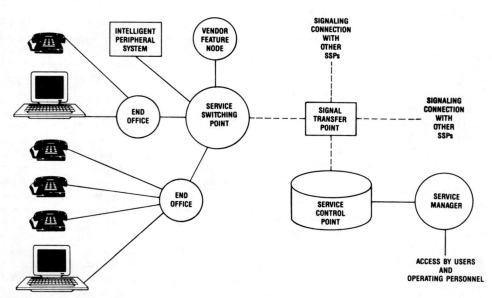

Fig. 4. Intelligent network architecture.

interface), and is accessed by the user through the switch. Only minimal hardware and software changes are required in the switch to support the service. Also, the VFN can be designed to have the capacity and characteristics required for only the new service, and not all the other services that the existing switch software and hardware provide.

Service Control

In the Intelligent Network architecture plan, many of the call-processing and decision-making functions which reside in switching systems in the present network have been distributed among these new network elements.

The SSP function, for example, resembles the call-processing logic in today's network switches. It performs an analysis of the user input to determine what service capability has been requested. If the switch is equipped to fully provide the service, it does so. However, if the user request includes information specifying a service capability which the switch cannot, by itself, deliver (e.g., 800 Service), the SSP requests the help of the SCP to process the service request, and uses the common channel signaling network to send the information supplied by the user to the SCP for analysis.

The SCP will interrogate information stored in databases to determine how the service request is to be satisfied. This could result in the establishment of a circuit-switched connection from the user to an Interexchange Carrier. Another possibility is that the SCP could effect the establishment of a connection from the user to a service vendor system providing voice message storage, or perhaps a data service.

In either of these cases, the SCP might first establish a connection from the user to an intelligent peripheral system providing a voice announcement (or a data message) to the user as an indication of state of the service request.

Technology Implications

These simple examples illustrate the following very important aspects of the Intelligent Network architecture:

• Compared with present switching systems, the required switch functionality is somewhat simplified, at least conceptually. It is reduced to a set of fundamental call processing, switching, and signaling tasks common to all the services. A given service would use a subset of these fundamental tasks, and a particular subset may be used by several services, although perhaps in a different order, or with different controlling parameters specified by the SCP for the service. In this way, the software of the switch need not be upgraded to implement a new service, since the switch would already have the software for the fundamental tasks.

• The switch is directed by the SCP through the necessary fundamental tasks in completing a service request in all cases where it cannot do so on its own, whereas today's switches generally are programmed to provide certain services, and only those, and to reject all other service requests. The significance of this architectural feature is that many new services can be made available to users without having to upgrade the end-offices themselves, but only the SCP.

• The use of intelligent peripherals provides for the delivery of new, user-oriented, service progress information to end-users without the installation of new hardware and software in all switching systems.

• The SM provides for the distribution of new services to a large population of users, terminated on several switches, by making software changes at one place (the SM) rather than at every switch. As ISDN access lines are introduced, customers using those lines will be able to interact directly with the SM to customize network capabilities to create services satisfying their own particular requirements.

The availability of standard interfaces for connecting service nodes will provide opportunities for the implementation and economic distribution of new services using the underlying capabilities of the network. This is perhaps one of the most important aspects of the Intelligent Network architecture plan. Its implications in terms of technology, the economics of new services, and the achievement of some of the goals of U.S. public policy are profound.

Signaling Network Architecture

Obviously, the architecture described here requires efficient signaling among the network elements. This is provided by a common-channel signaling network (CCS) which includes digital transmission terminals at every node interconnected with high-speed digital transmission links and signal transfer points which route the signaling messages through the network. Also, the network design will include safeguards to ensure the security of network control information and prevent a user from compromising network integrity and the quality of service provided. This requires that the network be implemented using interfaces between network elements which are well defined, and comply with the principles of an "open architecture" concept.

By means of this signaling network, any element can send and receive signaling information to every other element. The information exchanged among the network elements is carried in messages that are specified according to a layered structure in which the control functions and procedures at each end of a link, and at each signaling point, are unambiguously defined. Signaling System No. 7 (SS7) is the internationally standardized protocol that will be used in this network. Transaction Capabilities (TC) in the SS7 protocol specification provides a facility for the exchange of requests and responses between systems to remotely invoke operations in other systems. TC is the latest addition to the SS7 protocol, and it will support a variety of services that require transactions to be exchanged among the network elements in the Intelligent Network architecture plan. In TC, a relatively small set of non-service-specific capabilities is provided to be used by a large variety of services.

The specifications for the messages and procedures of the SS7 protocol are given in recommendations that are being developed in national and international standards-making bodies. These specifications are being developed for application in signaling networks supporting service of both the present voice and data networks, as well as future networks in which voice and data services are closely associated, and even integrated. This, of course, is the other new architecture plan that is shaping the future of telecommunications networks today, the Integrated Services Digital Network—ISDN.

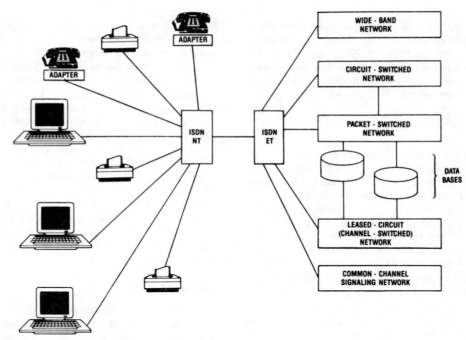

Fig. 5. ISDN architecture.

INTEGRATED SERVICES DIGITAL NETWORK ARCHITECTURE

Fig. 5 depicts the Integrated Services Digital Network Architecture. User terminals, supporting both voice and nonvoice services are connected through a network termination (NT) to a digital access line to an exchange termination (ET) at the end-office. The ET provides access to the circuit-switched, packet-switched, leased-circuit networks to obtain the same services available in today's networks. In addition, the ET will provide access to the common-channel signaling capabilities of the Intelligent Network and, in the future, also to wide-band network facilities supporting video and higher speed data capabilities.

An important characteristic of the ISDN access is that control signaling information is carried in a separate channel from the signals carrying other information for end-to-end communication. As discussed below, this access line may be implemented in several ways, using different technologies, but in each implementation it is capable of carrying signals for several different services simultaneously. The service is limited principally by the bandwidth of the access line technology chosen by the user.

Integrated Access

The implications of the integration of several services onto a single user access are substantial. First, the use of the same access line for multiple services yields cost advantages for both the network provider and the user. Second, the bandwidth provided on this access line will allow several services to be accessed by the user simultaneously, without their interfering with each other. This integration will allow the user to create connections in which a combination of capabilities are used, constituting a "multimedia" service, such as voice and data, or voice and image interconnections between two or more terminals.

Also, the service integration in a single access line is accomplished with a standardized, service-independent interface, allowing the implementation of a wide range of terminal capabilities. This permits the user to upgrade his terminal without changing the interface or access line to support the services his terminal may be equipped to provide.

Although Fig. 5 shows only a single access line serving each user terminal, there currently are two standardized interfaces. These are depicted in Fig. 6, which shows the "Basic Access" and "Primary Access" interfaces. The Basic Access interface is provided using twisted wire pairs covering a range of up to 18 kft from the Central office, or, as shown in the figure, from the remote terminal of a digital carrier system. The Basic Access provides two so-called "B" channels operating at 64 kbits/s which are used to carry customer information from a terminal through the circuit-switched or channel-switched network to another terminal. The basic access also includes a "D" channel operating at 16 kbits/s carrying customer signaling information and customer packet data for transmission through the packet-switched transport network.

Fig. 6 also shows the other standard interface, the "Primary Access." This interface provides 23 B channels and a 64-kbit/s D channel which, except for the bit rate, has the same logical characteristics as the 16-kbit/s D channel in the Basic Access. The figure shows that a Primary Access line may terminate in the Network at either an ISDN switch or a digital cross-connect system. In the first case, the B channels are circuit-switched although they may also be reserved for nonswitched services if the switch is equipped to provide that service. In the second case, where the access line terminates on a digital cross-connect, some B channels may be reserved for nonswitched services and the rest are connected to the switch for circuit-switching. In both cases, the channels reserved for nonswitched services

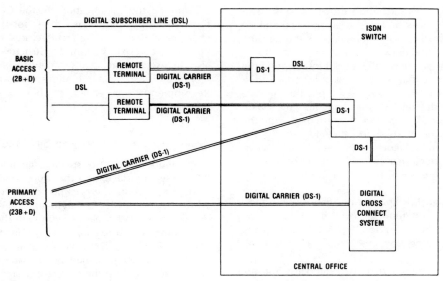

Fig. 6. ISDN access architecture.

are connected to the "channel-switched" transport network.

Technology supporting the Primary Access is presently available. It is the technology of the existing T-carrier digital transmission systems that have been applied in the interoffice transmission systems for the past 20 years. More recently, this technology has been applied in the feeder distribution plant as digital Subscriber Line Carrier (SLC) systems, on access arrangements for customers with special requirements, and to terminate directly at digital PBX systems on customer premises.

Technology for the Basic Access is currently being developed, and is expected to be available commercially by the end of 1987. Two approaches to the bidirectional transmission of the 2B + D channel signals on the two-wire loop have been the focus of intense study for the past several years. As these signals comprise in excess of 144 kbits/s in each direction, the challenge of achieving error-free transmission on the existing loop plant is formidable indeed. One approach, called "Time Compression Multiplexing" (TCM), utilizes unidirectional transmission alternately from network to subscriber, and then from subscriber to network. This results in a bit rate on the loop in excess of two and one-half times the net bit rate transmitted in each direction. The signal processing required at each end of the loop in this case is relatively straightforward, but this approach places energy in a wide spectrum on the loop, and it is very difficult to achieve the objective 18-kft loop range.

The other technology being considered for the ISDN Basic Access uses simultaneous transmission of the 2D + D signal in both directions on the loop with sophisticated digital signaling processing at both ends to discriminate the received signal from the transmitted one at each end. This technique, known as "Hybrid Echo Cancellation" (HEC), uses considerably less spectrum then TCM but requires more sophisticated signal processing in the terminal equipment. Currently, the HEC approach is favored due to the expectation that continued progress in semiconductor technology will yield cost-effective solutions while the conservation of spectrum will allow the achievement of the loop range objective. Standards for the design parameters of HEC devices is the subject of intense efforts in the national and international standards-setting bodies.

Wide-Band ISDN Access

Eventually, there will be a third standard ISDN access interface based on the availability and application of fiber-optic transmission technology for the user access line. This interface will support all of the services which the Basic and Primary accesses support, and will add to those video and image services utilizing wide-band transmission and switching capabilities in the network. Pressures for this extension of the narrow-band ISDN architecture stem principally from the desire of users to access these services, and the present availability of this technology in service-specific network configurations, such as local area networks and cable TV. From a services perspective, these two examples represent extremes in terms of the definition of the user interface and the standards that would govern the development of technology for it. Also, optical-fiber and semiconductor technology continue to evolve, and consensus has not yet been achieved among industry experts on which of the available choices will (or should) prevail in arriving at a standardized wide-band ISDN interface to satisfy the widest possible range of user needs.

Clearly, evolutionary forces will have to be accommodated including the need to provide metropolitan area networks interconnecting LANs at higher bandwidth than narrow-band ISDN, as currently planned, will support. Some of these networks are already being implemented, so the need for reaching consensus on these issues is especially urgent. Given this state of the developments in this area, further discussion of wide-band ISDN and the underlying technical matters is beyond the scope of this paper.

Message-Oriented User Signaling

Equally important as access integration in the ISDN architecture plan is the carriage of all customer signaling in a separate channel from those that are used to carry other customer information. This separation allows the process of signaling between the user and the network to take place without interfering with user communications through the network. This allows the network, for example, to notify the user of the arrival of a new call without interrupting the communications on one that is already established. Furthermore, in this case the busy user could possibly send a signaling message to the network specifying how the "waiting" call is to be handled, without interfering with the communications on the established call. Alternatively, the "busy" user could send a signaling message to the "waiting" caller to the effect that he will return the call, or enquiring as to the subject of his call, also without interrupting the existing call.

On this separate signaling channel is organized into messages, with network and terminal control functions and procedures specified in a "layered" model similar to the Open Systems Interconnection model used in data communications.

Although national and international standards for the messages and procedures are still evolving, the departure of this approach from the signaling presently used for telephony has special significance in terms of the services that can be supported by this network architecture. Current telephony signaling uses very simple signals; namely, on-hook, off-hook, and the digits 0–9 as the symbol alphabet for passing information between the user and the network. This implies very limited functionality in the terminal equipment, and calls for the user to represent a service request in only very elementary terms. By contrast, the ISDN D-channel signaling messages allow the user to exploit sophisticated digital technology in terminals for communicating with the network to specify addresses (perhaps using the name of the person the connection is to be made with), network service capabilities, and network control functions. It also allows the network provider to implement services relying on the presence of that technology in the terminal equipment, resulting in opportunities for users, terminal providers, and network providers to implement solutions that cannot be supported with today's networks.

STANDARDS

Implementation of the ISDN and Intelligent Network architectures is critically dependent on the adoption of technically sound specifications for the user interfaces, the interfaces between network elements, and the protocols to be followed in communicating across these interfaces in the provision of services. Much progress has been made in the development of national and international standards for the ISDN user interfaces and access signaling protocols, and for common channel signaling protocols during the past several years. This effort has been concentrated in the International Telegraph and Telephone Consultative Committee (CCITT) Study Groups XI and XVIII, culminating in the adoption of Recommendations in 1984. These Recommendations have provided an excellent starting point for the further development of the specifications.

This work is continuing both in CCITT as well as in Committee T-1 sponsored by the Exchange Carriers Standards Association, and which is accredited by the American National Standards Institute (ANSI). The activities of this group, which has representation from all segments of the telecommunications industry in the U.S. have been focused on refining the specifications in the CCITT Recommendations to assure that a basis exists for the earliest possible implementation of these architectures, consistent with the emerging competitive industry structure in the U.S. It has also focused on achieving timely resolution of some key technical issues that are unresolved in the 1984 CCITT Recommendations. A few of these are discussed briefly below.

Access Signaling and Network Signaling

The work in CCITT to specify signaling procedures at the ISDN user interface, and the work to specify the network signaling procedures was carried out in two different groups, which had somewhat different aims. The network signaling development had already begun with the goal of achieving a signaling system to support only telephony service several years before the ISDN access signaling work started. As a result, much progress had been made in developing this specification before there were any significant results from the study of ISDN. When these results did emerge, they called for some changes to the network signaling specifications. Reconciliation of the differences is a major goal in both CCITT and T-1 at this time.

Stimulus and Functional Signaling

In drafting its 1984 Recommendations, the CCITT included so-called "stimulus" procedures for terminals having relatively low functionality, such as a simple digital telephone, and "functional" procedures for ISDN terminals implemented with sophisticated logic, memory, and control capabilities. In planning initial implementation projects, controversy has arisen among implementors over the merits of such a "bifurcated" approach. It is argued on one hand that this creates ambiguity in the specification, complicating the design of both terminal and network systems. This position is countered with arguments that this introduces flexibility and is necessary until specific services are defined by network providers and users which then can be that basis for specifying only one approach. It is also suggested that this approach allows for economic evolution of the terminal products and networks.

User Access Technology

One area where there has been great progress in the development of ISDN interface standards relates to the physical interface specification. Based on extensive technical evaluation of alternative technologies consensus is developing that the use of the HEC technique at both ends of the loop is feasible, and is preferred over the alternatives available at this time. This technology is now proved to be capable of supporting reliable transmission of the line rate required for 2B + D access on greater than 95 percent of the existing nonloaded plant in the Bell Operating Company networks in the U.S. Furthermore, research indicates that this is expected to improve rapidly with continuing advances in semiconductor technology. The key technical issues remaining to be solved in this area include the specification of the line code, and parameters of the echo-canceler designs to allow devices supplied by different vendors to be applied at the two ends of the access line.

Internetworking

Although the basic network architecture concepts for ISDN and the Intelligent Network are now established, much work is needed in the area of interconnecting networks. The development of the specifications for ISDN interfaces has been dominated by concern for the user interface, with the result that many aspects of the interconnection of ISDNs, and the interconnection of an ISDN with another (non-ISDN) network have not yet received sufficient attention among the standards groups. The competitive nature of the United States telecommunication industry makes this an area of urgent study.

Performance

Performance is another area where the development of effective standards is in the area of performance. In this context, performance certainly has the traditional meanings relating to signal quality and the allocation of impairments to the various parts of the user-to-user connection. It also has another, possibly more important, meaning relating to service functionality and the extension of access and control of network capabilities to users and services providers outside of the traditional framework of the network. The standardization of protocols governing this access and control may not be adequate, and there may be need for assuring that the various implementations of those protocols are equivalent, or equal, on the basis of performance measures related to the requirements of emerging industry policy principles.

Conclusions

The two new network architecture concepts which are expected to be the foundation for the telecommunications networks of the 1990s are ISDN and the Intelligent Network. In the planning of these architectures, a few major goals are of paramount importance:

- the achievement of a flexible network structure in which functionality is distributed among the network components in a way which supports the timely and economic introduction of new services in response to user needs;
- the establishment of industry standards at the interfaces between network elements such that service suppliers can choose among a set of available systems products in building their networks and avoid dependence on a small set of suppliers;
- the development of standard user interfaces supporting signaling procedures which can provide the user with increased control of, and access to, services to satisfy his needs.

These goals, and achievement of them, constitute another concept which networks of the future must realize. This is the Open Network Architecture concept. The Intelligent Network and ISDN architectures exploit the inexorable growth of the application of digital technology in both user equipment and network systems. They also effectively respond to the commercial, sociological, and economic pressures of the "Information Age" by providing a basis for increased competition among the industry participants, and rich opportunities for the user to extract greater utility and benefit from the telecommunications infrastructure. While there are many technical challenges in both the design of the networks and the development of effective standards, the direction is clear. This paper has attempted to provide a high level description of the key elements of these concepts.

Acknowledgment

The author gratefully acknowledges the contributions of many colleagues in the Network Architecture Planning Center and the Services Planning and Implementation Center at Bellcore to the work and concepts reported in this paper. In particular, W. Gifford is deserving of special mention for his pioneering work on ISDN and for thoughts shared in many stimulating discussions on principles of network architecture.

Customer Installations for the ISDN

G. Robin

Essential characteristics of future systems

I NTERNATIONAL standards on the Integrated Services Digital Networks (ISDN's) are actively being prepared by the CCITT, and many Recommendations will be approved by the 1984 Plenary Assembly. Most of this activity is devoted to the definition of services offered by the ISDN, the architecture of public networks, and the universal ISDN terminal interface.

Considering planned applications, it is clear that the first customers of ISDN will be business users. For some years, the penetration of ISDN in residential subscriber installations will be marginal, and most ISDN terminals will be connected to private installations serving business users, PABX's, intercom systems, or local area networks. Surprisingly enough, very little has been done so far to identify the problems of private ISDN systems, to specify their requirements, and even further, to propose solutions. The aims of this paper are to indicate what can be expected from the introduction of ISDN in the business environment, and to identify the essential characteristics of future systems serving this market.

Examples given in this paper are based on the European digital hierarchy (primary rate-2.048 Mb/s) which is more familiar to the author, but the same conclusions could equally be applied to the North-American hierarchy (primary rate-1.544 Mb/s).

ISDN Services

The ISDN will provide access to a wide spectrum of services over a single type of digital interface, with a common basic access procedure. Several types of services are being considered with reference to the seven-layer model of the Open-Systems Interconnection (OSI).

1) Bearer services use the connection types defined for the ISDN and characterized by protocol layers 1–3 of the reference model. Currently, the following services are defined and offered by a public ISDN:

- 64-kb/s circuit-switched service with transparent information transfer;
- 64-kb/s circuit-switched service, with nontransparent information transfer (as dictated by specific requirements of voice transmission);
- 64-kb/s leased circuit service, both transparent and nontransparent;
- virtual circuit-switched service established on a B channel (access to a packet-switching network);
- virtual circuit-switched service established on the D channel;
- connectionless packet service on a D channel (datagram type service proposed for telemetry and alarms).

Optionally, the customer installation may provide additional low layer functions (ALLF), also belonging to layers 1–3. These are commonly

Reprinted from *IEEE Communications Magazine*, April 1984, pages 18-23.

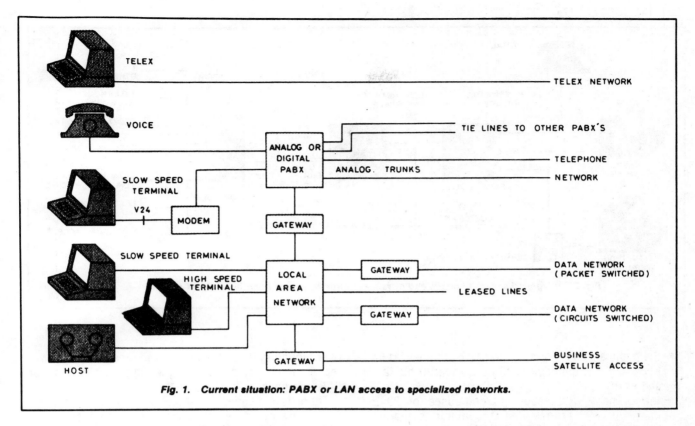

Fig. 1. Current situation: PABX or LAN access to specialized networks.

referred to as "facilities" or "supplementary services," for example, abbreviated dialing and call transfer.

2) Services which require all seven protocol layers and are sometimes called "value-added services"—such as information storage and retrieval, and message store and forward. Such functions are necessary for business users and may be integrated into the communications system, or segregated in a dedicated system interworking with it.

3) Telecommunications services are fully specified for all protocol layers, including functions performed by the terminals. They make use of bearer services, ALLF, and high layer functions. Examples of such services are Telephony and Teletex, which are standardized by CCITT.

From this analysis, it can be seen that, as a minimum, a subscriber installation for the ISDN must provide some or all of the bearer services since higher-layer functions may reside in the terminals or in specialized (private or public) servers accessed over the ISDN.

The current definition of ISDN also includes wideband services, but ISDN exchanges are only able, in a first step, to switch 64-kb/s channels. Digital information at rates higher than 64 kb/s can be carried by an aggregate of several 64-kb/s B channels, but digit sequence integrity is not provided by the network, and some capacity taken from the $n \times 64$-kb/s rate will be needed for that purpose by the terminal. However, enhancements of the ISDN must be anticipated and customer installations have to consider possible future requirements for wideband switched services.

The bandwidth required by video services eventually will be of interest; solutions offering such connections, possibly over a separate connecting matrix but using the same control channel and protocol, have to be investigated.

Interfaces for Business Installations

External Interfaces

Currently, businesses have separate accesses to several telecommunications networks: telephone lines, telex lines, data lines accessing packet-switched or circuit-switched networks, as well as leased lines for various purposes such as tie-lines to other PABX's constituting a private network (Fig. 1). This diversity does not permit optimization of the use of transmission or switching resources. Moreover, some accesses are not switched within the customer premises, and therefore a single terminal is connected to an external line: this is the case for telex, a telecommunications service which ends at the local mailroom and must be relayed by conventional internal mail for final distribution, thereby losing some of its speed and efficiency advantages.

In the case of ISDN, all these external interfaces will be replaced by a single one at the *T* reference point (Fig. 2). Depending on the size and traffic requirements, two types of access are currently candidates for standardization:

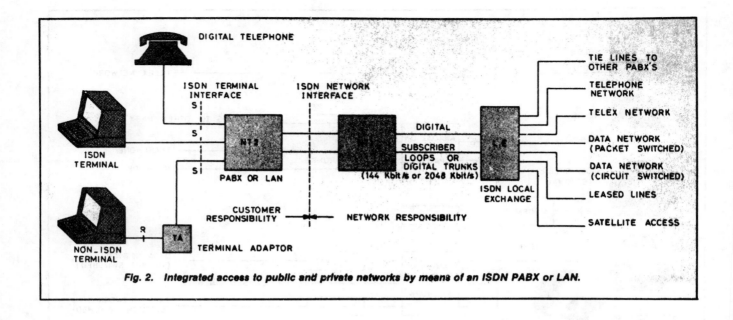

Fig. 2. Integrated access to public and private networks by means of an ISDN PABX or LAN.

- the basic access is composed of 2 B channels at 64 kb/s and 1 D channel at 16 kb/s (overall bit rate 144 kb/s);
- the primary rate access is composed of 30 B channels at 64 kb/s and 1 common signaling channel at 64 kb/s (2.048 Mb/s); another channel (channel 0) is reserved for synchronization and maintenance.

Some European administrations, considering that there is a gap between 2 and 30 B channels, are proposing an intermediate multiplex bit rate. Multiplexes carrying 9 or 6 B channels have been proposed, the latter providing a better definition of size ranges. However, at this point in time, there is little chance of agreement for an international standard on this subject.

As already mentioned, these access types can provide real or virtual circuits for all services, and therefore there is no point in providing both analog and ISDN trunk lines to a digital PABX. The necessary condition for installation of an ISDN PABX is that the area is served by an ISDN local exchange or remote unit. This condition will not be met everywhere at once when the ISDN service is opened. However, it is likely that telephone operating companies will make efforts to serve business areas reasonably soon. It is also possible to connect an ISDN PABX by means of foreign exchange lines to an ISDN exchange located in a neighboring area, using standard digital transmission systems. In this manner, a small number of ISDN exchanges could be sufficient to provide ISDN service to business users over a large territory.

The ISDN access can also offer permanent real or virtual circuits which are established over the local exchange, eliminating the need for leased lines and manual patching at the distributing frames. The D channel signaling system is designed to carry user-to-user signaling in a standardized manner, and this will replace the many different signaling schemes used for PABX tie-lines.

There is, however, a difficulty when the first ISDN PABX's are introduced in a private network which cannot be converted to ISDN in its entirety, due to economic or logistic reasons or to lack of digital access capability in some parts of the network. Therefore the coexistence of ISDN and conventional tie-lines will remain a necessity during the transition period.

Internal Interfaces

The universal interface specified by CCITT (S/T interface) is meant to be both the public exchange "basic access" and the terminal interface; therefore, ISDN PABX's will offer the S interface on all *digital* extension lines.

The S interface is designed to operate over a distance of about 1000 m in a point-to-point configuration, and over a distance of 150 m in a multipoint (passive bus) configuration. The overall bit rate across this interface is 192 kb/s. The distance of 1000 m should be sufficient for most centralized PABX installations. The basic S interface is 4-wire, in contrast with the 2-wire analog loop. Solutions exist for 2-wire digital transmission and indeed will be used for public exchange lines, and also perhaps for extension lines; it is expected that these 2/4-wire transmission end systems will become fairly cheap when specialized VLSI circuits become available (either for burst-mode or echo canceler types of transmission). There will be a trade-off between 2-wire transmission requiring these end systems and direct 4-wire transmission; the break-even distance will depend on local circumstances such as the existence of available spare cable capacity. It is likely that, in many cases, 4-wire transmission will prove more economic for the majority of extensions, and that only a small number of 2-wire interfaces will be provided for the longer loops.

There are three possible ways to use the S/T internal interface for PABX extensions:

1) As a T interface, that is, in a point-to-point configuration controling a network termination NT. The NT permits connection of a number of terminals on a passive bus of 150-m length;
2) As an S interface serving a single terminal. In this case, the full 1000-m distance can be covered;
3) As an S interface serving a passive bus directly. Such a passive bus can connect up to 8 terminals.

As the S interface serving an extension will be given a single directory number (network address), several modes of operation are possible for incoming calls:

1) In the case of point-to-point operation, a single terminal is addressed unambiguously.
2) In the case of multipoint operation, the problem of designating a particular terminal is solved in the following ways:

   - if the terminals are all different—such as a telephone, a teletex terminal, and a videotex terminal—the discrimination will be made by the service indicator which is carried by the incoming call message. Only the terminal compatible with the designated service will answer;
   - if several terminals designed for the same service

are present on the bus, anyone may receive the call and the first to answer gets it. This is similar to the present operation of parallel phones;

   - it is also possible to designate by additional digits (called "sub-address") a particular terminal.

It should be noted that only terminals for manual answer produce "ringing" in the usual sense, and automatic terminals connected to the same extension receive calls without disturbing the user of the line.

Thus, a typical ISDN extension line will be all that is needed to provide all services associated with "the office of the future," including a telephone, an interactive terminal for videotex and data base access, a teletex, a telewriting pad, fire or smoke detectors, temperature controls, and so on. . . All these terminals are served from a single "S interface" and it is clear that in most cases a gain of pairs will be obtained, despite the fact that the S interface is 4-wire. It can be expected that existing wiring in a building may be used without the installation of additional cables.

In typical cases, an S interface can serve both the manager and his secretary for telephone service. In this case, a common directory number is assigned, and a general rule can be established such as subaddress "1" for the manager and "2" for the secretary.

As can be seen, the ISDN internal interface may be used in a flexible manner and ensures total compatibility for all terminals fitted with the S interface. Existing

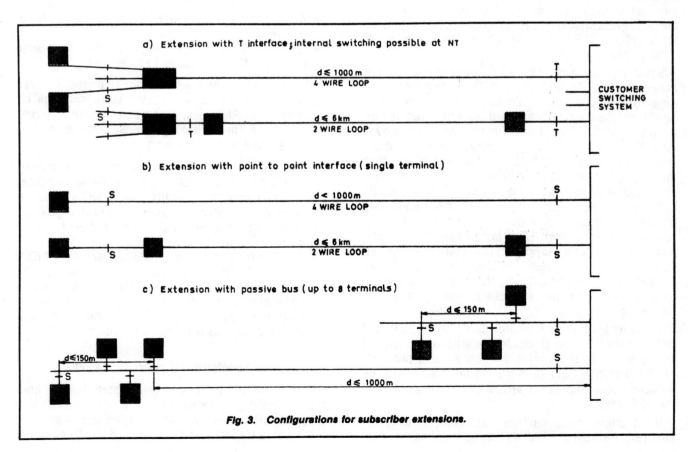

Fig. 3. Configurations for subscriber extensions.

terminals may still be used with an adaptor box. However, it is likely that extensions requiring only telephone service can be provided more cheaply with the conventional analog interface serving already-installed telephone sets. A possible strategy could be to convert to digital operation only those offices where several services or specific ISDN facilities are required. This conversion could be made throughout the life of the PABX by replacing interface cards or small modules. Therefore, the ISDN PABX should be quite flexible in accommodating various proportions of analog and digital extension lines.

Traffic and Dimensioning

There is very little information regarding actual usage of new services such as videotex or teletex in the business environment. In particular, the statistics obtained so far in experiments are mostly related to communications over the public network. However, it is likely that the largest part of the traffic in a working community will be local. It seems reasonable to assume that each external B channel is the equivalent of a telephone trunk. Considering the information-carrying capacity of a 64-kb/s channel, it is clear that nonvoice calls will have a shorter average duration than telephone calls—a 64-kb/s facsimile machine will transmit one A4 page in less than 5 s. Consequently, the dimensioning of the ISDN PABX should be based on current Erlang figures for telephony PABX's, or less, when referring to one B channel per line.

Some models have been built to characterize the packet traffic over the D channel—signaling messages for setting up and clearing down circuit-switched calls do not represent a significant load since the average rate is only a few b/s. The interactive type of services over virtual circuits, such as videotex, could be more significant. However, several studies have shown that the average bit rate per user will not exceed 200–400 b/s. This means that the load of the D channel is between 0.01 and 0.025 Erlang. This traffic is, by definition, very bursty and therefore should be characterized by two values—the maximum instantaneous bit rate, and the long-term average. The instantaneous bit rate defines the transmission delay for a given size packet, and the capacity of the buffers necessary to statistically multiplex packets over links to the packet network.

This average bit rate indicates that 1 64-kb/s channel can concentrate the D channel traffic of about 100 extensions with reasonable efficiency. As a practical rule, 1 B channel per primary digital system will be sufficient to multiplex the virtual circuit services from the 30 D channels.

If no intermediate multiplex structure is adopted, small and medium size PABX's (say up to 250 extensions) could be connected by a number of basic access lines, and larger machines by 2-Mb/s primary multiplex systems. If an intermediate structure such as 6 B + D is used, a more balanced range is obtained with small systems of up to 60 lines connected by basic access lines; medium systems from 60 to 500 lines connected by this intermediate multiplex; and only large systems (above 500 lines) connected by the primary multiplex.

Signaling

Out-of-band signaling is used in the ISDN. Messages carried by the signaling channel belong to two categories:

- user to network signaling, to establish and release calls and control additional facilities;
- user to user signaling, carried transparently by the network (for example, using CCITT No. 7 signaling between end exchanges) and used, for instance, to control calls on tie lines between PABX's.

When the customer installation is connected via basic access lines (2 B + D), the D channel of each basic access carries signaling related to the B channels of the same access. Such a PABX is thus treated exactly as a number of individual lines, with the possibility of hunt groups.

When the customer installation is connected by means of multiplexes, two kinds of signaling channels may be used:

- A D channel operating at 64 kb/s, using the LAP-D protocol;
- An E channel, also operating at 64 kb/s, but using the CCITT No. 7 message transfer part.

It is intended that both will use the same network layer protocol, that is, the D-channel Level-3 protocol would be the same as the PABX user part (yet to be standardized) of CCITT No. 7. The rationale to choose between the two solutions could be the following.

- The LAP-D protocol channel would be associated with the B channels carried by the same multiplex, with no changeover facility. Such PABX's would be connected to subscriber units of the local exchange. In this way, the subscriber unit handles the same protocol for all its subscribers;
- the E channel protocol would be used for PABX's connected directly to the group stages of the exchange, that is, like a subscriber unit. The E channel could benefit from the security and reconfiguration features of CCITT No. 7. The group stage in this case uses No. 7 for all its liaisons.

Telecommunications Services and Facilities

As indicated above, the customer installation has both analog and digital extensions. The analog extension lines will offer only the basic telephone service, and analog data transmission over a modem. This type of transmission is likely to be replaced by digital transmission when more of the network is converted to digital and more digital terminals are available.

The digital extension lines will offer all possible services of the ISDN, either in isolation or in combination (for example, voice and facsimile). Each service will include a number of additional facilities; some will simply be carried over from the current telephone and data transmission practice, for example:

from telephony: abbreviated dialing, call transfer, conferences;
from data networks: closed user groups, called party billing, user class negotiation.

In addition, a number of new facilities will be permitted due to the large message capacity of the D channel signaling system (in conjunction with common channel interexchange signaling in the public network). Such facilities are:

- calling party number display during ringing or camp-on busy,
- message display,
- charging information during a call, and
- date and time distribution.

As indicated above, most incoming calls will be treated by direct inward dialing, as many nonvoice services are entirely automatic. Operators will remain useful for telephone service, complemented by direct access to a videotex directory system. Interfacing an operator position is very simple in an ISDN PABX, since the universal S interface carries two speech channels and one control channel, corresponding to the usual requirement for an operator (inward and outward speech, and control).

Operation and maintenance functions will be unified for all services and greatly enhanced by the various loop test facilities provided by the S interface.

Conclusion

A customer installation designed for connection to the ISDN will provide all the services which are currently offered by two separate systems—a telephone PABX providing mostly voice services plus limited enhance-ment for digital data, and a local area network providing only data communications.

In addition, the ISDN system will benefit from the synergistic effect of combining voice and data services, used simultaneously to communicate with one or several parties. The adherence to the OSI reference model and to worldwide standards will ensure that all around the world, terminals and host systems can communicate freely, without the need for gateways, front ends, and other devices which were the result of shortsighted decisions and lack of communication between the data processing and telecommunications industries. Nothing has been said so far of the configuration of the customer installation—a subject of heated debates between the supporters of star (PABX's), ring, and bus systems (LAN's). In our opinion, this aspect of implementation is much less significant than the principles defined in the OSI architecture and the ISDN standards. As we have seen it, a small number of terminals (8, for example) can be connected over the universal S interface to a passive 4-wire bus of restricted length. A number of such buses can be tied in a star configuration to a switching controller. This unit may be the central part of a small or medium PABX, or may be a decentralized switching unit within a larger distributed system. In this case, the distributed clusters are connected by a high-bandwidth medium such as a coaxial or fiber-optic ring.

The French administration has made prospective studies of ISDN introduction in its network. Present plans are based on the successful outcome of the current CCITT Study Period in 1984, field trials between 1984 and 1986, and on the opening of the commercial service in 1987. At that time, about 50% of the French network will be digital. Within approximately ten years, all existing PABX's should be converted or replaced to allow for ISDN operation.

Gérard Robin is in charge of specification and standardization aspects of new telecommunications products at Laboratoire Central de Télécommunications, Vélizy, France. He is an expert in CCITT Study Groups XVIII and XI, dealing with the Integrated Services Digital Network project. Prior to that assignment, he managed network planning studies and initial work on the ISDN enhancement of ITT System 12 at ITT—Europe. ∎

CCITT Recommendations on the ISDN: A Review

MAURIZIO DÈCINA, FELLOW, IEEE, AND ERIC L. SCACE

Abstract — This paper gives a brief overview of the I-Series of Recommendations on the integrated services digital network (ISDN) developed by the International Telephone and Telegraph Consultative Committee (CCITT). This set of Recommendations is printed in the CCITT "Red Book." The I-Series Recommendations represent the first major step towards worldwide harmonization of the fast-growing digital network capabilities in support of multiservice (voice, data, and image) user applications. Over 25 different Recommendations set up standard guidelines and features for service capability, network architecture, and user–network interfaces in the ISDN.

INTRODUCTION

ISDN represents a focal point for the international standards activities of CCITT. The synergy between digital network evolution and signal processing technology advances is indeed stimulating the development of new multiservice (voice, data, and image) communications applications. CCITT is emerging as a necessary catalyst among telecommunications operators, industry, and research organizations to allow worldwide harmonization of these fast-growing communication capabilities. This paper presents a snapshot of the results obtained by the CCITT standardization activities in shaping the emerging ISDN. In particular, reference is made to the set of Recommendations on ISDN approved by the CCITT Plenary Assembly at the end of 1984. These Recommendations have been designated as the I-Series Recommendations.

The following sections of this paper are devoted to a description of the structure of the I-Series Recommendations and to an outline of the most important standard features in the ISDN concerning service capability, network architecture, and user–network interfaces. Among these aspects, emphasis is given to standard user-interface arrangements and protocols, since they are key factors in permitting a coordinated evolution of both network facilities and user applications according to technology advances and marketplace demand.

Manuscript received October 1, 1985.
M. Dècina is with the Central Research Laboratory, Italtel, 20019 Settimo Milanese, Milan, Italy, on leave from the University of Rome, Rome, Italy. He participates in the CCITT standardization process on behalf of the Italian Administration.
E.L. Scace is with the GTE Telenet Communications Corporation, Vienna, VA 22180.
IEEE Log Number 8407246.

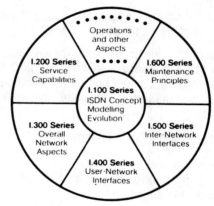

Fig. 1. Structure of the I-Series Recommendations.

FRAMEWORK OF I-SERIES RECOMMENDATIONS

An ISDN is a network, in general evolving from a telephony integrated digital network, that provides end-to-end digital connectivity to support a wide range of services, including voice and nonvoice services, to which users have access by a limited set of standard multipurpose user–network interfaces.

This concept requires a family of CCITT Recommendations. Fig. 1 shows the general structure of the I-Series Recommendations that cover:

- the concept and principles of an ISDN
- service capabilities
- overall network aspects and functions
- user–network interfaces
- internetwork interfaces
- maintenance principles.

The I-Series provide principles and guidelines on the ISDN as well as detailed specifications of the user–network and internetwork interfaces. They also contain suitable references to Recommendations of the other CCITT Series so that specific topics can continue to be standardized in the already established Series (for example: signaling protocols the Q-Series, data protocols in the X-Series).

At present, Recommendations have been prepared only on the first four topics listed above. Standards for internetwork interfaces and maintenance principles will indeed be developed during the present Study Period (1985–1988). During this period of time a consolidation is also expected for the present user–network interface standards (an accelerated procedure for updating such Recommendations is expected in 1986), while further Recommendations will be developed to cover more in-depth general principles, service capabilities, and overall network aspects.

It is worth pointing out that the user–network interfaces are at the core of the standardization process during the present Study Period and that the international debate on this topic involved several CCITT Study Groups, in particular Study Group XVIII for the general principles and the Layer 1 Recommendations, Study Group XI for the Layers 2 and 3 Recommendations, and Study Groups VII and XVII for the support of existing data interface Recommendations.

I.100 SERIES—GENERAL CONCEPTS

The I.100 series serves as a general introduction to the concept of ISDN. This series also hosts information common to other I-Recommendations, such as a vocabulary of terms specific to ISDN standards. For the user and network designer thirsting for detailed information on ISDN's and their interfaces, the I.100 series fails to quench the thirst. Nevertheless, important fundamentals are found here.

Recommendation I.130 lays out the "attribute method" for describing the services available through an integrated services network. Each ISDN service is characterized by specific values assigned to each descriptive attribute; e.g., 64 kbit/s circuit-mode demand service suitable for carrying human speech of a nominal 3.1 kHz bandwidth, etc. The meaning of each attribute value is defined here.

I.200 SERIES—SERVICES

The I.200 Series Recommendations provide a classification and method of description of the telecommunication services supported by an ISDN. They also form the basis for defining the network capabilities required by an ISDN. Services provided by ISDN's have been categorized into two broad families (see Fig. 2):

- "bearer services" and
- "teleservices."

Bearer services provide the means to convey information (speech, data, video, etc.) between users in real time and without alteration to the information "message." However, alterations may occur in the method of representing the information for certain services; e.g., transcoding between speech digital encoding laws when a speech bearer service is being provided by an ISDN. In contrast, the "unrestricted information transfer" bearer services make absolutely no changes in the digital bit sequence and grouping transported by the network (within quality-of-service limits); users employing this service for carriage of speech must provide their own transcoding, if required.

Telecommunication Service			
Bearer Service		Teleservice	
Basic Bearer Service	Basic + Supplementary Bearer Services	Basic Teleservice	Basic + Supplementary Teleservices
X 200 Layers 1-3		X 200 Layers 1-7	

Fig. 2. Services supported by an ISDN.

Many bearer services are fairly completely described in Recommendation I.211. In addition, certain bearer services have been characterized as "essential" and are intended to be provided internationally. Other bearer services are listed as "additional," meaning that they may be available on some networks, and might also be available internationally.

Teleservices use bearer services to move information around, and in addition employ "higher layer functions." These higher layer functions correspond to the transport through applications layers (4–7) of the Reference Model for open systems interconnection (OSI), described in the Recommendation X.200 series and related ISO International Standards. The characterizations of teleservices is still in a rather immature state. The CCITT has agreed that these higher layer functions can be provided by the public ISDN in some cases, while in other cases a teleservice can be provided by an independent service provider. The actual situation in a particular country depends on national regulations.

The I-Series Recommendations do not identify particular agreed-upon teleservices per se. A candidate set of teleservices are the "message handling services" such as electronic mail and voice mail. Other examples exist, but at this stage of the study the examples differ widely, depending on which CCITT participant is speaking.

Both bearer services and teleservices may be used in conjunction with "supplementary" services. A supplementary service is an additional service feature which can never stand alone; it is always used with some other bearer or teleservice.

A simple example of a supplementary service is reverse charging (also know as "collect call" in some countries). A user never asks for reverse charging all by itself; reverse charging always accompanies some kind of call, and thus, some sort of bearer service or teleservice is being provided.

Again, the characterization of supplementary services is not yet well understood. Certain supplementary services are already defined in CCITT Recommendations for dedicated service networks; e.g., for telex, telephony, or public data networks. These will be part of the supplementary service set for ISDN's. It is expected that many other new supplementary services will be defined early in the 1985–1988 Study Period.

The work on ISDN service definition highlights an artificial impediment to writing standards about ISDN. In the past, each set of service-specific network standards used similar, but not identical, terminology to describe their services and supplementary services. Part of the discussion on services is spent learning what people from different network backgrounds mean when they use a particular term such as "call redirection." Frequently, subtle but important differences are uncovered. As a result, new terminology springs up with specific meaning for an ISDN environment.

This leads to an important result: the description of service features and network functions that allow implementation in a service-independent manner. Again, the simple example is the ability of the users to signal for reverse charging the same way, and for the network to provide reverse charging in the same way, regardless of whether the reverse-charged call is speech, 64 kbit/s transparent information, or a packet-mode virtual circuit. Common functions allow elements of supplementary services, bearer services, and teleservices to be requested and provided in new combinations which may be more specifically tailored to the user's requirements at that instant.

I.300 SERIES—NETWORK ASPECTS

The main Recommendations in this series are I.320—Protocol Reference Model, and I.330—Numbering and Addressing Principles.

The purpose of the I.320 protocol reference model is the same as for other reference models: to allocate functions in a modular fashion that facilitates the definition of telecommunications protocols and standards. The I.320 protocol reference model is not a replacement of the Open Systems Interconnection (OSI) Reference Model, described in the Recommendation X.200 series and in a number of related International Standards produced by ISO. Rather, it extends the principles of the OSI model beyond the point-to-point, user–network–user, in-band signaling, data communications environment for which X.200 was originally produced.

Three major extensions to OSI are included in the Red Book I.320 model:
- separation of signaling and management operations from the flow of application information within a piece of equipment;
- definition of communication contexts which may operate independently from each other; and
- application of the above to internal network components.

The separation of signaling, management, and application operations within a piece of equipment allows each set of operations to be independently modeled. The information flow between applications in user equipment continues to be modeled with the OSI protocol reference model. Signaling operations involve protocols used to establish, modify, and disestablish communications services. Management operations deal with the management of internal equipment resources; management entities may also communicate with each other to manage distributed resources.

This separation leads to the definition of independent communications contexts. Each context can be modeled individually and use independent protocols.

In addition, all communications within the ISDN can be effectively modeled. A uniform protocol model applies to communications between users, between a user and the local service network, between equipment within the network, and between equipment responsible for managing network resources on an aggregated basis.

Although the I.320 protocol reference model shows promise, additional work is needed to continue to extend OSI principles into other areas, such as
- multipoint and asymmetric connections;
- nondata services (e.g., Where does one place a voice transcoder function?);
- changeover/changeback of signaling links; and
- communications for maintenance and network management.

These form the basis for work in the next CCITT Study Period.

Recommendation I.330 lays out principles for numbering and addressing in ISDN's. "Numbering" here refers to the number(s) allocated out of the public numbering plan to a public numbering plan to a public network subscriber. An expansion of the international public switched telephone network numbering plan, hinted at by CCITT Recommendations in the E-Series over the last 12 years, for use in ISDN's is described in Recommendation I.331/E.164. The ISDN numbering plan contains numbers with a provisional maximum length of 15 digits; the telephone country code system is retained.

In the ISDN, however, *addresses* contain not only the *number* out of a public numbering plan, but also additional *subaddress* information. This subaddress may be used by private communications facilities such as LAN's, PABX's, or other private networking arrangements. The subaddress may also be used by the end-user equipment (e.g., to distinguish between various OSI service access points within a single computer system).

I.330 also lays out other important principles:
- independence between address information and service specification (one cannot assume that addresses with a particular structure imply terminals which support a specific service);
- independence between address information and user-specified routing such as transit network selection (any transit network selection information must be provided separately);
- preference for single-stage procedures when interworking (e.g., when a telephony service user on an ISDN calls a public switched telephone network subscriber, single-stage operation means the originator needs to supply only the destination telephone number while two-stage selection would require the originator first to call an interworking gateway, obtain a "second dial tone," and then provide the true destination number);
- defined relationships between ISDN public numbers and various types of user–network interfaces, including interfaces to mobile users.

Other I.300 Series Recommendations illustrate how ISDN bearer services, teleservices, and supplementary services can be resolved into elementary network connections (of various defined types) and other additional functions.

During the 1985–1988 Study Period, performance objectives will be drawn together into new I.300 Series documents. It is worth pointing out, indeed, that the CCITT emerges as the only international body capable of establishing reliable and effective performance objectives (error rate, delay, call quality, voice quality, etc.) for user-to-user digital connections on a worldwide scale (the longest terrestrial connection envisaged by CCITT is 25 000 km long!). On the other hand, the critical issue of establishing performance objectives within local user networks (LAN's and PBX's) to allow their worldwide interconnection via open carrier networks still remains unsolved.

I.400 SERIES—USER-NETWORK INTERFACES

This series details the interface between user equipment and ISDN's.

This user–network interface lies within the customer premises. Recommendation I.411 outlines a functional model for the subscriber. These functions include:

• termination of the digital subscriber loop transmission system at the subscriber premises (Network Termination 1);

• on-premises distribution and switching (Network Termination 2), which embraces everything from passive, bridged wiring through an international private network;

• terminal equipment with ISDN user–network interfaces (Terminal Equipment 1); and

• terminal equipment with other interfaces (Terminal Equipment 2) and associated adapters (Terminal Adapter). ISDN user–network interfaces are recognized between each functional grouping.

Currently, the CCITT Recommendations do not specify a set of digital subscriber loop transmission systems; this has been left up to the network provider to select. However, standards for digital subscriber loop transmission systems are being developed in North America in response to recent regulatory decisions made by the United States Federal Communications Commission. The CCITT also has a study question on these systems for its 1985–1988 Study Period.

Only three ISDN user–network interfaces are described:

• basic, with an information-carrying capacity of 144 kbits/s simultaneously in both directions;

• primary rate, with an information-carrying capacity of 1536 kbits/s simultaneously in both directions; and

• primary rate, with an information-carrying capacity of 1984 kbits/s simultaneously in both directions.

When the overhead for framing and housekeeping is included, the aggregate transmission rates are 192, 1544, and 2048 kbits/s, respectively.

The number of different interfaces has been minimized in order to improve the economies of scale. For example, a digital telephone may never require all the information-carrying capacity available through a basic interface; how-

ever, a large spectrum of user equipment requirements can be met by the basic interface. Correspondingly, large manufacturing quantities will help reduce the cost penalty of an interface which might be oversized for a particular application.

The 1985–1988 Study Period may see the addition of one, or perhaps two, additional user-interface sizes with a capacity of tens or hundreds of Mbits/s (wide-band ISDN).

Although just a few interface types are employed within the subscriber's premises between functional groupings, the ISDN provider may choose to use digital subscriber transmission systems of other intermediate sizes. The choice of the transmission system depends on the customer's service requirement and the relative economics of implementing and operating each system.

Each of these interfaces is channelized. Recommendation I.412 defines the various types of channels, including:

• B channel: 64 kbits/s; carries circuit-mode or packet-mode user information such as voice, data, facsimile, and user-multiplexed information streams;

• D channel: 16 (basic) or 64 (primary rate) kbits/s; carries packetized signaling information to control the establishment, modification, and disestablishment of calls and services; idle time on this channel may be used for packet-mode user data and telemetry;

• H channels: 384 ($H0$), 1536 ($H11$), or 1920 ($H12$) kbits/s; carries circuit-mode or packet-mode user information such as voice, data, video/image, and user-multiplexed information streams.

I.412 also defines how channels may be packaged together within basic and primary rate interfaces. The channel structure for a basic interface is $2B + D$. Examples of primary rate structures are $23B + D$, $30B + D$, $5H0 + D$, $2H0 + 11B + D$, and $H11$.

The physical characteristics of these user–network interfaces are defined in Recommendations I.430 and I.431. These characteristics include electrical characteristics, permissible wiring configurations, connectors, powering arrangements, and time-division multiplex structure. The basic interface may support up to eight user equipments bridged on the same wiring. In this passive bus arrangement, each B channel is allocated to a specific user equipment for a period of time, as directed by the D-channel signaling from the network. The D channel is shared by all equipment; if multiple equipments simultaneously have information to transmit, I.430 provides a contention resolution procedure. The primary rate interfaces are only used in a point-to-point configuration.

Recommendations I.440–441, also known as Q.920–921, define the data-link-level procedure used on the D channel. This procedure extends the LAPB link-level protocol of Recommendation X.25. The link-level address field is expanded to two octets; this allows multiple, parallel, independent links to exist simultaneously on the D channel. The links may carry different types of information (signaling, X.25 packets) or terminate in different user equipments. The D-channel link-level procedure also uses several additional HDLC commands and responses.

Recommendations I.450–451/Q.930–931 outline a flexible signaling procedure for establishing, modifying, and disestablishing calls. While the Red Book version of these Recommendations is focused on elementary call control, the protocol is easily expanded to support more complex calls, standardized supplementary services, and network-specific services and supplementary services. This expansion is a focus of current intensive work in Study Group XI; an enhanced set of Recommendations may be adopted under the CCITT accelerated procedures in 1986.

These call control procedures may be exercised in two styles:

• "functional," when both the network and user equipment share knowledge about the call state and other details of the service(s) being provided;

• "stimulus," when the user equipment retains little information about a call's state, and may be restricted to handling a single call at a time.

Stimulus operation is appropriate for some manually operated user terminal equipment, such as a digital telephone. Functional operation may also be used by manually operated equipment, and is especially convenient for more autonomous user equipment, such as PABX's, LAN gateways, computers, and integrated workstations.

Furthermore, these signaling procedures may also be used within private networking arrangements. For example, I.451 may be used to control calls between two PABX's over a leased inter-PABX trunk group.

Note that these D-channel signaling procedures are common to all user–network interfaces, regardless of size and configuration. The requirement for the network to have a detailed knowledge of the configurations and capabilities of user equipments has been eliminated. As a result, users may expand, reduce, and change their equipment without prior notification to the network administration and without updating lengthy tables in the serving network switch(es).

Although not part of the I-Series, the intra- and internetwork signaling protocols described in the Q- and X-Series are also fundamental to the implementation of an ISDN. Common channel signaling techniques are generally assumed to be used within and between ISDN's, with most attention focused on Signaling System No. 7 (Q.700 series). However, more work is required to align the procedures and capabilities of Signaling System No. 7 with the user–network interface signaling protocol. Interworking arrangements with existing networks must be more clearly defined as well.

Several other Recommendations in the I.400 block provide standardized methods for rate adapting classical X.1 and V.5 data rates into B channels, for submultiplexing B channels and corresponding circuit-mode bearer services, and for interworking with digital network facilities providing only restricted 64 kbit/s information transfer rates. This restriction is nearly universal in the existing North American transmission system plant, which treats the octet 00000000 as an error.

Recommendation I.461/X.30 also defines the operation of a terminal adaptor for converting X.21 circuit-switched interfaces into the ISDN basic user–network interface. Recommendation I.462/X.31 defines how existing packet-switched public data network services can be provided in or through an ISDN to a packet-mode user terminal. This user terminal may be an X.25 DTE; in this case the Recommendation also outlines how the corresponding terminal adaptor function between the X.25 interface and the basic ISDN user–network interface operates. Packet-mode user terminals which use the X.25 packet-layer protocol, but attach directly to an ISDN user–network interface, are also covered by Recommendation I.462.

FURTHER STUDY AREAS

Although space has been reserved in the outline of the I-Series for Recommendations on internetwork interfaces and on maintenance principles, no Recommendations were prepared for the Red Book. This remains an open issue for further study in the 1985–1988 Study Period.

CONCLUSION

The CCITT Red Book I-Series Recommendations provide a wealth of guidance for early implementation of ISDN's, but should not be viewed as a design specification. Further work is being undertaken in additional areas, including:

• definition of supplementary services;

• definition of teleservices;

• harmonization and enhancement of Signaling System No. 7 ISDN User Part interexchange signaling protocols and the ISDN user–network signaling protocols;

• definition of principles and procedures for testing, maintenance, and other operational and administrative aspects of ISDN's; and

• agreement on international tariffing principles for ISDN's.

The experience gained during the development of trial quasi-ISDN's in a variety of countries will help resolve these open issues. Many issues should be resolved in time for a larger set of ISDN Recommendations to be approved, via CCITT accelerated procedures, in 1986.

Each organization participating in the work of the CCITT on ISDN's has benefited, and will continue to benefit, from these international agreements. The pool of collective intelligence and talent brought by those participating in CCITT's ISDN work far exceeds that which any individual organization can hope to assemble.

With today's hardware and software technology, any given technical problem has a vast number of potential solutions. It would be relatively uneconomical for each organization to develop its own, unique solution. In providing telecommunications for an increasingly interdependent, mobile society, the total costs of a plethora of such unique solutions are borne by the users. In contrast, standards provide a common solution to many similar problems. The widespread use of a standard yields economies of scale. The individual talents of both network-provider and users' organizations are then freed to focus on other problems.

This is especially true for ISDN's. Furthermore, the development of standardized, flexible, internationally available, and economical "building blocks" of digital information transport and processing services has exciting implications for the information society of the future.

Maurizio Dècina (M'83–SM'85–F'86), for a photograph and biography, see this issue, p. 314.

Eric L. Scace received the B.S. degree in atmospheric sciences from Cornell University, Ithaca, NY, where he did research on operational tornado and hurricane forecasting using computer modeling techniques.

He is the Senior Network Architect for ISDN at GTE Telenet Communications Corporation, Vienna, VA. Over the past ten years, he has participated in CCITT Study Groups II, VII, XI, and XVIII activities on packet-switching public data networks and on ISDN's. He currently chairs ANSI T1D1.2, a North American standards group on ISDN switching and signaling protocols, and is Vice-Chairman of ANSI T1 on telecommunications. He has also worked with the National Bureau of Standards and General Electric on data communications matters. During this time, he participated in the International Civil Aviation Organization's Automated Data Interchange Systems Panel, and in ISO and ANSI activities on data communications. He developed for the U.S. National Weather Service a fully computerized network for the collection and distribution of meteorological data, maps, forecasts, and warnings.

The Evolution of Telecommunications Technology

RICHARD VICKERS AND TOOMAS VILMANSEN

Invited Paper

The telephone network today is the result of many years growth. Technology has enabled this growth to occur, and simultaneously allowed a steady reduction in the real cost of service and an increase in the level of performance. Technology will continue to facilitate the economic growth of the network, and with the evolution to digital will enable a much larger variety of information to be carried. Managing this broad range of capabilities requires more than traditional hardware and even software technologies. The real leverage will come from a broader class of "soft" technologies that goes beyond language and operating systems to encompass system architecture, protocols, and databases. This paper will address the potential evolution of the network with respect to both hard and soft technologies. The trends in these technologies and in the network itself are leading towards a high-performance transport layer, with added intelligence layers to define the transport capabilities for a variety of services.

INTRODUCTION

Since the invention of the telephone, a hundred years ago, the network required to interconnect users has been growing steadily. In a simple model, the cost of providing service to a single subscriber would increase proportionally with the number of subscribers served by the network. In practice, this cost has decreased in real terms. At the same time, responsiveness and performance have both increased. This apparent contradiction is the result of the applying technology to control the costs of the network. Automatic switches and Direct Distance Dialing have reduced the number of manually switched calls. Multiplexed transmission systems have reduced the number of physical facilities and at the same time provided much improved performance. Today the carriage medium of the network is being transformed from analog (4 kHz) to digital (64 kbit/s).

Digital switching and transmission offer the user quality of sound independent of distance. The potential though goes beyond voice. The digital channel, because of its intrinsic freedom from impairment, offers the potential for the carriage of a plethora of different types of information.

Manuscript received March 18, 1986; revised June 27, 1986.
The authors are with the Advanced Network Technology Department, Bell-Northern Research, Ottawa, Ont., Canada K1Y 4H7.

As well, digital technology provides simplified maintenance, reduced space requirements, and diminishing first costs due to its mass availability.

In parallel with the introduction of digital carriage, the application of digital computing technology is providing the means to automate many of the network functions. As well as controlling the basic switching function, computers are used to provide many of the necessary ancillary operations necessary to run a network. Automated record keeping, network surveillance, billing, and other administrative tasks actually require more lines of code than does basic call processing.

The technologies being placed in the network today are leading to a more versatile and responsive network. The digital base, together with software systems and evolving end-to-end signaling protocols will, in future, provide users with services as sophisticated as voice and data conferencing, as utilitarian as data connections, and as easy to install as a standard telephone set.

This paper will address the trends in base technologies, and their application to network components. The final section discusses the potential evolution of the network, based on current trends in the network architecture and capabilities, and the technologies that will be required for that evolution to occur.

I. BASE TECHNOLOGIES

The key to extracting the potential of the digital base lies with software—or more generally "soft" technologies that include protocols, network architecture, and network planning. By comparison with the hard technologies, soft technologies are still in their infancy. By the 1990s, however, they will emerge as dominant in the network.

A. Soft Technologies

The key technology is the basic system architecture. It is influenced by several forces: the applications to be supported, the architecture of the hardware in the system,

336

performance requirements, the design principles and development methodology adopted, and support systems. These factors all change with time, and therefore the network must incorporate features that allow it to evolve gracefully.

The nature of telecommunications systems has resulted in software architectures that are characterized by long service life, the ability to support extensions to the system, and graceful evolution. The modularity and reusability of software modules protects the large existing software investment. The architecture must also allow for the incorporation of advances in hardware technology to support growth and reduce cost, without major software change. At the same time, new services should be capable of being added without disruption to existing applications.

Current telecommunication system designs use layering principles to decouple the system software architecture from the underlying hardware technology. The operating system provides the interface between the hardware and the software. This approach allows the hardware and software to evolve independently, as long as the interfaces are maintained. This reduces costs by allowing for the introduction of new technology as it becomes feasible. This basic principle of layering can also be applied within the software, allowing an independent evolution path for each layer.

Today's telecommunications operating systems all employ some form of concurrent processing. This allows for software architectures in which cohesive processing activities can be modeled as separate processes that only need to know about and communicate with related processes. The operating system provides the primitives for communication and synchronization, and hides the details of processor sharing and scheduling from the component processes. This approach provides flexibility in partitioning of functions, and facilitates the addition of new applications. It also improves software design productivity because developers need only be concerned with their own specific applications.

Languages impose additional discipline on the development of software architecture. A current trend is towards languages that support formalized interfaces between software modules. This permits the use of software architecture based on loosely coupled software modules that can evolve individually without affecting the entire architecture of the system.

The falling cost of processing elements and storage, along with the ability to loosely couple parts of the system, has underpinned the evolution of telecommunications systems with distributed intelligence that allows for choice in partitioning software architecture. In the past, each intelligent component had its own operating system with interfaces that allowed communication with other components. The current trend, however, is towards fully distributed operating systems. Such systems hide complexity from designers by letting the operating system handle interprocess communication and synchronization. They also facilitate the migration of functionality between processing components if this is dictated by performance needs and becomes cost-effective.

Even if a network does not have a fully distributed operating system, standardized operating systems and communications protocols for processors will facilitate the distribution of intelligence and reusability of software components across systems.

Current telecommunications systems incorporate a large software component. Modern switching systems typically contain millions of lines of code, generated by hundreds of programmers over several years. Such systems also require substantial ongoing maintenance over their life span of ten to twenty years. Much of that maintenance involves the addition and modification of specific network features.

Telecommunications system manufacturers have developed comprehensive tools to support the development, maintenance, and ongoing evolution of their products. Software development tools are available for coding, compiling, linking, testing, and debugging software. Administrative tools support the management of change, the generation of system loads from appropriate versions of software modules, and the incorporation of new modules into a library of existing modules.

In the telecommunications environment, several factors will influence the evolution of software tools. There is an ongoing need to reduce development and maintenance costs by increasing the productivity of software engineers over the total software life cycle. There is also a need to implement services in accordance with the requirements of customers, and to meet rapidly evolving demands for new services. Although different tools address different phases of the software life cycle, there is an increasing emphasis on the requirements phase because it has been shown that lower cost software can be delivered if design issues are addressed in a more rigorous fashion at the beginning.

In checking for the consistency and completeness of design requirements, limited use has been made of formal specification languages. One example of this type of tool is the CCITT's Specification and Description Language (SDL) which has been proposed for telecommunications applications. This introduces more formalization to the definition of system requirements and this could evolve along with the automation of other phases of the design process to significantly reduce development time and cost.

Another emerging software development technique is rapid prototyping. It demonstrates the capability that the end user expects at an early stage in the design process, thus allowing user feedback to influence the final result. Thus far, rapid prototyping has had limited application in telecommunications systems because the environments to support this type of development are not generally available. Furthermore, the services being developed are complex and require interactions among many elements. If these systems become more common, however, they carry the potential of designing products that better meet user needs at lower overall cost because problems can be detected at the front end of the development process.

Automated testing of software is another tool that continues to reduce the cost of maintaining networks. System maintenance is a long-term activity that is a significant contributor to overall life cycle costs. Static analysis, dynamic analysis, and symbolic execution are some of the testing approaches that are being developed and applied to this task.

Telecommunication systems are being developed with high-level languages such as CHILL, PROTEL, C, and variants of PASCAL. It is likely that languages will evolve towards even higher levels of abstraction. One possibility is to use CAD/CAM principles to develop graphics-oriented languages. Languages that simplify programming by hiding complexity, have already been shown to increase the productivity of programmers. In addition, the skill required to generate applications would decrease significantly, thus addressing the present shortage of trained real-time programmers.

Another way of improving the productivity of programmers is to design reusable software. Software should be designed in modules with well-defined interfaces. Also, the programmer should be made aware of what reusable modules are already available.

Advanced software tools allow for the creation and introduction of increasing amounts of intelligence in telecommunications networks. Among the most promising forms of artificial intelligence are expert systems which are now ready to make the transition from university research to industrial application. Expert systems are benefiting from the appearance of specialized hardware and software environments, development costs are decreasing, and there is a growing base of expertise in this area.

Expert system technology could be used to automate and streamline many of the functions in network operations, administration, and control. These functions include equipment and network configuration, monitoring and maintenance systems, circuit provisioning, and testing. Expert systems can also be developed to automate parts of the development cycle.

Some testing and diagnostic systems are already in service. Applications will become more widespread when the expert is able to transfer his knowledge by interacting directly with the expert system rather than going through an intermediary. Successful transfer of expert knowledge may depend on natural language interfaces. Presently, systems can understand and generate questions about restricted problem domains. As capabilities are enhanced, the technology can be applied not only to acquire knowledge from experts, but also to allow casual users to access systems that currently demand significant expertise. With this type of enhancement, users of telecommunications systems could, for example, access databases to customize their capabilities and to get appropriate management information.

Database and database management system design are well-defined disciplines. Current hardware supports large databases at acceptable levels of performance. Database systems designed according to hierarchical, network, and relational models are used extensively in commercial applications.

Current telecommunications systems use database technology, for example, in the management of system software. In order to provide customers with the appropriate services, networks require customer record databases. Separate databases support ongoing operations, administration, and maintenance. The databases in telecommunications systems must meet stringent criteria for response time, reliability, consistency, efficient recovery, and security. Methods for achieving these requirements have already been successfully implemented and future database systems will be able to build on this technology.

The evolution of services will increase the importance of database technologies. Decreasing processing and storage costs are leading to distributed database management systems. The increasing ability to provide customized services will lead to more sophisticated customer service databases requiring higher transaction rates to meet a service request.

Users will be able to interact with databases to modify service profiles. Medium to large distributed database management systems will be accessed by many users of varying skills. The databases will have to meet very stringent performance criteria regarding anticipated response time, reliability, and security.

Database management systems will eventually evolve towards knowledge management systems by merging expert systems with database management systems. Such systems will permit extraction of information that is not explicitly stored but can be inferred by combining database facts with rule-based knowledge. Such a development will simplify the user's interaction with the database.

Developments in software, intelligence, and database management have been matched by rapid progress in computing technology which is evolving to meet a demand for more processing power. Microprocessing elements are continually being enhanced to improve performance and achieve lower costs. New computing structures are being developed to overcome the limitations of traditional serial processing structures.

Computing technology applied to telecommunications supports switching systems, network control systems, network management systems, administration, and system development. Each of these areas can require different processing capabilities, and increasingly complex functions will demand significant increases in processing capacity. Tradeoffs of cost and performance will favor specific hardware architectures, such as fully distributed systems for switching products.

The processing power of chip-based microprocessors is approaching that of mainframes. Thirty two bit microprocessors are now available and reduced instruction set computer (RISC) architectures are also being actively investigated. The latter are simple to implement in VLSI, they offer significant cost/performance gains over conventional microprocessors, and they promote the distribution of processing.

The exploitation of parallelism promises dramatic increases in processing power. Ongoing research is investigating computer structures with multiple processing and storage elements, communicating through a connection network. Examples include multiple processors featuring local memory with a cube or bus connection network, and multiple processors connected to multiple global memory elements through a multistage interconnection network. A further refinement is a flexible architecture in which the memory and processor can be reconfigured for specific applications. Although these systems are inherently parallel, compilers must be developed to exploit this parallelism.

Telecommunications switching systems can derive great benefit from parallel architecture because, typically, they handle simultaneous calls requiring different levels of processing. The multiprocessor system satisfies existing demands for additional capacity and provides a flexible path for further expansion as service requirements evolve.

There are several specific computing technologies which could be applied to telecommunications. Supercomputers and array processors are examples of commercially available systems which are optimized for numerical processing. These systems could serve in telecommunications resource control applications. Another available technology is the database machine, which is optimized for rapid query of large databases and which could satisfy the performance requirements of telecommunications system databases.

The architecture necessary for knowledge processing machines is more futuristic. One example is the inference machine investigated for the Japanese fifth generation computer project. Such a technology could support telecommunications system users, developers, and operators.

At a more fundamental level, alternative computation mechanisms are being investigated. Data flow and reduction mechanisms may be alternatives to the traditional control flow approach. Although the application of data flow techniques to switching system call processing has been investigated, much of this work is in the research stage.

Because of the range of computing requirements in telecommunications, many alternative computer structures could eventually have diverse application. The challenge will be to match the application to the appropriate technology and then adapt it to the needs of telecommunications.

Some of the emerging technologies discussed have more impact than others on the future network. The migration of functions such as database management, surveillance, and network configuration towards a centralized function will require much complexity in the supporting software systems. Database technology together with expert systems will become increasingly important as a result. In the development of new features, a software architecture providing well-defined modules and interfaces is essential for the preservation of existing software and to facilitate integration of the feature. Software tools and languages will reduce the manual effort involved in developing features, and in fact ensure the performance of the finished product.

B. Hard Technologies

The evolution of computing structures is associated with the development of necessary base hardware technologies. Because of their more general application, computing structures are a more dominant influence on the evolution of "hard" technologies than other network components.

The current base hardware technology for virtually all mass computing applications is CMOS. CMOS provides high integration densities, moderate speed, and moderate power consumption. The speed achieved by CMOS with planned processes will be of the order of 100 MHz. Any increases in CMOS speed are achieved by manipulating device geometry, particularly linewidths. However, because linewidths are expected to reach a minimum at 0.2 μm, this will ultimately limit the speed of the technology.

Silicon bipolar in the form of emitter coupled logic (ECL) dominates devices with speeds of hundreds of megabits. It is a very mature technology, but uses considerably more power than CMOS. It is not yet possible to achieve a combination of high speed, high complexity, and moderate power consumption, so architectural compromises such as parallel connections are necessary.

Transistor–transistor logic (TTL) currently falls between CMOS and ECL, with lower speed and lower power consumption than the latter. TTL technology has evolved towards lower power and higher speed. For example, the low power Schottky family has now replaced standard TTL in general applications.

In comparison with CMOS, gallium arsenide (GaAs) technology has the most potential for very high logic speeds (Fig. 1). Medium scale integration (MSI) is already available in gallium arsenide, operating at speeds up to 2 Gbits/s. The technology is still maturing, and rapid increases in

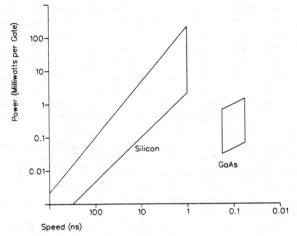

Fig. 1. Speed/power envelope for CMOS and GaAs.

complexity are expected as die manufacture is perfected. Its greatest potential in the network is for high-speed interfacing and switching. It is also a candidate material for the integration of electronic and photonic technologies, together with other materials such as indium gallium arsenide phosphide (InGaAsP).

Interfacing to copper pairs requires tolerance to high voltages (up to 350 V). A different process is required for these applications. Because of silicon's inherent properties, dielectric breakdown between tracks can occur at relatively low voltages, depending on track spacing. There are a number of techniques for providing high voltages, the most promising of which is dielectric isolation. In this technology, an insulating barrier is placed between components to provide high voltage tolerance. Using this technique, fairly complex integrated circuits can be built that tolerate up to 500 V.

As each process evolves, speed and power differences between them will diminish. These will no longer be the selection criteria for a given process. Fan out and line driving capabilities are the domain of bipolar, while component density and complexity are the chief strengths of CMOS. In the future, custom chips will be able to support a variety of processes, using the best features of each.

Very large scale integration (VLSI) today consists of up to 10^5 gate equivalents. This level of integration offers a large degree of functional sophistication in single integrated circuits. With the development of computer-aided engineering (CAE), complex functions can be integrated even with relatively low production volumes. VLSI will continue to be one of the major technology enablers of the future. It has begun to influence telecommunications, but its capabilities are only just starting to be exploited.

In telecommunications, as elsewhere, LSI and VLSI first appear in mass applications. Switch line cards are a good example of where the most advanced custom LSI has been used to good effect, and where the complexity available in VLSI will soon be exploited.

In line card applications, technology allows for some major architectural changes. In early channel banks, the voice (analog) to digital conversion was performed by equipment shared between a number of line cards which were totally analog. Analog technology, however, is not as amenable to integration as digital technology. It also has intrinsic limitations on crosstalk, and noise addition. The first local digital switches were equipped with per-channel codecs on the line cards, moving the analog/digital boundary further towards the periphery. This was possible because of the decreasing cost of codecs, and led to simplified architecture and improved reliability. With the Integrated Services Digital Network (ISDN) the conversion will move right to the subscriber.

Line card designers are simplifying the analog portion of the codec to migrate analog functions into the digital domain. Hence oversampled sigma–delta codecs with digital filtering are starting to appear because digital technology is now easier to implement and more predictable than analog technology.

The density, or feature definition parameter of integrated circuits has a physical limit. Below 0.2 μm, the silicon semiconductor material has insufficient dielectric strength to sustain current–voltage levels. Technology, however, has not yet reached that limit. The most advanced production tools are masks produced by electron-beams writing directly onto the die, and achieving linewidths below 1 μm. Production using this technique though will depend on much higher writing speeds.

Usable die size has increased considerably over the past few years with 350 mil square dies now becoming commonplace. Usable die size depends on both the size and the density of defects on the wafer as well as on the definition of the features being etched. It follows that the smaller the feature size the smaller the usable die size for a given yield. It might be felt that this would ease pressure for smaller feature definition, but in practice the quality of silicon wafers is keeping sufficiently ahead of lithography techniques to allow good yields on more complex chips.

Wafer-Scale Integration (WSI) is the ultimate limit of die size. Technology is not yet capable of achieving that level of integration at micrometer feature sizes. There are, however, some potential intermediate steps using silicon wafers as high-density interconnect systems between other chips. The usable capacity of integrated circuits may be dependent upon the ability to interconnect them, particularly when parallel input/output is used to achieve throughput.

Logic speed has increased less rapidly than complexity (Fig. 2). Speed increases in CMOS and bipolar technologies were achieved almost as by-products of the drive towards feature size reduction. The need to increase speed for its own sake is just starting to appear in mass applications. One stimulus to higher speeds might be new computing structures, requiring much faster memory read/writes. In traditional Von Neumann structures, processing speed limits the usefulness of fast access memory.

Evolution happens by a series of pushes on particular technologies. As the certain technologies advance, they pull other related technologies along with them. For example, as the complexity of circuits increases, the associated

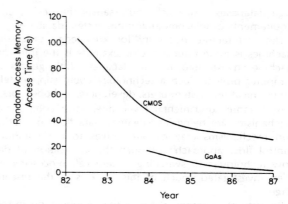

Fig. 2. Random-access memory speed trends.

packaging technology and ability to interconnect also have to develop. This is true both at the integrated circuit level and on the macro scale of circuit boards and backplanes.

To increase throughput, designers have increased the number of parallel paths. This has resulted in a large increase in pin counts for both integrated circuits and printed circuit board edge connectors. To increase the packing density, printed circuits boards are now made with multiple layers. Tracks have become finer and track spacing smaller as technology and production tolerances improve.

In addition to the trend to parallel connections, the density and therefore complexity of both integrated circuits and printed circuit boards has increased, leading to a further increase in interconnection. The combination of parallel paths, increasing path lengths, and greater complexity increases the relative space devoted to interconnection. This leads to diminishing returns for increases in chip area or printed circuit board component density (Fig. 3). By

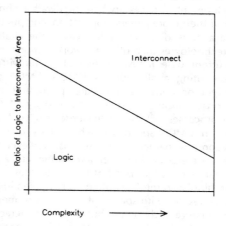

Fig. 3. Decrease of packaging efficiency with complexity.

mounting chips directly on silicon wafers as opposed to conventional printed circuit boards leads to higher packing density. The reason is that the silicon wafer can provide fine tracks (< 25 μm) and can therefore provide dense interconnection between chips.

Within a chip there is generally some form of power amplifier or matching impedance on the pads to ease interconnection. It consumes a high proportion of the power passing into the device. Similarly, in intercard connections, power is required to minimize the effects of crosstalk and electron magnetic interference (EMI), and to match the impedance of the transmission. In both cases, power dissipation and increasing component densities have led to more widespread use of forced-air cooling as opposed to simple convection cooling.

One potential development in interconnection lies with integrated optics. High-speed optical serial connections can replace parallel connections to reduce complexity and power dissipation. Integrating optics and electronics on the same device is a major challenge, but the potential of such devices to interface at gigabit rates through fiber or free space is driving a lot of interest and research into such devices.

As a packaging technology, optical transmission has advantages over electrical transmission at frequencies in excess of 10 MHz. Electrical connections require proper termination and impedance matching which limits their usefulness in applications such as distribution and bussing where multiple terminations have to be matched. Optical transmission does not suffer from the same problems until the multiple gigabit frequency range is reached. Because of their inherent ease of termination matching, and their waveguide characteristics, highly effective directional couplers can lead to practical bus systems at bit rates in the gigabit per second range.

This type of interconnect technology carries the potential of increasing interconnect capacity on boards or backplanes to hundreds of gigabits per second. Optoelectronic interconnection may well be one of the most important applications for gallium arsenide electronics.

Optical fibers have extremely high intrinsic bandwidth capacity, far in excess of what can be utilized today. Conventional on/off modulation techniques are now used in commercially available systems to 565 Mbits/s, with 4 Gibts/s achieved in the laboratory. The first step towards using more of the bandwidth is in the use of wavelength division multiplexing. This requires controlled frequency lasers and narrow-band optical filters. As in any frequency multiplexed system, potential capacity is limited by the passband and rolloff of the frequency response at the skirts. Current state of the art is 20-Gbit/s total transmission using 10 wavelengths.

In the more distant future the use of homodyne or heterodyne coherent systems promises a further increase in fiber efficiency, but the impetus to employ such techniques will only come as a response to a greater demand for bandwidth—with the 2-Gbit/s technology expected by the end of the decade, a 36-fiber cable has a two-way capacity of at least half a million telephone conversations. The pressure for higher capacities will largely be dependent on the proliferation of high-bandwidth services.

Optical fiber technology has dramatically reduced the overall cost of short haul transmission. This cost can be divided into carriage cost and termination cost. Carriage costs dominate termination with conventional copper and radio based systems because of the inherent bandwidth limits of the carriage medium and, in the case of copper, the need for intermediate repeaters. With fiber, the accessible bandwidth is so high that cost per bit for carriage becomes much less significant. The termination costs for

multiplexing to the high speeds necessary to achieve low-cost carriage now dominate. The multiplexing cost for a high-speed channel will be less than or equal to that for a low-speed channel resulting in a significant reduction in the cost differential between transmitting high- and low-speed channels. Transmission cost is no longer the limiting factor on getting broad-band service to the customer.

Despite advances in optical transmission, optical switching remains largely in the laboratory with the exception of specialized applications such as optical protection switching. The cost of optical switching is not yet competitive with electronic switching and even when it does become competitive, more research is needed to make the systems useful. Photonic cross points exhibit some crosstalk and attenuation of the signal, making multistage switching unattractive. Pure photonic switching will not be feasible until direct optical regenerators become commercially available.

The conclusion to be drawn is that in the immediate future, optical switching will remain limited to areas such as direct protection of optical components and cables. It will be some years, if ever, before a commercially viable pure optical switch, competitive in size or cost to electronic switches, will become available.

Optical technology holds more immediate promise in the area of information storage. Conventional compact disks provide a large bulk storage medium with current capacity on commercially available devices in the order of 550 Mbytes. That is equivalent to two comprehensive encyclopedias, or about 2000 novels. This technology, however, only allows for reading from the disk: writing is still complex and expensive, although Sony Corporation has announced it is developing a "write" system for domestic use to be available in the next five years.

In contrast to magnetic materials, optical disks provide a much more robust medium. They are unaffected by changes in magnetic field and potentially resistant to relatively high temperatures. Mechanical damage, such as surface scratches, have to be severe before errors occur.

Optical storage could have important implications in a telecommunications network. In a read-only form, optical storage could be used in applications such as routing and directory databases. Eventually, with a write capability, this storage medium could be applied to detailed billing records.

Another area of technological development is signal processing. In the telecommunications network, it will be primarily applied to video signals. Video signals are susceptible to reduction because of the large amount of redundant information both within a single picture and in motion from frame to frame. Differential coding techniques are particularly effective, especially where the background is uniform and changes from frame to frame are small. Video teleconferencing is one area where these techniques have been applied.

Full motion video, however, presents more challenge. Scene changing, panning, and zooming all limit the effectiveness of frame to frame differential encoding.

An uncompressed NTSC signal requires in excess of 100 Mbits/s for high-quality transmission. Minimal processing is required to reduce this signal to 90 Mbits/s, which is still of broadcast quality with minimal degradation. A reduction to 45 Mbits/s is possible, but this results in degradation that is detectable under controlled conditions. Below this rate, the limitations of processing become more obvious, although if one accepts considerable constraints on the picture dynamics, codecs can be developed to fit the 64-kbit/s bandwidth required for speech.

Speech remains the high-volume traffic on the network. A speech signal resembles white noise in that it manifests unpredictability and low susceptibility to reduction. Pulse code modulation (PCM) at 64 kbits/s is used for speech transmission, with the signal already processed to some extent by using a logarithmic companding law. Recent developments in processing and silicon technology have enabled cost-effective algorithms to be implemented at rates of 32 kbits/s with no noticeable degradation, and at 16 kbits/s with some degradation. Because of the nature of the algorithms at 16 kbits/s there is a tradeoff between signal quality and processing delay. Quality is therefore degraded. Together with the high-speech channel capacity of fiber systems there is therefore less incentive to reduce bandwidth for speech than for video.

The amount of processing applied to signals will be a tradeoff between quality, cost of transmission, and cost of processing. Reductions in the cost of transmission will tip the balance away from signal processing. This is illustrated qualitatively in Fig. 4 where the total cost of operation is mapped against three different transmission costs. The

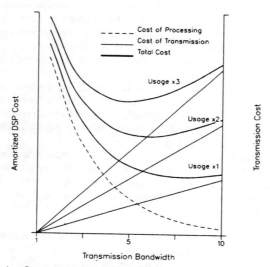

Fig. 4. Cost sensitivity of DSP with usage.

diagram assumes that transmission cost is proportional to bandwidth and that the cost of processing increases exponentially with a reduction in bandwidth.

There are several other very specific applications of digital signal processing that will have volume application in the network. The introduction of digital line cards requires signal processing, both in analog and digital domains, to recover signals in the presence of high, correlated interference. With voice, signal processing can provide both simple functions such as level control, and complex functions such as automatic loop routining and per-channel echo cancellation. In terminals, voice recognition and

synthesis will provide one of the means for a user-friendly interface.

There has been a continuing shift in emphasis from analog to digital technology. Telecommunications has followed this shift and new developments are nearly all built on a digital base. Digital transport is now starting to reach the subscriber, and together with the processing power and intelligence being placed in the network, will give the opportunity to carriers to offer a much wider range of service.

II. IMPLICATIONS OF TECHNOLOGY ON THE TRANSPORT NETWORK

The Integrated Services Digital Network, or ISDN, is the basis for enhanced service offerings. The first stage of ISDN implementation is the conversion of the basic transport medium from a 4-kHz analog channel to a 64-kbit/s digital channel, with a new, powerful end-to-end signaling system. The rationale for this change is the premise that digital channels are more versatile and ultimately cost less than analog channels.

ISDN will provide a universal access method for all types of service using 64-kbit/s channels. With a pure ISDN network there would be no requirement for the large variety of interface hardware currently used in the provision of voice-grade special services, for example; there would ideally be only one type of line card.

The vision is to provide ubiquitously a single transport layer with performance sufficient for all services. From the user's perspective, the vision is for a single jack in the wall that will provide all his communications needs. A look at how technology will allow the network to evolve to meet those needs must necessarily begin with the device that delivers the services—the customer's terminal.

A. Customer Terminals

Terminal capabilities can evolve more rapidly than those of the network. Terminals are already at the stage where they could make use of greater bandwidth and more powerful signaling. Both these factors will become available as the network evolves, but at present terminals themselves have to possess some intelligence to improve their own functionality.

Dial Pulse signaling and its successor DTMF are relatively simple protocols. However, because of the limited instruction set, and the necessity to work in an analog environment they are slow. Some terminals use local intelligence and storage to add features and improve the user interface; for example, Speedcall, Ring Again, and Conference capability. The next generation signaling to be introduced with ISDN is considerably more powerful, but with that power comes complexity. Terminals will have to be designed to buffer the human user from the complexity of the protocols. To make an analogy, programming in ISDN primitives is like programming a computer in an assembly language; it is restricted to the expert user. The majority of users require processing power to simplify the user interface. In the same way that the majority of users of personal computers run pre-packaged software so the terminal vendor will have to provide software to interface the user to a powerful, but fundamentally unfriendly network. This will require an additional level of terminal intelligence not generally present today.

Current generation microprocessors are capable of handling 32-bit words at rates around 2 million instructions per second (MIPS). The network, at 64 kbits/s, is today incapable of passing information at a comparable rate. There exists, therefore, a bandwidth gap between the capabilities of the terminal and the network. This gap provides a potential role for digital signal processing (DSP). Some of the processing power of the terminal can be allocated to reducing the redundant information in the signal prior to transmission. Not all types of information can be compressed, but of the information that can be the graphic and pictorial files can be subject to quite substantial reduction. A high-resolution image, if mapped on a pixel basis, requires many megabytes of uncompressed digital data. Most images have high degrees of redundancy; for example, a line drawing contains a background with no variation in color or density. This requires very little information for its definition and yet may cover 95 percent of the area of the image. Similarly with pictorial images, variation between one pixel and its neighbors may be very small, and some form of differential coding will require less information for equivalent definition.

Substantial gains can be made in the transmission of information by processing the information prior to transmission. Given unrestricted bandwidth in the network the amount of processing performed is a pure cost tradeoff. Given a bandwidth gap between the capabilities of the terminal and the network, DSP will be used to achieve higher speed transfer of information.

There will be a necessary growth in the intelligence in the terminal for the reasons stated above. In addition it will be shown that with new network architectures there will be mobility of functions between the terminal and the network. Part of the terminal will therefore be a constituent of the network architecture, and highly dependent upon its evolution.

B. Access

By far the largest proportion of a carrier's investment is in outside plant, particularly the loop. The current copper loop plant currently provides service up to 4.8-kbit/s data, or in special cases up to 56 kbits/s. With ISDN this will increase to 144 kbits/s, but above this rate new installation is required. Because of its intrinsic inertia, wide-scale change in the access network will be slow. To look at the introduction of broad-band services, the access network can be segmented on the basis of needs, and each segment addressed independently. The capability of delivering high bandwidth to the subscriber is being implanted into new installations by some changes in distribution.

Subscriber carrier systems are now being deployed on a wide-scale basis, and increasingly via optical fiber feeder cables. This brings high bandwidth closer to the subscriber, and potentially the ability to offer bit rates beyond ISDN basic service. In such systems also the variety of line cards required is being reduced by adding some degree of programmability to cover different requirements on the associated copper loops.

Within the year ISDN basic service will be available from all the major switch manufacturers. It is the first step towards a network which will provide for all communications needs, and will finally cap the proliferation of different varieties of line cards.

Current ISDN implementations use two-wire echo-canceling hybrid technology. The cancellation algorithms are complex, and in North America are made even more complex by the widespread use of bridge taps (unterminated stubs). These bridge taps cause long, high-amplitude echoes which required the development of more complex cancellation algorithms and circuitry.

The introduction strategy for basic rate ISDN will be for business applications. Its introduction to residential environment is dependent on the buildup of quantity production of the necessary silicon chips, and the resulting reduction of cost in providing the service to the point that it is comparable to basic telephone service.

As an intermediate state between basic rate access and fiber-based broad band, DS-1 (1.544-Mbit/s) service is being provided in both channelized and unchannelized forms. With remote carrier terminals, the distance to the customer premises can be sufficiently short to allow repeaterless connections on copper loop. Examples of services using such facilities are PBX-CO voice trunks and video teleconference.

Radio provides a means of deploying DS-1 service to customers. It can be deployed much more rapidly than either fiber- or copper-based systems, and in addition can be used at high bandwidths (tens of megahertz). In this mode it should be regarded as a temporary expedient until permanent (fiber, copper) can be placed. Its permanent application requires coordination and could lead to congestion in the allocated frequency bands. The other major access application of radio is mobile radio, which by definition allows mobility of users' terminals. The use of cellular radio techniques allows efficient use of available spectrum and is capable of supporting large numbers of users. It may be the progenitor of a network by which a person or terminal's logical address is independent of physical location.

The operating companies are starting to deploy fiber to their major customers. Fiber has the property of long reach with high bandwidths. Major customers will find applications for the bandwidth that are currently not cost-effective, such as video teleconferencing and bulk data transfer. Because of the high value of such services there will be downward pressure for supply to smaller and smaller businesses. These services can be performed at 1.5 Mbits/s, but in the case of video, at considerable terminal complexity, and in the case of bulk data, with delay. The rate of penetration will be largely dependent on the tariff structure for such services.

Fiber loop has the potential of giving much higher bandwidth services, with little increased complexity in control. This is broad-band ISDN. All the necessary hard technologies already exist. Suitable switching mechanisms for high-bandwidth services have been announced, although they are not yet in the form of product. Dedicated fibers are being placed into major business customer premises, so there will be adequate bandwidth. Significantly, only about 10 percent of investment in access is in loops to business premises. This makes it relatively easy and strategically advantageous to place fiber in these applications.

The demand for fiber to residence is far more difficult to

predict. There are some moves to integrate delivery of communications services to the home into a single vehicle such as fiber. This would only apply to new construction. The retrofitting of fiber to existing homes will require new services of high value to the user. Some television-based services may not be cost-competitive with other delivery systems such as video cassettes, although there are services that are more effectively delivered via fiber such as High Definition TV (HDTV). A service such as electronic catalog shopping might provide a trigger. The information in a catalog is equivalent to hundreds of megabits. To scan this information requires transmission bandwidths that are many tens of megabits per second. If services such as this are of value, and can be provided cost-effectively, it could provide a demand for fiber loops to residences.

There are some parallels between the revolution in personal computing, and the bandwidth requirements for services using the network. In computing, increases in functionality or improvements to the man–machine interfaces have resulted from an exponential increase in low-cost computing power. If the analogy to the network is accurate, bandwidth requirements will also grow exponentially as the users learn how to apply it to improve the quality and utility of the information being transferred. To seed the process excess bandwidth has to be capable of being placed at little or no cost penalty over existing methods of providing the basic service.

Fiber loop systems are currently not cost-competitive with copper loop for telephony applications in residential environments. Although intrinsic material and placement costs are similar, fiber splicing and cabling costs remain high. High-speed optoelectronics and multiplexing are not yet cost-effective. The cost of interfacing will drop with the introduction of low-cost lasers or LEDs. Carrier systems with outside plant remotes will aid the economics of fiber loop, although some of the savings obtained apply equally to copper loop.

The cost of multiplexing signals on to the fiber is the dominant cost factor of the interface. This cost in turn is directly related to the number of channel appearances (channel cards) and the complexity of the processing required for each. In the business environment both the number of channels and their complexity will be higher than in the residential environment. The bit rate of the service bearer could be fixed to both business and residential users. This leads to common interface hardware. If a technology/cost breakpoint is used the rate may be as high as 500 Mbits/s. The user, though, would only pay for the amount of bandwidth and number of channels in use. The same arguments apply to radio, but because of the necessity to conserve spectrum, the bearer rate for radio may be around 50 Mbits/s.

In summary, the vision is for unification of the physical loop interface, the practice will be a limited number of standard interfaces. The access evolution is technologically straightforward, the next question is how to switch the information once it reaches the central office.

C. Switching

Telephone switching has undergone a technology revolution from analog-based space switching to digital space and time switching. The switching is 64 kbits/s based and performed under the control of a common processing element. The basic structure of switches has evolved from a single processor to the distribution of some functions into the peripherals. The control and switch matrix are generally still physically separate. Telephone switches are in a continuous state of evolution, and are today capable of handling a wide range of service other than basic telephony. High-speed digital data and the support of highly featured business telephone sets are examples of such capabilities.

In addition to telephony switches, other more specialized switches continue to be deployed for services with different requirements, but because of the dominance of voice service in terms of traffic, the telephone switch will continue to evolve to carry more diverse services for the foreseeable future. It is the evolution of this switch that will be the theme of this section.

Today's networks have evolved around the ability to provide basic telephone service at low cost. To provide new services at low cost they have to fit within the capabilities of the telephony network. The alternative is to provide an overlay. This has led to packet networks and switches, where fast connection time is important, and channel switches (cross-connects), where services have long holding times and perhaps higher bandwidth requirements. The result is that three types of switches exist in today's network, each type tailored to a different service. The switches are generically characterized as follows:

Network	Characteristics
Telephony	Channel Rate 4 kHz (64 kbits/s) holding time—minutes traffic volume—large message rate—medium
Packet	Channel Rate $\leqslant$ 9.6 kbits/s holding time— < 1 s traffic volume—low message rate—high
Special Services	Channel Rate—up to 1.5 Mbits/s holding time— > 1 h traffic volume—medium → high message rate—low

The telephony switch crosspoints each typically handle tens of calls per hour. The volume of traffic, and therefore the number of crosspoints, is very high. Today's switches require high-capacity processing capability as with common control; a single processing structure has to manage all crosspoints. The throughput of the core of today's digital switch is of the order of 5 Gbits/s. Multistage switches have propagation delays across the switch of the order of a millisecond. The path setup is dependent on traffic load but is generally a few tens of milliseconds from the termination of signaling.

The packet switch is an order of magnitude smaller in terms of throughput than the voice switch. Because the holding times are generally an order of magnitude shorter, the total number of path setting and clearing operations is of a similar order. Packet switches have flow control, which results in a relatively long delay through the switch. This delay is load-dependent but is at least as long as the packet because of the error control scheme used. Routing is somewhat simpler than in a conventional switch and there are few embedded features. Processor requirements are generally less demanding than a circuit switch.

The channel switch used for special services is designed to hold crosspoints for long periods of time, ranging from hours to years. The number of crosspoints tends to be smaller than that on a telephony switch. The basic instruction set is very simple (e.g., connect channel 15 port 3 to channel 2 port 16) resulting in a low processing requirement. The process of path setup is today by preprogrammed network intelligence, or manual control. This is evolving to customer control of his own facilities.

The reason for the existence of the channel switch is that it performs the switching function at much lower cost than a conventional switch. It has generally an order of magnitude fewer channels than a message switch and requires call handling rate three orders of magnitude lower. The crosspoint bit rate may be of higher bandwidth, leading to applications for facility management, grooming, and restoration in addition to special service switching.

To compare the three types of switches, an envelope of performance can be developed (Fig. 5). The axes represent

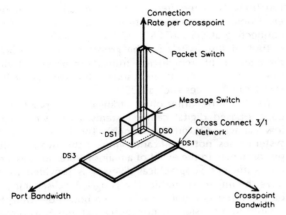

Fig. 5. Switch characteristics.

three aspects of the switch which give some comparison of the difference in capabilities. The port size on DS-0 level cross-connects and message switches today is generally DS-1, and on packet switches 56 kbits/s. The DS-1 level channel switch (CCD3/1) will have DS-3 ports. There is approximately an order of magnitude difference between cross-connects and message switches, and the same between message switches and packet switches. The crosspoint speed is highest in the 3/1 channel switch (DS-1) and lowest in the packet switch.

The development of digital and software technologies has allowed telephone switches to evolve to provide more functionality and features. In the course of this evolution, processing loads have increased considerably over those required for basic telephone service. To provide for this increased load, distributed processing architectures are incorporated into current digital switches. Future developments will provide nonblocking switch elements and integrate packet and channel switching capabilities. Network intelligence and databases are required to coordinate and integrate the different network elements. Initial implementations use physically separate switching mechanisms, and use intelligence to give the user the appearance of a single,

high-performance network. A common switching mechanism with a much higher capture ratio of service requirements would help further integration.

Two emerging technologies may evolve to provide this mechanism: fast packet switching and fast circuit switching. The way they operate is consistent with the current migration of intelligence to peripherals and networks rather than residing centrally in switches. With both these technologies, call processing software is taken out of the individual switches and becomes a network function. Some functions such as contention resolution become embedded in hardware or firmware, and are removed from the software load. Other functions (such as error recovery) are performed on a network-wide basis instead of a link basis and are therefore not replicated across the network.

Such mechanisms provide the potential of sufficient performance to embrace the capabilities (Fig. 6) of all existing types of switches at reasonable cost. These switches though cannot be considered as stand-alone devices. The real advantages only accrue when they are combined to form a network. A detailed discussion of their potential role is therefore given in the discussion of transport networks. Today, such switches are in the exploratory phase, but can be expected in product form in the mid-1990s. With their

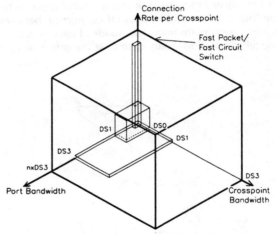

Fig. 6. Fast packet/fast circuit switch characteristics.

introduction, the need for overlay networks for new services will be diminished. The vision of a single switch for all services may be reality.

D. Transmission, Cross-Connects, and Interfaces

Transmission systems were the first to employ digital techniques on a widespread basis. The penetration of digital systems in metropolitan environments is over 80 percent, and the circuits are in theory available to any traffic in digital form. In practice, some segregation is required to guarantee error performance on data circuits.

Transmission has traditionally evolved independently from switching. There are standard interfaces between them, giving points of demarcation for independent engineering, operations, etc. The interfaces were originally two or four wires per voice circuit, with additional signaling. In the digital domain, it is on a digroup (24-channel) basis with in-channel signaling. This gives a pair-gain of 12 at the interface.

As has been noted in the previous section, in large switching systems the costs are shifting away from the switching mechanism towards the peripherals. One major cost element in the periphery is interfacing the switch to transmission systems. The cost is in terms of both hardware and administration. The use of higher capacity interfaces helps to contain the costs and, at the same time, provides for less wiring in the central office.

To achieve maximum benefit from high-capacity interfaces requires some conversion between different requirements of transmission and switching. To understand the problem, a brief explanation of today's network topology is necessary.

Today's networks consist of several layers. The structural network consists of cables, microwave radios, and the associated rights of way. The physical network consists of the cross sections between nodes. The logical network is the interconnection of nodes.

In today's network, the physical layer is a simple interconnection (Fig. 7), whereas the logical layer is an almost fully interconnected mesh (Fig. 8). In a metropolitan environment this results in large cross sections in the physical and structures network, but much smaller cross sections in the logical network. The physical connection between two nodes may contain many thousands of circuits, made up of logical trunks from many routes. At the office, at each end

of the physical connection, the constituent logical trunks must be split out of the whole and routed according to their logical destinations. In a digital environment, the logical trunk cross sections are administered at a modularity of DS-1 (24 channels). Cross-connect panels operating at this modularity provide the routing and engineering flexibility between the logical and physical connections.

For maximum efficiency in the use of switch core hardware the modularity of interfaces in numbers of channels should be in powers of two. Current digital switches have their cores designed in this fashion. The modularity of transmission, on the other hand, has evolved with the need to conserve bandwidth, and has resulted in bundles of $24 \rightarrow 96 \rightarrow 672$. In the days of voice-frequency interface between switching and transmission this difference in modularity was unimportant, but with multiplexed interfaces format and modularity conversion is necessary.

With DS-1 (24-channel) modularity, the conversion may be performed efficiently at the periphery of the switch. With DS-3 (672-channel) modularity, a different network functionality is required. That functionality is that of the electronic cross-connect with interfaces at DS-3 and cross-connecting at DS-1 (DCS 3/1). Using this device, the bulk output of the switch can be groomed and electronically directed to different logical trunks and physical facilities. Interfacing can be at high speed, but still generally via standard interfaces such as DS-3.

The existing inter-office transmission plant is still copper-based digital carrier systems, each with relatively low capacity (24 or 48 channels). The failure of a single system does not significantly impact the overall network performance. In the event of a major failure, such as a cable cut, sufficient geographical diversity is provided to minimize the impact on traffic. The geographical separation can be as simple as running cable down both sides of a street.

Fiber transmission, as previously mentioned, is providing massive capacities in terms of number of channels. The capacity of a single fiber cable is sufficient to cover foreseen growth on the vast majority of physical routes. The placement of a second cable for diversity is generally not cost-effective. To utilize the capacity of fiber systems, operating companies are collapsing their physical topologies to backbone fiber routes, and concentrating their traffic on to these routes (Fig. 9). This does not provide the same level

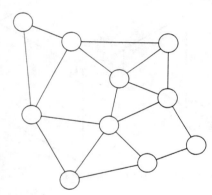

Fig. 7. Physical network.

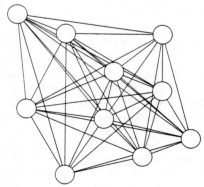

Fig. 8. Logical network.

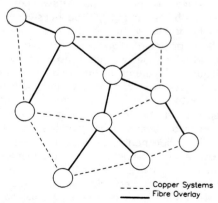

Fig. 9. Fiber deployment.

of survivability as the copper-based network. For example, a single 565-Mbit/s system carries 8 thousand 64-kbit/s circuits, and a cable may carry a number of such systems. If the cable is cut, the repair may take several hours; the resultant impact on the network performance can be severe. With today's manual cross-connect technology the re-routing of all special services would be impossible. The availability of such circuits would be reduced to unacceptable levels. The use of channel switches to perform the cross-connect function is essential to restore the failed circuits in an acceptable time. Network intelligence is also required to know where sufficient spare capacity for restoration exists. With such restoration systems, and careful network design, the need for geographic diversity on fiber is minimized. This illustrates by example how new network design and architecture rules have to evolve to make application of new technologies effective.

Network design for survivability involves consideration of how much of the available capacity of the fiber system needs to be restored in the event of failure. This, together with the total traffic requirement, will limit the capacity requirement of fiber systems. With today's traffic levels there are probably very few requirements for systems with greater than the 565-Mbit/s systems currently being installed. The real requirement for higher capacity systems will only happen with the introduction of broad-band services.

The introduction of electronic cross-connects eliminates one separate stage of multiplexing at each end of a physical facility (Figs. 10 and 11). Standard interfaces between systems are still by means of asynchronously multiplexed signal formats. The next stage of development is a direct synchronous interface. With a traditional asynchronous multiplex approach much of the multiplexing functionality has to remain, buried in both the switch and cross-connect. This is because the individual DS-1s and DS-0s are not readily identifiable in the DS-3 bit stream. The simplification of the interface is to some extent dependent on the simplification of the transmission or interface formats, al-

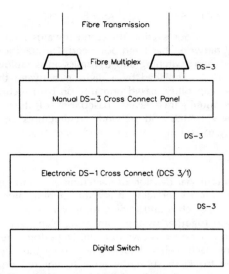

Fig. 11. Transmission termination—electronic cross-connect.

though application of VLSI to conventional formats reduces the cost advantage from the original predictions.

Syntran was developed by Bellcore to fulfill this function for the DS-3. Some characteristics of the asynchronous DS-3 have been retained for transmission compatibility. Syntran therefore is based on a 672-channel structure, and is not optimal for switch interfacing. It does, however, introduce the concept of total synchronism within the network, which can ultimately lead to further savings in interface costs.

The ideal format is synchronous with a simple interleaving multiplex scheme. Such schemes are typified by Bellcore's SONET, intended as a public standard. This type of a format is designed to simplify interfacing at the expense of transmission bandwidth efficiency.

Formats such as SONET are designed for direct interfacing between termination and transmission. With such systems the electrical interface can be replaced by a direct optical interface. The electronics interfacing the electrical DS-3 signal to the fiber are integrated into the cross-connect or switch. With the reach being achieved by optical fiber systems outside repeaters become unnecessary. Transmission is effectively collapsed to the fiber cable itself and into the switches and cross-connects (Fig. 12).

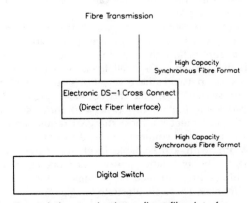

Fig. 12. Transmission termination—direct fiber interface.

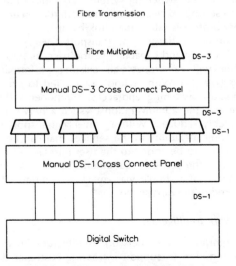

Fig. 10. Transmission termination—manual cross-connect.

In the previous section the trend towards less complex physical networks has been described. The traditional division between switching and transmission is rapidly disappearing to be replaced by a division between the basic transport capability, based primarily on hard technologies, and the intelligence layers, based on soft technologies. With new switch and network architectures comes the flexibility to add more features and functionality.

A. Transport

The traditional division of switching and transmission is more than just interfacing, it is a fundamental architectural split. The switches and transmission were stand-alone devices held together by the numbering plan. Each element had autonomy over its own domain. To perform the routing functions each switch had to know only the destination outgoing trunk, or subscriber line. It required no knowledge of network topology. Within a switch, however, there has been a migration from distributed control—in step systems each selector need only known the next selector group—to common control, whereby a central processor knows all the paths through the switch. This migration of intelligence is continuing beyond the boundaries of the switch to centralized network intelligence. Together with network-wide databases, this allows features to be added on a network-wide basis. With common channel signaling network features, such as Enhanced 800 calling, can be offered. A major development planned is Common Channel Signaling System #7 (CCS#7), a universal call processing system. It is the one element required to make separate transport networks appear as a single entity to the user.

Three types of switching systems are currently required; namely, message, packet, and channel, each with different characteristics tailored to their application. These switch types are arranged to give three distinct types of transport service. The performance envelopes are different for the different logical networks, and a set of parameters can be used to characterize networks.

For voice service delay is a noticeable impairment, and significant delay variation can be intolerable. Packet data are insensitive to delay; message integrity is more important. For packet data, recovery is performed on a link basis resulting in the long and variable delay characteristic of packet networks.

The other parameter of importance is connection speed. This is the speed with which a user can establish and tear down a connection through the network. As a rule of thumb this should be a small fraction of the connection time. Packet networks have very fast setup times whereas the "non-switched" special services have long lead times.

To provide a single transport network for all services, the network must meet the most stringent requirements of all services in terms of connection speed, switching delay, and bandwidth, with a reasonable level of accuracy and lost messages. The two latter performance objectives can be met with end-to-end recovery schemes, albeit at the expense of delay. These are illustrated in Fig. 13.

The ultimate goal is for "transparent transport" where

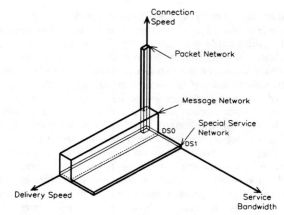

Fig. 13. Network performance envelope.

connection time, blocking, and transmission performance become insignificant. Current partitioned networks are evolving towards such a ubiquitous solution, and by the addition of an overall network intelligence layer, the network will have the appearance of ubiquity to the user.

The two generic technologies with the potential for sufficiently high performance are Fast Packet Switching and Fast Circuit Switching. Both of these technologies use much simplified network protocols to minimize the amount of processing required for setting paths. Both systems also make use of distributed hardware processing to gain very rapid path setup. Simplistically, the main difference between the two technologies is in the way the information is packaged for transmission. With the fast packet approach, the available bandwidth of the transmission pipe is divided in the time domain such that all of the bandwidth is used by all of the circuits for part of the time. With fast circuit switching, the bandwidth of the pipe is multiplexed on a channelized and framed basis such that part of the bandwidth is available to all of the users for all of the time. In practice the technologies use hybrid techniques, biased appropriately.

There are great similarities between performance capabilities of the two switching systems: both have intrinsically fast path setup and cleardown, and both rely on end-to-end recovery protocols rather than link-by-link.

The universal transparent transport layer provides extreme flexibility and capability to the user. Its performance envelope will encompass all existing performance envelopes (Fig. 14). With technology applied to simplify the transmission, switching, and access components, the physical size and the administration of such a network will be considerably reduced. With a universal transport network there is high flexibility in the services that can be offered. Software has to be added to the user interface to tailor the transport to the service requirements so that its invocation is by means of a simple man–machine interface. This will be covered in the next section.

B. Network Intelligence

Telecommunications network architectures in the 1990s will be required to support increasing customer demands for new services. Also, these architectures will allow service

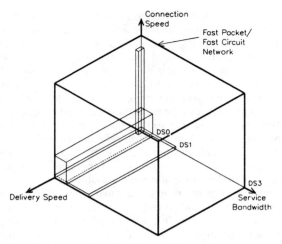

Fig. 14. Fast packet/fast circuit network characteristics.

providers the flexibility to assess new service viability through rapid deployment of technology and/or market trials on a network basis. It is to be expected that these new services will, to some extent, be defined or evolved empirically, leading to the need for capabilities to modify and change service parameters. These needs, in aggregate, point to a requirement for opening up network architectures to allow the service provider or end user to customize features to his own requirements. To provide this capability with rapid response requires a flexible open-ended architecture.

The network today contains several switching technologies, but with switch modernization, software-controlled switches are now dominant. The capabilities of these switches are determined by the programs loaded in them. The carrier modifies capabilities by procuring new software loads from the switch vendor.

The current network has been built up on the basis that switches and transmission systems are independent items, meeting at standard interfaces. Switch architecture is self-contained; all processing requirements supplied locally. Features have been implemented on a switch basis, the switch vendor writing and supplying the appropriate software. This is the traditional view of network architecture.

To develop the ability to provide network-wide features, an alternative view of the architecture is required. In this view the architecture can be considered in layers (Fig. 15). The core is the transport, consisting of the individual

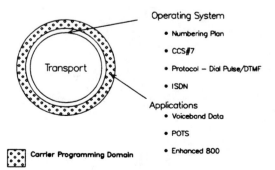

Fig. 15. Network functional structure—1990.

switches and transmission systems. The next layer provides the operating system.

With step-by-step and crossbar switches, the operating system was the numbering plan. The protocol used for messaging was dial pulse, both internal to the network and to the subscriber. Programming of the operating system was by hardwire connection, features were difficult if not impossible to engineer. With the advent of more software into switches the flexibility increased. Tasks such as number change, which previously required a physical wiring change, became electronic. Some capability to provide enhanced calling features was present. The flexibility of the network was still limited, and the programming of the features still a switch vendor responsibility.

Recent developments provide an overall network intelligence layer, which supplies the necessary shared resource and coordination functions for provision of features on a network-wide basis. This is a major step in the development of architectures which will lead to the desired, responsive network. Other steps, though, are necessary.

One such step is the development of software architecture. Partitioning of functions into well-defined blocks has been at the expense of less efficient use of real time, and real time has precedence. The rapid development of computing power, the ability to partition and distribute computing tasks, and the implementation of repetitive time-sensitive tasks in hardware will reduce the need for real-time optimization. More highly structured functional architectures will be possible.

An extension of the layering principle is the first step in partitioning. An example of such layering is given in Fig. 16.

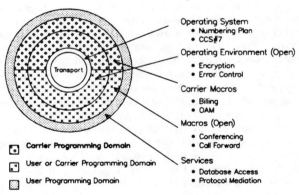

Fig. 16. Network functional structure—2000.

The inner layer provides the raw transport. It may consist of different types of switching and transmission systems. The next layer is the kernel which will match the switching mechanics to a common standard. The kernel will be addressed by a set of call control messages, an example of which could be "Set Path Port A to Port B." The kernel will generate the appropriate switch-specific messages to set the path. It will also pass out basic call metrics, such as bandwidth and duration and messages on the operational status of the network elements.

Outside the kernel will be a series of functions which generate and use the kernel messages. These functions may be proprietary to the operating company, or open to user programming. Examples of the former are the generation of bills and restoration under fault conditions, and of the latter are functions for establishing private networks and the invocation of error control.

This set of primitive commands will constitute a language on which networking applications can be built. It will provide all the basic capabilities for setting paths, measuring traffic, counting messages, and other necessary functions. These primitives can be built into application-specific macros which can be restricted to operating company applications such as billing and network administration, or to more general applications, such as Call Transfer and Conferencing.

Within each layer, emphasis will be placed on the definition and standardization of the software modules and their interfaces to minimize undesirable interaction with other modules. This will result in more flexibility to transfer and distribute modules; if the functions and interfaces are defined, physical location and implementation both become of secondary importance. Of more importance is that protected software domains can be defined.

In this example, it is possible to define programming domains. The kernel is programmed by the switch vendor, the billing will be defined by the carrier. Provided the stability and performance of the network remain under control of the carrier, it may be possible to allow end user to access the primitives and configure his own network; the programming and customizing of features by the user is somewhat easier.

The type of network structure envisioned above leads to more graceful and rapid evolution of services and the network itself. The network will be more responsive and more versatile. The use will be protected from a more complex interface by software "personality" packages resident either in the terminal, in the line card, or as a shared function in the network itself. This personality package will invoke those network functions for the service called: error correction for data, echo suppression, and controlled loss for voice. Features such as Speedcall lists will also be resident in the personality package. The sophisticated user will be able to write his own customized package to fine tune the network to his own requirements.

The result of this type of structure is the potential to establish, on demand, virtual networks and closed user groups by invoking both network shared resources and dedicated customized resources. Together with the high-performance transport layer, it can provide the user with the ability to program new requirements under his own control. However, as with any sharing system, there have to be some controls implanted to prevent one user's demands affecting other users. These controls could include some restriction on access, to prevent one user monopolizing transport capabilities, and maybe restrictions on shared processing. The mechanisms, though, are not well understood and will be largely empirical. There will be a continuing need for network modeling and testing to provide the necessary basis for updating traffic models and engineering.

The network and network components of the 1990s will continue to take advantage of developments in technology. The hard technologies are reducing the cost of providing the basic transport functions, and at the same time giving increased performance and more versatile switching mechanisms. Some of the simple repetitive functions performed in software today will revert to hardware implementations with lower cost VLSI. This will give the ability to collapse more and different types of traffic on to a common transport network. Integration will initially be achieved logically by the addition of intelligence layers, with physical integration dependent on the generation of switching mechanisms of sufficient capacity and capability.

The prime contribution though will be from the soft technologies. Network architectures will provide more versatility and more functionality to exploit the capabilities of the high-performance transport layers. The ability of both operating companies and users to respond to new demands will be dramatically improved by the ability to write their own, custom software functions, facilitated by the use of standard, powerful high-level languages. SDL and CHILL provide some of the early steps in this process.

The major challenge is also in the soft technologies. Allowing users to configure their own networks will place demands on shared network resources, both physical and processing. The needed technologies for the separation of functions into well-defined blocks are understood; the ability to isolate the performance of the network from the demands of different users is not. The opening of the network protocols to the end user must be preceded by sufficient testing to provide reasonable guarantees on the ability to maintain performance for all network users.

The vision of a single, responsive, high-performance network is technologically achievable before the end of this century. In fact, the constituent fundamental hardware and, to some extent, software technologies exist today. The challenge is to put in place the necessary standards and structures to allow the superficially contrary requirements of technical innovation and standard building blocks to coexist. That is the ultimate ISDN.

Design of an Integrated Services *Packet* Network

JONATHAN S. TURNER, MEMBER, IEEE

Abstract—The Integrated Services Digital Network (ISDN) has been proposed as a way of providing integrated voice and data communications services on a universal or near-universal basis. In this paper, I argue that the evolutionary approach inherent in current ISDN proposals is unlikely to provide an effective long-term solution and I advocate a more revolutionary approach, based on the use of advanced packet-switching technology. The bulk of this paper is devoted to a detailed description of an Integrated Services *Packet* Network (ISPN), which I offer as an alternative to current ISDN proposals.

AN integrated voice and data packet communications system has several advantages over existing methods.

• It uses a common set of switching and transmission facilities for both voice and data communication. This is less costly than current systems that use separate mechanisms.

• It allows voice communication to be done using less than 25 percent of the bandwidth currently needed, without sacrificing quality. This allows major savings in long-distance transmission costs. It also allows customers to carry on two or three simultaneous voice conversations along with a substantial amount of data traffic over a standard copper loop.

• It provides much higher performance data communication and at lower cost than current systems. This is largely due to the integration of data communication with voice, which allows one to take advantage of the economies of scale possible in the large systems needed for a national telephone network.

The system is based on high-performance packet switches which are large and fast enough to effectively support both voice and data communication on a large scale. Each packet switch can have a raw throughput of up to 1.5 Gbits/s, allowing it to support as many as 50 000 simultaneous voice conversations using a 32 kbit/s voice encoding scheme. The one-way cross-network delay in a worst-case connection in a national network in the U.S. can be limited to about 150 ms. The key elements of the design are as follows.

• The use of high-speed digital transmission facilities (1.5 Mbits/s) with excellent error performance.

• Simple link protocols. In particular, there is no flow

Manuscript received April 15, 1986. This work was supported by a Bell Communications Research Grant. This paper was presented at the ACM 9th Data Communications Symposium, Whistler, B.C., Canada, September, 1986.

The author is with the Department of Computer Science, Washington University, St. Louis, MO 63130.

IEEE Log Number 8609694.

control or error correction done at the link level, which eliminates the need for state information in the link level protocol processors.

• A predominantly connection-oriented service. This allows the routing of most packets to be handled by a very simple method and facilitates bandwidth allocation and overload control. A connectionless (datagram) service can also be supported.

• Hardware implementation of basic switching and protocol functions. Switching is done using a large self-routing network containing roughly 1300 custom VLSI chips. All per-packet protocol functions are handled by protocol processors (one for each link) consisting of one custom controller chip plus one large memory chip.

• Implementation of higher level protocol functions (including error correction and flow control) on an end-to-end and application-dependent basis.

The governing philosophy behind the design is that the communications network should provide transport of information at the highest possible level of performance, but nothing else. The network achieves generality, not by providing every service a user might conceivably require, but by providing only those services that every user requires. The system can be implemented with currently available technology at a cost that is comparable to that of conventional telephone switching systems.

THE TROUBLE WITH ISDN

The Integrated Services Digital Network has been heralded as the mechanism that will usher in the Information Age. ISDN, it is said, will facilitate the development of new communications services, including a wide range of data services and maybe even video. It will allow such services to be implemented on a large scale at a cost most customers can afford, and spur the transformation to a "postindustrial society."

While I share the long-range goals of ISDN proponents, I do not believe that current ISDN plans can take us very far. The fundamental problem is that current plans (see [1], for example) implicitly assume a network model based on circuit-switched voice and a combination of circuit and packet-switched data. This assumption is evident in current standardization efforts which focus on a transmission format for digital subscriber lines that provides two 64 kbit/s circuit-switched channels plus a 16 kbit/s packet-switched channel. While this plan can provide a limited data communications capability (it is certainly a vast improvement over the current situation), it is too inflexible to satisfy long-term needs. What happens, for example, to an application that requires a 75 kbit/s channel? Do we implement it using the two 64 kbit/s channels or one of them and the packet-switched channel? Neither option is particularly attractive.

The source of the trouble is the reliance on circuit switching, which requires that the available bandwidth be divided up into fixed-size channels. The channel size is a permanent feature which cannot be changed easily, if at all. There is no reason to think that 64 kbits/s is a particularly useful channel size. The only reason for choosing it is that current telephone switching systems are based on that size. In fact, the driving force behind current ISDN plans is to preserve the investment in existing equipment while ''evolving'' to a more flexible network.

Unfortunately, it will not work because it is inherently a hybrid approach. The only thing integrated about ISDN is its name. Two (or more) switching networks are required to support ISDN as it is currently envisioned. Manufacturers may talk about their integrated architectures for ISDN, but what they mean is that they plan to put a packet switch and a circuit switch in the same box and call it an integrated system. This is not their fault—what else can they do? Packet switching and circuit switching are very different communications methods. They require different switching and transmission facilities and all the proposed schemes for combining them are little more than packaging.

What to do then? Is a hybrid network really necessary or is there an integrated solution that can satisfy the needs of both voice and data and remain flexible enough to meet satisfy new requirements as they arise? I claim that there is such a solution, but it requires abandoning circuit switching and moving to new network designs based on packet switching.

WHY PACKET SWITCHING?

Here are three reasons that make packet switching an attractive method for providing integrated voice and data services.

• *Adaptability to Changing Traffic:* Packet switching naturally provides the user with exactly the bandwidth required. As new services are developed with different bandwidth requirements, packet-switching systems can adapt to the changing conditions easily. Circuit-switching systems cannot.

• *Integrated Internal Architecture:* As outlined above, current ISDN plans require separate switching networks for different types of information. Packet switching can provide both an integrated customer interface and a single network solution for a wide range of communications needs, leading to substantial cost savings in switching systems and system administration.

• *Transmission Efficiency:* Many data services are characterized by bursty communications patterns which make poor use of conventional circuit-switched facilities. For example, interactive data users typically use only a few percent of the bandwidth available to them. Although

it is less widely recognized, voice is also bursty. In the average telephone conversation, less than 40 percent of the available bandwidth is actually used. Packet switching can exploit this burstiness to double the number of conversations that can share a single transmission facility.

Given all these advantages, why has packet switching not been used extensively for voice? To answer this, we must take a closer look at the technical specifications and costs of conventional circuit switches and packet switches and see how they compare.

Circuit switching has been the technology of choice in the telephone network since its origin about 100 years ago. During that period, circuit switches have evolved from manually operated switchboards to small but automatic step-by-step switches to larger panel switches to the computer-controlled electronic switching systems that currently dominate the scene. The current systems are very large—local switches can provide service to over 100 000 customers; large toll switches support as many as 50 000 simultaneous voice calls corresponding to a raw bandwidth of 6 Gbits/s. The network as a whole is also very large. There are approximately 10^8 telephones in the United States at present and over 10 000 local switching offices. The performance of the switching systems is quite impressive—information passing through a modern digital switch is typically delayed no more than a few milliseconds, and new connections can be established in a fraction of a second. Equally impressive is the cost—while there is considerable variation, per-line equipment costs for modern digital telephone systems is typically in the $100–200 range.

Just as circuit switching has long dominated the telephone network, so has packet switching dominated the data communications scene, principally because of the advantages cited earlier. Packet switching is exemplified by the ARPANET, which was the first major example of a large-scale data communications network. In the ARPANET, the endpoints of the communication are typically large time-shared computers, called hosts. Communication is provided by packet switches, each of which typically connects to a few hosts and several other packet switches. There are currently about 300 hosts in the ARPANET and 100 packet switches. The packet switches are implemented using general-purpose computers (usually minicomputers), although more recent versions have been supplemented with front-end communications processors to reduce the load on the main processor. The transmission facilities used by the ARPANET include low-speed modem connections (1–10 kbits/s) and higher speed digital facilities (56 kbits/s). The throughput of the packet switches is generally under 1 Mbit/s and delays can be substantial (50–100 ms per switch). The cost of packet switching as provided by the ARPANET is quite high since each packet switch supports a small number of hosts. Commercial data networks do better, but there remains a large gap between per-host costs in data networks and per-line costs in the telephone network.

This comparison explains the conventional wisdom that

packet switching is poorly suited to the needs of telephony and the resulting conclusion that an ISDN implementation must include a circuit-switching component to support voice. The fallacy in the conventional wisdom is that the disadvantages attributed to packet switching by this comparison are not due to any inherent properties, but are side effects of the conventional implementations. The requirements and design constraints that shaped the development of the ARPANET and commercial data networks were completely different from those that shaped the telephone network. The scale, the performance requirements, and the cost sensitivities are all very different. It is because the *needs* differ that the resulting systems differ so in their technical specifications. In the remainder of this paper, I will attempt to correct the widely held, but erroneous view that packet switching is poorly suited to the needs of voice by describing a system capable of supporting voice and data communication on a large scale and at a cost that is competitive with conventional telephone systems.

ISPN ARCHITECTURE

Let us begin by deciding what general properties a packet-switching system must have if it is to be a suitable vehicle for providing voice and data communication on a large scale. First and most obviously, it must be big—comparable in raw bandwidth to conventional telephone systems. Second, it must be fast—long end-to-end delays are annoying in voice connections, and hence unacceptable. While opinions differ on the exact amount of delay that can be tolerated, most experts would agree that 100 ms is acceptable, while more than 500 ms is not. Third, it must be inexpensive. Any system that seeks to replace circuit switching must be able to provide voice services at a competitive cost. It is not enough to offer a better product at a higher cost—for a system to succeed on a large scale, it cannot significantly increase costs for basic services.

How can we achieve the requisite scale, performance, and economy in a packet-switching network? Well, the telephone network is one place to look for the answer. One of the first things one notices is that the telephone network makes extensive use of high-bandwidth digital transmission facilities. Modern digital switching systems are designed to interface directly to 1.5 Mbit/s transmission facilities carrying 24 voice channels in a 64 kbit/s format. Interfacing costs are cheap since they are shared by the 24 channels and the bit error rates are excellent. Conventional packet-switching systems, on the other hand, interface to much smaller channels, and must be prepared to cope with the much higher error rates present in modem connections. A second thing one notices about the telephone network is the amount of special-purpose hardware used to provide the switching function. While telephone systems contain general-purpose computers, their function is primarily connection establishment. The

computers do not move the bits. That function is handled by large specialized networks. In contrast, most conventional packet switches include a general-purpose computer that must do some processing on every packet and can quickly become a bottleneck.

These observations suggest that a high-performance packet switch should 1) interface directly to high-speed digital transmission facilities and 2) use special-purpose hardware to perform all per-packet processing. Now, those familiar with link level protocols used in conventional packet networks may recognize this as a challenging task. Typical protocols are quite complex, requiring extensive state information in the protocol processors and complicated error recovery procedures. This complexity almost demands a programmable processor (a hardware implementation would never be completely debugged), but a microprocessor-based implementation is unlikely to be fast enough to keep up with a 1.5 Mbit/s link, and also may be too expensive. Fortunately, there is a way out—simplify the protocol. Current packet-switching protocols were designed for hostile environments—the low speeds and high error rates typical of modem connections. In a network composed entirely of high-speed digital facilities, much of the complexity of typical protocols can be eliminated from the lower protocol layers. In particular, error correction and flow control can be taken out of the link layer and provided on an end-to-end basis rather than a link-by-link basis. This eliminates the need for extensive state information at the ends of each link and the synchronization and recovery procedures required to maintain it, and in turn makes possible the construction of inexpensive, high-performance protocol processors. In addition, if these functions are provided on an end-to-end basis, they can be provided selectively. Hence, delay-sensitive applications like voice can avoid the performance penalties they can cause.

Returning to the question of scale, how large must our packet switches be? In order to compete with conventional telephone switches, they should be capable of supporting at least 50 000 simultaneous voice conversations. How many conversations can a 1.5 Mbit/s link carry? When operated in a circuit-switching mode using the 64 kbit/s digital encoding method currently employed, the answer is 24. When operated in a packet-switching mode, that number jumps to about 50. One can obtain another factor of two improvement by using newer voice encoding methods. 32 kbit/s adaptive differential pulse code modulation (ADPCM) is a good choice since it provides comparable quality to the method currently used, but requires only half the bandwidth. This leads to a figure of 100 voice channels per 1.5 Mbit/s link, implying that our packet switch must terminate 1000 full-duplex links if it is to support 50 000 simultaneous voice conversations. This, in turn, means that our switch must include some mechanism capable of receiving over two million packets per second, and sending each one out on the appropriate outgoing link.

NETWORK OVERVIEW

Based on the discussion in the previous section, we can begin to develop an architecture for the ISPN. The major components are shown in Fig. 1. The packet switches (PS) each terminate up to 1000 high-speed links (HSL). Using 32 kbit/s ADPCM coding for voice, they can support over 50 000 simultaneous voice conversations. Residential customers connect to the network over medium-speed links (MSL), which operate at 100 kbits/s. High-speed access can be provided to business customers. The customer premises interface (CPI) can take a variety of forms, depending on the kind of service required. At the low end of the spectrum would be a simple controller providing service for a single telephone and implemented using a single-chip 8 bit microcomputer. Customers wanting several phones and data communication would require a more complex controller. Businesses would typically have an interface to a private branch exchange (PBX) or local area network (LAN). The network interfaces (NI) provide concentration, accounting data collection, and network protection. A configuration designed for residential customers would support about 500 customers and would be connected to a local packet switch by four HSL's. Thus, one PS could have as many as 125 NI's and support over 60 000 custom rs.

NI's can also be designed to provide a conventional line interface as supported by current telephone systems. Even higher concentration ratios can be supported in this case— up to 2000 customers could be supported on a single NI connected to its host PS by four HSL's. While this configuration does not exploit the advanced capabilities of the system, it does provide a mechanism for easing the transition from a circuit-oriented network to a packet-oriented one. In a similar fashion, NI's can be designed to interface to the current telephone network. In this case, the NI would interface directly to up to 960 digital trunks.

The network provides two communications services.

- *Point-to-Point Channels:* These are two-way channels joining pairs of customers. The customer establishes a channel by sending a connection request message to the network specifying the destination and the average bandwidth needed. If the network accepts the connection, it in effect guarantees that the customer can expect to have the requested bandwidth available. The network may refuse to accept the connection if there are not adequate resources available. The connections do not provide perfectly reliable information transport. In particular, the network does not provide mechanisms for error correction and flow control. The network can provide connections at any speed up to 1.5 Mbits/s (although larger connections are more likely to be blocked). The bandwidth requirement may be asymmetric—that is, it may depend on the direction of transmission. No distinction is made between voice and data connections. A voice connection is simply one with an average bandwidth of about 12 kbits/s.

- *Datagrams:* These are individually addressed packets, not associated with a preestablished connection. The

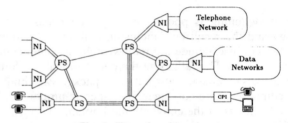

Fig. 1. Network architecture.

network makes an effort to deliver them, but does not guarantee delivery.

The communications protocols are divided into three levels—the *link level*, the *network level*, and the *customer level*. The interrelationships among the different levels are indicated in Fig. 2. There is a single link level protocol, two network level protocols (one for connection-based communication and one for datagrams), and a variety of customer level protocols. The customer level protocols are not discussed in detail here, but the intention is that these protocols would be application-dependent and built on top of one of the two network level protocols. One of these would be a telephony protocol. Another might be an internet protocol such as the DARPA IP to facilitate communication among different data networks. Still another might be a connection-oriented internet protocol.

The link level protocol provides frame delimiting, link transparency, error detection, packet timing, and congestion control, but not error correction. There are four fields used by the link protocol, the frame type field (FTYP), the priority field (PRI), the time stamp field (TS), and the frame check field (FC). The FTYP field identifies the frame as a test frame, a datagram, or a frame belonging to a connection. The priority field (PRI) contains a customer-specified priority. The network preferentially discards low-priority frames to alleviate short-term overload conditions in the network. The network uses the time stamp field (TS) to record the delay encountered by the packet as it crosses the network. More precisely, it records the number of milliseconds the packet spent in each PS and NI it passed through (see [11]). This is important for applications such as voice that are sensitive to delay variations and need a mechanism to remove the timing "jitter" that packet networks can introduce. It can also be used in distributed programming applications for clock synchronization. The frame check field (FC) uses a 16 bit cyclic redundancy code to detect errors in the frame. Frames with errors are simply discarded.

There are two protocols at the network level, one for datagrams and one for connections. Connection packets contain a 4 bit packet type field (PTYP) and a 12 bit logical channel number field (LCN). The packet type field (PTYP) identifies each packet as either a data packet or a control packet and contains a congestion control subfield, used to inform the NI's and customers of internal network congestion. They also contain an information field, which can have any length up to 144 bytes. Control packets are used to establish and control connections and contain a

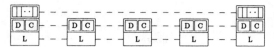

Fig. 2. Protocol structure.

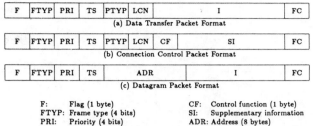

F	FTYP	PRI	TS	PTYP	LCN		I		FC

(a) Data Transfer Packet Format

F	FTYP	PRI	TS	PTYP	LCN	CF		SI	FC

(b) Connection Control Packet Format

F	FTYP	PRI	TS		ADR		I		FC

(c) Datagram Packet Format

F: Flag (1 byte)   CF: Control function (1 byte)
FTYP: Frame type (4 bits)   SI: Supplementary information
PRI: Priority (4 bits)   ADR: Address (8 bytes)
TS: Time stamp (1 byte)   I: Information field
PTYP: Packet type (4 bits)   FC   Frame check sequence (2 bytes)
LCN: Logical channel number (12 bits)

Fig. 3. Packet formats.

control function field (CF) and a supplementary information field (SI).

PACKET SWITCH DESIGN

The network is built using large high-performance packet-switching systems, each terminating up to 1023 HSL's. The structure of such a packet switch is illustrated in Fig. 3, which shows a small version with 15 HSL's.

The system is controlled by a control processor (CP) which performs all connection control functions, plus administrative and maintenance functions. The CP is a large general-purpose computer. Its role is analogous to that of the control processor in large telephone switching systems such as the No. 4 ESS [2].

Each HSL is terminated by a packet processor (PP), which performs the link level protocol for all packets and the network level protocol for data transfer packets. It also forwards connection control packets to the CP and datagram packets to the datagram routers (DR).

The heart of the switch is the switch fabric (SF) which consists of a large binary routing network. The important property of such networks is that the path each packet takes through the network is determined by successive bits of its destination address. The figure shows paths from two different SF input ports to output port 1011. Note that at the first stage, the packets are routed out the lower port of the nodes (corresponding to the first ''1'' bit of the destination address), at the second stage they are routed out the upper port (corresponding to the ''0'' bit), and in the third and fourth stages they are routed out the lower ports. The self-routing property is shared by a variety of interconnection patterns, including the so-called delta, shuffle-exchange, and banyan networks (see [5]). The PS uses a ten-stage binary routing network with 1024 ports.

The datagram routers (not shown) are special-purpose devices used to route datagrams. The number of DR's can be engineered to suit the traffic. Each occupies a port on the SF, replacing one PP. Since most of the traffic is expected to be connection-oriented, the number of DR's required should be modest.

PACKET PROCESSING

When a packet is received by a PP, it is placed in a buffer with several additional header fields added. The destination field (D) identifies the destination port on the SF. The source field (S) identifies the port where the packet arrived. The length field (LNG) gives the packet length in bytes. The switch packet type field (SPTYP) is used to identify various packet types within the PS. The arrival time field (AT) gives the time at which the packet arrived at the PS (this is used for processing the TS field).

For data transfer packets, the destination port is determined by the PP using the packet's LCN field and the PP's logical channel translation table (LCXT). Each entry in the LCXT contains an outgoing port number and a new LCN. The outgoing port is placed in the D field of the packet and the new LCN goes in the LCN field. The packet is then sent on to the switch fabric, which uses the D field to route the packet to the proper outgoing port as described earlier. When a packet arrives at the outgoing PP, it is buffered and then transmitted on the HSL with the extra header information stripped off.

The contents of the LCXT's is controlled by the CP, which can read and write the LCXT using special control packets sent to the PP's through the SF. Thus, the connection establishment process includes the sending of messages from the CP to the two PP's selected for the connection, updating their LCXT's appropriately.

SWITCH FABRIC

The basic operation of the switch fabric has already been described. The SF's highly regular and parallel structure allows the construction of very large switches, without the bottlenecks that can arise in bus or ring-based interconnection networks.

The nodes of the SF operate as miniature packet switches. Each node has a buffer at each input port capable of holding one maximum length packet. The data paths joining the nodes are bit serial and operate at 12 Mbits/s. This gives the SF an 8 : 1 speed advantage over the external HSL's. Thus, if all the external links are operated at an occupancy of 85 percent, the internal links will have an average occupancy of less than 11 percent. (This is assuming just one switch plane active. When both are active, the average occupancy is less than 6 percent.) There is also an upstream control lead joining each pair of adjacent nodes. This is used to implement a simple hardware flow control mechanism, which prevents buffer overflows within the SF.

A more detailed look at the switch node appears in Fig. 4. It consists of two input controllers (IC) and two output controllers (OC). The IC's contain a buffer large enough to hold one packet and a controller implemented as a state machine. The OC's are simple state machines that arbitrate requests for their ports.

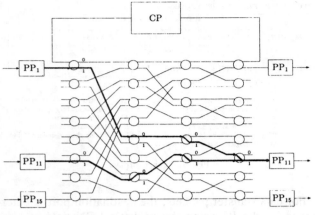

Fig. 4. Packet switch structure.

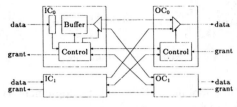

Fig. 5. Switch node.

When a packet is received, the IC determines the proper outgoing port by examining the appropriate bit of the packet's destination field, then requests permission to use that port. If the desired port is immediately available, the packet is sent to it directly, bypassing the buffer. Hence, a packet can pass through a switch node after experiencing a delay of just a few bit times. In fact, this is the normal case due to the relatively low occupancy of the internal data paths.

If the desired port is not available, the packet is shifted into the buffer. As soon as the desired port becomes available, the packet is sent out—even if not all of the packet has been received. If the port is still not available when the end of the packet is received, the IC holds its *grant* lead low to prevent the arrival of new packets. The grant lead is reasserted as soon as the desired link becomes available, allowing a new packet to enter the buffer, while another is leaving.

PACKET PROCESSORS

The structure of the packet processors (PP) is shown in Fig. 5. It is organized around a 16 kbyte RAM, which contains four packet buffers and the logical channel translation table (LCXT). The principal buffers are the receive buffer (RCB) used for packets received from the HSL on their way to the switch, and the transmit buffer (XMB) for packets going from the switch to the HSL. The link test buffer (LTB) and the switch test buffer (STB) are small buffers that provide loop-back paths for testing the HSL and switch, respectively.

Access to the memory is provided through the address controller (ADC), which contains read and write pointers for the buffers and arbitrates memory access.

The receive circuit (RCV) receives the incoming packets from the HSL, removes the flag field, discards packets with errors, adds the extra header fields, initializes the length (LNG) and arrival time (AT) fields, converts from bit serial to 8 bit parallel format, and writes the packet to the RCB through the ADC.

The output circuit (OUT) takes packets from the RCB,

performs the logical channel translation described above, and sends the packets onto the switch in bit serial format. The input circuit (IN) takes packets from the switch and writes them to the XMB.

The transmit circuit (XMIT) takes packets from the XMB, performs the time stamp calculation, strips the extra header information, adds the flag field, and transmits the packets on the HSL.

The switch interface (SI) connects to the duplicated switch planes, normally routing packets to and receiving packets from the active plane. The SI can also send and receive packets from the standby plane—this is used for testing.

The PP can be implemented using two chips—one custom controller chip and one memory chip. The simplicity of the protocols makes this possible. There is no need to buffer unacknowledged frames that may need to be retransmitted as in conventional link level protocols that perform error correction and flow control. Similarly, there is no need for the recovery and synchronization procedures that are required by such protocols.

One problem with binary routing networks is that they can become congested in the presence of certain traffic patterns. This is illustrated in Fig. 6, which shows a traffic pattern corresponding to several *communities of interest*. In this pattern, all traffic entering the first four inputs is destined for the first four outputs, all traffic entering the second group of four inputs is destined for the second group of four outputs, and so forth. Note that with this pattern, only one fourth of the links joining the second and third stages are carrying traffic. Thus, if the inputs are heavily loaded, the internal links will be hopelessly overloaded and traffic will back up. In a 1024 × 1024 network, there are ten stages and the links between the fifth and sixth stages can, in the worst case, be carrying all the traffic on just 32 of the 1024 links.

This problem can be solved by using two networks instead of one. One of these is called the routing network (RN) and is a standard binary routing network. The other network sits in front of the RN and is called the distribution network (DN). It has the same structure as the RN, but instead of routing packets based on their destination address, it attempts to distribute packets evenly across all its output ports. This is done by having each switch node route packets alternately out its two ports. The alternate-port strategy is modified if one or both ports is unavailable. In this case, the first port to become available is used. This approach breaks up any communities of inter-

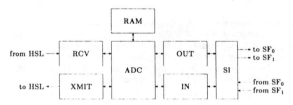

Fig. 6. Packet processor.

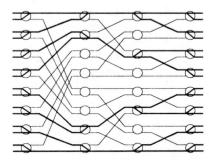

Fig. 7. Congestion in binary routing networks.

est and makes the combination of the DN and RN robust in the face of pathological traffic patterns.

One drawback of the DN is that it doubles the number of stages in the SF, thus roughly doubling the packet delay and the circuit complexity. This loss can be recovered by using larger switch nodes for the RN and DN. Instead of 2×2, we can use 4×4 nodes. With the larger nodes, we can construct a network with half the number of stages required with 2×2 nodes. This also halves the delay and the circuit complexity.

Summary

This paper has described the design of a high-performance packet-switching network capable of supporting both voice and data communication on a large scale and at a cost to the user comparable to current telephone service. The key features of the design are the use of high-speed digital transmission facilities and simple link level protocols, a predominantly connection-oriented service, hardware implementation of all per-packet functions, and implementation of higher level protocol functions on an end-to-end and application-dependent basis. I claim that the advanced packet-switching technology described here is inherently better suited for the provision of advanced communications services than the circuit-switching technology which underlies current ISDN proposals.

Many aspects of the packet-switching system described here have been patented by AT&T Bell Laboratories. See [14]-[20] for further details.

Acknowledgment

The work described here was carried out between 1981 and 1983 when I was a member of the Technical Staff at Bell Laboratories, Naperville, IL. Many individuals contributed ideas. I particularly want to thank H. Andrews, D. Creed, B. Hoberecht, W. Montgomery, and L. Wyatt who helped turn the original idea into a workable design. I also want to thank B. Cardwell and E. Nussbaum for recognizing the potential and providing the backing to develop it. Special thanks go to M. Dècina, whose refusal to accept half-baked ideas forced me to make sure that mine were always well-cooked.

References

[1] D. J. Aoki, D. H. Florin, S. D. McKenna, G. R. Welsh, and P. E. White. "Digital feature evolution in the 5ESS™ digital switch," in *Proc. GLOBECOM 83*, 1983, pp. 601-605.

[2] *Bell Syst. Tech. J.*, vol. 56, Sept. 1977 (Special Issue devoted to the No. 4 ESS).

[3] M. Dècina, "Performance requirements for integrated voice/data networks," *IEEE Trans. Commun.*, vol. COM-30, pp. 2117-2130, Sept. 1982.

[4] D. M. Dias and J. R. Jump, "Packet switching interconnection networks for modular systems," *Comput.*, vol. 14, pp. 43-54, Dec. 1983.

[5] T.-Y. Feng, "A survey of interconnection networks," *Comput.*, vol. 14, pp. 12-30, Dec. 1983.

[6] J. G. Gruber, "Performance requirements for integrated voice/data networks," *IEEE J. Select. Areas Commun.*, vol. SAC-1, pp. 981-1005, Dec. 1983.

[7] W. L. Hoberecht, "A layered network protocol for packet voice and data integration," *IEEE J. Select. Areas Commun.*, vol. SAC-1, pp. 1006-1013, Dec. 1983.

[8] Y.-C. Jenq, "Performance analysis of a packet switch based on a single-buffered banyan network," *IEEE J. Select. Areas Commun.*, vol. SAC-1, pp. 1014-1021, Dec. 1983.

[9] M. Kasahara, H. Shimizu, and M. Imaizumi, "Basic concept of the digital network for the information network system," in *Proc. GLOBECOM 83*, 1983, pp. 969-973.

[10] P. Kermani and L. Kleinrock, "Virtual cut-through: A new computer communication switching technique," *Comput. Networks*, vol. 3, pp. 267-286, 1979.

[11] W. A. Montgomery, "Techniques for packet voice synchronization," *IEEE J. Select. Areas Commun.*, vol. SAC-1, pp. 1022-1028, Dec. 1983.

[12] R. Rettberg, C. Wyman, D. Hunt, M. Hoffman, P. Carvey, B. Hyde, W. Clark, and M. Kraley, "Development of a voice funnel system: Design report," Bolt Beranek and Newman, Rep. 4098, Aug. 1979.

[13] J. S. Turner and L. F. Wyatt, "A packet network architecture for integrated services," in *Proc. GLOBECOM 83*, 1983, pp. 45-50.

[14] J. S. Turner, "Packet load monitoring by trunk controllers," U.S. Patent 4 484 326, Nov. 20, 1984.

[15] ——, "Packet switching loop-around network and facilities testing," U.S. Patent 4 486 877, Dec. 4, 1984.

[16] ——, "End-to-end information memory arrangement in a line controller," U.S. Patent 4 488 288, Dec. 11, 1984.

[17] ——, "Interface facility for a packet switching system," U.S. Patent 4 488 289, Dec. 11, 1984.

[18] ——, "Packet error rate measurements by distributed controllers," U.S. Patent 4 490 817, Dec. 25, 1984.

[19] ——, "Fast packet switch," U.S. Patent 4 491 945, Jan. 1, 1985.

[20] ——, "Fast packet switching system," U.S. Patent 4 494 230, Jan. 15, 1985.

[21] C. J. Weinstein and J. W. Forgie, "Experience with speech communication in packet networks," *IEEE J. Select. Areas Commun.*, vol. SAC-1, pp. 1022-1028, Dec. 1983.

Jonathan S. Turner (M'77) received the B.S.C.S. and B.S.E.E. degrees from Washington University, St. Louis, MO, in 1977, and the M.S. and Ph.D. degrees in computer science from Northwestern University, Evanston, IL, in 1979 and 1981.

From 1977 to 1983 he worked for Bell Laboratories, in Naperville, IL, first as a member of the Technical Staff and later as a Consultant. His work there included the development of maintenance software and design of system architectures for telephone switching systems. From 1981 to 1983 he was the principal system architect for the fast packet switching project, an applied research project which established the feasibility of integrated voice and data communication using packet-switching technology. He has been awarded seven patents for this work and has several others pending. He is now an Associate Professor of Computer Science at Washington University, where he continues his research on high-performance communications systems. His research interests also include the study of algorithms and computational complexity, with particular interest in the probable performance of heuristic algorithms for NP-complete problems.

Dr. Turner is a member of the Association for Computing Machinery and the Society for Industrial and Applied Mathematics.

Concluding Remarks

An overview of this nature would not be complete without some prognostications on future private branch exchanges (PBX) directions. Accordingly, in what follows, a number of observations will be provided regarding the state and direction of PBX technology from the perspective of early 1988.

PBX packaging. As was noted several times previously, the dominant PBX manufacturers have clung to the standard 19 inch rack, multiple cabinet format, and, for larger systems, the forced air environment. Improvements have been made in implementation details, such as universal slot design (allowing any line card to be used in any slot), automatic disconnect and/or backup for failed components, and auto-dial access and/or automatic intervention of remote maintenance services. The dominant makers have not, however, made the drastic change in packaging seen in the computer arena, where, for example, the DEC VAX line has been repackaged in the form of the much more compact Micro-VAX line.

As was noted in Chapter 5, smaller makers such as CXC, Ztel, and Ericsson have led in setting new standards for PBX repackaging. Items as diverse as portable cellular phones, laptop personal computers (PCs), and the Sony Walkman have shown that when a manufacturer drastically reduces the size of an electronic item, it frequently changes as well the way people use it. The tendency toward physically distributed, networked PBXs makes compactness and office-level environmental requirements a definite plus; striving to offer a product sufficiently modular that is economically competitive across a wide market, from under 100 to over 20,000 lines, adds an additional impetus toward creative packaging.

Full software functionality regardless of size. In earlier PBX lines where processor and memory capacities were more restrictive, many PBX features, data transmission being a typical case, were available only on the larger—and invariably more expensive—models. Also, manufacturers, to be simultaneously competitive in the small, medium, large, and very large markets were forced to field switches whose base hardware and software were dissimilar. This dissimilarity limited upgrades, and the incompatibilities between PBXs of a single manufacturer made for embarrassing and uncomfortable situations for both the sales representatives and the customer. AT&T, an early player and the long-time leader in the PBX arena, has been particularly hobbled by the incompatibilities between its S/25, S/75, and S/85 offerings.

With the newer designs, of which the NEC 2400 is one of the best examples, there has been a concerted effort to implement a uniform software suite on a PBX line that extends from small to very large systems. This software strategy offers several benefits. It allows the customer features previously available only on the high-end switches. It allows the customer to impose a standard feature set companywide, even when the physical environments vary widely. Last, it provides the customer several upgrade paths, ranging from straightforwardly augmenting a central system to installing one or more satellite switches, whichever is preferable.

Physical distribution/multisystems networking. As was noted in chapter 5, all the new PBX designs were designed to be capable of operating as a single node, star-configured network, or in a multi-node networked configuration ("a network of stars"). The chief advantages of physical distribution and multisystems networking are in fault toleration and economics.

Providing multiple central office paths—perhaps even to multiple central offices—reduces one source of catastrophic failure. Providing multiple nodes, each with UPS, reduces the effects of a power failure in a single building to the building affected. For example, Rolm, in a clever approach to extending both the capacity and the fault-tolerance of their CBX II 9000 without physical repackaging, developed a network operating system. This allowed Rolm to field a single "logical" system of over a dozen nodes and effectively offer multiple PBX nodes where the competition was forced to bid more expensive central office equipment. (With some repackaging, this approach has been followed with the successor product, the IBM 9751 CBX Voice/Data Controller.)

The economic advantages of physical distribution and multisystems networking are more difficult to access, but the ability to closely match hardware to specific user requirements, and offer equipment with minimal demands for space, air conditioning, and UPS definitely improves a makers' competitive position, other factors being equal.

ISDN capable. The PBX will be a front-line player in integrated services digital networks (ISDN), and migration to ISDN, likely extending over several decades, will pose considerable complexity for the local telecommunications manager. As was noted in chapter 6, ISDN services are end-to-end (user to user). Given all the equipment in the calling path that must be compatible, the earliest implementations of ISDN are likely to be in private, electronic tandem networks (ETN). In this limited and ETN context, PBX changes will be the accommodation of ISDN primary service (trunks) and basic services (the PBX-to-user loop and station equipment). Although new line cards and software extensions will be necessary, the most traumatic—and expensive—aspect will be the new station equipment capable of controlling the integrated services.

All the while, there will be substantial minority, perhaps a majority of users, who, for reasons of cost, may not be ISDN equipped yet must continue to be supported with multiple, but pre-ISDN services. Even for the ISDN equipped, a situation will exist similar to today where different generations of telephony technology coexist (rotary dial, touch tone, touch tone/direct dial, rotary dial/direct dial) will be common. In today's deregulated, post-divestitature environment, there may be sparsely populated and remote locations that even 20 years hence do not offer ISDN. As a result, off the well-known highways of a company ETN or major city calling patterns, ''service negotiation,'' essentially an electronic exchange of relative technological capabilities, will be the norm with the least technologically advanced partner determining the nature of the service. To the telecommunications manager falls the task of managing this changing environment (via PBX equipment modifications) in the light of his or her company's calling patterns, service requirements, and cost elasticities. It is a daunting prospect.

Finally, were a designer creating a PBX system for the next decade, what would the high-level design specify? One such specification appears next:

- Via modular design and packaging options, cost competitive from 50 to 20,000 lines,
- Air-cooled, capable of operating in office environments,
- Upwardly scaleable and nonblocking, to 20,000 lines,
- A single operating system and full function applications software suite capable of running on all size configurations,
- Hotel, central office, and tandem switching applications packages capable of running on all size configurations,
- A networking operating system capable of integrating a large number of individual nodes into a single ''logical'' system of 20,000 lines,
- All software written in a high level language capable of being ported to next-generation microprocessor families,
- Full support of ISDN services and the capability of running them in conjunction with pre-ISDN operating modes,
- Providing functional equivalence for all software features offered in one's current installed base, and
- Supporting all station equipment in one's current installed base.

Glossary

Abbreviated Dialing: A PBX feature which allows users to dial calls by using one or two digits. The system translates the abbreviated dial number into an associated destination number for completion of the desired call.

Acceptance Test: Actions made to prove that a system fulfills agreed upon criteria; for example, that the processing of specified input yielded expected results.

Access Arrangement: Used in reference to the interconnection of customer-provided data modems or automatic calling units in which data access arrangement service includes the provision of a data access arrangment with appropriate loop conditioning (including adjustments for loop loss) to meet data requirements.

Access Charge: A charge by the local telephone company, for use of the local company's exchange facilities, and/or interconnection with the telecommunications network.

Access Code: A digit that must be dialed prior to dialing an outside call. The access code is required so that the PBX can determine the correct plan for the call.

Access Line: A telecommunication line that continuously connects a remote station to a data switching exchange. A telephone number is associated with the access line.

Account Code: A two to four digit number that must be entered before (or after) a call is dialed. When account codes are used, they are recorded in a data base for subsequent use by a PBX or user-provided billing program.

Acoustic Coupling: A method of coupling a data terminal equipment or similiar device to a telephone line by means of transducer that utilizes sound waves to or from a telephone handset or equivalent.

Actual Work Time: The average time an operator requires to handle a call. This corresponds to the expected value (mean value) of the holding time distribution used in the Erlang C model.

Adaptive Channel Allocation: A means of multiplexing where the information capacities of channels are not predetermined but are instead assigned on demand.

Add-On: Circuitry or system that is attached to a computer to increase memory or performance.

Advanced Mobile Phone Service: A dial mobile telephone service whose concept includes a cellular system of low power radio stations and a mobile telephone switching office. The use of low power radio signals, the ability to have numerous cells in an area, and the switching capacity combine to significantly increase mobile telephone capacity.

Alarm: A visible or audible signal to alert personnel of the existance of an abnormal condition.

Alerting (Alerting Signal): A signal sent to a customer PBX, or switching system, to indicate an incoming call. A common form is the signal that rings a bell in the telephone set being called.

Algorithmic Routine: A program or routine that directs a computer specifically toward a solution of a problem in a finite number of distinct and discrete steps as contrasted to trial-and-error methods, that is, heuristic methods or routines.

All-Number Calling (ANC): The system of telephone numbering that uses all numbers and replaces the two-letter plus five-number (2L + 5N) numbering plan. ANC offers more usable combinations of numbers than the 2L + 5N numbering plan and is becoming the nationwide standard.

Alpha-Numeric Display: A visual display device provided on many attendant consoles and telephone sets to display call status information.

Alternate Routing: A means of selectively distributing traffic over a number of routines ultimately leading to the same destination.

American Standard Code for Information Interchange: See ASCII.

Amplifier: A device that, by enabling a received wave to control a local source of power, is capable of delivering an enlarged reproduction of the essential characteristics of the wave.

Analog Data: Data represented by physical quantity that is considered to be continuously variable and whose magnitude is made directly proportional to the data or to a suitable function of the data.

Analog Switch: Switching equipment designed, designated, or used to connect circuits between users for real-time transmission of analog signals.

Analog-to-Digital Converter (A/D): A functional unit that converts analog signals to digital data.

Answerphone: A unit for automatically responding to telephone calls, recording messages for playing back later.

Application Package: A series of interrelated routines and subroutines designed to perform a specified task.

Architecture: A specification that determines how something is constructed, defining functional modularity as well as the protocols and interfaces that allow communications and cooperation among modules.

Archiving: The storage of backup files and associated journals, usually for a given period of time.

Area Code: A three-digit number identifying one of 152 geographic areas of the United States and Canada to permit direct distance dialing on the telephone system.

ASCII: Ámerican National Standard Code for Information Interchange. The standard code, using a coded character set consisting of seven-bit coded characters (eight bits including parity check), used for information interchange among data-processing systems, data communications systems, and associated equipment. The ASCII set consists of control characters and graphic characters.

Asynchronous: Transmission in which time intervals between transmitted characters can be of unequal length. Synchronization of the data bits in the character is provided by stop and start elements at the beginning and end of each character.

Assembly Language: A computer programming language whose statements may be instructions or declarations. The instructions usually have a one-to-one correspondence with machine instructions.

Attached Processor: A processor affixed to a central processor, often sharing its memory.

Attempts per Circuit per Hour: An indication of calling pressure.

Attendant Camp-On with Tone Indication: A PBX feature that allows an attendant to hold an incoming call in a specified waiting mode when the desired station is busy and send a camp-on tone indication to the busy station. When the busy station becomes idle, it is automatically rung and connected to the waiting party. Camp-on warning tone is heard only by the internal station. Music on hold may be provided to the waiting party.

Attendant Console: A desk-top position from which the attendant handles and distributes calls by pushbutton keys.

Attendant Lockout: This PBX feature denies an attendant the ability to re-enter an established trunk/station connection unless recalled by the station.

Attendant Loop Release: This PBX feature enables an incoming console loop to become available for another incoming call as soon as the attendant directs the first call to a station even if the station does not answer.

Attenuation: A decrease in magnitude of current, voltage, or power of a signal in transmission between points. It may be expressed in decibels or nepers.

Audible Alarm: An alarm that is activated when predetermined events occur that require operator attention or intervention for system operation.

Audible Indication Control: A PBX console feature; the console audible indication may be increased or decreased in volume, to conform with ambient noise levels. Console audible may be inhibited completely for operations requiring visual cautions only.

Audio Response: A form of output that uses verbal replies to inquiries. The computer is programmed to seek answers to inquiries made on a time-shared on-line system and then to use a special audio response unit that elicits the appropriate prerecorded responses to the inquiry.

Authorization Code: A three to nine digit number entered before (or after) a toll call is dialed. The use of authorization codes is usually established by class of service. When predefined authorization codes are used, it is not necessary to dial the authorization code. When variable authorization codes are used, the PBX software checks the authorization code for validity before the call is placed.

Autoanswer: An instrument which automatically answers calls via a telephone network.

Auto Dialer: A device permitting automatic dialing of calls via the telephone network.

Automatic Call Distributor (ACD): A system for automatically providing even distribution of incoming calls to operator or attendant positions; calls are served in the approximate order of arrival and are routed to positions in the order of their availability for handling a call.

Automatic Equalization: The process of automatically compensating for linear distortion. This is generally accomplished by an adaptive transversal equalizer.

Automatic Number Identification (ANI): The automatic identification of a calling station, usually for automatic message accounting.

Automatic Recall: A PBX feature whereby calls processed by the attendant that have exceeded a programmed duration in a hold, unanswered, or carried-on condition will recall to the console.

Automatic Switchover: An operating system that has a standby machine that detects when the on-line machine is faulty, and once this determination is made, switches this operation to itself.

Back Plane: The area where the boards of a system are plugged, synonymous with mother board.

Balanced Circuit: A circuit that is terminated by a network whose impedance balances the impedance of the line so that the return losses are infinite.

Band: In data communication, the frequency spectrum between two defined limits.

Bandwidth: The maximum number of data units that can be transferred along a channel per second.

Bandpass Filter: A device that allows signal passage to frequencies within its design range and which effectively bars passage to all signals outside that frequency range.

Base Band: The frequency band occupied by one or more information signals that either modulate a carrier or are transmitted at base band frequency over a suitable medium.

Basic Service: At one time this phrase meant plain old Telephone service (P.O.T.S.) The most common usage now is that provided by the FCC in the Second Computer Inquiry: basic service is the common carrier offering of transmission capacity for the movement of information between two or more points. Most authorities on the issue have agreed that the term basic service covers both exchange and interexchange offerings.

Baud Rate: The transmission rate that is in effect synonymous with signal events, usually bits per second.

Binary: Refers to a numbering system using a base (radix) of 2. The two digits used in the binary numbering system are 0 and 1.

Binary Synchronous Transmission: A data communication protocol using synchronous signaling as opposed to asynchronous signaling.

BISYNC: See Binary Synchronous Transmission.

Bit: An acronym for "binary digit." A single digit in a binary number. The bit can have one of two values: 0 or 1.

Bit Error Rate: The frequency at which errors occur during the transmission of digital information.

Bit Transparency: Refers to the transmission or reception of data without altering the data.

Bits Per Second: The speed at which digital information is transmitted. Also expressed as bit rate.

Blocking: The inability to interconnect two idle ports because all possible paths between them are in use.

BOCs: The 22 operating companies of the Bell System.

Bootstrap (BOOT): A technique or device designed to bring itself into a desired state by means of its own action; for example, a machine routine whose first few instructions are sufficient to bring the rest of itself into the computer from an input device.

bps: See bits per second.

Bridge: The connection of one circuit to more than one port.

Broadcast: The dissemination of information to several receivers simultaneously, usually through electromagnetic signals.

Buffer: A routine or storage used to compensate for a difference in rate of flow of data, or time of occurrence of events, when transferring data from one device to another.

Bus: One or more conductors used for transmitting signals or power.

Busy Hour: The peak 60-minute period during a business day when the largest volume of communications traffic is handled.

Busy Lamp Field (BLF), Flexible: A PBX console feature; the console is equipped with a BLF, which provides visual indications of busy or idle station conditions for a particular group of a hundred station lines selected by the attendant.

Byte: An eight bit binary string operated upon as a unit.

Cable: One or more conductors found within a protective sheathing. When multiple conductors exist, they are electrically isolated from each other.

Cable Fill: The percentage of pairs in a cable sheath actually assigned and used.

Cable Vault: An area, generally on the lower level of a telephone company building, where cables enter the building.

Cache: In a processing unit, a high-speed buffer storage that is continually updated to contain recently accessed contents of main storage. Its purpose is to reduce access time.

Call: In data communication, the action performed by the calling party, or the operations necessary in making a call, or the effective use made of a connection between two stations.

Call Back: This feature allows a calling station, upon encountering a busy station, to have his call automatically completed when the called station becomes idle. Users of conventional stations will hook flash and dial a feature access code.

Call Detail Recording: A PBX option that produces a detailed history of calls made through the PBX system. Data recorded can include the number of the station making the call, area code and a number called, type of trunk used (DDD, WATS, FX), the date,time, duration, and any special billing or authorization code.

Call Forwarding, All Calls: When activated by a customer, all calls to that line are automatically routed to another line designated during activation.

Call Forwarding, Busy Line: This PBX service feature permits a call to a busy station to be forwarded to a designated station or to the attendant after a programmed ringing interval.

Call Forwarding, Don't Answer: This PBX feature permits calls to unattended stations to be forwarded to a designated station or to the attendant after a programmed ringing interval.

Call Hold: By using this feature, PBX station users can hold any call in progress, freeing their line to initiate a second call, or perform a second feature.

Call Pickup, Group: This PBX feature allows a station to intercept calls directed to another station within its group. Ringing station number need not be dialed. Group registration is performed through the maintenance administration terminal.

Call Processing Indication: Visual and audible indications are provided to the attendant relating to the status of calls being processed.

Call Progress Messages: A series of prompts (audible or visual) that aids a PBX user in making or receiving a call.

Call Transfer, All Calls: Allows the station user to transfer established calls to another station, without attendant assistance.

Call Transfer, Attendant: Station users may transfer established calls to the attendant for further processing.

Call Waiting: A PBX feature that signals the called party that another call is waiting.

Camp-On: A method of holding a call for a line that is in use and of signaling when it becomes free.

Card Cage: A chassis or frame that holds a central processor, memory cards, and interfaces.

Carrier: A high frequency that is modulated with voice or digital signals for bulk transmission over cable or radio circuits.

Carterfone Decision: A decision made by the Federal Communications Commission in 1968 to the effect that telephone company customers should be permitted to connect their own equipment (e.g., data modems) to the public telephone network provided that this interconnection does not adversely affect the telephone companies' operations or the utility of the telephone system to others. Prior to this decision, only telephone company-provided equipment could be connected to the network.

CBX: See Computerized Branch Exchange.

CCITT: Consultative Committee for International Telegraph and Telephony. This organization develops international communications usage standards.

CCS: See Century Call Seconds.

CDR: See Call Detail Recording.

Cellular Radio: See Advanced Mobile Phone Service.

Central Office: A location where switching of subscriber telephone lines is accomplished.

Centralized Automatic Message Accounting (CAMA): A process using centrally located equipment, including a switchboard or a traffic service position, associated with a tandem or toll switching office, for automatically recording billing data for customer-dialed extra-charge calls originating from several local central offices. A tape record is processed at an electronic data processing center.

Central Office Code: A three-digit identification under which up to 10,000 station codes are subgrouped. Exchange area boundaries are associated with the central office code which accordingly has billing significance. Note that several central office codes may be served by a central office.

Central Processing Unit: The part of a computer system that performs arithmetic operations, controls instructions, timing signals and other logic operations.

Centrex: Central office telephone equipment serving subscribers at one location on a private automatic branch exchange basis. The system allows such services as direct inward dialing, direct distance dialing, and console switchboards.

Century Call Seconds: A unit of traffic measurement meaning 100 call seconds. Example: a telephone call of 5 minutes equals 300 call seconds, or 3.0 CCS.

Channel: A path along which signals can be sent, for example, data channel, output channel.

Channel Bank: Channel terminal equipment used for combining (multiplexing) channels on a frequency-divison or time-division basis. Voice channels are combined into 12- or 24- channel groups.

Channel Service Unit (CSU): An AT&T unit that is part of the AT&T nonswitched digital data system.

Checksum: A summation of digits or bits according to an arbitrary set of rules. Primarily used for checking purposes.

Chip Microprocessor: LSI circuits residing on a single silicon chip, able to perform the necessary activities of a computer; popularly called "computer on a chip."

Circuit: In data communications, a means of two-way communication between two data terminal installations.

Circuit Board: A board to which is affixed the circuitry of a microprocessor.

Circuit Grade: The information-carrying capability of a circuit, in speed or type of signal. The grades of circuits are broad band, voice, subvoice, and telegraph. For data uses, these grades are identified with certain speed ranges.

Circuit Switch: A method of switching in which a path or circuit is established between origin and destination and held available for as long as required by the users, whether information is passing or not.

Class of Service: Used to describe the type of calls and features a line or trunk group is permitted.

Cladding: A sheathing or covering, usually of glass, fused to the core of an optical fiber.

Class 5 Office: A local central office that serves as the network entry point for station loops and certain special-service lines. Other offices, classes 1, 2, 3, and 4, are toll offices in the telephone network.

Clock Pulse: A synchronization signal provided by a clock.

Coaxial Cable: A transmission line in which one conductor completely surrounds the other, the two being coaxial and separated by a continuous solid dielectric or by dielectric spacers. Such a line radiates no external field and is not susceptible to external fields from other sources.

Codec: Coder/Decoder. An electronic circuit that accepts analog (voice) input and converts it to digital data and accepts digitial data and converts it to analog.

Common Carrier: A company that furnishes communication services to the public under a charter from a regulatory body.

Common Channel Interoffice Signaling (CCIS): An electronic means of signaling between any two switching systems independent of the voice path. The use of CCIS makes possible new customer services, versatile network features, more flexible call routing and faster call connections. By interfacing with communications processors that control data bases incorporated into the system, CCIS can be used to store and provide access to large amounts of information for vast numbers of terminals.

Communications Act of 1934: Establishes a national telecommunications goal of high quality, universally available telephone service at reasonable cost. The act also established the FCC and transferred federal regulation of all interstate and foreign wire and radio communications to this commission. It requires that prices and regulations for service be just, reasonable, and not unduly discriminatory.

Computerized Branch Exchange: A term used to describe a programmable software-controlled business telecommunications system.

Computer Network: A complex consisting of two or more interconnected computing units.

Conditioning: The addition of equipment to leased voice-grade channels to provide minimum values of line characteristics required for data transmission.

Conference: A PBX feature that allows a user to join parties in conversation.

Conference Bridge: The PBX hardware used to provide the conference feature.

Connection: An association established between functional units for conveying information.

Connections per Circuit per Hour (CCH): An indication of holding time of calls.

Consultation Hold, All Calls: This PBX service feature allows a station user to hold any incoming or outgoing public network call, tie line call, or intra-office call, and, on the same line, originate a call to another station within the system. After consultation, the station user may add the second station to the original call (add-on). After add-on, that station user may return to the original call if the second station goes on hook or may transfer the call to the second station.

Consultive Committee International Telegraph and Telephone (CCITT): An advisory committee established under the United Nations within the International Telecommunication Union (ITU) to recommend world-wide standards.

Contention: A line-control scheme in which stations on a line compete for the use of that unused line; the station that is successful in gaining control of the line is able to transmit.

Country Code: The one-, two-, or three-digit number, that, in the world numbering plan, identifies each country or integrated numbering plan in the world. The initial digit is always the world-zone number. Any subsequent digits in the code further define the designated geographical area (normally identifying a specific country). On an international call, the country code is dialed before the national number.

CPU: See Central Processsing Unit.

CRC: See Cyclic Redundancy Check.

Cream Skimming: A situation in which market suppliers can selectively choose to serve only the more profitable areas (or the "cream") of the communications market. For example, cream skimming occurs when an other common carrier (OCC) offers private-line service along high-volume, lower-cost routes, while ignoring areas in which it costs more to provide service. The common carriers are required by law to service all routes, even the high-cost, low volume routes. Since the established telephone common carriers charge uniform prices based on average costs over all their routes, their lower-cost routes are vulnerable to cream-skimming competition.

Cross-Connect: A connecting point for wiring the PBX to the outgoing trunks. Also a connecting point for the wiring of the PBX ports to the system.

Cross Plan Translation: The conversion of 10 digit telephone numbers to 7 digits, or the reverse.

Crosstalk: Unwanted coupling from one signal path to another. Faint speech or tone heard in one circuit, coming from an adjacent circuit.

Customer Premises Equipment: All telecommunications terminal equipment located on the customer premises, both state and interstate, and encompassing everything from black telephones to the most advanced data terminals and PBXs.

Cyclic Redundancy Check: An error detection scheme in which the check character is generated by taking the remainder after dividing all serialized bits in a block of data by a predetermined binary number.

Data Above Voice: A transmission system carrying digital data on a portion of the microwave radio spectrum above the frequency used for voice transmission.

Data Access Arrangement: Equipment that permits attachment of privately owned data terminal equipment and telecommunication equipment to the network.

Data Communications Equipment: The equipment that provides the functions required to establish, maintain, and terminate a connection, as well as signal conversion and the coding required for communication between data terminal equipment and data communications equipment. A modem is an example of DCE. Contrasted with DTE.

Data Encryption Standard (DES) Algorithm: A cryptographic algorithm designed to encipher and decipher data using 64-bit cryptographic key, as specified in the Federal Information Processing Standard Publication 46, January 1, 1977.

Datagram: In packet switching, a self-contained packet that is independent of other packets, that does not require acknowledgement, and that carries information sufficient for routing from the originating data terminal equipment (DTE) to the destination DTE without relying on earlier exchanges between the DTEs and the network.

Data in Voice: Transmission where digital data displaces voice circuits in a microwave channel.

Data Link Layer: In open system architecture, the layer that provides the functions and procedures used to establish, maintain, and release data link connections between elements of the network.

Data Phone: A unit that permits data to be transferred over a telephone line.

Data-Phone: Both a service mark and a trademark of AT&T and the Bell System. As a service mark, it indicates the transmission of data over the telephone network. As a trademark, it identifies the telecommunication equipment furnished by the Bell System for transmission services.

Dataphone Digital Service (DDS): AT&T's private-line service, filed in 1974, for transmitting data over a digital system. The digital transmission system transmits electrical signals directly, instead of translating the signals into tones of varied frequencies, as with the traditional analog transmission system. The digital technique provides more efficient use of transmission facilities, resulting in lower error rates and costs than analog systems.

Data Under Voice: An arrangement for transmitting 1.544 megabit-per-second data streams in the bandwidth portion of the base band used for voice channels on existing microwave systems.

Data Terminal Equipment: The equipment comprising the data source, the data link, or both. Examples of DTE are computer terminals such as CRTs or computer systems data communications ports. Contrasted with DCE.

Data Terminal Ready: A RS-232C signal that indicates a DTE device is in a condition to transmit or receive data.

Day/Night Class of Service: Station class of service can be automatically altered (in accordance with programmed system data) when the system is in the night mode.

dBm: Decibel based on 1 milliwatt.

dBrnC: A power level in dB relative to a noise reference of -90 dBm, as measured with a noise meter, weighted by a special frequency function called C-message weighting that expresses average subjective reaction to interference as a function of frequency.

Decibel (db): A unit for measuring relative power. The number of decibels is 10 times the logarithm (base 10) of the ratio of the measured power levels; if the measured levels are voltages (across the same or equal resistance), the number of decibels is 20 times the log of the ratio.

DCE: See Data Communications Equipment.

DDCO: See Direct Dial Central Office.

Dedicated Channel: A channel that is not switched.

Dedicated Circuit: A circuit designated for exclusive use by two users.

Delay Dial: An off-hook signal returned to the calling end of a trunk as soon as the called end receives a connect (off-hook) signal followed by an on-hook signal to indicate that is ready to receive digits.

Delta Modulation: Conversion of an analog signal, such as voice, to a digital format in which the amplitude difference between successive samples of the analog signal is represented by a set of digits coded to express the quantized amplitude difference. In its simpliest form, the quantized magnitude of the amplitude difference can have only one value other than zero.

Demarcation: A connecting point between the PBX and Telco or any other carrier or vendor. Also see Main Distribution Frame (MDF).

Demark: See Demarcation.

Diagnostics: Programs and procedures used to check and isolate equipment malfunctions.

Dial: To use a dial or push-button telephone to initiate a telephone call. In telecommunication, this action is taken to attempt to establish a connection between a terminal and a telecommunication device over a switched line.

Dial Access to Attendant: This PBX feature allows a station user to access the attendant by dialing the operator call code.

Dial Plan: A description of the number of digits required to place a call. The number of digits required varies according to whether the call is on-net, local, or long distance. Local or long distance dial plans have different numbers of digits depending upon the use of the account and/or authorization code.

Dial Pulse: An interruption in the loop of a calling telephone. The interruption is produced by the breaking and making of the dial pulse contacts of a calling telephone when a digit is dialed. The loop current is interrupted once for each unit of value of the digit.

Dial Tone: An audible signal indicating that a device is ready to be dialed.

Dial Tone Delay: A measure of time required to provide dial tone to customers. This measures one aspect of the performance of a switching system.

Dial Tone Response: A tone indicating the switching equipment is ready to receive dial signals.

DID: See Direct Inward Dialing.

Digital Data System (DDS): A national wide private-line synchronous data communications network formed by interconnecting digital transmission facilities and providing special maintenance and testing capabilities. Customer channels operate at 2.4, 4.8, 9.6, 56, or 1544 kilobits per second.

Digital Error Rate: The frequency of error occurrence in the transmission of data.

Digital Switching: A process in which connections are established by operations on digital signals without converting them to analog signals.

Digital Transmission: A mode of transmission in which all information to be transmitted is first converted to digital form and then transmitted as a serial stream of pulses. Any signal—voice, data, television—can be converted to digital form.

Direct Dial Central Office: A PBX station feature that allows a station user to access the DDD network by dialing the access code and receiving a central office dial tone.

Direct Distance Dialing: A telephone service that enables a user to directly dial telephones outside the user's local area without the assistance of an attendant.

Direct Inward Dialing (DID): Dialing of a call directly to a station without the assistance of an attendant.

Direct Inward System Access: A PBX option that allows a user located off-premises to dial into the PBX from outside the system, to obtain a PBX dial tone, and make calls.

Direct Outward Dialing (DOD): A PBX station user may gain access to the exchange network without the assistance of an attendant by dialing an access code and receiving a second dial tone from the central offfice. The user may then proceed to dial the desired exchange or network number.

Distinctive Ringing: A PBX feature; incoming calls to a station may be identified as external, intra-system, or feature by the type of audible signal (on-off duration) provided to the station.

Dominant Carrier: In legislative proposals to rewrite or amend the Communication Act of 1934 it is used to describe a carrier having control over a substantial portion or subportion of the telecommunications market. A dominant carrier would be subject to special restrictions, usually including a requirement to establish a fully separated subsidiary for offering other than basic services. FCC Docket 79-22 defines a dominant carrier as one having significant market power. This includes AT&T and all the independent telephone companies in the voice market, and Western Union in the domestic record market, which are subject to more stringent rules regarding tariffs and regulatory oversight.

Do Not Disturb: A PBX station feature under user control that prevents calls from ringing at the station. Calls can originate from a station in a Do-Not-Disturb state.

Drop (Subscribers): The line from a telephone cable to a subscriber's building.

DTE: See Data Terminal Equipment.

DTR: See Data Terminal Ready.

Dual Tone Multi-Frequency: A method of signaling by using a voice transmission path. This method employs 12 distinct signals (the 12 buttons on the dialing pad) each composed of two voice band frequencies, one from each of the two geometrically spaced groups designated as the low group and the high group. The selected spacing assures that no two frequencies of any group combination are harmonically related.

EBCDIC: Extended binary-coded decimal interchange code. A coded character set consisting of eight-bit coded characters.

E&M Signaling: A signaling arrangement characterized by the use of separate paths. E (ear) is receive and M (mouth) is transmit.

Echo: An attenuated signal derived from a primary signal by reflection at one or more impedance discontinuities and delayed relative to the primary signal.

Echo Suppressor: A device that detects speech signals transmitted in either direction on a four-wire circuit and introduces loss in the opposite direction of transmission for the purpose of suppressing echos.

Echo Suppressor Control: A PBX feature that disables the echo suppressor used on voice-grade lines when the circuit is used for data.

Economy of Scale: As the need for increasing capacity in switching and transmission facilities develops, owing either to growth or concentration, the cost per unit of capacity may decrease because of two factors: (1) fixed start-up costs that are spread over an increasing number of units, and (2) technological advantages that can be achieved when designing for large capacity.

EIA: Electronic Industries Association.

Electronic Switching System: A common carrier communication switching system that uses solid state devices and other computer-type equipment and principles.

Emergency Maintenance: Maintenance specifically intended to eliminate an exisiting fault, which makes continued production work unachievable.

End-to-End Responsibility: The principle that assigns to communications common carriers complete responsibility for all the equipment and facilities involved in providing a telecommunications service from one end of a connection to the other. These responsibilities included design control, installation, operation, and maintenance.

End-to-End Signaling: A mode of network operation in which the originating central office (or station) retrains control and signals directly to each successive central office (or PBX) as trunks are added to the connection. This contrasts with operation in which each office takes control in turn, called link-by-link signaling.

Engineered Capacity: The highest load for a trunk group or a switching system at which service objectives are met. In general, for a switching system, carried load is equal to offered load below engineered capacity, but is less than offered load above engineered capacity.

Envelope: A group of binary digits formed by a byte augmented by a number of additional bits which are required for the operation of the data network.

Equalization: The procedure applied to transmission media or channels in order that the amplitude and phase (or envelope delay) characteristics of a signal to be transmitted are preserved at the receiving end of the connection. (If a channel introduces phase or frequency offset, equalization does not preserve the wave form of the signal.)

Equivalent Four-Wire System: A transmission system using frequency division to obtain full-duplex operation over only one pair of wires.

Erlang: A dimensionless unit of traffic intensity used to express the average number of calls under way or the average number of devices in use. One Erlang corresponds to the continuous occupancy of one traffic path.

Erlang B: One of the basic traffic models and related formulas used in the Bell System. The assumptions are Poisson input, negative exponential holding time, and blocked calls cleared. Used for trunk engineering.

Erlang C: One of the basic traffic models and related formulas used in the Bell System. This is the queuing model with assumptions of Poisson input, negative exponential holding times, and blocked calls delayed. The queuing discipline may be arbitrary but is usually approximately first come, first served. Used for common-control engineering.

Error: A discrepancy between a computed, observed, or measured value or condition and the true, specified, or theoretically correct value or condition.

Error-Checking Code: A general term for all error-correcting and all error-detecting codes.

ESS: See Electronic Switching System.

Exchange: A telephone switching center.

Exchange Classes: Class 1, Regional Center; Class 2, Sectional Center; Class 3, Primary Center; Class 4, Toll Center, Class 5, End Office.

Expert Programs: Computers acting as intelligent assistants, providing advice and making judgments in specialized areas of expertise.

Facsimile: The transmission of images (photographs, maps, diagrams, and other graphic data) by communication channels. The image is scanned at the transmitting site, transmitted as a series of impulses, reconstructed at the receiving station, and duplicated on paper.

Fallback: The use of a backup module in a redundant system during degraded operation.

Fault: An accidental condition that causes a functional unit to fail to perform in a required manner.

Fault Tolerant: A program or system that continues to function properly in spite of a fault or faults.

Federal Communications Commission (FCC): A board of commissioners appointed by the president under the Communications Act of 1934, having the power to regulate all interstate and foreign electrical telecommunications systems originating in the United States.

FAX: See Facsimile.

Feature: Service and equipment that is common to PBX systems as a standard.

Fiber Optic Cable: A transmission line made of transparent glass fibers bundled together parallel to one another. The length of each fiber is much greater than its diameter.

Final Group: A trunk group that acts as a final route for traffic. Traffic can overflow to a final group from high-usage groups that are busy. Traffic cannot overflow from a final group.

Foreign Exchange (FX): A trunk line provided by a common carrier from an exchange other than the local exchange.

Foreign Numbering Plan Area: Any NPA outside the boundaries of the home NPA.

Four-Wire: A two-way transmission circuit that uses four conductors, one pair for the transmission direction and the other pair for the receiving direction.

Frame: A set of consecutive digit time slots in which the position of each digit time slot can be identified by reference to a frame alignment signal.

Front-End Processor (FEP): A processor that can relieve a host computer of certain processing tasks, such as line control, message handling, code coversion, error control, and application functions.

Full-Duplex: Transmission in two directions simultaneously.

FX: See Foreign Exchange.

Giga: Ten to the ninth power, 1,000,000,000 in decimal notation.

Grade of Service: The proportion of calls, usually during the busy hour, that cannot be completed owing to limits in the call-handling capability of a component in a network.

Glare: Call blocking by the simultaneous seizure of a trunk at both ends.

Ground: The point of reference in an electrical circuit; considered to be at nominal zero potential and other potentials within the circuit are compared to it.

Ground Start Trunk: A ground start trunk detects ground applied on the tip conductor at the central office as an indication of incoming seizure. To initiate an outgoing seizure, the trunk circuit grounds the ring conductor through a maximum local resistance of 550 ohms at the interface.

Guard Band: A frequency band between two channels of a data transmission device, left unused so as to prevent interference between channels.

Half-Duplex: Transmission in two directions, but in only one direction at a time.

Handset: A telephone mouthpiece and receiver in a single unit that can be held in one hand.

Handshaking: An exchange of predetermined signals when a connection is established between two data-set devices.

HDLC: See High-Level Data Link Control.

Hertz (Hz): A unit of frequency equal to 1 cycle per second.

High-Level Data Link Control: A bit oriented line protocol used for transmission of data between stations.

Hit: A transient disturbance to a data communication medium.

Hold: The ability to maintain an established connection while performing another function.

Holding Time: Total time when a given channel is occupied for each transmission or call; consisting of operating time and conversation time.

Host Interface: The interface between a network and a host computer.

Hotline: A PBX feature that allows a line to dial a predefined number automatically upon off-hook. Any line within the system can usually be class-marked as a hotline.

Howler Tone Sending: This PBX feature causes stations that are left off-hook in an undetermined condition to be released from the system. A distinctive tone may be sent at a louder than normal level (''howler''), before the station is released. Placing a ''locked -out'' station on-hook for a few seconds will cause service to the station to be restored.

Hundred Call Seconds: A unit of traffic used to express the average number of calls in progress or the average number of devices in use. Numerically, it is 36 times the traffic expressed in Erlangs.

Hunting: Searching activities performed in switching systems to find the called line or the next available line in an equivalent group.

IDF: See Intermediate Distribution Frame.

IEEE: The Institute of Electrical and Electronics Engineers, Inc.

Immediate Dial: A type of trunk that sends digits immediately after seizing.

Inband Signaling: Signaling that uses the same path as a message and in which the signaling frequencies are in the same band used for the message.

Incoming Call Identification: This PBX feature allows an attendant to identify the type of service or trunk group associated with a call arriving at the console by the visual indication provided for each trunk type.

Independent Telephone Company: A telephone company not affiliated with the Bell System and having its own ''independent'' territory.

Individual Trunk Access: A PBX feature whereby the attendant(s) may access specific trunks within a trunk group by dialing an access code.

Insertion Loss: The insertion loss of a transmission system (or component of the system) inserted between two impedances, ZT (transmitter) and ZR (receiver), is the ratio of the n power measured at the receiver before the insertion of the transmission system to the power measured after insertion. Insertion loss is normally expressed in decibels (dB).

Inside Plant: With respect to cable and wire, all fixed ground cable plant extending inward from the main distribution frame, central office equipment, teletypewriters, and so on, including the protectors and associated hardware on the telephone central office main distribution frame.

Installation Charge: A one-time charge, due upon installation of customer-premises equipment, that is used to help recover the actual expenditures.

Intercepting: The routing of a call or message that is placed for a disconnected or nonexistant telephone number or terminal address to an operator position or to a specially designated terminal.

Interexchange Channel: A channel connecting two different exchange areas.

Interface: A common boundary between two system or pieces of equipment where they are joined.

Intermediate Distribution Frame: The IDF provides connections between the Main Distribution Frame (MDF) and individual station wiring to larger pair cable from the IDF to the MDF.

International Direct Distance Dialing: Non-operator calls that terminateoutside the North America integrated DDD network.

International Organization for Standardization (ISO): An organization established to promote the development of standards to facilitate the international exchange of goods and services and to develop mutual cooperation in areas of intellectual, scientific, technological, and economic activity.

International Record Carrier (IRC): Carrier providing overseas/international telecommunications services, other than voice communications (e.g., teletypewriter, facsimile, and data).

Interoffice Trunk: A direct trunk between local central offices in the same exchange.

Isochronous: Having a regular periodicity.

Jack: A connecting device to which a wire or wires of a circuit may be attached and which is arranged for the insertion of a plug.

Jitter: Short-term variation of the significant instants of a digital signal from their ideal positions in time.

K: An abbreviation for the prefix kilo, that is 1000.

Kbps: Kilo (thousand) bits per second.

Key-Pulsing Signal: In multifrequency signaling, a signal, keyed by the operator, that is used to prepare the distant equipment for receiving digits.

Key Telephone Systems: An arrangement of key telephone stations and associated circuitry, located on a customer's premises, providing combinations of certain voice communications arrangements such as multiline pickup, call line status lamp signals, and interconnection among stations without the need for connections throught the central office or PBX facilities.

KIPS: - Kilo instructions per second.

Lamp Check: A PBX console feature; each console is equipped with a key that illuminates all LEDs and sounds the console audible signal simultaneously allowing the attendant to confirm proper operation of each indicator.

Large Scale Integration (LSI): The process of integrating large numbers of circuits on a single chip of semiconductor material.

LATA: See Local Access and Transport Areas.

Least Cost Routing (LCR), 3/6 Digit: To select the most economical route available for each outgoing call, based on calling station, class of service, available facilities, time of day, and other conditions. The PBX system can be programmed to perform dialed digit translation (addition or deletion) to comply with the requirements of the selected facility (WATS, FX, TIE DDD, etc.). Station users dial a single access code for all outgoing calls (typically 9). Trunk queuing, speed calling, and other features may be used in conjunction with LCR.

Light-Emitting Diode (LED): A semiconductor chip that gives off visible or infrared light when activated.

Line Concentration: Matching a large number of input channels with a smaller number of output channels, the latter performing at higher speed.

Line Level: The signal level in decibels (or nepers) at a particular position on a telecommunications line.

Line Lockout: This feature causes stations that are left off-hook to be released from the system. Service will be restored to the station once the receiver is placed on-hook for a few seconds.

Line Loop Resistance: The metallic resistance of the local loop.

Line Noise: Noise originating in telecommunication line.

Line Protocol: Rules for controlling the sequence of transmission on a synchronous line, explaining bidding for a line, methods for positive and negative acknowledgments, request for retransmission, receiver and transmitter time-out constraints, and other necessary controls for an orderly movement of message blocks from terminal to terminal.

Local Access and Transport Area (LATA): An AT&T concept where calls between points within a LATA are handled entirely by the local telephone company.

Local Area Network (LAN): A system linking together computers, word processors, and other electronic office machines to create an inter-office or inter-site network. These networks usually provide access to external networks, for example, public telephone and data transmission networks, or information retrieval systems.

Local Call Description Tables: Part of a PBX's alternate routing package, which represents a picture of the local call area for routing purposes.

Local Exchange: An exchange in which trunks terminate. Also called end office.

Local Service Area: In telecommunications, the area containing the telephone stations that a flat-rate customer may call without incurring toll charges.

Loop: A channel between a customer's terminal and a central office. The most common form of loop is a pair of wires.

Loop-Back Test: A test in which signals are looped from a test center through a data set or loop-back switch and back to the test center for measurement.

Loop-Start Trunk: Employs standard DC loop signaling and ring-down control of two-wire central office, OPX telephone, and foreign exchange lines.

Main Distribution Frame (MDF): The MDF is located near the PBX cabinets and provides connections between the cabinets, trunks, and intermediate distribution frames.

Make Busy: Conditioning a circuit, a terminal, or a termination to be unavailable for service. When unavailable, it is generally necessary that it appear busy to circuits that seek to connect to it.

Manual Exchange: An exchange where calls are completed by an operator.

Mark: In digital transmission, the binary state "one" when any one of the following conditions exist: (1) loop closed, (2) current flowing, (3) tone one, (4) positive voltage, (5) line to ground.

Master Clock: The primary source of timing signals used to control the timing of pulses.

Master Clock Frequency: The number of pulses per second produced by the master clock.

Mastergroup: An assembly of 10 supergroups occupying adjacent bands in the spectrum for the purpose of simultaneous modulation and demodulation.

MCI Decision: This decision by the FCC in 1969 expanded competition in the intercity private-line market. It allowed Microwave Communications Inc. (MCI) to become the first specialized common carrier and to provide interstate intercity private-line service in competition with the Bell System. In its MCI decision, and its 1971 SCC decision (Specialized Common Carriers), the FCC said it expected the SCCs would provide novel and innovative services not readily available from the telephone companies under conditions that do not threaten the technical or economic viability of the telecommunications network.

Mean-Time-Between-Failures (MTBF): For a stated period in the life of a function unit, the mean value of the lengths of time between consecutive failures under stated conditions.

Mean-Time-To-Repair (MTTR): The average time required for corrective maintenance.

Media: In transmission systems, the structure or path along which the signal is propagated, such as wire pair, coaxial cable, wave guide, optical fiber, or radio path.

Megabyte: One million bytes. Also see Byte.

Meet-Me Paging: This PBX feature allows a station user to transfer a call to a "meet-me" circuit for intercept by the desired party.

Meet-Me Paging, Attendant: This PBX feature allows the attendant to transfer incoming calls to a "meet-me" circuit, for intercept by the desired party.

Mesh Network: A network configuration in which there are two or more paths between any two nodes.

Message: In telephone communications, a successful call attempt that is answered by the called party and followed by some minimum period of connection.

Message Circuit: A long-distance telephone circuit used in furnishing regular long-distance or toll service to the general public. The term is used to differentiate these circuits from circuits used for private-line service.

Micro: A prefix denoting 1 millionth.

Milliwatt: A signal having a power output of 1 thousandth of a watt and a frequency of 1 thousand Hertz used by maintenance personnel for testing for one-way transmission.

MIPS: Million instructions per second.

Miscellaneous Trunk Access: This PBX service feature allows a station user or a terminating tie line party to originate an outgoing call to such trunk groups as FX, WATS, CCSA.

Modem: Equipment that combines the functions of modulator and demodulator for transmission/reception of data. Connects data equipment to telephone lines. Also called data set.

Modem Pool: A group of modems available to a PBX data user on a switched basis when the requirement is to transmit data over an analog facility (trunk).

Modulation: The process by which some characteristics of one wave is varied in accordance with another wave or signal. This technique is used in modems to make business machine signals compatible with communications facilities.

Module: A packaged functional hardware unit designed for use with other components.

Multifrequency Pulsing: An in-band interoffice address signal method in which 10 decimal digits and five auxiliary signals are each represented by selecting two frequencies out of the following group: 700, 900, 1300, 1500, and 1700 Hz.

Multiplexing: In data transmission, a function that permits two or more data sources to share a common transmission medium such that each data sources has its own channel.

Music On Hold: This PBX feature provides music to held station/trunk parties or camped-on stations.

"Nailed" Data Connection: A permanently established connection between two PBX data interfaces. This connection is defined in the PBX data base, and the switch paths are established when the call processing software is initiated.

Nano: Prefix denoting 1 thousand millionth (10-9).

Nationwide/Statewide Cost Averaging: A method of averaging costs upon which uniform prices are set for telephone service so that subscribers using more costly-to-serve, lightly trafficked routes—such as those between small communities—receive the same service for the same price as subscribers on lower-cost highly trafficked metropolitan routes.

Negative Acknowledge Character (NAK): A transmission control character transmitted by a station as a negative response to the station with which the connection has been set up.

Network: An interconnected group of nodes.

Network Congestion: A network condition when traffic is greater than the network can carry, for any reason.

Network Numbering Plan: A PBX option that supports a user's private telephone numbering scheme.

Network Topology: The schematic arrangement of the links and nodes of a network.

Night Connection, Fixed: This PBX feature provides arrangements to route incoming exchange network calls, normally directed to the attendant, to a preselected stations(s) when the attendant positions are placed into the night mode of not attended. These calls answered at the night station(s) can be transferred to another station within the system by means of call transfer, all calls.

Night Connection, Flexible: This feature provides arrangements to route incoming exchange network calls, normally directed to the attendant, to any station or stations when the regular attendant positions are not attended. These calls answered at the night stations can be transferred to another station within the same system.

NNX Code: The first three digits of a seven-digit telephone number. The three digits represent the central office.

NNX Exclusion Tables: A data base listing of special routing within an NPA.

Noise: Random variations of one or more characteristics of any entity such as voltage, current, or data.

Non-Blocking System: Indicates that the internal network of the switching system is such that the total number of available transmission paths is equal to the number of ports, therefore, all ports; have simultaneous access through the network.

Nonswitched Line: A telecommunications line on which connections do not have to be established by dialing.

Nonvolatile Storage: A storage device whose contents are not lost when power is removed.

Numbering Plan: A uniform numbering system wherein each telephone central office has a unique designation similar in form to that of all other offices connected to the nationwide dialing network. In the numbering plan, the first three of 10 dialed digits denote area code; the next three, office code; and the remaining four, station number.

Numbering Plan Area (NPA): Any of the 215 geographical divisions of the United States, Canada, Bermuda, the Caribbean, Northwestern Mexico, Alaska, and Hawaii within which no two telephones have the same seven-digit telephone number. Commonly called Area Codes.

Numbering Plan Area Tables: The NPA tables are a part of a PBX's uniform routing option. The tables contain the route guide numbers for specified area codes.

Offered Load in Erlangs: The average number of calls that would have been in progress if there had been no delay or blocking.

Off-Hook: A term meaning that an instrument's voice frequency circuit is closed.

Off-Net: Beyond the user's private dialing network.

Off-Premises Extension: An extension telephone located off-premises from the PBX site.

Ohm: The unit of measurement for resistance. One ohm is the amount of resistance in a circuit having a potential difference of 1 volt and a current of 1 ampere.

On-Net: Within the user's private dialing network.

On-Hook: A term meaning that an instrument's voice frequency circuit is open. The hand-set portion of the instrument is in its cradle.

Open System Architecture: A model that represents a network as a hierarchical structure of layers of functions; each layer provides a set of functions that can be accessed and that can be used by the layer above it.

Option: Service or equipment offered a user that is not included in the standard PBX system.

Original Equipment Manufacture (OEM): A manufacturer of equipment that may be marketed by another manufacturer.

Other Common Carrier (OCC): Includes specialized common carriers (SCCs), domestic and international record carriers (IRCs), and domestic satellite carriers which are authorized by the FCC to provide communications services in competition with the established telephone common carriers.

Outgoing Trunk Queuing: Upon encountering a trunk busy condition for outgoing calls, station users may access a queue for service. The queued station will be automatically called back by the system and connected to the desired trunk when available. If used in conjunction with least cost routing, the system will dial out the stored number when the station answers.

Out-of-Band Signaling: A method of signaling that uses the same path as voice-frequency transmission and in which the signaling is outside the band used for voice frequencies.

PABX: See private automatic branch exchange.

Packet: A group of bits including data and control elements switched and transmitted as a composite whole. The data and control elements and possibly error control information are arranged in a specified format.

Packet Assembly/Disassembly (PAD): A functional unit that enables data terminal equipment (DTEs) not equipped for packet switching to access a packet-switched network.

Packet Data Switching: A data transmission process, employing addressed packets, in which a channel can be used to transfer packets to many destinations on a time-multiplexed basis.

Paging Access: A PBX feature allowing access to voice paging equipment on a trunk level access.

Paired Cable: A cable made up of one or more separately insulated twisted pairs or lines, none of which are arranged with others to form quads.

Parity Bit: A binary digit appended to a group of binary digits to make the sum of all the digits either always odd (odd parity) or always even (even parity).

PBX: See Private Branch Exchange.

PBX Tie Trunk: A trunk between two PBXs.

PBX Trunk: A line that connects a PBX and a central office. Sometimes called a PBX line in central office terminology.

PCM: See Pulse Code Modulation.

Peak Load: Denotes a higher-than-average quantity of traffic; usually expressed for a 1 hour period and as any of several functions of the observing interval, such as peak hour during a day, average of daily peak hours over a 20-day interval, maximum of average hourly traffic over a 20-day interval.

Peer: In network architecture, any functional unit that is the same layer as another entity.

Peg Count: A count of all calls offered to a trunk group, usually measured for an hour. As applied to units of common-control switching systems, peg count or carried peg count means the number of calls actually handled.

Phase Modulation: Modulation in which the phase angle of a carrier is the characteristic varied.

Picosecond: One trillionth of a second.

Point-to-Point Connection: A connection established between two data stations for data transmission. The connection may include switching facilities.

Polling: Interrogation of devices for purposes such as to avoid contention, to determine operational status, or to determine readiness to send or receive data.

Post Telephone and Telegraph Administration: A generic term for the Government-operated common carriers in countries other than the United States and Canada.

Power Failure Transfer: Calls normally terminated to the attendant positions are routed during a commercial power failure to predesignated single line telephones. This feature is used when reserve power is not equipped or when the battery reserve is depleted. Critical service, both incoming and outgoing to pre-assigned stations, is maintained by this feature.

Preventive Maintenance (PM): Maintenance specifically intended to prevent faults from occurring. Corrective maintenance and preventive maintenance are both performed during maintenance time.

Private Branch Exchange: A private telephone exchange that provides for the transmission of calls internally and to/from the public telephone network.

Private Line: Depreciated term for nonswitched line.

Protocol: A formal set of conventions governing the format and relative timing of message exchange between two communicating processes.

Protocol Conversion: A PBX option that converts dissimilar format and relative timing of message exchange (protocol) between communicating systems.

Public Network: A network established and operated by communication common carriers or telecommunication administrations for the specific purpose of providing circuit-switched, packet-switched, and leased-circuit services to the public.

Public Switched Network (PSN): Any switching system that provides a circuit switched to many customers. In the United States there are four: Telex, TWX, telephone, and Broad band Exchange.

Pulse Code Modulation: A process in which an analog signal is sampled; the magnitude of each sample is quantized with respect to a fixed reference, and the quantized value is converted to a digital signal.

Queuing: A PBX option that stacks requests for outgoing circuits and processes them in order of reception.

Rack: A frame or chassis on which a microprocessor is mounted.

Regeneration: The process of recognizing and reconstructing a digital signal so that the amplitude, wave form, and timing are constrained within stated limits.

Regional Center: A control center (class 1 office) connecting sectional centers of the telephone system together. Every pair of regional centers in the United States has a direct circuit group running from one center to the other.

Ring: One side of a standard telephone line or one lead of a two-wire/four-wire circuit.

Ring 1: One side of a standard four-wire telephone line or one lead of a four-wire circuit.

Ringdown Interface: A private line two-wire interface, also called Loop Start Trunk.

Ringer: A device, usually part of a telephone set, that responds to a 20-Hz signal to produce a ringing sound. Ringers separate from the associated telephone sets are sometimes installed.

Rotary Dial Pulsing: A method of transmission of the called number to a distant office that uses a series of on-hook and off-hook signals for each digit.

RS-232C: An IEEE standard that specifies the electrical characteristics for the interconnection of data terminal equipment (DTE) and data communication equipment (DCE) employing serial binary data interchange.

RS-422: An EIA standard that specifies the electrical characteristics of a balanced voltage digital interface circuit, normally implemented in integrated circuit technology, that can be employed for the interchange of serial binary signals between DTE and DCE or in any point-to-point interconnection of serial binary signals between digital equipment.

RS-423: An EIA standard that specifies the electrical characteristics of an unbalanced voltage digital interface circuit.

RS-449: An EIA standard that specifies the general purpose 37-position and nine-position interfaces for data terminal equipment and data circuits terminating equipment employing serial binary interchange.

SDLC: See Synchronous Data Link Control.

Seize: To gain control of a line in order to transmit data.

Signal-to-Noise Ratio (S/N): The relative power of the signal to the noise in a channel.

Simplex: Refers to operation of a channel in one direction only, with no capability for operating the channel in the other direction.

SMDR: See Station Message Detail Record.

Source Code: The program in a language prepared by a programmer. This code cannot be directly executed but first must be compiled into object code.

Space: In digital transmission, means the binary state zero when any of the following conditions exist: (1) current or tone is off, (2) the loop is open, (3) there is a negative voltage, (4) line to ground.

Special Assembly: Services provided on a special, nontariffed basis. The rate for these special assemblies is based directly on the cost to provide the service. A frequently provided special assembly is a candidate for a tariff filing.

Specialized Common Carrier: An intercity communications common carrier, other than an established telephone common carrier, authorized by the FCC to provide private-line communications services in competition with established telephone carriers. Specialized common carriers usually provide service to the high-density, low-cost intercity private-line routes.

Speed number: The use of three or four digit numbers to dial frequently called numbers as a convenience.

Splitting: This PBX feature allows an attendant to consult privately with one of the calling parties while placing the other party on hold.

Standard Telephone Equipment: Refers to the standard 2500 telephone.

Start-Stop: A time and framing technique used in data-transmission systems, especially teletypewriter systems. Data are transmitted in the form of serial characters, each composed of a start element, information bits, and a stop element, with fixed timing of all of these. Characters are sent asynchronously.

Station Code: The final four digits of a standard 7- or 10-digit address. These digits define a connection to a specific customer's telephone(s) within the larger context of a NPA and central office code. The term "main station code" is an equivalent expression. In the past, a line number and a party letter often were combined to provide station identification. With the discontinuance of party letters, the four numerics have assumed the role of station identification.

Station Hunting, Circular: This PBX service feature allows a call to be directed to an idle station line in a prearranged group when the called station is busy. The hunting sequence is undirectional, and station numbers can be in either consecutive or nonconsecutive order.

Station Hunting, Secretarial: This PBX service feature allows a station to be a common last choice in one or more hunting groups.

Station to Station Calling: This PBX service feature permits a station user to directly dial another station within the same system without operator assistance.

Station to Station Calling, Operator Assistance: This PBX feature allows a station user to call another station within the same system with the assistance of the attendant.

Station Message Detail Record: A standard station message detail record (SMDR) format used by AT&T for traffic and cost reporting.

Statistical Multiplexing: A method of transmitting a number of signals over a common path. The service rate of each signal is a function of its percentage of use as compared with the other signals.

Stop/Go Dialing: An off-hook signal that is sent to the originating end to stop outpulsing is the "stop" signal, and the on-hook signal that is sent to the originating end for a resumption of outpulsing is the "start" signal.

Supervision: A signal indication that is initiated by the called end changing from the on-hook to the off-hook state after the ringing cycle has begun or a circuit is not longer in use, given by each end changing from the off-hook to the on-hook state.

Switched Connection: A mode of operating a data link in which a circuit or channel is established to switching facilities, as, for example, in a public switched network.

Switched Line: A telecommunication line in which the connection is established by dialing.

Switch Hook: A switch on a telephone set, associated with the structure supporting the receiver or handset. It is operated by the removal or replacement of the receiver or handset on the support.

Switch-Hook Flash: A momentary depression and release (off-hook to on-hook to off-hook) of the plunger of the handset cradle.

Synchronous: Transmission in which the data characters and bits are transmitted at a fixed rate with the transmitter and receiver synchronized. This eliminates the need for start-stop elements, thus providing greater efficiency. Compare Asynchronous.

Synchronous Data Link Control (SDLC): A communications line discipline associated with the IBM Systems Network Architecture (SNA) and offering a number of advantages to users of data networks.

System: An assembly of components united by some form of regulated interaction to form an organized whole.

T1: The 1.544 Mbps North American Transmission Standard used by carrier data transmission systems.

Talk Off: False operation of in-band signaling receivers caused by customer speech simulating the supervisory tone for a sufficiently long interval (usually more than 10 ms) to cause accidental release of the connection.

Tandem Office: In general, an intermediate switching system for interconnecting local and toll offices. All toll offices are tandem offices. A more specific meaning of local tandem or metropolitan tandem is an office that connects class offices to other class 5 offices or to other tandem offices within a metropolitan area.

Tandem Switching: Using an intermediate switch to interconnect circuits from the switch of one serving central office to the switch of a second serving central office in the same exchange.

Tariff: The published rate for a specific unit of equipment, facility, or type of service provided by a telecommunications facility. Also, the vehicle by which the regulating agencies approve or disapprove such facilities or service. Thus, the tariff becomes a contract between the customer and the telecommunication facility.

T-Carrier: A series of transmission systems using pulse code modulation technology at various channel capacities and bit rates.

Telecommunication: Communication over a distance, such as by telegraph or telephone.

Teleconference: A meeting between people who are remote from each other but linked by a telecommunications system.

Telephone Dialer: Under program control, a circuit that divides the output of an on-chip crystal oscillator, providing the tone frequency pairs required by a telephone system. The tone pairs are chosen through a latch by means of a BCD code from the bus.

Telephone Recording Statement: A device that enables telephone conversations to be recorded on a dictation machine.

Telephone Set: The terminal equipment on the customer's premises for voice telephone service. Includes transmitter, receiver, switch hook, dial, ringer, and associated circuits.

Telephony: Transmission of speech or other sounds.

Tera: Ten to the twelfth power, 1,000,000,000,000 in decimal notation.

Terminal: A device, usually equipped with a keyboard and a display device, capable of sending and receiving information over a link.

Telex: An international teletypewriter exchange service requiring format and protocols for international transmission.

Tie Line: A trunk that is connected between two private branch exchanges.

Time-Division Multiplexing (TDM): A multiplexing approach to allocate a communications channel for a stated short period to a number of differing units.

Timing Jitter: In digital carrier systems, an accumulative relative timing discrepancy between digitial signal elements. The most common causes are transmission media with nonuniform delay versus frequency characteristics and imperfect timing recovery in digital line regenerators.

Tip and Ring Conductors: The two conductors associated with a two-wire cable pair. The terms tip and ring derive their names from the physical characteristics of an operator's cord-board plug, in which these two conductors terminated in the days of manual switchboards. Use of the tip and ring has extended throughout the plant. The cord-board plug also had a sleeve, and the name is occasionally used for a third conductor associated with tip and ring.

Toll: In public switched systems, a charge for a connection beyond an exchange boundary, based on time and distance.

Toll Center: A central office where channels and toll message circuits terminate. While this is usually one particular central office in a city, larger cities may have several central offices where toll message circuits terminate; a class 4 office.

Touchtone Dial: A push-button pad and associated oscillator circuitry used to transmit address (or end-to-end data) signals from customer stations by means of in-band tones. Each decimal digit, plus a maximum of six additional signals, is uniquely represented by selecting one frequency from each of two mutually exclusive groups of four. The dial is ordinarily powered from the central office.

Traffic Measurement: This feature provides the system traffic usage record in CCS.

Three-Way Calling: This PBX service feature allows a station user to add another station within the same PBX system to an existing connection for a three-party conference. Subsequently, transfer is accomplished when the original party hangs up.

Transient Error: An error that occurs once or at unpredictable intervals.

Transmission Facility: An element of physical telephone plant that performs the function of transmission; for example, a multipair cable, a coaxial cable system, or a microwave radio system.

Trunk: A circuit that connects the PBX to a central office or other system. There are several types of trunks, including CO trunks, FX trunks, tie trunks, and WATS trunks.

Trunk Group: Those trunks between two points, both of which are switching centers, individual message distribution points, or both, and which use the same multiplex terminal equipments.

Twisted Pair: A pair of wires used in transmission circuits and twisted about one another to minimize coupling with other circuits. Paired cable is made up of a few to several thousand twisted pairs.

Two-Wire Circuit: A transmission circuit used for transmit and receive. The two conductors are used for signal and ground.

Unattended Mode: A mode in which no operator is present or in which no operator station is included at the system generation.

Unbalanced Line: A transmission line in which the magnitudes of the voltages on the two conductors are not equal with respect to ground; for example, a coaxial line.

Undetected Error Rate: The ratio of the number of bits incorrectly received but undetected or uncorrected by the error-control device, to the total number of bits, unit elements, characters, and blocks that are sent.

Uniform Call Distribution: A PBX feature that distributes incoming calls to stations in a sequential or equalized order.

Uninterruptible Power Supply: An auxiliary power unit using stored energy to provide continuous power within specified voltage and frequency tolerances; usually consists of a power supply that charges a battery, and an inverter that provides AC power from the battery. In the event of a power failure, the power supply is inoperative, but the batteries and inverter will provide power for a period of time.

UPS: See Uninterruptible Power Supply.

usec: Microsecond; 1/1,000,000 of a second.

Value Added Common Carrier: A common carrier that itself does not establish telecommunications links, but leases lines from other carriers. It can establish a computer-controlled network offering specific telecommunications services.

Value Added Network (VAN): A data network operated in the United States by a firm that obtains basic transmission facilities from the common carriers; for example, the Bell System, adds "value" such as error detection and sharing and resells the services to users. Telenet and Tymnet are examples of VANs.

V-H Coordinates: Numerical coordinates that define the location of a rate center. A simple calculation using the V-H coordinates for two rate centers gives the airline mileage between the two points for use in determining charges for toll calls and private lines.

Video Conferencing: Teleconferencing where participants see and hear others at remote locations.

Virtual Circuit: In packet switching, those facilities provided by a network that give the appearance to the user of an actual connection.

Vocoder: A contraction of Voice-Operated Coder. A device used to compress the frequency bandwidth requirement of voice communications.

Voice Band: The 300 Hz to 3400 Hz band used on telephone equipment for the transmission of voice and data.

Voice Message System: An electronic system for transmitting and storing voice messages that can later be accessed by the individual to whom they are addressed.

Voice Response Unit: A voice synthesizer used to generate voice messages.

WATS: See Wide Area Telecommunication Service.

Wide Area Telecommunication Service: A trunk line tariffed for communication at reduced rates. For example, incoming (800), outgoing, continental United States, and intrastate.

Word: An ordered set of characters that is the normal unit in which information can be stored, transmitted, or operated within a computer.

X.25: An international standard recommended by the CCITT for the interface between data terminal equipment and data circuit-terminating equipment operating in the packet mode on public data networks.

Annotated Bibliography

Bibliographies

National Technical Information Service, *Digital Private Branch Exchange - Digital PBX* (1975-January, 1986) (Citations from the INSPEC Data base). Contains 187 citations on PBXs from the INSPEC data base.

National Technical Information Service, *Private Branch Exchange* (PBX) (March, 1985-March, 1986) (Citations from the INSPEC Data base). This update contains an additional 106 citations.

National Technical Information Service, *Private Branch Exchange* (PBX) (March, 1985-March, 1986) (Citations from the INSPEC Data base). This updated bibliography contains 235 citations of which 48 are new entries to the previous edition.

Terminology

Rosenberg, J. M., *Dictionary of Computers, Data Processing and Telecommunications* (New York: John Wiley & Sons, 1984).

National Communications System, Office of Technology and Standards, *Glossary of Telecommunications Terms*, Federal Standard 1037A (Washington, D.C.: General Services Administration, Office of Information Resources Management, June 1986).

General

Aldermeshian, H., "ISDN Standards Evolution," *AT&T Technical Journal* (January-February 1986), pp. 19-25. A discussion of the ISDN primary and basic rates interfaces and the specification problems remaining.

Anderson, G.M., "Transition to the AT&T-IS Integrated Private Network Architecture," *IEEE Journal on Selected Areas in Communications* (July 1985), pp. 600-605. Describes the evolution of analog and digital ETN networks—and the role that PBXs play in them—to ISDN.

AT&T Information Systems, *Reference Manual DIMENSION System 85 Data Management, Reference Manual DIMENSION System 85 Voice Management, and Reference Manual DIMENSION System 85 Hardware* (Indianapolis, Ind: Western Electric, 1983). AT&T's massive documentation set for System 85; it's size and complexity is such that one is never really sure that one has it all or that the various pieces are current.

Baxter, L.A., Berkowitz, P.R., Buzzard, C.A., Horenkamp, J.J., and Wyatt, F.E., "System 75: Communications and Control Architecture," *AT&T Technical Journal* (January 1985), pp. 153-173. Describes the workings of AT&T's System 75 PBX and provides a detailed introduction to the concepts, terms, and control architecture of a modern PBX.

Bhushan, B. and Opderbeck, H., "The Evolution of Data Switching for PBX's." *IEEE Journal on Selected Areas in Communications* (July 1985), pp. 569-573. The authors argue that the fundamental natures of voice and data traffic are such that data traffic should be handled by virtual calls and packetization, and separately within the switch by a dual bus architecture. This separation scheme is used on GTE's OMNI PBX.

Bir, D., Eng, J., and Hoo, R., "The Evolution of the SL/1 PBX," *Telesis* (1984 one), pp. 20-27. Recounts the software and hardware evolution of the very successful Northern Telecom Sl-1 PBX, introduced in 1975, up to 1984.

Browne, T.E., "Network of the Future," *Proceedings of the IEEE* (September 1986), pp. 1222-1230. Presents a view of the ISDN future, where synergistic relations between flexible and distributed network structures, industry standards at the interfaces, and standard user interfaces and signaling procedures will generate new information services for the "Information Age."

Bhusri, G.S., "Consideration for ISDN Planning and Implementation," *IEEE Communications Magazine* (January 1984), pp. 18-32. Takes the long view on ISDN service provisioning. Initially he argues that high-bandwidth, supplemental overlay networks will be necessary during "first-generation" ISDN to meet the needs of high-end business customers. Later, with a high performance, packet-switched, "second generation" ISDN, ISDN will be able to meet the needs of all its constituencies.

Casey, L.M., Dittburner, R.C., and Gamage, N.D., "FXNET: A Backbone Ring for Voice and Data," *IEEE Communications Magazine* (December 1986), pp. 23-28.

Colby, W., "Is Voice Messaging Still on Hold?," *Infosystems* (June 1985), pp. 85-90. Boosts voice messaging systems but contains some useful cost-of-service numbers.

Coover, E.R., "Voice-Data Integration in the Office: A PBX Approach," *IEEE Communications Magazine* (July 1986), pp. 24-29. The author starts with assumptions about a typical office and argues that a PBX approach to voice/data integration offers the best path to eliminating equipment duplication, implementing a uniform wiring plan, and achieving device interconnection. He then discusses a number of ways a PBX can be used to achieve these objectives.

Coover, E.R., and Kane, M.J., "Analog vs. Digital Station Equipment: The Not-So-Obvious Economics," *Telephony* (August 10, 1987), pp. 76-81. The authors argue that the costs of moves and changes in a dynamic environment justify the higher purchase costs of digital station sets.

Coover, E.R., and Kane, M.J., "Notes from Mid-Revolution: Searching for the Perfect PBX," *Data Communications* (August 1985), pp. 141-150. The authors contend that, sooner or later, small, energy efficient, distributed PBXs will change the look—and cost structure—of the PBX market.

Creadon, R.D., "Voice-Data PABX Makes Data Analysis Easy for E-Tech," *Telephony* (March 2, 1987), pp. 28-32. A case study application involving a Siemens Saturn II PBX for E-Tech, a Florida systems and training analysis house. Of particular importance to E-Tech was voice and data over twisted pair wiring, cost optimization for long distance calls, and, in the Florida location, the ability of a PBX to operate normally over a broad temperature range.

CXC, *General Description: ROSE Business Communications System* (Irvine, Calif.: CXC Corporation, 1985). An introduction to one of the most interesting of the radical, distributed designs. Note that many functional capabilities are designated "future."

Data Communications, Editors, "Who Switches Data Along with Voice? PBX Users, Increasingly," (February 1987), pp. 77-81.

Datapro Research Corporation, "PABX Systems," *Report CS15-610* (March 1982), pp. 101-109.

Decina, M., and Scace, E.L., "CCITT Recommendations on the ISDN: A Review," *IEEE Journal on Selected Areas in Communications* (May 1986), pp. 320-325. Provides an overview of both the CCITT's role in the standards process and the CCITT's I-Series of Recommendations on ISDN. (The I-Series of Recommendations outline the most important standard features of ISDN regarding service capability, network architectures, and user-network interfaces.)

Densmore, W., Jakubek, R.J., Miracle, M.J., and Sun, J.H., "System 75: Switch Services Software," *AT&T Technical Journal* (January 1985), pp. 197-212. Describes the software that runs the AT&T System 75 PBX. In considerable detail the reader is carried step by step through a call and the large number of cooperating processes are described.

Dhawan, A., "One Way to End the Brouhaha Over Choosing an Optimal LAN," *Data Communications* (March 1984), pp. 287-296.

Drake, S.R., "The Role of the PBX in the Automated Office," *Computerworld*, Special Edition on Office Automation, (June 23, 1982).

Ebert, I.G., "The Evolution of Integrated Access Towards the ISDN," *IEEE Communications Magazine* (April, 1984), pp. 6-11. Reviews the factors involved in positioning the public access network, the link from the customer to the first network node, to provide ISDN services.

Ehreth, D., and Jackson, A., "New Switch Offers Integrated Services," *Telephony* (March 25, 1985), pp. 30-37. Describes Harris' 20/20 switch. Of particular interest is its dual use design (tandem switching and PBX), its orientation toward both the North American and European markets, and it parallel subsystem approach to reducing interprocessor communications.

Enomoto, O., and Kadota, S., "The Role of Information Managment System in Business Communication," *IEEE Communications Magazine* (July 1986), pp. 37-43. Details the very wide range of computer and communications services offered by NEC's 2400 Information Managment System PBX and how these services can be provided intra-office, inter-office, and inter-corporationally.

Ericsson Information Switching Systems, *MD110 General Description* (Garden Grove, Calif.: Ericsson, November 1987). A sparse and almost minimalist description of one of the most interesting of the new, distributed PBXs.

Falconer, W.E., and Hooke, J.A., "Telecommunications Services in the Next Decade," *Proceedings of the IEEE* (September 1986), pp. 1246-1261. The authors speculate on the impact of ISDN in the next decade, particularly in regard to the creation and proliferation of ISDN-enabled services and their effects on "first world" life and workstyles.

Feiner, A., "Architecture, Design, and Development of the System 75 Office Communications System," *IEEE Journal on Selected Areas in Communications* (July 1985), pp. 522-530. A summary of the papers on various aspects of AT&T's System 75 PBX that appeared in the January 1985, AT&T Technical Journal.

Fritz, J.S., Kaldenbach, C.F., and Progar, L.M., *Local Area Networks:* Selection Guidelines (Englewood Cliffs, N.J.: Prentice Hall, 1985). The digital PBX appears as an "alternative technology" in chapter five in this volume on LAN selection.

Fromm, F.R., "Centrex is Not Dead Yet!," *Telephony* (April 6, 1987), pp. 40-43. Fromm argues that the Centrex market remains strong, that add-on features through software upgrades will hold the customer base until ISDN arrives, and that Siemen's EWSD central office is well positioned to meet future needs under ISDN.

Goeller, Jr., L. F., and Goldstone, J. A., "The ABCs of the PBX," *Datamation* (April 1983), pp. 178-196. An introductory article, particularly strong on describing the progress of a call.

GTE, *The GTD-4600-E Digital PBX* (McLean, Va.: GTE, April, 1984).

Gupta, G., Liu, P., Sevcik, R., "The SL-1 PBX Integrates Voice and Data Switching," *Telesis* (1984 one), pp. 28-34. Describes, feature by feature, each of the data capabilities added to Northern Telecom's SL-1 PBX between 1980 and 1984.

Hagen, J. G., "Networking: A Multilevel View," *Telecommunications* (May 1985), pp. 48-52. Brief article hyping IDSN services, and seeing ISDN as a new transport network for the information society, serving as an "economic engine, driving and shaping future growth worldwide."

Harris, *20-20 Integrated Network Switch* (Novato, Calif.: Harris Corporation, undated).

Harris, F.H., Sweeney, Jr., F.L., and Vonderohe, R.H., "New Niches for Switches," *Datamation* (March 1983), pp. 109-114. Describes the cutover experiences of a large (approximately 8,500 stations) InteCom PBX for voice and data services for the University of Chicago campus.

Heather, J. A., "Integrated PBX's: The Solution to Information Management Needs," *Telecommunications* (February 1985), pp. 36-39. Argues that the newest PBX designs—using Northern Telecom's Meridian SL as example—provide, by physically incorporating multiprocessing and high speed buses and supporting both circuit and packet switching, the multimedia answer for the integration of information in the office.

Horwitt, E., "Surveying the PBX Path," *Business Computer Systems* (May 1985), pp. 56-65. Summarizes why, for most offices, a PBX approach to office automation combining both voice and data services is preferable.

InteCom, Inc., General Description, *IBX S/80 Integrated Business Exchange* (Allen, Texas: InteCom, Inc., February, 1986).

Janakiraman, N., "An Overview of Recent Developments in the Designs and Applications of Customer Premises Switches," *IEEE Communications Magazine* (October 1985), pp. 32-45. Overview of recent developments—"office of the the future," new technology, ISDN, and deregulation—and how they are changing the design and marketing of PBX systems.

Johnson, H.W., "An Inside Look at the New ROLM CBX II Architecture," *Business Communications Review* (January-February 1984), pp. 9-19. Not only does Johnson's paper provide ROLM's view of the office and where it is headed but details how ROLM has evolved a 16-bit single-node PBX with 75 Mbps bus introduced in 1975 into a 32-bit multinode PBX with a 295 Mbps bus while maintaining upward compatibility.

Junker, S.L., and Nolle, W.E., "Digital Private Branch Exchanges," *IEEE Communications Magazine* (May 1983), pp. 11-17.

Karavatos, W.P., "The Next Generation in Business Communications," *Telecommunications* (August 1983) Describes Ztel's Private Network Exchange or "PNX." It consists of a triple ring (voice, data, spare) IEEE 802.5 LAN with two kinds of node processors (ring, appplications) and offers both circuit-switched voice and packet-switched data.

Kennedy, T. S., Pezzutti, D. A., and Wang, T. L., "System 75: Project Development Environment," *AT&T Technical Journal* (January 1985), pp. 269-285. A description of a product development project plan and its execution, which happens to be the one used with AT&T's System 75 PBX.

Kinck, C.A., "Just When You Thought It Made Sense to Get Rid of Centrex...," *Data Communications* (March 1986), pp. 118-129. Surveys the numerous techniques by which the BOCs are adding PBX-like features to central office switches to stay competitive.

Lake, C. J., Shanley, J. J., and Silverstein, S. M., "System 75: GAMUT: A Message Utility System for Automatic Testing," *AT&T Technical Journal* (January 1985), pp. 305-320. The subject is testing message-based systems; GAMUT was the automated testing tool used by AT&T developers with System 75.

Levin, D., "Comparing Local Communications Alternatives," *Data Communications* (March 1985), pp. 243-256. Compares three LAN technologies, PBXs, baseband, and broadband and makes rather negative assessments of PBX-based and CSMA/CD-baseband roles.

Levin, D., "Private Branch Exchanges: The Best Time to Shop May Be Right Now," *Data Communications* (August 1987), pp. 100-122. Surveys the very crowded low end of the PBX market; the article contains some very useful diagrams and summary information.

Loverde, A. S., Frisch, H. D., Lindemulder, C. R., and Baker, D., "System 75: Physical Architecture and Design," *AT&T Technical Journal* (January 1985), pp. 175-195. Detailed account of the physical design features of an orthodox PBX product, AT&T's System 75.

Lu, K. S., Price, J. D., and Smith, T. L., "System 75: Maintenance Architecture," *AT&T Technical Journal* (January 1985), pp. 229-249. Long but good discussion of the principles of PBX maintenance, with the AT&T System 75 as the case in point.

McFarland, M. A., and Miller, J. A., "System 75: Introduction Activities and Results," *AT&T Technical Journal* (January 1985), pp. 321-332. A good, programmatic review of how to bring out a complex product, in this case the System 75 PBX.

McKay, F. G., "New Wave Coming in Data/Voice Switching," *Telephone Engineer & Management* (October 1984), pp. 71-74. An overview of GTE-sponsored research in burst and packet switching, PBX-on-a-chip, LAN-linked distributed switches, and other new technologies.

Mehta, S., "Who Needs A LAN?," *LAN Magazine* (June, 1987), pp. 24-28. Written by a cable LAN enthusiast, the author admits that for average office environments a data PBX is hard to beat for price and features.

Mier, E. E., "PBX Trends and Technology Update: Following the Leaders," *Data Communications* (September 1985), pp. 82-96. Concentrates on the "big three" of the PBX market—AT&T, Northern Telecom, and Rolm—and, in a series of tables and schematic diagrams, shows the various ways they have responded to user demands for data services. Also included is a user rating chart of PBX's and Centrex.

Mitel, *SX2000 General Description and SX2000 Features and Services Guide* (Boca Raton, Fa: Mitel, 1984). Mitel's comparatively concise system manual for its SX2000.

Muller, N. J., "Enhanced Data Switches May Outshine LAN, PBX Alternatives," *Data Communications* (May 1987), pp. 185-193. Discussion of the functionality and features offered by modern data PBXs, especially in the areas of control, security and administration.

Murray, C. and Carr, W., "Meridian DV-1: A Fast, Functional and Flexible Data Voice System, *IEEE Communications Magazine* (December 1986), pp. 36-42. Describes Northern Telecom's new Meridian DV-1 PBX and provides a nice contrast, both in basic architecture and in functional objectives, with the Bir et. al., article on the SL-1.

NEC, *NEAX2400 IMS General Description* (Melville, N.Y.: November 1986).

Nerys, C. F., "PBXs: Switching to A New Generation," *Today's Office* (July 1985), pp. 20-24. Short paper with brief product descriptions on two new "fourth generation PBXs, CXC's Rose and Ztel's PNX, and two older PBXs that have been recently augmented, AT&T's S/85 and Northern Telecom's Meridian SL.

Northern Telecom, *SL-1 Business Communications System, Generic X11 Feature Document Business Features* (Santa Clara, Calif.: Northern Telecom, March 1983), *Electronic Switched Network Feature Document* (Santa Clara, Calif.: Northern Telecom, March 1984), and *SL-1XN System Description* (Santa Clara, Calif.: Northern Telecom, April 1984). Another of the massive documentation school, NT provides very large basic documents and endless feature supplements. Because much of this material is not dated, especial care must be taken that materials are current.

Patrick, M.W., "The Heat Is on for Phone Switches That Do a Lot of Fast Shuffling," *Data Communications* (March 1985), pp. 227-236. Argues that the 19.2 Kbps asynchronous/64 Kbps synchronous data service offered by many PBXs provides sufficient throughput for most office applications and that twisted pair media is clearly less expensive than alternatives.

Pedersen, T. J., Ritacco, J. E., and Santillo, J.A., "System 75: Software Development Tools," *AT&T Technical Journal* (January 1985), pp. 287-304. A description of the Unix-based software development environment employed for developing Sytem 75 software.

Pitroda, S. G., "Telephones Go Digital," *IEEE Spectrum* (October 1979), pp. 51-60.

Robin, G., "Customer Installations for the ISDN," *IEEE Communications Magazine* (April 1984), pp. 18-23. Focuses on ISDN services and the arrangments necessary to bring them to business environments. Uses the seven-layer OSI reference model to distinguish between "bearer services" (OSI levels 1-3) and "value-added services (OSI levels 1-7).

Roca, R. T., "ISDN Architecture," *AT&T Technical Journal* (January-February 1986), pp. 5-17. Discusses the evolution of AT&T's CCIS network, and AT&T's DMI and DCP protocols, toward the CCITT-defined ISDN standard.

Rolm, CBX II Business Communications System (Santa Clara, Calif.: Rolm, October 1983) and *9000 System Data Communcations Feature, System Administrator's Guide* (Santa Clara, Calif.: Rolm, September 1983). Rolm's system manuals are both copious and corpulent and are supplemented by masses of slippery supplements. Special care must be taken to ascertain whether materials are current.

Sagaser, B. C., "Use Integrated PBXs and X.25 in Today's Networks; Don't Wait for ISDN," *Data Communications* (May 1986), pp. 253-261. Argues strongly that the combination of data-capable PBXs with X.25 interfaces and public data networks can deliver many of the services promised by ISDN today.

Sager, G. R., Melber, A. J., and Fong, K. T., "System 75: The Oryx/Pecos Operating System," *AT&T Technical Journal* (January 1985), pp. 251-268. On message-implemented, real-time operating systems; the fact that it runs a PBX (AT&T's System 75) is almost incidental.

Sharma, R., "Some Architectural Considerations for Local Area Networks," *8th Conference on Local Computer Networks* (Washington, D.C.: IEEE Computer Society Press, 1983), pp. 6-13. Believes that PBXs are emerging as the natural telecommunications service centers for office workers. Also contains useful sumations of various work breakdown studies for office workers.

Sikes, A., "The Data PBX Solution Local For Networks," *Teleconnect* (December 1985), pp. 112-125.

Stallings, W., "Integrated Services Digital Network," in William Stallings (ed.), *Data and Computer Communications* (New York: Macmillan, 1985), pp. 6-23. A thorough introduction to ISDN circa 1984.

Storer, T. S.. "Voice/Data Integration in Small Offices," *Telephony* (October 8, 1984), pp. 32-40. Describes the capabilities of the Telenova 1, a small-office oriented PBX handling from 10-100 stations.

Talley, D., *Basic Electronic Switching for Telephone Systems* (Rochelle Park, N.J.: Hayden, 1982). Basically a description of the old Bell System analog central office switches although a new chapter (9) has been added on "Digital Telephony."

Tang, W. V., "ISDN: New Vistas in Information Processing," *IEEE Communications Magazine* (November 1986), pp. 11-16. Identifies three fundamental principles embodied in ISDN (digitization with high bandwidth, worldwide standards, & integrated functionality/services) and speculates on their likely effects on information-intensive business enterprizes.

Thomson, D., "Helping Data Managers Find Their Voice," *Data Communications* (May 1985), PP. 111-123. A very useful paper on PBX administration with its emphasis on cost minimization.

Thurber, K. J., and Freeman, H. A., "An Introduction to Integrated Voice/Data PBX Systems," in Thurber and Freeman's *Local Computer Networks*, 2nd edition

(Washington, D.C.: IEEE Computer Society, 1986), pp. 209-215. A nice introduction to PBXs in several possible LAN roles, with discussions of coding, switching, and blockage.

Trewhitt, J., "Central-Office LANs: BOCs' Plans for Interim ISDNs," *Data Communications* (January 1987), pp. 58-64.

Turner, J. S., "Design of an Integrated Services Packet Network," *IEEE Journal on Selected Areas in Communications* (November 1986), pp.1373-1379. A critique of the current ISDN plan. Argues that ISDN is unnecessarily dependent on a circuit switching mentality and will result in a less-than-satisfactory hybrid circuit/packet switched solution. Argues that for adaptability to changing traffic, a truly integrated internal architecture, and transmission efficiency, a high performance packet network would have been the better choice.

United Technologies Lexar, *UTX-5000 Voice/Data Switching System General Description* (Westlake Village, Calif.: Lexar Corp., February 1984). The system manual for an interesting PBX—one that employs delta modulation voice coding—whose corporate parent seems to have changed almost yearly.

Vickers, R., and Vilmansen, T., "The Evolution of Telecommunications Technology," *Proceedings of the IEEE* (September 1986), pp. 1231-1245. Begins with a wide-ranging assessment of achieveable software and hardware technolgies and ends with a visionary, post-ISDN telecommunications future dominated by highly intelligent and adaptable "soft technologies."

Western Union, *Vega System General Description Manual* (McLean, Va.: Western Union, March 1984). Western Union marketed CXC's "Rose" as the Western Union "Vega." Their materials closely follow CXC's.

Woodland, H. K., Reisner, G. A., and Melamed, A. S., "System 75: System Management," *AT&T Technical Journal* (January 1985), pp. 213-228. A fairly detailed account of the system management features available on the AT&T System 75 PBX.

Yaski, E., "Is There A PBX in Your Future?," *Datamation* (March 1983), pp. 100-104. The first of a four-part series on PBXs, this short paper evaluates PBXs for data transfer.

Ztel Technical Publications, *Ztel PNX System Description* (Wilmington, Mass.: Ztel, Inc., May 1984). The modest-sized system manual for the ahead-of-its-time Ztel PNX.

Author Biography

Edwin R. Coover was born in Washington, D.C., in 1942. He earned B.A. and M.A. degrees from the University of Virginia (Charlottesville) and then served in Vietnam as a Marine Corps Communications Officer. After a Ph.D. at the University of Minnesota (Minneapolis), he taught at Indiana University (Bloomington) and Warsaw University (Poland). Upon his return to the Washington area, he worked for Rapidata, Electronic Data Systems, and the Network Analysis Corporation. For the last five years he has been with the MITRE Corporation in McLean, VA., where he is a Lead Engineer. He and his family live in Chevy Chase, MD. Dr. Coover is a member of the Institute of Electronic and Electrical Engineers, the Association for Computer Machinery, and the American Statistical Association.

THE IEEE COMPUTER SOCIETY
A member society of the Institute of Electrical and Electronics Engineers, Inc.

Computer Society Press Publications

Monographs: A monograph is a collection of original material assembled as a coherent package. It is typically a treatise on a small area of learning and may include the collection of knowledge gathered over the lifetime of the authors.

Tutorials: A tutorial is a collection of original materials prepared by the editors and reprints of the best articles published in a subject area. They must contain at least five percent original materials (15 to 20 percent original materials is recommended).

Reprint Book: A reprint book is a collection of reprints that are divided into sections with a preface, table of contents, and section introductions that discuss the reprints and why they were selected. It contains less than five percent original material.

Technology Series: The technology series is a collection of anthologies of reprints each with a narrow focus on a subset of a particular discipline.

Offices of the Computer Society

Headquarters Office
1730 Massachusetts Ave. NW
Washington, DC 20036-1903
Phone: (202) 371-0101
Telex: 7108250437 IEEE COMPSO

Publications Office
10662 Los Vaqueros Circle
Los Alamitos, CA 90720
Membership and General Information: (714) 821-8380
Publications Orders: (800) 272-6657

European Office
13, Avenue de l'Aquilon
B-1200 Brussels, Belgium
Phone: 32 (2) 770-21-98
Telex: 25387 AVVALB

Asian Office
Ooshima Building
2-19-1 Minami-Aoyama, Minato-ku
Tokyo 107, Japan

Purpose

The Computer Society strives to advance the theory and practice of computer science and engineering. It promotes the exchange of technical information among its 90,000 members around the world, and provides a wide range of services which are available to both members and nonmembers.

Membership

Members receive the highly acclaimed monthly magazine *Computer*, discounts on all society publications, discounts to attend conferences, and opportunities to serve in various capacities. Membership is open to members, associate members, and student members of the IEEE, and to non-IEEE members who qualify as affiliate members of the Computer Society.

Publications

Periodicals. The society publishes six magazines (*Computer, IEEE Computer Graphics and Applications, IEEE Design & Test of Computers, IEEE Expert, IEEE Micro, IEEE Software*) and three research publications (*IEEE Transactions on Computers, IEEE Transactions on Pattern Analysis and Machine Intelligence, IEEE Transactions on Software Engineering*).
Conference Proceedings, Tutorial Texts, Standards Documents.
The society publishes more than 100 new titles every year.
Computer. Received by all society members, Computer is an authoritative, easy-to-read monthly magazine containing tutorial, survey, and in-depth technical articles across the breadth of the computer field. Departments contain general and Computer Society news, conference coverage and calendar, interviews, new product and book reviews, etc.

All publications are available to members, nonmembers, libraries, and organizations.

Activities

Chapters. Over 100 regular and over 100 student chapters around the world provide the opportunity to interact with local colleagues, hear experts discuss technical issues, and serve the local professional community.
Technical Committees. Over 30 TCs provide the opportunity to interact with peers in technical specialty areas, receive newsletters, conduct conferences, tutorials, etc.
Standards Working groups. Draft standards are written by over 60 SWGs in all areas of computer technology; after approval via vote, they become IEEE standards used throughout the industrial world.
Conferences/Educational Activities. The society holds about 100 conferences each year around the world and sponsors many educational activities, including computing sciences accreditation.

European Office

This office processes Computer Society membership applications and handles publication orders. Payments are accepted by cheques in Belgian francs, British pounds sterling, German marks, Swiss francs, or US dollars, or by American Express, Eurocard, MasterCard, or Visa credit cards.

Ombudsman

Members experiencing problems—late magazines, membership status problems, no answer to complaints—may write to the ombudsman at the Publications Office.